Photoactive Functional Soft Materials

Photoactive Functional Soft Materials

Preparation, Properties, and Applications

Edited by Quan Li

WILEY-VCH

Editor

Quan Li
Kent State University
Liquid Crystal Institute and Chemical Physics Interdisciplinary Program
1425 Lefton Esplanade
OH 44242
United States

Cover
© zhou shu/Getty Images

Library of Congress Card No.: applied for

British Library Cataloguing-in-Publication Data
A catalogue record for this book is available from the British Library.

Bibliographic information published by the Deutsche Nationalbibliothek
The Deutsche Nationalbibliothek lists this publication in the Deutsche Nationalbibliografie; detailed bibliographic data are available on the Internet at <http://dnb.d-nb.de>.

Print ISBN: 978-3-527-34482-6
ePDF ISBN: 978-3-527-81674-3
ePub ISBN: 978-3-527-81676-7
oBook ISBN: 978-3-527-81677-4

Typesetting SPi Global, Chennai, India

Printing and Binding C.O.S. Printers Pte Ltd, Singapore

Printed on acid-free paper

10 9 8 7 6 5 4 3 2 1

Contents

Preface

Elegant and enabling light-triggered processes and phenomena abound in nature. For example, photoisomerization of the retina enables our vision so that we can see and appreciate the vivid natural environment that surrounds us. Similarly, the photoisomerization of phytochromes in plants regulates some critical physiological processes leading to heliotropic and phototropic motions. The movement of the sunflower toward sunlight is a convincing example. As always, such natural phenomena have tremendously inspired scientists and engineers to develop functional photoactive materials and systems for fundamental scientific studies as well as for technological applications. Among various photoactive materials, photoactive soft materials based on biomaterials, polymers, liquid crystals, gels, colloids, etc. that can be driven by light stimulus have recently emerged as advanced functional materials that facilitate a plethora of applications in multiple realms of science and technology. These enabling materials exploit the gamut of distinguished characteristics of light such as convenience of application, noninvasive nature, and fast and orthogonal energy delivery in addition to precise temporal and spatial resolution with remote and digital controllability. Through irradiation by light of tunable wavelength, intensity, and polarization, the existing properties of photoactive soft materials can be readily modulated and/or new properties can be prescribed and, when necessary, certain attributes can be silenced so that the materials can exhibit desired functions suitable for specific applications. Moreover, the functionalities of photoactive soft materials can be wirelessly programmed and their assembly and disassembly into hierarchical structures can be facilely controlled to yield reconfigurable materials and systems. The alignment and reorientation of photoactive soft materials can be driven by light thereby enhancing their properties and performances for switchable applications. In photoactive soft materials, light-driven molecular events have been cooperatively translated leading to dynamic and controllable macroscopic photophysical and photochemical changes that have been harnessed for the purpose of enabling different applications. In this book, the design, synthesis, properties, and applications of functional photoactive soft materials are covered. Topics dealing with distinct aspects of polymers, biomaterials, liquid crystals, gels, colloids, etc. have been included in this work. Authoritative accounts from an international team of active and leading experts describing the various functional aspects of photoactive soft materials have been combined in a unified manner in this book. The primary focus of this book is on the design and fabrication of photoactive functional soft materials for application in materials science, nanophotonics, nanotechnology, optomechanics, soft robotics, biological and biomedical applications, switchable surfaces and interfaces, microfluidics, etc. The assessment of different properties of photoactive materials and their performance in devices have been highlighted and the major challenges as well as future prospects are outlined in the chapters.

This book is not an attempt to exhaustively cover all the interesting and relevant topics on photoactive functional soft materials since it is simply not possible to do so in a concise manner and within a single volume book. Instead, the book focuses on the recent developments of the most fascinating themes about photoactive functional soft materials: preparation, properties, and applications. The various chapters covering trendy topics are as follows: Soft Materials Driven by Photothermal Effect and Their Applications (Chapter 1), Photoresponsive Supramolecular Polymers (Chapter 2), Light-Driven Self-Organized Liquid Crystalline Nanostructures Enabled by Chiral Molecular Switches or Motors: From 1D to 3D Photonic Crystals (Chapter 3), Photochemical Chirality Induction and Inversion in Soft Materials (Chapter 4), Soft Photoactuators in Microfluidics (Chapter 5), Liquid Crystal Polymer Networks and Elastomers for Light-Fueled Robotics (Chapter 6), Light-Driven Phase Transitions in Liquid Crystals and Their Applications (Chapter 7), Photomechanical Soft Nanocomposites: Synergies between Soft Matrix and Energy Conversion Additives (Chapter 8), Photoresponsive Polyolefins (Chapter 9), A Photoresponsive Multi-Bilayered Film for a Tunable Photonic Crystal (Chapter 10), Photoinduced Liquid Crystal Domain Engineering for Optical Field Control (Chapter 11), Azobenzene Polymers as Photoactive Materials for Shape Changes of Micro/Nano-objects (Chapter 12), Light-Controlled Encapsulation and Release Enabled by Photoresponsive Polymer Self-Assemblies (Chapter 13), and Photoresponsive Soft Materials Based on Reversible Proton Transfer (Chapter 14). In each chapter, the state of the art, along with future prospects in the respective fields, is discussed and highlighted by the experts. The overarching goal of this book has been to reflect on the significant recent developments and current approaches toward the realization of functional photoactive soft materials.

The book has been organized with the hope that it would be accessible to undergraduates and graduate students and would be a ready guide for researchers and engineers both in academia and industry. The comprehensive account of the topics covered in various chapters is expected to offer useful information, and provide insight into the exciting concepts and enabling techniques for both early career and experienced researchers in the fields of materials science, materials engineering, polymer science, liquid crystals, chirality induction, colloid and interface science, microfluidics, biological and biomedical science, chemical engineering, organic chemistry, photonics, optoelectronics, optomechanics, soft robotics, imaging and sensing, nanotechnology, and energy efficient materials. Moreover, it is anticipated that the book would stimulate new ideas and interest which could drive future developments and forward march of this emerging endeavor.

I would like to express my gratitude to Martin Preuss at Wiley-VCH for inviting us to bring this exciting and enabling field of research to a wide audience, and to all our distinguished contributors for their dedicated efforts. Also, I am indebted to my wife Changshu and my sons Daniel and Songqiao for their great support and encouragement.

April of 2018

Quan Li
Kent, OH, USA

1

Soft Materials Driven by Photothermal Effect and Their Applications

Hari K. Bisoyi[1], Augustine M. Urbas[2], and Quan Li[1]

1 Kent State University, Liquid Crystal Institute and Chemical Physics Interdisciplinary Program, 1425 Lefton Esplanade, Kent, OH 44242, USA
2 Wright-Patterson Air Force Base, Materials and Manufacturing Directorate, Air Force Research Laboratory, Dayton, OH 45433, USA

1.1 Introduction

We are surrounded by soft materials. Most organs of the human body are made of soft materials. One defining characteristic of soft materials is that they can be readily deformed by external stimuli and stress. Even the thermal energy available at room temperature is sufficient to cause significant deformation in soft materials [1–13]. de Gennes in his 1991 Nobel lecture emphasized "flexibility" as a defining characteristic of soft materials. Colloids, polymers, liquid crystals (LCs), gels, and biological materials are categorized as soft materials. Being pervasive in nature, soft materials have been a source of inspiration for scientists and engineers for the design and fabrication of stimuli-responsive smart materials and systems for scientific investigations to understand the basic principles and technological application in devices and health care. Accordingly, a wide variety of engineered soft materials have been developed, which play a crucial role in modern technology. Soft nanotechnology, soft robotics, and soft lithography are examples that reflect the far-reaching implications of soft materials in biology and engineering [14–22]. Soft materials constitute one of the most stimulating interdisciplinary frontiers of modern science and are built on diverse experimental and theoretical foundations. Although soft materials in neat state have found a wide range of applications, fabrication of composites of soft materials by integrating other functional materials has recently become a common practice to realize advanced and tunable materials with enhanced properties for high-tech applications. Incorporation of photothermal agents into the matrices of soft materials is one such emerging approach.

Photothermal agents absorb light and convert it into heat, which is referred to as the photothermal effect. Over the past decade, many different types of photothermal agents have been reported, including inorganic nanomaterials (e.g. noble metal nanoparticles and carbon-based materials) (Figure 1.1) and organic

Photoactive Functional Soft Materials: Preparation, Properties, and Applications, First Edition.
Edited by Quan Li.

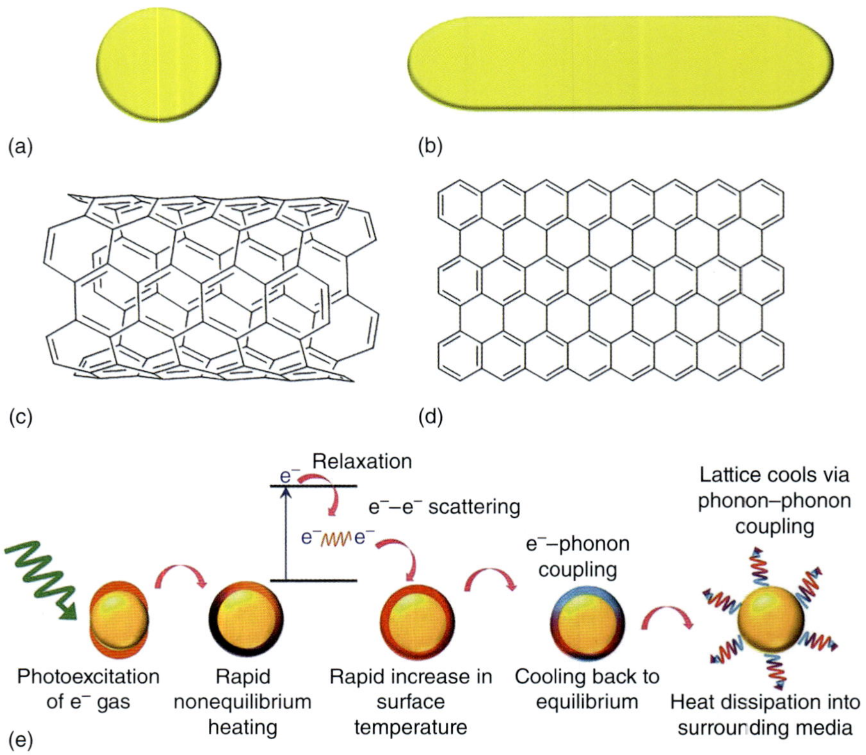

Figure 1.1 Commonly employed photothermal agents for driving soft materials by light irradiation. (a) Gold nanoparticle, (b) gold nanorod, (c) carbon nanotube, and (d) graphene. (e) Schematic describing the principle of photothermal light to heat conversion by plasmonic nanostructures. Source: Qin and Bischof [23]. Copyright 2014. Reproduced with permission from The Royal Society of Chemistry.

compounds or materials (e.g. indocyanine green and polyaniline) [23–54]. Different classes of photothermal agents are associated with their own advantages and disadvantages. Near infrared (NIR)-absorbing photothermal agents have earned a special position owing to their suitability in biological applications since biological tissues are largely transparent to NIR light. They have been applied in photothermal therapy and drug delivery remotely actuated by NIR light. Local heat produced by these agents raises the temperature and causes cell death in cancer treatment. Photothermal agents have also been used to trigger and accelerate drug release in biomedical applications. A useful photothermal agent is required to exhibit strong absorption of light and efficient transduction of light into heat. Gold nanoparticles (GNPs), gold nanorods (GNRs), carbon nanotubes (CNTs), graphene, and iron oxide nanoparticles have been employed as photothermal agents owing to their high photothermal conversion efficiency [55–67]. Similarly, conjugated polymers and dye molecules have also been utilized as photothermal agents. These organic compounds convert absorbed light through a non-radiative relaxation process. Plasmonic particles such as gold nanoparticles produce heat subsequent to light absorption through a distinct

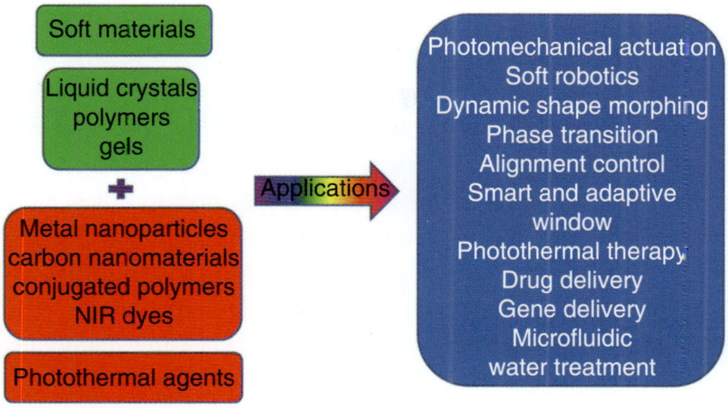

Figure 1.2 Different demonstrated applications of the hybrid systems that have been fabricated by combining soft materials and photothermal agents.

mechanism. Light absorbed by GNPs excites the electrons in the plasmonic band. These excited electrons relax through electron–phonon interaction by colliding with the gold lattice. This collision produces heat, which is transferred to the surrounding medium through phonon–phonon coupling, resulting in increasing the surrounding temperature (Figure 1.1e).

Recently, remote driving of soft materials by combining with photothermal agents is an emerging endeavor that reaps the benefits of both classes of promising materials. Accordingly, both inorganic and organic photothermal agents have been incorporated into the matrices of soft materials. The functional composite materials have been driven by light where the photothermal agent absorbs the light and converts it into heat, thereby increasing the temperature. The remotely triggered local temperature increase causes various physical and morphological changes including phase transitions. The occurrence of such changes in the soft matter matrix has been exploited for different applications (Figure 1.2). In this chapter, we present the developments on soft materials driven by photothermal effect of various photothermal agents. The combination of well-known soft materials such as LCs, polymers, and gels with inorganic nanoparticles, carbon nanomaterials, and organic dyes is discussed.

1.2 Liquid Crystals Driven by Photothermal Effect

Liquid crystals (LCs) represent a state of matter that appears between the crystalline solid state and isotropic liquid state [68–75]. This thermodynamically stable state of matter has been recognized as the fourth state of matter after solid, liquid, and gas. In this state of matter, the constituent elements, i.e. molecules, macromolecules, and molecular aggregates, simultaneously exhibit order and mobility, which renders it very intriguing and fascinating. The presence of order makes this state anisotropic whereas molecular mobility facilitates stimuli responsiveness. LCs are commonly known for their extensive applications in liquid crystal display (LCD) devices. However, they have been applied in numerous non-display applications [76–86]. In addition to their technological applications,

LCs are interesting for fundamental studies to understand self-assembly and self-organization of matter in multiple dimensions and over different length scales [87–97]. LCs serve as model supramolecular systems. They have been regarded as an important class of soft materials owing to their large response to small stimuli [98–108]. Moreover, they play very critical roles in living systems. LCs offer a significant contribution in nanoscience and nanotechnology. LCs have been classified into two broad classes: thermotropic and lyotropic. In thermotropic materials, the occurrence of LC phases is dependent on temperature. In lyotropic systems, the LC phase formation is primarily determined by the concentration of solutes in appropriate solvents although temperature also plays a role in the appearance and stability of LC phases.

Thermotropic and lyotropic LC phases undergo phase transition in response to the effect of temperature change. The temperature of the systems can be changed by directly heating or cooling in contact with heat sources. However, temperature variation by remote control and noncontact methods is appealing in many instances where application of direct heat is either detrimental or impractical. In this context, the use of light energy to vary the temperature of thermotropic and lyotropic LCs is very promising. In order to achieve this goal, it is often necessary to have a component in the system that can absorb the light energy and efficiently convert it into heat, thereby increasing the temperature of the medium. This role of converting light energy into heat has been played by certain metal nanoparticles, carbon nanomaterials, and organic compounds that exhibit photothermal effect. Thus, LC materials can be effectively driven by light using such photothermal agents in their matrices. The temperature variation by photothermal effect can either cause phase transitions or modification of physical properties of LC phases. Therefore, a variety of photothermal agents including inorganic nanoparticles and organic compounds have been employed for driving LC materials for fundamental studies and different applications.

Sun et al. have prepared gold-nanocrystal-doped LC elastomer microparticles and studied their light-driven shape morphing [109]. Soft lithography and polymerization techniques were used to fabricate microcylinders from the LC elastomer nanocomposite. The processing conditions were such that weakly undulating director orientation along the long axis of the microcylinders was obtained. Upon irradiation with infrared (IR) laser, the dispersed GNPs efficiently produce heat and transfer it to the surrounding in the matrices. This local heat produced upon exposure to IR light is able to modulate the ordering in the liquid crystalline microparticles. Both reversible and irreversible changes in the shape of the microcylinders have been observed using scanned beams (Figure 1.3).

Fong et al. doped GNRs into a lyotropic LC matrix [110, 111]. Through NIR laser irradiation, they could achieve control over the nanostructure of the mesophase (Figure 1.4). Since lyotropic mesophases are used for drug and gene delivery applications, control over their nanostructures could facilitate on-demand delivery of encapsulated agents. Integration of GNRs and driving phase transitions by photothermal effect in lyotropic mesophases are promising way to control the nanostructure of the mesophases. Accordingly, GNRs were dispersed into the lyotropic mesophase matrix and the reversible changes in their nanostructures under NIR laser irradiation were investigated by different techniques.

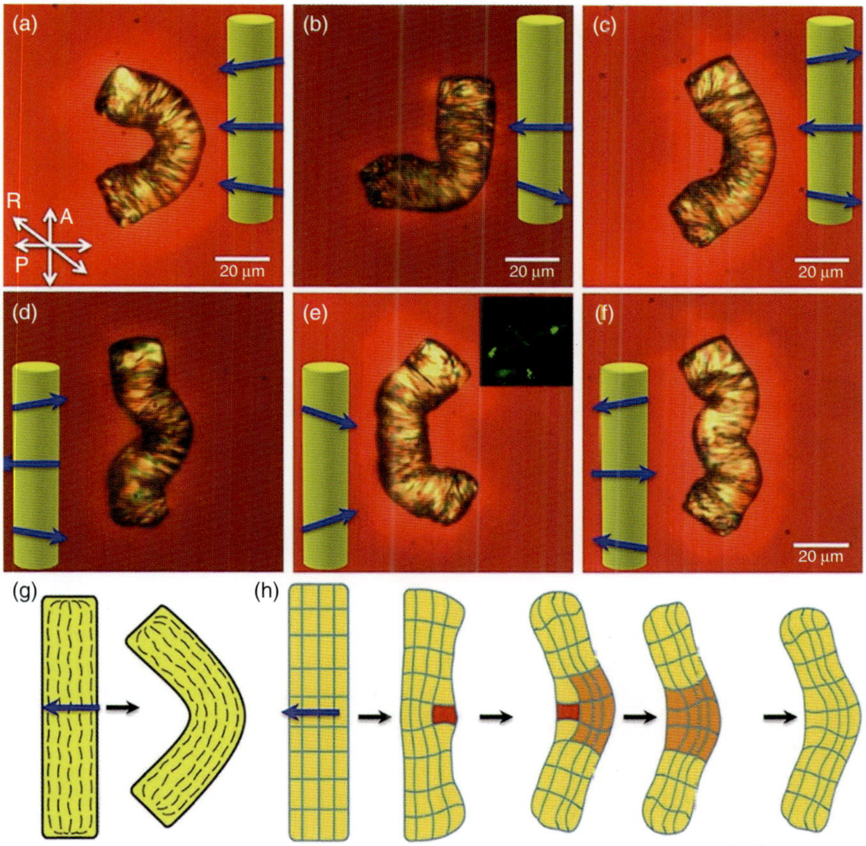

Figure 1.3 Examples of robust reversible morphing of liquid-crystalline elastomer (LCE) microparticle shapes by means of unidirectional laser beam scanning along blue arrows shown in the insets of (a–f). The inset in (e) shows a typical CARS-PM image of a microparticle that was bent using a scanning laser beam. (g) Schematics of transformation of weakly undulating director field ($n(r)$) (shown by dashed lines) as the particle is bent due to scanning. (h) Schematic drawings show the effect of local manipulation of LCE orientational ordering by a scanned laser beam along the direction marked by the blue arrow; the scanning causes nonreciprocal unidirectional motion of a molten region (red) and in modification of the LCE polymeric network in the vicinity of the "hot" scanning region (orange), eventually leading to the stable modification of the particle shape that persists after the laser is turned off. Source: Sun et al. 2012 [109]. Copyright 2012. Reproduced with permission from the American Institute of Physics.

Moreover, the effects of lipid composition, aspect ratio, and concentration of GNRs as well as laser pulse duration on the response of the systems were studied.

Fong et al. further undertook studies to understand the role of photothermal effect in non-lamellar LC systems [112]. They set out to reveal the photothermal response of known and new lamellar and non-lamellar lyotropic phases by correlating the thermal properties of the materials and the photothermal energy conversion by the photothermal agent. Photothermal heating was carried out by irradiating the composite systems with NIR laser and the consequent structural

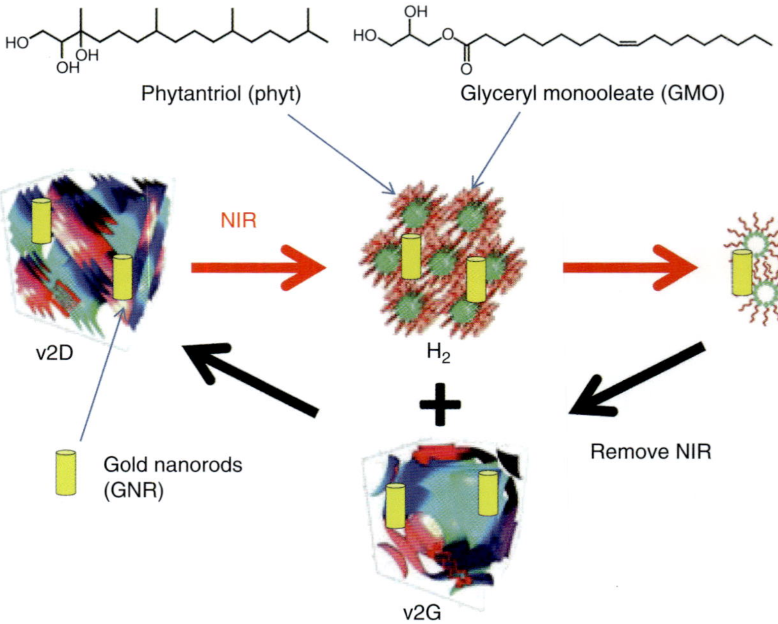

Figure 1.4 (Top) The molecular structures of phyt and GMO. In excess water, the two lipids transition from the double diamond bicontinuous cubic phase (v2D) to the inverse hexagonal phase (H_2), and then the fluid isotropic inverse micellar phase (L_2) with increasing temperature. (Bottom) The observed phase transition of the phyt-GNR LC upon laser irradiation. On irradiation, the phyt-GNR LC transitioned from v2D to H_2 to L_2, and on cessation transitioned back from L_2 to H_2 + the nonequilibrium gyroid bicontinuous cubic phase, v2G, before returning to v2D. Source: Fong et al. 2012 [110]. Copyright 2012. Reproduced with permission from The American Chemical Society.

changes of the LC phases were monitored by X-ray scattering studies. The effect of GNRs on the hexagonal phase was studied by electron microscopy techniques. Detailed understanding of such lyotropic nanocomposites is critical for their use in drug delivery applications remotely triggered by photothermal effect.

Pezzi et al. studied the photothermal effect of GNPs in thermotropic nematic liquid crystal (NLC) phases both experimentally and theoretically [113]. They characterized the photothermal effect of GNPs doped into thermotropic LCs and subsequently modeled the observations. Theoretical modeling of the photoinduced temperature variation in the LC nanocomposites was performed using the thermal heating equation that is applicable to an anisotropic medium such as the host nematic phase. Predictions of the theory were verified by conducting photothermal experiments on an LC nanocomposite fabricated by dispersing GNPs into a nematic host. Thermographic comparison between the nanocomposite and pure host clearly shows that GNPs are responsible for the observed photothermal effect under resonant optical frequency (Figure 1.5). This comprehensive investigation contributes toward the understanding of the physical phenomenon of heat propagation at the nanoscale.

Liu et al. observed rapid and reversible actuation of LC elastomer micropillars containing GNPs [114]. In this study, a new experimental protocol has been

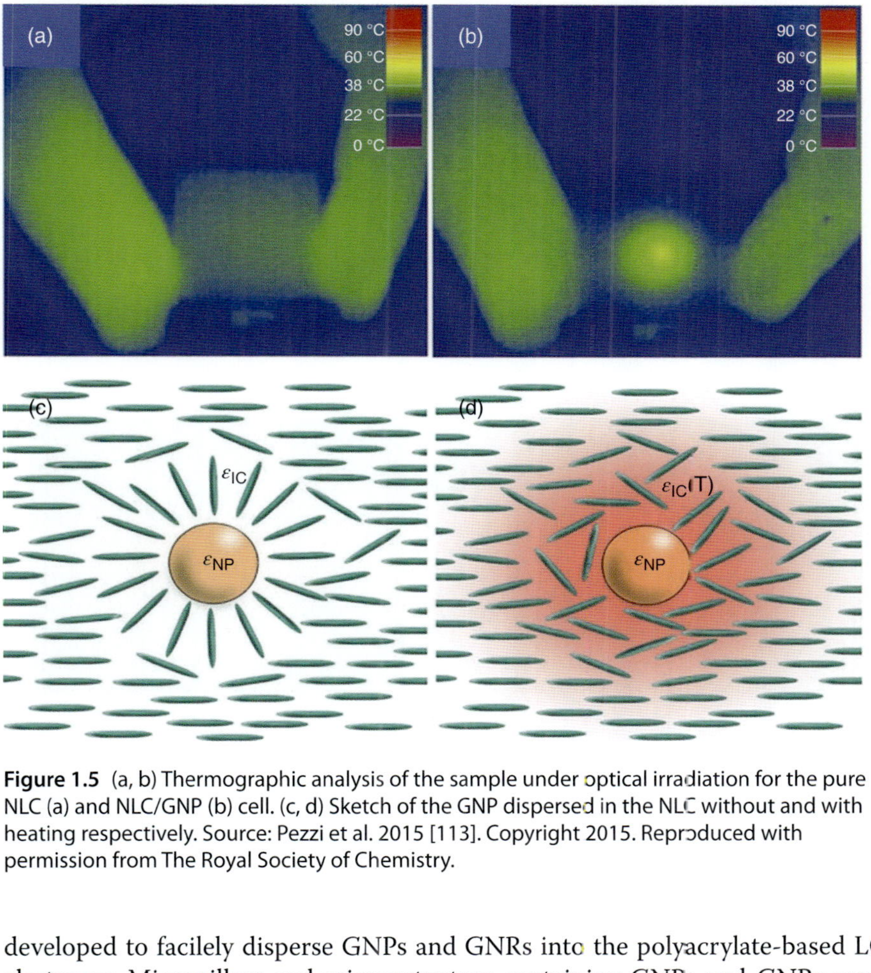

Figure 1.5 (a, b) Thermographic analysis of the sample under optical irradiation for the pure NLC (a) and NLC/GNP (b) cell. (c, d) Sketch of the GNP dispersed in the NLC without and with heating respectively. Source: Pezzi et al. 2015 [113]. Copyright 2015. Reproduced with permission from The Royal Society of Chemistry.

developed to facilely disperse GNPs and GNRs into the polyacrylate-based LC elastomer. Micropillars and microactuators containing GNPs and GNRs were fabricated from the plasmonic polymer nanocomposites. The effect of the incorporated GNPs was investigated by a battery of characterization methods including microscopy, spectroscopy, thermal analysis, and scattering experiments. Photothermal actuation behavior of the composites and the pure elastomer host were studied and compared. It was found that GNR is a good candidate for the development of LC elastomer-based high-performance photothermal actuators. Yang et al. prepared GNR-containing LC elastomer composites using thiol-click chemistry [115]. These elastomeric composites exhibited responsiveness to NIR and underwent nematic to isotropic phase transition under NIR irradiation.

We designed and synthesized mesogen-functionalized plasmonic GNRs and dispersed them into thermotropic NLC hosts [116]. Mesogen functionalization of the nanorods enhanced their compatibility in the LC matrix and enabled their homogeneous dispersion. Upon irradiation with NIR laser, the LC nanocomposite exhibited nematic to isotropic phase transition (Figure 1.6). However, when the light source was turned off the mixture returned back to the nematic phase. Thus, reversible phase transition was observed in the nanocomposites. It was noticed

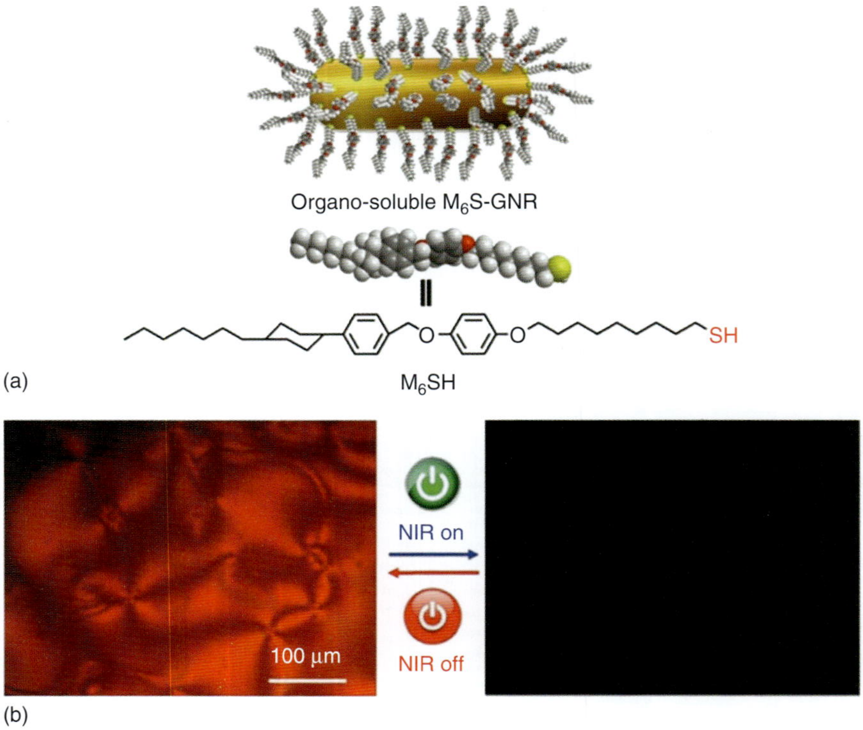

Figure 1.6 (a) Mesogen-functionalized organosoluble GNR. (b) Phase transition of the 0.5 wt% M_6S-GNR/E7 composite at DT = 3 1C with NIR laser irradiation took approximately two seconds from the nematic to the isotropic phase whereas it took approximately one second to return from the isotropic to the nematic phase when the NIR light was turned off. Source: Gutierrez-Cuevas et al. 2015 [116]. Copyright 2015. Reproduced with permission from The Royal Society of Chemistry.

that composites containing higher concentration of mesogen-functionalized GNRs underwent faster phase transitions.

Pezzi et al. found that GNPs embedded in NLCs exhibit a photothermal effect that is dependent on the refractive index of the medium [117]. They implemented a theoretical model to account for the observed photoinduced temperature variations in GNPs dispersed in NLCs. Experimental results of GNP LC composites have been compared with the theoretical predictions. de Sio et al. have studied plasmonic heating of GNRs in chiral LCs [118]. It was found that the presence of GNRs desensitizes the shift of the cholesteric liquid crystal (CLC) reflection band to temperature. Suitable cell design enables one to "turn off" the wavelength shift of the peak reflection, thereby turning the system into a pure amplitude measurement tool.

We have achieved handedness inversion in a GNR containing a CLC phase driven by NIR light irradiation [119]. Mesogen-functionalized GNRs were dispersed into the CLC phase formed by a chiral mesomorphic compound. Upon NIR irradiation, the sample temperature increases due to the photothermal effect of GNRs, which brings about pitch changes in the cholesteric system and thus a change in the selective reflection. Interestingly, the cholesteric system exhibits handedness inversion upon increasing the temperature (Figure 1.7). Red, green, and blue reflection colors were obtained both in the right- and left-handed states of the CLC. This NIR-light-driven plasmonic system would be

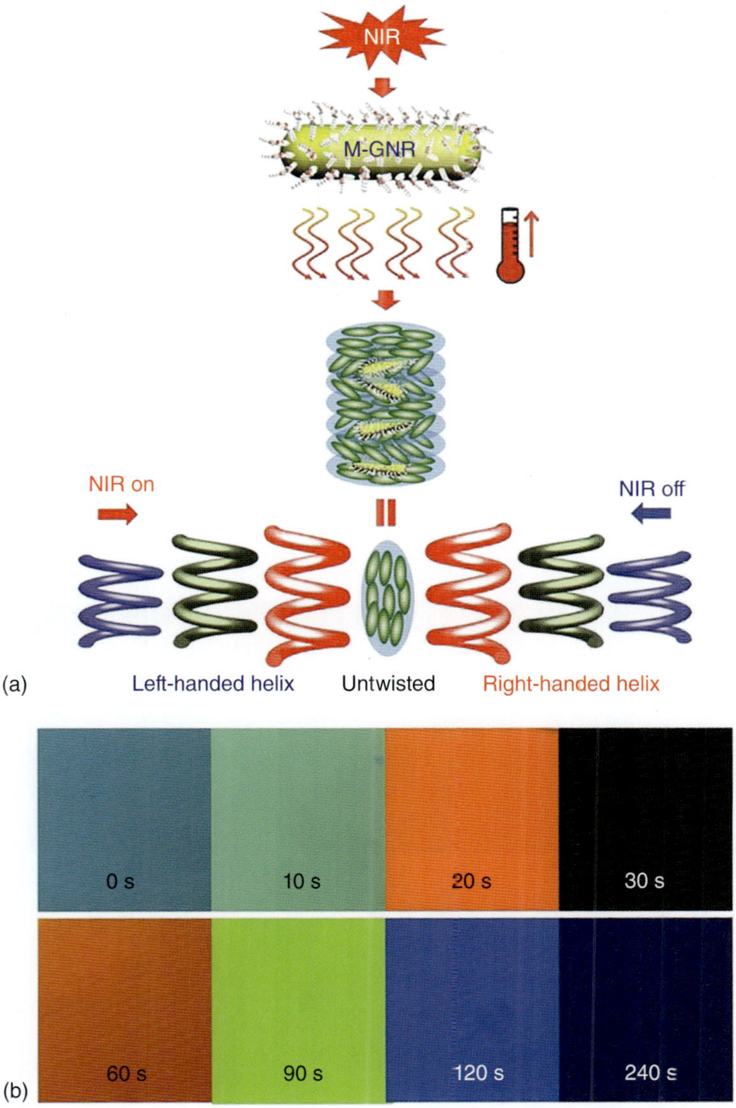

Figure 1.7 (a) Schematic illustration of NIR-light-directed handedness inversion in mesogen-grafted GNR (M-GNR)-embedded helical superstructure resulting from the "photothermal effect" of the plasmonic nanorods. (b) Dynamic photonic reflection colors of self-organized helical superstructure loaded with 0.2 wt% 770 nm M-GNRs in a 10-μm-thick planar cell upon NIR laser irradiation at 808 nm, taken under a crossed polarized optical microscope in reflection mode. Source: Wang et al. 2016 [119]. Copyright 2015. Reproduced with permission from Wiley-VCH Verlag GmbH & Co. KGaA.

promising in applications where both left- and right-handed circularly polarized light with tunable wavelengths are required.

We have also utilized the photothermal effect of GNRs to study phase transitions in LC blue phase composites [120]. Mesogen-functionalized GNRs were designed, synthesized, and dispersed into the cubic matrix of a blue phase. Upon irradiation with NIR light, the composite exhibited phase transition from a body-centered cubic lattice to a simple cubic lattice before turning into the isotropic phase. The NIR-light-driven phase transition has been found to be reversible (Figure 1.8). Interestingly, the lattice constant of the blue phase could be tuned by NIR light irradiation, which has produced red, green, and blue reflection colors from these photonic nanostructures.

Palermo et al. have demonstrated optical control of the plasmonic heating effect in NLCs employing the photoalignment technique [121, 122]. Here, they have exploited the dependence of the refractive index of NLCs on the molecular orientation. By varying the molecular alignment of the NLCs, control over the rate of plasmonic heating by the GNPs has been achieved.

Yang et al. have reported IR-light-driven reversible actuation in LC elastomer nanocomposites containing CNTs [123]. They used poly(*p*-phenyleneethynylene)-functionalized single-walled CNTs as fillers in the LC elastomer matrix. Such functionalization was found to enhance the dispersibility of CNTs in the elastomer matrix. The NIR response of the composite was found to increase upon high loading of CNTs in the composites. Ji et al. have developed a protocol for the dispersion and alignment of CNTs in LC polymers and elastomers [124].

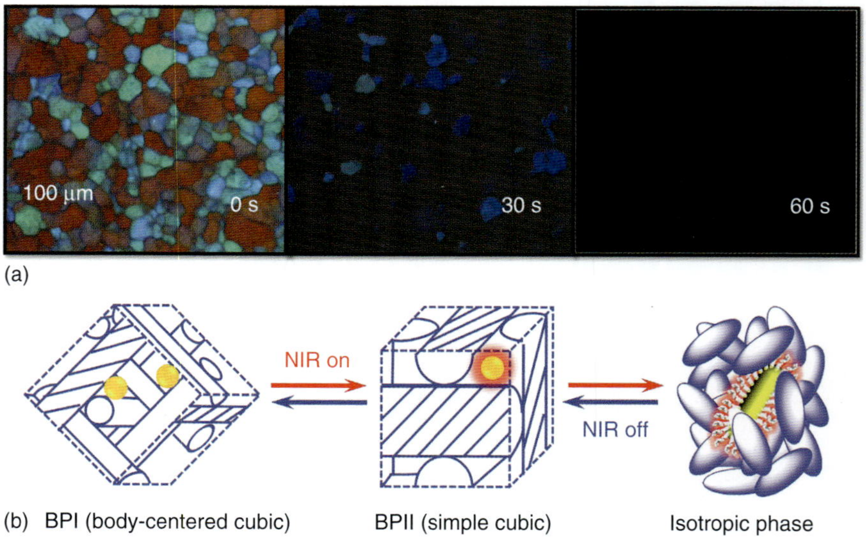

Figure 1.8 (a) Typical polarized optical microscope (POM) images of blue phases (BPs) doped with 0.3 wt% M-GNRs in 10-mm-thick untreated cells upon irradiation with an 808-nm NIR laser (2 W). (b) Schematic illustration of the NIR-light-tunable self-organized soft photonic superstructures. The yellow dots represent the cross section of GNRs in the BP matrices. Source: Wang et al. 2015 [120]. Copyright 2015. Reproduced with permission from The Royal Society of Chemistry.

They have taken advantage of the photothermal property of the embedded CNTs to achieve remote-controlled actuation of the elastomeric nanocomposites. Li et al. demonstrated actuation of LC elastomer nanocomposites containing CNTs driven by IR as well as white light [125]. Similar observations have been reported by Marshall et al. [126]. Direct sunlight has also been employed to drive LC elastomer nanocomposites containing CNTs [127].

Kohlmeyer and Chen fabricated and demonstrated wavelength-selective IR-light-driven hinges based on LC elastomers [128]. Composites of the elastomers were prepared by dispersing either CNTs or NIR dyes. The elastomeric nanocomposites exhibit fast and reversible bending upon NIR light irradiation. Bilayers films and inchworm walkers have been realized using these nanocomposites (Figure 1.9).

Li et al. reported bulk nanocomposites from LC elastomers and single-walled CNTs [129]. Light-driven bulk actuation in the composites was demonstrated under uniform as well as asymmetric irradiation. Wang et al. have synthesized polysiloxane-based LC elastomers containing azobenzene moieties [130]. Nanocomposites prepared from these elastomers can be driven either by NIR or by UV irradiation. NIR laser irradiation was found to cause shrinkage of the nanocomposite films.

Yu et al. fabricated polymer-dispersed LC–graphene oxide nanocomposites and prepared thin films using the solution casting method and mechanical stretching [131]. In the composites, graphene-oxide-doped poly(vinyl alcohol) acted as the matrix whereas a low molar mass room temperature NLC formed

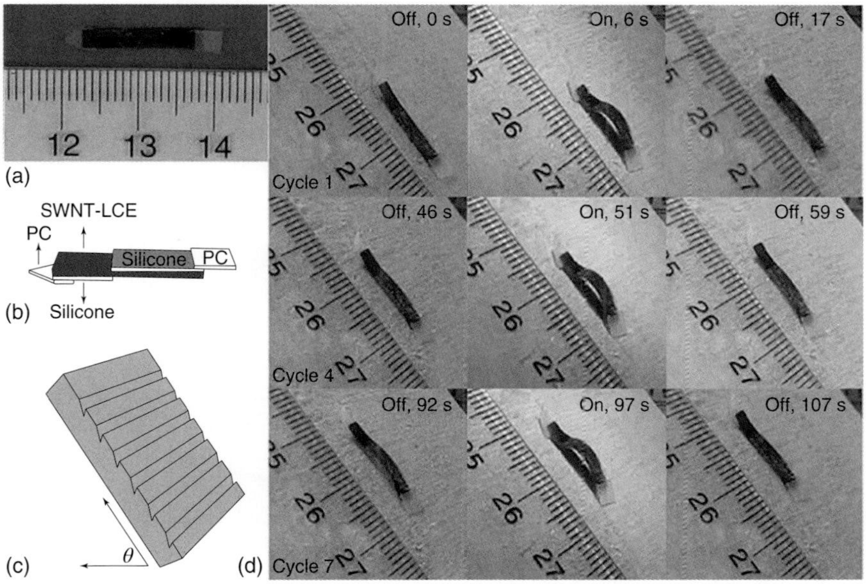

Figure 1.9 (a) Photograph and (b) scheme of an inchworm walker device consisting of an asymmetric SWNT-LCE/silicone bilayer film. (c) Scheme of a ratcheted wood substrate. (d) The inchworm walker crawling up the wood substrate at a 50° incline in response to on and off cycles of CW NIR light (28.2 mW mm^{-2}). Source: Kohlmeyer and Chen 2013 [128]. Copyright 2013. Reproduced with permission from Wiley-VCH Verlag GmbH & Co. KGaA.

a separated phase. The nanocomposite films exhibited photomechanical response upon irradiation with visible light (Figure 1.10a–c). The origin of this response has been attributed to arise from the photothermal effect of graphene oxide.

Yu et al. further demonstrated light-driven tumbler movement in polymer graphene oxide nanocomposite films [132]. Reversible and stable photomechanical behavior was obtained in these nanocomposites due to the combination of shape memory effect of the polymer host and photothermal effect of the graphene oxide guest. Control over such complicated motion could lead to devices with improved performances.

Quinn et al. have dispersed graphene nanoparticles into lyotropic LC phases [133, 134]. They observed that upon NIR light irradiation, the lyotropic system undergoes phase transition from cubic to micellar through hexagonal phase (Figure 1.10d). The cubic phase recovers when the NIR source is removed. The occurrence of reversible phase transition has been attributed to the local heating of the matrix through photothermal effect of the embedded graphene particles.

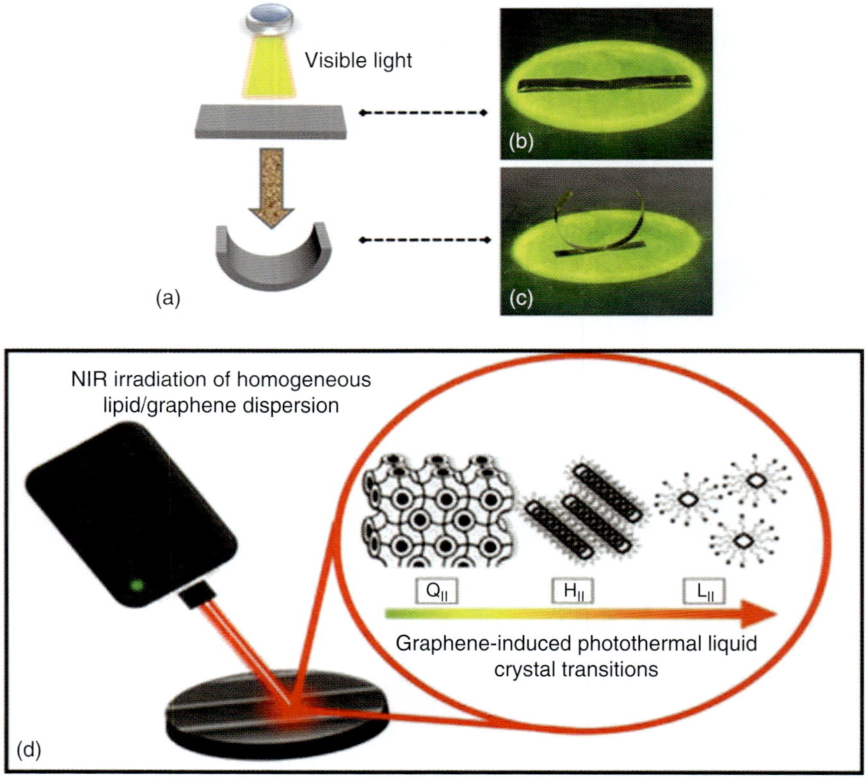

Figure 1.10 Visible-light-responsive behavior of the stretched polymer-dispersed liquid crystalline/graphene oxide (PDLC/GO) nanocomposite films. (a) Schematic illustration of photomechanical behaviors. (b) and (c) Photographs of the nanocomposite films before and after irradiation with visible light. (d) Schematic diagram showing the change in phase upon photothermal heating of the phytantriol–water LC. Source: Xiang et al. 2015 [131]. Copyright 2014. Reproduced with permission from The Royal Society of Chemistry. Yu et al. 2014 [101]. Copyright 2015. Reproduced with permission from The American Chemical Society.

Such hybrid systems could be employed for *in vivo* payload release by stimulation with NIR irradiation.

Yang et al. have prepared LC elastomer nanocomposites containing graphene nanosheets by *in situ* photopolymerization [135]. Photopolymerization and hot-drawing of the composites have yielded composites with highly aligned graphene in their matrix. They have investigated the photomechanical actuation in nanocomposite films driven by NIR irradiation (Figure 1.11).

Li et al. have observed substantial light-induced uniaxial contraction in LC elastomer nanocomposites containing graphene oxide nanoparticles [136]. Moreover, photoactuation of the nanocomposites has been observed under both UV and IR light irradiation.

Wang et al. have synthesized mesogen-functionalized graphene, which facilitates homogeneous dispersion and high compatibility in a chiral LC matrix [137]. Utilizing the stable LC nanocomposites, an adaptive window system has been demonstrated that responds to multiple stimuli (Figure 1.12). The optical transparency of the device could be modulated between transparent and opaque states by IR light irradiation by exploiting the photothermal effect of graphene sheets incorporated into the system. Here, light-induced heating of the film causes phase transition in the chiral LC, leading to transparency modulation. The chiral nanocomposites hold huge potential toward the fabrication of adaptive and energy-efficient windows for automotive and architectural applications.

Vallooran et al. have introduced magnetic nanoparticles into the matrix of lyotropic LCs and investigated the photothermal effect of the nanoparticles on the phase behavior of the system [138]. The magnetic nanoparticles greatly facilitate the alignment of the LC phase under magnetic field. Upon photoirradition, the system was observed to undergo LC to isotropic phase transition owing to the photothermal effect exhibited by the iron oxide nanoparticles. It was found that the effect of the photothermal effect could be tuned by adjusting

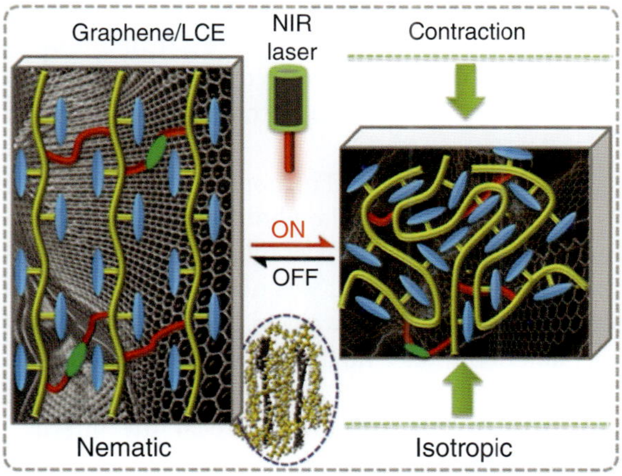

Figure 1.11 Schematic illustration of reversible photomechanical actuation in graphene/LCE nanocomposites upon an on–off switching of NIR light. Source: Yang et al. 2015 [135]. Copyright 2015. Reproduced with permission from Wiley-VCH Verlag GmbH & Co. KGaA.

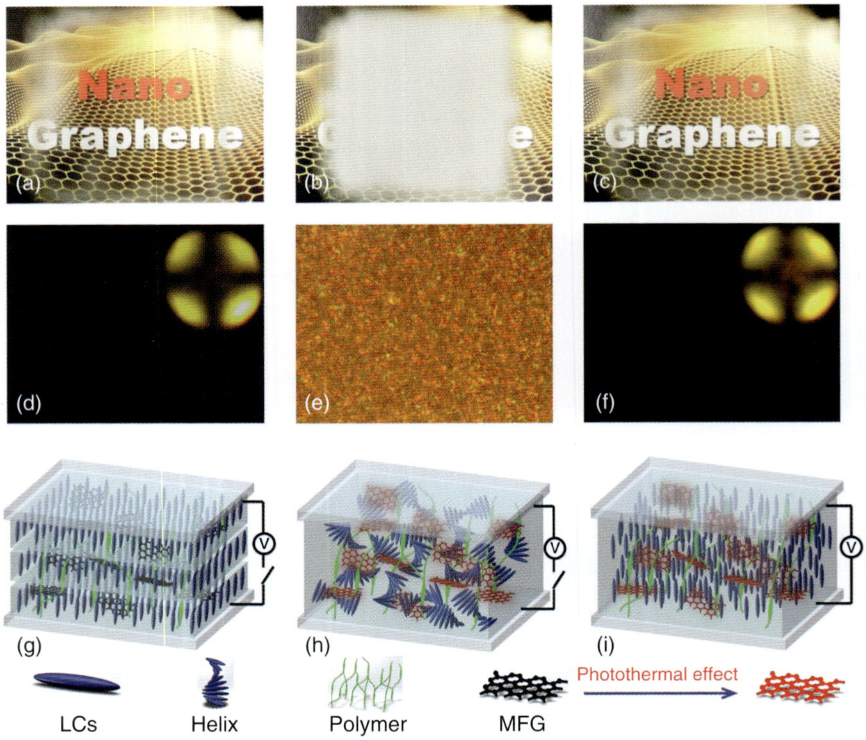

Figure 1.12 Polymer-stabilized MFG-containing chiral LC thin film. Photograph of the polymer-stabilized MFG-containing chiral LC film with 7 wt% R811: (a) transparent state at room temperature; (b) opaque state after NIR irradiations (2 W); (c) transparent state from the unwound cholesteric focal conic phase upon applying the electric field. POM images of the devices under different addressing conditions (d–f). Schematic illustrations of the corresponding configurations of LC phases (g–i) in the devices. Source: Wang et al. 2017 [137]. Copyright 2017. Reproduced with permission from Elsevier.

the concentration of the nanoparticles in the matrix. Moreover, the orientational order can also be modulated by changing the concentration of the particles. This hybrid mesophase system is responsive to both magnetic field and photoirradiation.

Ono and Kawatsuki have studied the photothermal response characteristics of host–guest LC systems [139]. The results showed the photoresponsive properties to be dependent on the change in refractive index of the system due to temperature variation caused by laser light absorption. It was noticed that refractive index change was proportional to laser beam intensity. The photothermal response of the system was reversible. They have further investigated photothermal effect in NLCs on a dye-doped polymer film [140]. The NLC molecules were found to undergo reorientation due to rise in temperature at the interface between the NLC layer and the polymer surface.

Guo et al. have fabricated and investigated NIR-absorbing elastomer composites containing mesogenic croconaine dye [141]. They designed and synthesized the first example of an NIR-absorbing liquid crystalline croconaine-based dye.

The LC nature of the NIR dye makes it more compatible in the elastomer matrix. Moreover, homogeneous distribution of the dispersed dye could be achieved in the soft polymer matrix. A monodomain LC elastomer composite was obtained by adopting the traditional LC alignment method in combination with photopolymerization of monomer mixtures. Films fabricated from the elastomer composite were able to show a fully reversible contraction and expansion cycle in response to NIR light irradiation. Liu et al. have dispersed polyaniline-conducting polymer nanoparticles into the matrix of LC elastomers and investigated their NIR response through photothermal effect [142]. The effect of doping concentration and the conjugation form of the polyaniline nanoparticles on the photomechanical response of the elastomer composite was studied. The elastomeric composite material was found to exhibit fully reversible muscle-like actuation and could lift up heavier weights under NIR irradiation owing to the photothermal effect of the polyaniline nanoparticles.

Gelebart et al. have reported on self-sustained mechanical oscillators that are produced by embedding NIR-absorbing dyes into LC networks [143]. They developed splay-aligned LC networks using commercially available LC monomers. To obtain self-sustained oscillation in the system by light irradiation they have added photostabilizers into the polymer network matrix that convert absorbed light into heat through non-radiative relaxation. The observation of mechanical self-oscillation has been linked to temperature variation over the film thickness. The oscillation of a beam-shaped device (Figure 1.13) is activated in a localized responsive zone that acts as a hinge. It was found that light spanning across UV to the NIR region could be employed for actuation.

Liu et al. synthesized NIR chromophore-functionalized LC elastomers that are capable of exhibiting fast response speed and promising mechanical property [144]. A croconaine dye-based cross-linker was used as the NIR-absorbing motif. They prepared uniaxially aligned main-chain LC elastomers using a two-step diene metathesis *in situ* polymerization and cross-linking. Under NIR irradiation, the elastomer system raises its temperature from 18 to 260 °C by efficient photothermal conversion due to the presence of homogeneously distributed and covalently linked NIR chromophore. The soft actuator fabricated from these NIR-absorbing elastomers was able to lift loads 5600 times heavier than its own weight.

Li et al. have doped polydopamine nanoparticles into the matrices of LC elastomers. The photothermal effect of polydopamine has been exploited in the systems for developing dynamic three-dimensional (3D) structures by light-irradiation-induced deformations [145]. The dispersion of polydopamine in the elastomer matrices turned out to be more facile than the dispersion of CNTs. The mechanical properties of the polydopamine-dispersed elastomers have been found to be superior to those of the pure host materials. Tian et al. have reported the fabrication and light-driven actuation of polydopamine-coated LC elastomers [146]. These coated systems exhibit rapid response and significant contraction because of the photothermal effect of polydopamine cladding. Interestingly, by surface scanning with NIR laser the polydopamine-coated elastomer films could be made to bend or roll up. A prototype robotic swimmer was designed and demonstrated using the elastomer films, which was able to swim

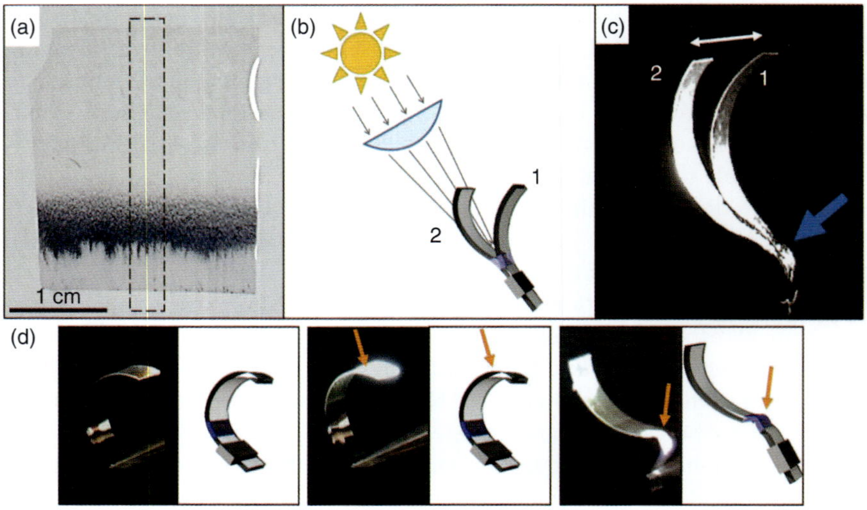

Figure 1.13 Self-oscillation obtained with the indigo hinge. (a) Picture of the nonactive film with the hinge of indigo component. (b) Schematic representation of the setup used to actuate the liquid crystalline network (LCN) film containing a stripe of indigo. The sunlight is focused by a lens and has sufficient intensity to bring the system into an oscillating mode, between positions 1 and 2. (c) Superposition of two pictures of the oscillating film in positions 1 and 2. Over time, the film is moving continuously between these two positions. The self-shadowing at the hinge is visible in this superposition. In position 1 the hinge is in the shadow while it is in the sunlight in position 2. (d) In the first picture, the film is in the dark at the resting state. In the second picture, the focus point of the light (indicated by the orange arrow) does not overlap with the indigo stripe and no bending is observed. In the last picture, the film is positioned so that the focused light reaches the indigo stripe. In that condition only, the film bends significantly and self-oscillates. Source: Gelebart et al. 2017 [143]. Copyright 2017. Reproduced with permission from Wiley-VCH Verlag GmbH & Co. KGaA.

on the water surface in response to NIR laser irradiation. The above-described research shows that polymer-based photothermal agents could be employed in the fabrication of optically driven artificial muscle for application in soft robotics and functional optomechanical devices.

1.3 Polymers Driven by Photothermal Effect

Biopolymers such as polysaccharides, polypeptides, and polynucleotides have played a vital role in the evolution and advance of life. Their criticality in life process is well appreciated by scientists and nonscientists alike. Synthetic polymers such as teflon, nylon, and polyesters have revolutionalized the way we use materials and have significantly influenced our modern life, starting from the comfortable clothes we wear to the luxury cars we drive. Very recently, supramolecular polymer materials have been developed by appropriate engineering of different noncovalent interactions. These dynamic and functional materials display a variety of promising properties and enable many applications that are either difficult or impossible to achieve by conventional polymer materials.

While polymers themselves are interesting, recently it has been recognized that their composites often exhibit remarkably superior traits and perform better in certain applications. Accordingly, many varieties of polymer composites have been developed and explored, and in the recent years polymer nanocomposites containing CNTs, graphene, and graphene derivatives have been paid special attention [147–152]. It has been noticed that carbon-based nanomaterials often enhance the optical, electrical, and mechanical properties of polymer nanocomposites and introduce new capabilities into the hybrid polymer systems.

CNTs, which absorb light and convert it into heat, have been dispersed into the matrices of various polymers to realize functional polymer nanocomposites. The polymer matrices can be driven by the photoinduced temperature increase [153]. CNTs have been used in thermoplastic elastomer matrices for efficient shape recovery by utilizing the photothermal effect of CNTs that facilitates local heating in the system upon light irradiation. Ahir et al. investigated IR-light-induced actuation in aligned polymer–nanotube composites [154, 155]. The polymer–CNT composites have been found to either expand or contract depending on the processing of the hybrid system. Rubber composites with multi-walled CNTs have been prepared and their reversible photomechanical actuation response has been investigated under NIR light irradiation. CNTs were aligned in the rubber matrix by applying stress to the composites. The degree of alignment of the nanotubes was found to influence the photomechanical response of the composites. These composites contract along the alignment direction of the nanotubes. Different polymer hosts were evaluated to gain greater insight into the photomechanical actuation triggered by NIR light.

Lu and Panchapakesan investigated the photomechanical responses of CNT/polymer actuators [156]. They compared the photomechanical response of bulk polymer nanocomposites and multilayer constructions containing single-walled CNTs and multi-walled CNTs. The photomechanical actuation behavior of the composites containing single-walled CNT and multi-walled CNT was related to nanotube alignment in the matrix and the presence of pre-strain. Comparable actuation properties were observed between multilayered constructions and bulk composites. Expansion and contractions were observed under small and large pre-strains respectively. Miyako et al. fabricated and studied CNT–polymer composites for light-driven microthermal control [157]. Here, polydimethylsiloxane (PDMS) was chosen as the soft polymer and noncovalently surface-functionalized CNTs were dispersed into its matrix. Noncovalent functionalization of CNTs is often preferred since this procedure does not significantly compromise the intrinsic electronic property of the nanotubes. A phospholipid and a protein were used for the surface modification of the nanotubes. A microchip was fabricated from the polymer nanocomposite and temperature distribution in a microchannel was investigated under NIR irradiation. Temperature control of an aqueous solution in the microchannel has been achieved, which makes this hybrid system suitable for various lab-on-a-chip applications.

Czanikova et al. developed photoactuating materials based on elastomers and modified CNTs [158]. The matrix of ethylene–vinylacetate was integrated with multi-walled CNTs. The nanotubes were surface functionalized with a pyrene cholesterol conjugate by the noncovalent approach to enhance its compatibility and dispersibility in the elastomer matrix. Nanotubes were uniaxially aligned in the composite by shearing forces. The photoactuation behavior of the soft and responsive nanocomposites was investigated by atomic force microscopy and nanoindentation techniques. Czanikova et al. have also reported single-walled CNT-based photoactuators in ethylene vinyl acetate copolymer [159]. Photoactuation and relaxation behaviors of the nanocomposites with aligned nanotubes were studied with respect to CNT concentration, light intensity, etc.

Tang et al. reported CNT-dispersed poly(ethyleneglycol) (PEG) composites for solar energy conversion and storage [160]. They prepared a ternary nanocomposite using multi-walled CNT, PEG, and silicon dioxide. The composite could absorb sunlight and efficiently convert it into thermal energy and can act as an energy storage system. Given the huge demand for energy conversion and storage devices and systems, this simple ternary nanocomposite containing a photothermal agent and a phase change material is very interesting for fundamental studies and technological applications. Ilcikova et al. developed acrylic thermoplastic elastomers that were dispersed with block-copolymer-grafted CNTs [161]. The nanocomposite of the triblock copolymer containing diblock-copolymer-grafted CNTs exhibited photoactuating behavior. Under red light exposure, the elastomeric nanocomposite showed fast and reversible actuation. CNTs and fullerene-based hybrid systems have been developed and applied as temperature indicators by taking advantage of the photothermal conversion characteristic of CNTs [162]. CNTs and GNPs have also been employed as photothermal agent in the shape, morphology, and wetting properties of fullerene-based microparticles [163].

Graphene, the two-dimensional (2D) allotrope of carbon, has gained enormous attention [164]. Impressive electronic, optical, and mechanical properties along with high surface area have rendered graphene an appealing material. Insertion of graphene into polymer matrices has yielded significantly improved physical properties.

Liang et al. fabricated graphene-based nanocomposites that could be triggered by IR light and can act as actuators [165]. They dispersed sulfonated functionalized graphene particles into thermoplastic polyurethane to prepare light-driven nanocomposite actuators. The nanocomposite not only exhibited excellent photoactuation but also helped in significantly improving the mechanical properties of the hybrid system. Upon irradiation with IR light the nanocomposite undergoes contraction and thereby could lift a 21.6 g weight by 3.1 cm with a force of 0.21 N. The mechanical properties of the nanocomposites were found to improve with increase in graphene loading in the system. Comparison of different kinds of graphene materials has revealed that the light-driven actuation behavior of the nanocomposites depends on the nature of the aromatic network of graphene as well as the homogeneity of its dispersion in the matrix. Loomis et al. have revealed layer-dependent mechanical responses of graphene composites to NIR light [166]. They fabricated graphene nanosheet dispersed polymer nanocomposites and prepared thin films from the hybrid system. Nanocomposites containing single layer graphene and multilayer graphene were made for comparison of their physical response to NIR light irradiation. Both elastic moduli and stress response of the single layer and multilayer graphene containing composites were observed to be different from one another. The difference in behavior of these composites has been attributed to the number of graphene layers in contact with the host polymer in the matrix. Loomis et al. have developed photothermal nanopositioners from graphene–elastomer nanocomposites [167]. The nanopositioners were fabricated from the nanocomposites made by dispersing graphene nanoplatelets into the soft and responsive matrix of PDMS. Upon light irradiation the grahene nanoplatelets absorb the photons and convert them into heat, thereby increasing the local temperature in the nanocomposite. They developed a two-axis stage with sub-micron resolution, which permitted two-axis positioning (120 nm resolution and about 5 $\mu m\,s^{-1}$ speed) driven by photothermal effect. Good actuation speed and appreciable actuator efficiency were obtained for the nanocomposites. However, there is still room for improvement in the performance of the photothermal nanopositioners.

Ansari et al. made optical actuators from graphene and triblock copolymer composites [168]. Freestanding thin films were prepared from the nanocomposites containing graphene platelets in the matrix of the block copolymer poly(styrene-b-isoprene-b-styrene). Solvent casting technique was used in the preparation of the freestanding nanocomposite films. It was found that under strained conditions the nanocomposite film contracts upon IR irradiation. However, under stress-free conditions the films were found to expand. Such films could find application in biomedical field in situations where remote-controlled actuation is preferred. Meng et al. observed enhanced photothermal effect of graphene conjugated polymer composites [169]. Poly(3-hexylthiophene)-conjugated polymer was covalently grafted onto reduced graphene oxide sheets to prepare the hybrid system. This hybrid system exhibited superior photothermal effect when compared with the system without covalent linking. This enhanced performance has been attributed to photoinduced energy transfer from the conjugated polymer backbone to the reduced graphene oxide surface. Using a thin layer of this hybrid material a photocontrolled electrical switch has been demonstrated.

Muralidharan et al. prepare nanocomposites from biocompatible natural polymer chitosan and reduced graphene oxide [170]. Using solvent casting technique they obtained nanocomposite-based actuators that exhibit photomechanical actuation under IR light illumination. The samples studied underwent contraction in length in response to IR exposure. Light-induced stress and strain were found to increase upon increase in reduced graphene oxide loading in the actuator matrices. In this system the applied pre-strain was found to reduce photomechanical stress of the actuators, which is an unusual behavior. Since photomechanical actuation in the nanocomposite is dependent on the concentration of graphene oxide and the applied pre-strain, the actuation behavior could be tuned by varying these two parameters. Leeladhar et al. demonstrated the fabrication of highly versatile photomechanical actuators using graphene platelets dispersed in PDMS matrix [171]. The actuators were prepared in a bilayer actuator where a thin layer of chromium was coated on one side of the polymer nanocomposite film. The bilayer actuators were found to exhibit fast and reversible actuation with a large deflection angle. Moreover, the actuator was found to exhibit photoresponse depending on the wavelength used. The large bending observed has been ascribed to the mismatch of the coefficients of thermal expansion between the two layers.

Robinson et al. developed biocompatible reduced-graphene-oxide-based hybrid materials that exhibit high absorption in the NIR region [172]. Single-layered reduced graphene oxide nanosheets were noncovalently surface functionalized with amphiphilic PEGylated polymer chains, which are able to show sixfold higher NIR absorption when compared to their nonreduced and covalently functionalized counterparts. Photoablation of U87MG cancer cells was demonstrated *in vitro* by employing reduced graphene oxide hybrids bearing a peptide with the Arg–Gly–Asp (RGD) moiety as photothermal agents. The reduced graphene oxide hybrid system, which is biocompatible and exhibits high NIR absorption, could be useful for photothermal therapy of cancer.

Feng et al. prepared polyurethane-based efficient actuators by incorporating functionalized reduced graphene oxide and CNTs into the polymer matrix [173]. These polyurethane hybrids containing the nanofillers exhibit good IR light absorption and consequent actuation. By adjusting the doping ratio of reduced graphene oxide and CNT, impressive IR-actuated stress recovery has been achieved in these multicomponent photothermal systems. Niu et al. have demonstrated a photomechanical soft actuator that has been fabricated from PDMS doped with graphene nanoplatelets [174]. The soft actuator is made up of a bilayer film: one layer is the graphene-dispersed PDMS and the other layer is pure PDMS. In this bilayer system, the mismatch between the coefficients of thermal expansion of the layers has been utilized to achieve controllable and reversible bending in response to NIR light irradiation. Different bending behavior has been registered by selecting the layer for exposure to light source. The occurrence of temperature gradient across the thickness of the layers has been assumed to be responsible for the distinct bending behavior observed.

Organic conjugated polymers capable of conducting ions and electrons have been developed and extensively investigated. Their applications in organic electronic and optoelectronic devices such as solar cells, field effect transistors, and light-emitting diodes have been well documented. Apart from these well-known applications, their suitability in various other areas has been assessed and they have been employed in high-tech applications such as biomedical science, biosensing, supercapacitor, chemical sensing, and fluorescence imaging. Owing to their extended conjugation length, many conducting polymers could absorb IR radiation and through nonradiative relaxation can produce heat. This energy transduction property of the conjugated polymers qualifies them as an attractive class of organic photothermal agents. Consequently, they have been used in both *in vitro* and *in vivo* photothermal therapy of cancer. These polymers could also be used as highly efficient drug delivery systems with large loading capacity. Thus, they could be useful in combination therapies. Recently conjugated polymers with large absorption coefficient, good photostability, bicompatibility, biodegradability, and acceptable photothermal conversion efficiencies have been realized and their application potential as photothermal agents has been evaluated.

Yang et al. reported organic nanoparticles based on the conducting polymer polyaniline and used them as photothermal agents in the ablation of cancer cells by light irradiation [175]. Polyaniline can exist in emeraldine base (EB) or emeraldine salt (ES) state. The EB form of polyaniline can be readily converted into the ES state by doping process (Figure 1.14). The doping process shifts the

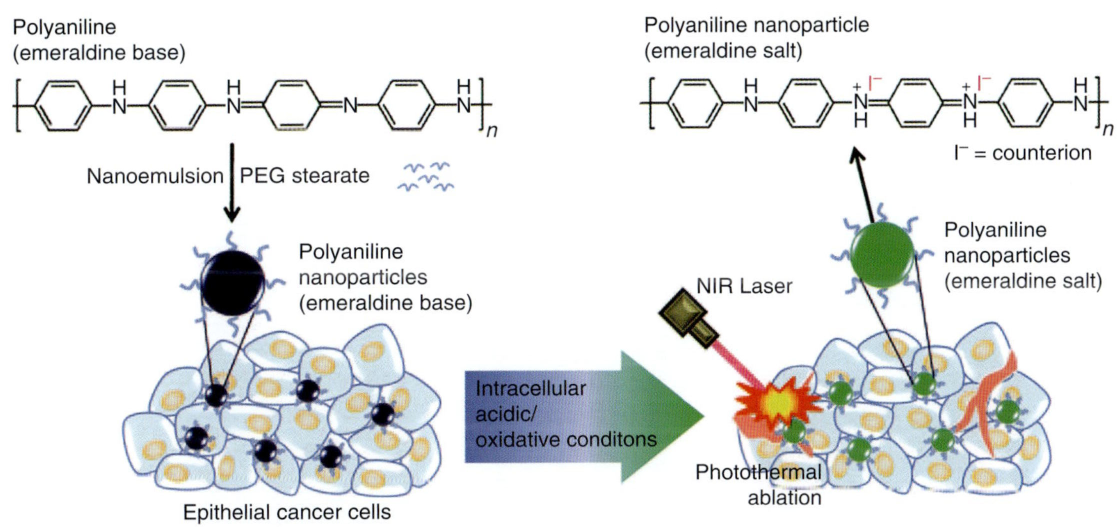

Figure 1.14 Schematic illustration of the preparation of organic photothermal agents based on polyaniline nanoparticles and their application for the photothermal ablation of epithelial cancer cells by NIR laser irradiation. Source: Yang et al. 2011 [175]. Copyright 2011. Reproduced with permission from Wiley-VCH Verlag GmbH & Co. KGaA.

absorption of polyaniline into the NIR region of the electromagnetic spectrum by decreasing the energy bandgap. Polyaniline is found to be capable of converting the absorbed NIR light into heat and is biocompatible. Therefore, it has been used as an organic photothermal agent in photothermal therapy of cancer. Water soluble and stable polyaniline nanoparticles have been prepared by nanoemulsion method and have been applied in the treatment of epithelial cancer cells using NIR light. Similar surface-coated polyaniline nanoparticles have also been used for both *in vitro* and *in vivo* photothermal therapy of cancer [176]. The surface-functionalized uniform polyaniline nanoparticles were found to possess high extinction coefficient and exhibit higher photothermal conversion efficiency. Low toxicity of these photothermal nanoparticles to living cells has been ascertained using a variety of analytical characterization techniques.

Geng et al. designed and synthesized biocompatible conjugated polymer nanoparticles for efficient photothermal tumor therapy [177]. They have judiciously engineered the polymer backbone through the coupling of donor and acceptor moieties. Polymer nanoparticles were fabricated using a suitable precipitation method in an encapsulation matrix. Both *in vitro* and *in vivo* cancer cell killing by NIR light irradiation has been demonstrated using these polymer nanoparticles with large absorption coefficient. Such carefully designed polymers could find use in photothermal cancer therapy.

Lyu et al. developed and used the photothermal effect of semiconducting polymer bioconjugates for activation of targeted neurons [178]. The semiconducting bioconjugates have been found to possess higher photothermal conversion efficiency than GNRs. They have demonstrated NIR-light-driven control of thermosensitive ion channels in neurons using the polymer bioconjugates. Owing to the fast energy conversion ability to heat and high spatial precision to the ion channel, safe and reversible activation of the intracellular calcium ion influx of neuronal cells has been accomplished using the polymer bioconjugates. This method is benign in that it does not involve genetic transfection. This organic nanoparticle-based photothermal technique is powerful and could be a potential alternative to optogenetics. Lyu et al. have demonstrated *in vivo* photothermal therapy using semiconducting polymer nanoparticles [179]. They carried out intraparticle molecular orbital manipulation to enhance the photothermal efficiency of the polymer nanoparticles for cancer therapy. For this purpose, they have used fullerene particles in combination with semiconducting polymers. The energy levels of the donor polymer and acceptor fullerene derivative have been so aligned that photoinduced electron transfer occurs efficiently and quenches the fluorescence in the system. This leads to nonradiative relaxation following light irradiation and thus produces heat. This approach to photothermal therapy could be extended to other binary composite systems. Cao et al. designed and synthesized an organic conjugated polymer and found that it possessed high efficiency in photothermal conversion [180]. The conjugated polymer prepared contains donor and acceptor entities in its structure and has a very narrow bandgap. The narrow bandgap enables the polymer to absorb NIR light. This photostable polymer with high extinction coefficient exhibits high photothermal conversion. The photothermal effect of the polymer has been used to drive temperature-dependent devices. The conjugated polymer could be applicable in biomedical technology due to its high absorbance in the biocompatible NIR window.

1.4 Gels Driven by Photothermal Effect

Gels are a promising class of soft materials. They have been extensively investigated in materials science and bioscience owing to their outstanding characteristics and a variety of potential and practical applications [181–185]. Gels do not flow and exhibit solid-like rheology. Stimuli responsiveness, adaptability, and self-healing of morphological defects are some of their excellent attributes. Gels contain a large volume of liquid (water or organic solvent), which is immobilized by a network of gelators. The gelators could be small molecules with low molecular weight or macromolecules with high molar mass. Hydrogels are made up of water as the solvent whereas organogels contain organic solvents. Gels can vary in the nature of network formation by the gelators. The gelators can form chemically cross-linked networks via covalent bonding or exhibit physically cross-linked networks through noncovalent supramolecular interactions. Different properties are observed in different types of gels. It should be noted that gels can also be fabricated from LC phases using suitable gelators. Unlike in hydrogels and organogels, the solvent in LC gels is an anisotropic fluid medium. Advanced nanocomposites have been developed by dispersing nanoscale particles into the matrix of gels. Nanocomposites of gels could exhibit superior properties than the pure host gels due to the synergy between complementary strengths of the two functional components in one system. Recently, distinct nanocomposite gels containing different photothermal agents have been attracting considerable amount of attention. Application prospects of such nanocomposites have been demonstrated in materials science, biomedical science, optomechanics, microfluidics, drug delivery, etc. Moreover, the gel nanocomposites are interesting hybrid systems to understand how collective properties arise from interaction between the components within the soft stimuli-responsive matrices.

Mebrouk et al. have designed and synthesized nickel bis(dithiolene)-based gelators containing pendant cholesteryl moieties [186]. They have fabricated gels using these metal complexes that form intertwined fibers. It was observed that the nickel bis(dithiolene) derivatives exhibit efficient absorption of NIR light comparable to GNPs and convert it into heat in the gel matrix. Taking advantage of this photothermal property, the stability of the gels could be modulated by NIR light irradiation. Such photothermal gels could offer new opportunities in materials science and nanotechnology owing to the facile realization of photothermal gelators.

Wang et al. have developed an injectable hydrogel that can be used to carry out photothermal therapy of cancer [187]. Interestingly, the hydrogel could be degraded on demand following therapy to clear out from the body, thereby reducing the risk of cytotoxicity that arises due to long-term retention of hydrogel in the body. The hydrogel is based on alginate-calcium, which immobilizes platinum nanoparticles containing dendrimers (Figure 1.15). The hydrogel is biocompatible and degrades upon injection of appropriate chelates. Here the

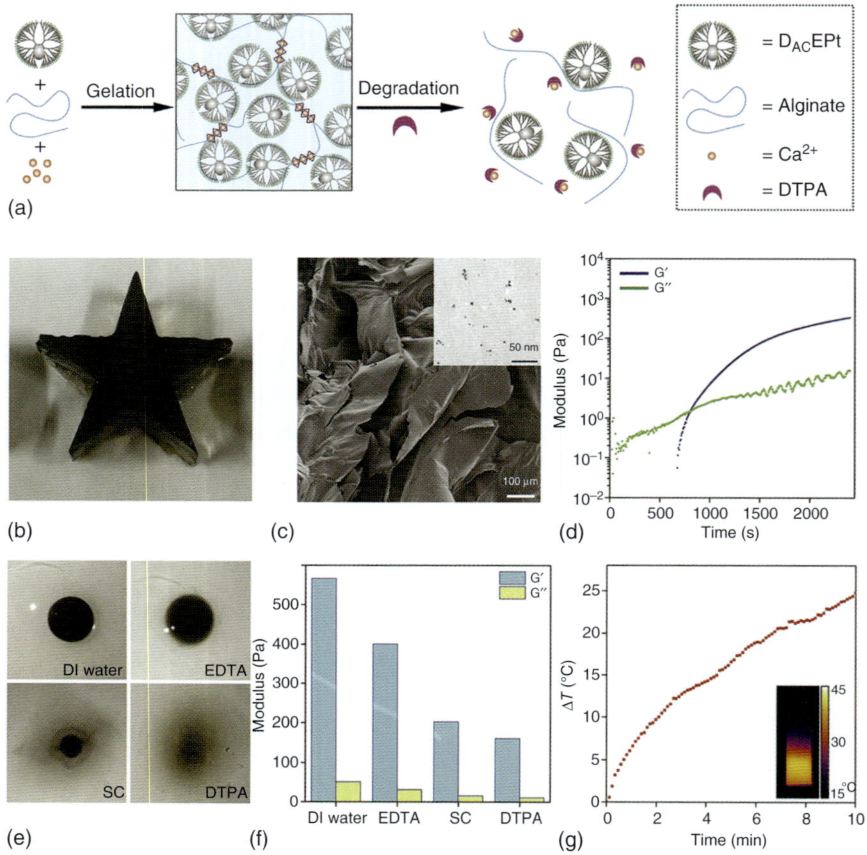

Figure 1.15 Schematic depicts the gelation and on-demand degradation of the gel. Source: Wang et al. 2016 [187]. Copyright 2016. Reproduced with permission from Elsevier.

dendrimer-encapsulated platinum nanoparticles act as photothermal centers and raise the surrounding temperature upon NIR irradiation. It has been observed that this hydrogel could ablate tumors when coated onto the tumor instead of being injected into it. This protocol of injection and on-demand degradation of the hydrogel in cancer photothermal therapy is attractive since it not only reduces the adverse effects but also drastically diminishes toxicity in the body.

Zheng et al. developed supramolecular hydrogels that are capable of exhibiting switchable size and dynamic mechanical properties in response to light of different wavelengths (Figure 1.16) [188]. The gel containing dendrimer-encapsulated platinum nanoparticles displays softening and stiffening in response to UV and NIR light irradiation. The gel is found to shrink in size upon NIR light irradiation; however, it recovers when the NIR source is put off. A photothermally driven high-strength double-network gel has been developed, which exhibits IR-light-induced actuation behavior [189].

Mohamed et al. have investigated the dynamics of hot electrons and phonons of GNPs that are inserted in gel matrices and irradiated by laser pulses [190].

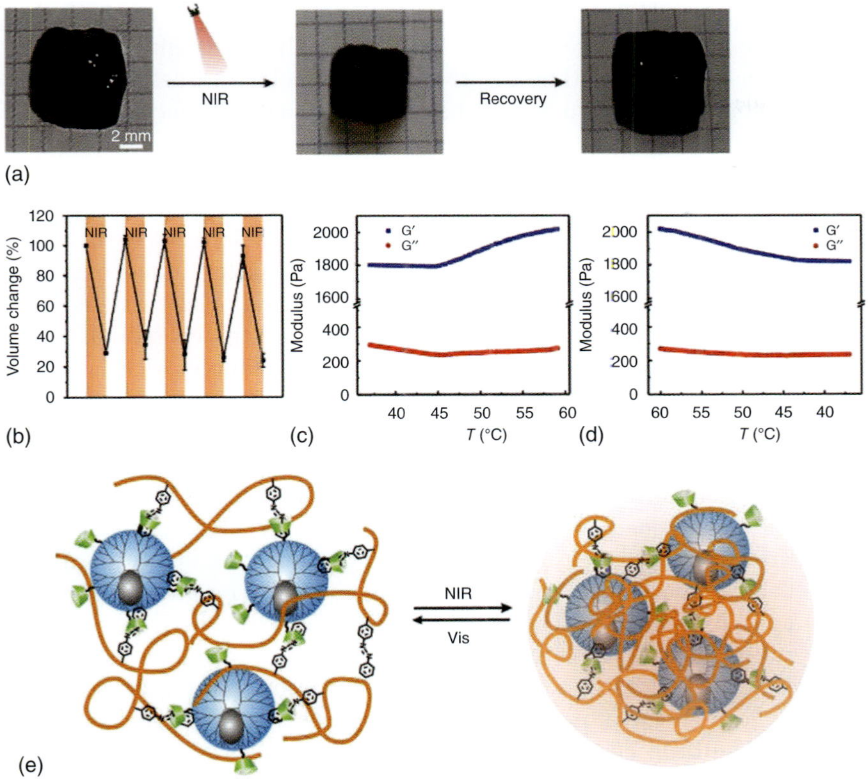

Figure 1.16 NIR-responsive behavior of the supramolecular gel. (a) Photographs of the supramolecular gel irradiated by an NIR laser. (b) Changes in the volume of the gel when an NIR laser is periodically turned on and off. (c, d) Stiffness of the gel at varying temperatures. (e) Mechanism of the NIR-responsive gel. Source: Zheng et al. 2017 [188]. Copyright 2017. Reproduced with permission from The American Chemical Society.

Electron–phonon and phonon–phonon coupling relaxation times have been found to increase in the hydrogel and organogel matrices compared to the aqueous solution. Kim and Lee have developed hydrogel-coated GNPs with different shell thicknesses [191]. Owing to the biocompatibility of the hydrogel polymer shell, these hybrid nanoparticles could serve as drug delivery vehicles. Moreover, the hydrogel shell of the nanoparticles could be easily thermally activated by exposure to light with wavelengths that fall in the plasmon absorption band of the GNP core. Kawano et al. designed and synthesized gel-coated GNRs [192]. These hybrid core–shell nanorods served as heating centers in targeted delivery in response to NIR laser irradiation. They show remote and reversible *in vivo* action upon NIR irradiation. These nanorods can be employed as probes for *in vivo* imaging techniques. *In vivo* fabrication of hydrogels needs delivery of two different solutions that would spontaneously cross-link upon mixing. However, in such processes the kinetics of gelation is difficult to control. To address this issue, Gramlich et al. carried out transdermal gelation using GNRs and NIR light irradiation [193]. The heat produced during NIR irradiation of the nanorods brings about polymerization leading to gelation. The characteristics of gels thus produced could be modulated spatially and temporally by changing the experimental parameters. This method is a promising protocol toward gelation through tissue using light that could penetrate tissue without harmful effects.

Stowers et al. have demonstrated dynamic tuning of the stiffness of a 3D hydrogel by light irradiation [194]. Hydrogels are widely used to unravel the role of stiffness on cell behavior since cell phenotype is greatly affected by the stiffness of the extracellular matrix. However, in biological processes the change in matrix stiffness is dynamic. In this context, a hydrogel with tunable stiffness over time could serve as a model system. For this purpose, they developed a system where alginate gel stiffness can be temporally modulated. The stiffness modulation of the gel was carried out by light-triggered release of calcium or a chelate from liposomes (Figure 1.17). Both dynamic stiffening and softening have been demonstrated in the system. This method could be extended to transdermal stiffness modulation by IR light irradiation.

Collins et al. have investigated charge modulation of a GNR in an ion gel device [195]. GNRs were added into an ion gel capacitor and changes in the electron concentration were studied using dark-field imaging spectroscopy.

Park et al. have developed a drug delivery system based on micro-organogels and loaded with GNRs [196]. They prepared the micro-organogels by emulsification technique using gelator fiber and vegetable oil, which served as a biocompatible drug carrier. A nonsteroidal anti-inflammatory model drug was encapsulated into the micro-organogel matrix for evaluation of the drug delivery properties. The drug delivery system released the model drug slowly in the absence of NIR light irradiation. However, upon exposure to NIR light the drug release was significantly accelerated owing to temperature increase of the GNR-containing drug delivery system that transforms the gel into liquid (Figure 1.18). The micro-organogel nanocomposite has the potential for on-demand drug delivery applications.

Sershen et al. have demonstrated independent control of microfluidic valves using nanocomposite hydrogels containing GNPs [197]. The hydrogel has been formed by poly(*N*-isopropylacrylamide), which exhibits reversible thermal phase changes. Fluid flow through the valves can be controlled by adjusting the intensity of the light source. Shiotani et al. have observed light-driven fast shrinking of a polymer hydrogel that was loaded with GNRs [198]. A model drug-loaded composite was prepared from the hydrogel and it was observed that the system rapidly releases the drug upon NIR laser irradiation. Moreover, irradiation region-specific release of the drug has been demonstrated by taking advantage of the high spatial resolution of light irradiation of the nanocomposite gel. Erdogan et al. have shown light-triggered control of degelation in peptide-based organogels embedded with GNPs [199]. They have used GNPs of different sizes and different shapes to evaluate their degelation capability under light irradiation. The control over degelation of biological molecules could offer promising possibilities in nanomedicine, drug delivery, nanodiagnostics, and tissue engineering. Zhou et al. have fabricated photothermally responsive hydrogel nanocomposites and developed micropatterned bilayer photoactuators that exhibit rapid and reversible bending and unbending movements upon

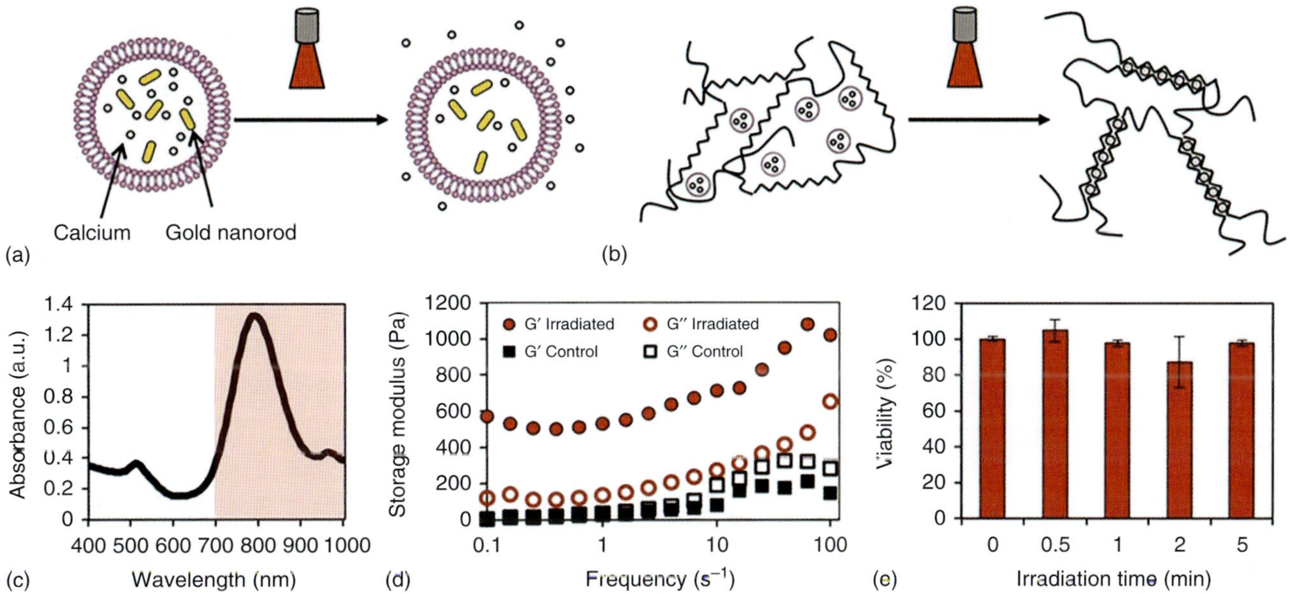

Figure 1.17 Mechanism of light-triggered release from liposomes. (a) Schematic of temperature-sensitive liposome loaded with gold nanorods and calcium. Upon irradiation, gold nanorods heat the lipid bilayer past its transition temperature, allowing calcium to permeate the lipid bilayer. (b) Irradiation of liposomes mixed with an alginate solution causes release of calcium and an increase in cross-linking density. (c) Absorbance spectrum of gold nanorods, demonstrating the narrow peak in NIR region. Shading denotes the optical window of highest penetration depth through tissue. (d) Rheometry of alginate: liposome solutions after irradiation. Irradiation causes release of calcium and gelation, shown by increased G′ over controls. (e) MTS assay demonstrates no loss of viability after irradiation of cell-seeded gels. Source: Stower et al. 2015 [194]. Copyright 2015. Reproduced with permission from National Academy of Sciences.

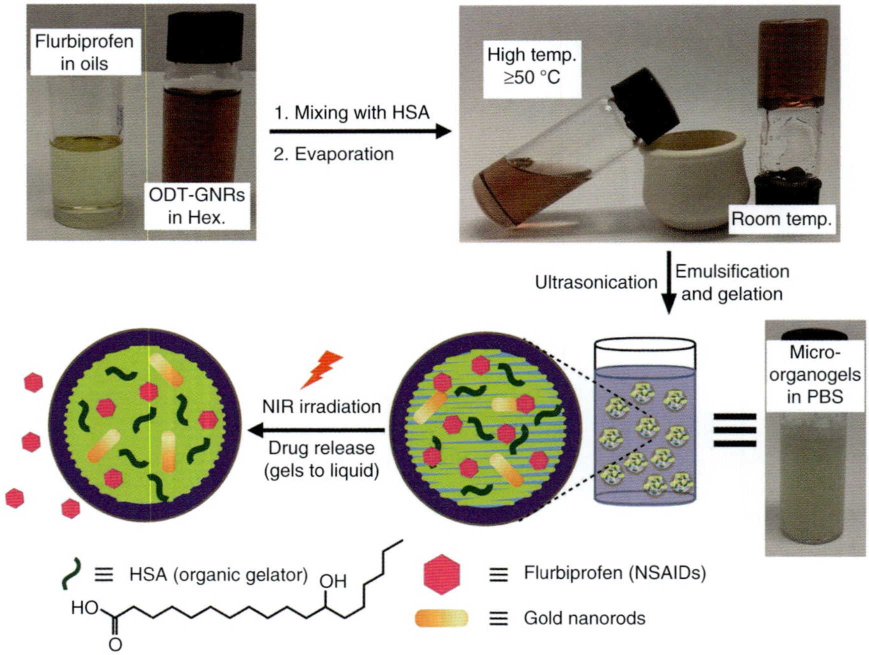

Figure 1.18 Schematic illustration for formation of GNR-containing micro-organogels and their drug release upon NIR irradiation. Source: Park et al. 2016 [196]. Copyright 2016. Reproduced with permission from The American Chemical Society.

exposure to visible light [200]. Light-waveguided through optical fiber directly into the photoactuator film has been found to be equally effective in causing actuation as free space light.

Lo et al. have developed a new light-responsive poly(*N*-isopropylacrylamide) hydrogel nanocomposite incorporating glycidyl-methacrylate-functionalized graphene oxide [201]. The hydrogel nanocomposites were produced using photopolymerization technique. The highly efficient photothermal conversion of glycidyl methacrylate-functionalized graphene oxide was found to cause large volume change of the nanocomposite hydrogels upon exposure to IR light. This large volume change has been exploited in the fabrication and operation of a microvalve that is able to control fluidic flow in the channel by remote actuation with IR light. Zhang et al. have fabricated CNT-containing hydrogel polymer composites and investigated their responsive behavior [202]. The polymer hydrogel was obtained using poly(*N*-isopropylacrylamide). Polymer hydrogel nanocomposites were fabricated by dispersing single-walled CNTs into the soft hydrogel matrix. Fast photomechanical actuation was observed in the hydrogel nanocomposites upon exposure to NIR light due to strong absorption by the CNTs. The response time of the actuators could be tuned by varying the loading level of CNTs in the matrix. Cubes and flowers capable of self-folding have been demonstrated using the polymer hydrogel nanocomposites. Zhang et al. fabricated graphene-oxide-containing polymer hydrogel nanocomposites capable of IR-light-driven actuation [203]. They have prepared a bilayer architecture using graphene-oxide-containing hydrogel and another hydrogel without graphene oxide. Upon IR irradiation the bilayer undergoes anisotropic bending motion due to the temperature increase in the hydrogel layer containing graphene oxide particles.

Zhu et al. synthesized a hydrogel nanocomposite containing graphene oxide and investigated its photothermal responsiveness [204]. The hydrogel was synthesized by *in situ* polymerization of *N*-isopropylacrylamide monomers in an aqueous solution dispersed with graphene oxide. The gel exhibits reversible phase transition upon irradiation with NIR laser. They have demonstrated the application potential of the nanocomposite gel in a microfluidic device for controlled fluid flow (Figure 1.19a–d). Such gels with superior photothermal sensitivity could find application in biomaterials and biomedical devices.

Fujigaya et al. have studied reversible volume change in CNT-containing gels upon NIR light irradiation [206]. These nanocomposite gels were found to possess excellent fatigue resistance during ON/OFF cycles. Li et al. have developed hydrogel nanocomposites by incorporating graphene oxide into a polymer matrix [207]. The hydrogels can be facilely triggered by NIR light irradiation. Using these functional gels as 3D scaffolds they demonstrated reversible cell capture and on-demand release by light exposure. Wang et al. have developed light-driven hydrogel actuators from genetically engineered elastin-like polypeptides doped with reduced graphene oxide particles [208]. Tunable motions with spatial resolution such as finger-like flexing as well as crawling has been demonstrated in these hybrid systems that can be controlled by varying different parameters of light source. Cong et al. have demonstrated water treatment using a thermoresponsive nanocomposite hydrogel containing graphene particles [209]. These mechanically robust hydrogel composites exhibited high adsorption capacity and good recyclability in treating water. Reduced-graphene-oxide-dispersed hydrogel composites have been used for the fabrication of bendable bilayer-type photoactuators that can be driven by visible light irradiation [210].

Satarkar et al. have reported the fabrication and investigation of remote-controlled microfluidic valves using magnetic nanoparticle hydrogels [211]. Gallium-based liquid metal alloys are recently attracting considerable attention in microfluidic devices as functional fluid microcomponents such as valves, heaters, and electrodes [212, 213]. Local switching of chemical patterns has been achieved by Yoon et al. employing the creased surface of a hydrogel [214]. The hydrogel was impregnated with iron oxide nanoparticles, which exhibit photothermal effect. Upon exposure to light the hydrogel temperature rises due to the photothermal conversion by the iron oxide nanoparticles, which causes volume change in the system. Light-controlled microvalves have been developed by Lee et al. using hydrogels incorporated with photothermal iron oxide nanoparticles [215]. For this purpose, they prepared the hydrogel composite that is capable of fast response and substantial volume change upon exposure to light. Movable position heating source [216], photoactuators from comb-type hydrogels [217], and stimuli-responsive hydrogel microfibers [218] have been obtained using photothermal magnetic nanoparticles dispersed into hydrogel matrices.

Hsiao et al. have used *in situ* formed NIR-absorbing micellar hydrogels in photothermal tumor ablation [205]. They have designed and synthesized a chitosan derivative functionalized with conjugated polymer polyaniline as side chains. This covalently linked hybrid polymer material is able to self-assemble into micelles. The micelles could be transformed into hydrogels by appropriately adjusting the pH of the micellar solution. Owing to the presence of the conducting polymer polyaniline in its structure, the micelles are able to absorb NIR light and convert it into heat. This photothermal property of the micelles was used in the selective destruction of cancer cells in light-exposed areas. The efficiency of the micellar hydrogel in photothermal therapy has been evaluated using a tumor-bearing mouse model (Figure 1.19e). This model study demonstrated

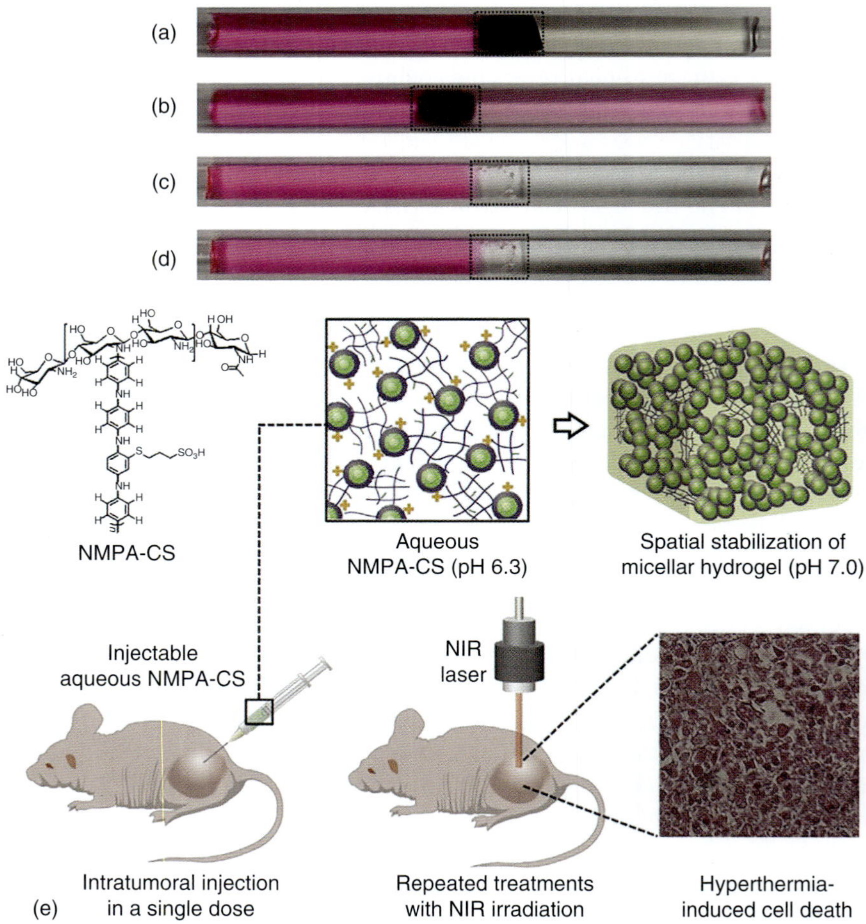

Figure 1.19 (a–d) Liquid microvalves made with the PNIPAM/GO hydrogel (a, b) and a PNIPAM hydrogel (c, d). The photographs show the microvalves before (a, c), after (b, d) NIR (808 nm, 2 W) irradiation for two minutes, where the positions of the hydrogels are marked by the rectangular boxes. The solutions encapsulated in the left and right sides are a red rhodamine solution and light yellow $NiSO_4$, respectively. Source: Gramlich et al. 2014 [193]. Copyright 2012. Reproduced with permission from Wiley-VCH Verlag GmbH & Co. KGaA. (e) Chemical structure of CS derivative that contained self-doped PANI side-chains and mechanism by which it photothermally treats tumors. Source: Hsiao et al. 2016 [205]. Copyright 2015. Reproduced with permission from Elsevier.

that biocompatible NIR-absorbing micellar hydrogels can be formed in the target site for minimal invasive photothermal treatment of cancer.

1.5 Summary and Outlook

The combination of soft materials and photothermal agents is emerging as an enabling endeavor. Functional soft composite materials fabricated by bringing these two classes of interesting materials have been demonstrated to act as actuators, therapeutic agents, drug delivery systems, microvalves, etc. Remotely triggered photoinduced phase transitions have been carried out by taking advantage of the photothermal effects of the photothermal agents. Inorganic metal (gold, platinum) nanoparticles, iron oxide nanoparticles, carbon nanomaterials (CNTs, graphene derivatives), and organic compounds (small molecule dyes and conducting polymers) have been employed as photothermal agents to drive LCs, polymers, and gels. The photothermal effect of GNPs has been used in drug delivery application of lyotropic LCs and they have enabled optical property modulation of self-assembled thermotropic 1D and 3D photonic crystals. Light-driven dynamic shape morphing and photomechanical actuations such as bending, twisting, folding, and oscillations have been achieved by combining photothermal agents with functional polymers and elastomers. Photothermally activated hydrogels have been employed in efficient water treatment and control of fluid flow in microfluidic devices. Smart and adaptive systems have been realized by dispersing photothermal agents into soft matter matrices. Synergy between the properties of soft materials and photothermal agents has been found to enhance the functionalities of the stimuli-responsive composites. In addition to physical mixing of photothermal agents with the soft matrices, they have also been covalently conjugated in the systems to fabricate better performing and robust materials. In many instances, the photothermal agents have been suitably functionalized to enhance their compatibility and homogeneous dispersion in the host matrices, thereby increasing their efficiency. However, phase separation at higher loading amounts remains an issue with the photothermal agents, which requires a great deal of attention. In drug delivery and photothermal therapy the long-term toxicity of the agents needs to be systematically evaluated before they could be clinically viable. Inorganic nanoparticles, being nondegradable, could accumulate in the cell and cause toxicity if they are not efficiently cleared from the body after they have performed their job. To address this issue, appropriate physical and chemical functionalization of inorganic nanoparticles is being developed and assessed. However, it needs extensive evaluation by materials scientists and clinicians for commercial applications. Organic photothermal agents have certain advantages over their inorganic counterparts; however, it is necessary to make a different variety of organic agents available for use. The covalent linking of conjugated polymers and NIR-absorbing dyes with soft matter components seems to be a promising approach to take advantage of the photothermal effect of organic agents; nevertheless, robust and versatile functionalization protocols that are facile and benign need to be developed and optimized to exploit the full potential of organic photothermal agents. The work accomplished in driving soft materials by photothermal effect and the applications demonstrated have laid the groundwork for further exploration. This area of research is in its early infancy and it is expected to make great strides in materials science, nanotechnology, bioscience, and beyond with the availability of appropriately engineered materials and processing techniques.

Acknowledgments

We thank the support from the US Air Force Office of Scientific Research (AFOSR), US Air Force Research Laboratory (AFRL), US Department of Defense Multidisciplinary University Research Initiative (DoD MURI), DoD-Army, US National Aeronautics and Space Administration (NASA), and the Ohio Third Frontier. We thank all Li's current and former group members as well as our collaborators, whose names are found in the references, for their significant contributions in this project.

References

1 de Gennes, P.-G. (1992). Soft matter. *Angew. Chem. Int. Ed. Engl.* 31: 842.
2 de Gennes, P.-G. (1992). Soft matter. *Rev. Mod. Phys.* 64: 645.
3 Hamley, I.W. (2007). *Introduction to Soft Matter: Synthetic and Biological Self-Assembling Materials*. Chichester, UK: Wiley.
4 Li, Q. (ed.) (2013). *Intelligent Stimuli Responsive Materials: From Well-defined Nanostructures to Applications*. Hoboken, NJ: Wiley.
5 Kleman, M. and Lavrentovich, O.D. (2003). *Soft Matter Physics: An Introduction*. New York: Springer-Verlag.
6 Selinger, J.V. (2016). *Introduction to the Theory of Soft Matter: From Ideal Gases to Liquid Crystals*. New York: Springer-Verlag.
7 Hirst, L.S. (2012). *Fundamentals of Soft Matter Science*. Boca Raton, FL: CRC Press.
8 Mezzenga, R., Schurtenberger, P., Burbidge, A., and Michel, M. (2005). Understanding foods as soft materials. *Nat. Mater.* 4: 729.
9 Marchetti, M.C., Joanny, J.F., Ramaswamy, S. et al. (2013). Hydrodynamics of soft active matter. *Rev. Mod. Phys.* 85: 1143.
10 Nagel, S.R. (2017). Experimental soft-matter science. *Rev. Mod. Phys.* 89: 025002.
11 Wang, L. and Li, Q. (2016). Stimuli-directing self-organized 3D liquid crystalline nanostructures: from materials design to photonic applications. *Adv. Funct. Mater.* 26: 10.
12 Bisoyi, H.K., Bunning, T.J., and Li, Q. (2018). Stimuli-driven control of the helical axis of self-organized soft helical superstructures. *Adv. Mater.* 30: 1706512.
13 Bisoyi, H.K. and Li, Q. (2016). Light-directed handedness inversion of self-organized helical superstructure: dynamic photoswitching the chirality of circularly polarized reflection. *Angew. Chem. Int. Ed.* 55: 2994.
14 Hamley, I.W. (2003). Nanotechnology with soft materials. *Angew. Chem. Int. Ed.* 42: 1692.

15 Quake, S.R. and Scherer, A. (2000). From micro- to nanofabrication with soft materials. *Science* 290: 1536.
16 Rogers, J.A. and Nuzzo, R.G. (2005). Recent progress in soft lithography. *Mater. Today* 8: 50.
17 Whitesides, G.M. and Lipomi, D.J. (2009). Soft nanotechnology: "structures" vs. "function". *Faraday Discuss.* 143: 373.
18 Whitesides, G.M. (2018). Soft robotics. *Angew. Chem. Int. Ed.* 57: 4258.
19 Li, Q. (ed.) (2018). *Functional Organic and Hybrid Nanostructured Materials: Fabrication, Properties, and Applications*. Weinheim: Wiley-VCH.
20 Wang, L. and Li, Q. (2018). Photochromism into nanosystems: towards lighting up the future nanoworld. *Chem. Soc. Rev.* 47: 1044.
21 Bisoyi, H.K. and Li, Q. (2014). Light-directing chiral liquid crystal nanostructures: from 1D to 3D. *Acc. Chem. Res.* 47: 3184.
22 Wang, Y. and Li, Q. (2012). Light-driven chiral molecular switches or motors in liquid crystals. *Adv. Mater.* 24: 1926.
23 Qin, Z. and Bischof, J.C. (2012). Thermophysical and biological responses of gold nanoparticle laser heating. *Chem. Soc. Rev.* 41: 1191.
24 Abadeer, N.S. and Murphy, C.J. (2017). Recent progress in cancer thermal therapy using gold nanoparticles. *J. Phys. Chem. C* 120: 4691.
25 Riley, R.S. and Day, E.S. (2017). Gold nanoparticle-mediated photothermal therapy: applications and opportunities for multimodal cancer treatment. *WIREs Nanomed. Nanobiotechnol.* 9: e1449. doi: 10.1002/wnan.1449.
26 Song, X., Chen, Q., and Liu, Z. (2015). Recent advances in the development of organic photothermal nano-agents. *Nano Res.* 8: 340.
27 Bao, Z., Liu, X., Liu, Y. et al. (2016). Near-infrared light-responsive inorganic nanomaterials for photothermal therapy. *Asian J. Pharm. Sci.* 11: 349.
28 Shi, C., Wu, J.B., and Pan, D. (2016). Review on near-infrared heptamethine cyanine dyes as theranostic agents for tumor imaging, targeting, and photodynamic therapy. *J. Biomed. Opt.* 21: 050901.
29 Huang, X., Jain, P.K., El-Sayed, I.H., and El-Sayed, M.A. (2008). Plasmonic photothermal therapy (PPTT) using gold nanoparticles. *Lasers Med. Sci.* 23: 217.
30 Yuan, A., Wu, J., Tang, X. et al. (2013). Application of near-infrared dyes for tumor imaging, photothermal and photodynamic therapies. *J. Pharm. Sci.* 102: 6.
31 Melamed, J.R., Edelstein, R.S., and Day, E.S. (2015). Elucidating the fundamental mechanisms of cell death triggered by photothermal therapy. *ACS Nano* 9: 6.
32 Huang, X. and El-Sayed, M.A. (2011). Plasmonic photo-thermal therapy (PPTT). *Alexandria J. Med.* 47: 1.
33 Wang, J. and Quiz, J. (2016). A review of organic nanomaterials in photothermal cancer therapy. *Cancer Res. Front.* 2: 67.
34 Huang, X., Zhang, W., Guan, G. et al. (2017). Design and functionalization of the NIR-responsive photothermal semiconductor nanomaterials for cancer theranostics. *Acc. Chem. Res.* 50: 2529.
35 Yang, K., Feng, L., Shi, X., and Liu, Z. (2013). Nano-graphene in biomedicine: theranostic applications. *Chem. Soc. Rev.* 42: 530.
36 Qiu, J. and Wei, W.D. (2014). Surface plasmon-mediated photothermal chemistry. *J. Phys. Chem. C* 118: 20735.

37 Fablan, J., Nakazumi, H., and Matsuoka, M. (1992). Near-infrared absorbing dyes. *Chem. Rev.* 92: 1197.
38 Qian, G. and Wang, Z.Y. (2010). Near-infrared organic compounds and emerging applications. *Chem. Asian J.* 5: 1006.
39 Xu, L., Cheng, L., Wang, C. et al. (2014). Conjugated polymers for photothermal therapy of cancer. *Polym. Chem.* 5: 1573.
40 Webb, J.A. and Bardhan, R. (2014). Emerging advances in nanomedicine with engineered gold nanostructures. *Nanoscale* 6: 2502.
41 Her, S., Jaffray, D.A., and Allen, C. (2017). Gold nanoparticles for applications in cancer radiotherapy: mechanisms and recent advancements. *Adv. Drug Delivery Rev.* 109: 84.
42 Jaque, D., Maestro, L.M., del Rosal, B. et al. (2014). Nanoparticles for photothermal therapies. *Nanoscale* 6: 9494.
43 Hu, Y., Li, Z., Lan, T., and Chen, W. (2016). Photoactuators for direct optical-to-mechanical energy conversion: from nanocomponent assembly to macroscopic deformation. *Adv. Mater.* 28: 10548.
44 Zhang, A., Li, A., Zhao, W., and Liu, J. (2018). Recent advances in functional polymer decorated two-dimensional transition-metal dichalcogenides nanomaterials for chemo-photothermal therapy. *Chem. Eur. J.* 24: 4215.
45 Dou, L., Liu, Y., Hong, Z. et al. (2015). Low-bandgap near-IR conjugated polymers/molecules for organic electronics. *Chem. Rev.* 115: 12633.
46 Cheng, L., Wang, C., Feng, L. et al. (2014). Functional nanomaterials for phototherapies of cancer. *Chem. Rev.* 114: 10869.
47 Jain, P.K., Huang, X., El-Sayed, I.H., and El-Sayed, M.A. (2008). Noble metals on the nanoscale: optical and photothermal properties and some applications in imaging, sensing, biology and medicine. *Acc. Chem. Res.* 41: 1578.
48 Watts, C.M., Liu, X., and Padilla, W.J. (2012). Metamaterial electromagnetic wave absorbers. *Adv. Mater.* 24: OP98–OP120.
49 Bayazitoglu, Y., Khedarmand, S., and Tullius, T.K. (2013). An overview of nanoparticle assisted laser therapy. *Int. J. Heat Mass Transfer* 67: 469.
50 Baffou, G. and Quidant, R. (2013). Thermo-plasmonics: using metallic nanostructures as nano-sources of heat. *Laser Photonics Rev.* 7: 171.
51 Cui, Y., He, Y., Jin, Y. et al. (2014). Plasmonic and metamaterial structures as electromagnetic absorbers. *Laser Photonics Rev.* 8: 495.
52 Baffou, G. and Quidant, R. (2014). Nanoplasmonics for chemistry. *Chem. Soc. Rev.* 43: 3898.
53 Xue, C. and Li, Q. (2015). *Gold Nanorods in Encyclopedia of Surface and Colloid Science* (ed. P. Somasundaran). Taylor & Francis.
54 Li, Q. (ed.) (2015). *Anisotropic Nanomaterials: Preparation, Properties, and Applications*. Heidelberg: Springer-Verlag.
55 Zhu, M., Baffou, G., Meyerbroker, N., and Polleux, J. (2012). Micropatterning thermoplasmonic gold nanoarrays to manipulate cell adhesion. *ACS Nano* 6: 7227.
56 Wang, Y., Black, K.C.L., Luehmann, H. et al. (2013). Comparison study of gold nanohexapods, nanorods, and nanocages for photothermal cancer treatment. *ACS Nano* 7: 2068.

57 Huang, X., El-Sayed, I.H., Qian, W., and El-Sayed, M.A. (2006). Cancer cell imaging and photothermal therapy in the near-infrared region by using gold nanorods. *J. Am. Chem. Soc.* 128: 2115.
58 Huschka, R., Zuloaga, J., Knight, M.W. et al. (2011). Light-induced release of DNA from gold nanoparticles: nanoshells and nanorods. *J. Am. Chem. Soc.* 133: 12247.
59 Jiang, K., Smith, D.A., and Pinchuk, A. (2013). Size-dependent photothermal conversion efficiencies of plasmonically heated gold nanoparticles. *J. Phys. Chem. C* 117: 27073.
60 Nedyalkov, N.N., Imamova, S.E., Atanasov, P.A. et al. (2011). Interaction of gold nanoparticles with nanosecond laser pulse: nanoparticle heating. *Appl. Surf. Sci.* 257: 5456.
61 Govorov, A.O., Zhang, W., Skeini, T. et al. (2006). Gold nanoparticle ensembles as heaters and actuators: melting and collective plasmon resonances. *Nanoscale Res. Lett.* 1: 84.
62 Hu, M. and Hartland, G.V. (2002). Heat dissipation for Au particles in aqueous solution: relaxation time versus size. *J. Phys. Chem. B* 106: 7029.
63 Baffou, G., Quidant, R., and Girard, C. (2009). Heat generation in plasmonic nanostructures: influence of morphology. *Appl. Phys. Lett.* 94: 153109.
64 Roper, D.K., Ahn, W., and Hoepfner, M. (2007). Microscale heat transfer transduced by surface plasmon resonant gold nanoparticles. *J. Phys. Chem. C* 111: 3636.
65 Chen, H., Shao, L., Ming, T. et al. (2010). Understanding the photothermal conversion efficiency of gold nanocrystals. *Small* 6: 2272.
66 Xue, C. and Li, Q. (2015). Anisotropic gold nanoparticels: preparation, properties, and applications. In: *Anisotropic Nanomaterials: Preparation, Properties, and Applications*, Chapter 3 (ed. Q. Li), 69–118. Heidelberg: Springer-Verlag.
67 Cong, B., Kan, C., Wang, H. et al. (2014). Gold nanorods: near-infrared plasmonic photothermal conversion and surface coating. *J. Mater. Sci. - Chem. Eng.* 2: 20.
68 Bisoyi, H.K. and Li, Q. (2014). *Liquid Crystals, Kirk-Othmer Encyclopedia of Chemical Technology*. New York: Wiley.
69 Li, Q. (ed.) (2014). *Nanoscience with Liquid Crystals: From Self-Organized Nanostructures to Applications*. Springer-Verlag.
70 de Gennes, P.G. and Prost, J. (1993). *The Physics of Liquid Crystals*. Oxford: Oxford University Press.
71 Chandrasekhar, S. (1992). *Liquid Crystals*. Cambridge, UK: Cambridge University Press.
72 Collings, P.J. and Hird, M. (1997). *Introduction to Liquid Crystals: Chemistry and Physics*. London, UK: Taylor & Francis.
73 Warner, M. and Terentjev, E.M. (2003). *Liquid Crystal Elastomers*. New York: Oxford University Press.
74 De Jeu, W.H. (ed.) (2012). *Liquid Crystal Elastomers: Materials and Applications*. Springer-Verlag.
75 Bisoyi, H.K. and Li, Q. (2016). Light-driven liquid crystalline materials: from photo-induced phase transitions and property modulations to applications. *Chem. Rev.* 116: 15089.

76 Li, Q. (ed.) (2012). *Liquid Crystals Beyond Displays: Chemistry, Physics, and Applications*. Hoboken, NJ: Wiley.
77 Wang, L., Gutierrez, K., and Li, Q. (2017). *Photochromic Chiral Liquid Crystal Sensors, Chapter 2 in Liquid Crystalline Sensors* (ed. A. Schenning, G. Crawford and D.J. Broer). CRC Press, Taylor & Francis Group.
78 Bisoyi, H. and Li, Q. (2016). Discotic liquid crystals for self-organizing photovolatics. In: *Nanomaterials for Sustainable Energy*, Chapter 6 (ed. Q. Li), 215–252. Heidelberg: Springer-Verlag.
79 Zheng, Z. and Li, Q. (2016). Self-organized chiral liquid crystalline nanostructures for energy-saving devices. In: *Nanomaterials for Sustainable Energy*, Chapter 14 (ed. Q. Li), 513–558. Heidelberg: Springer-Verlag.
80 Bisoyi, H.K. and Li, Q. (2015). Liquid crystalline anisotropic nanoparticles: from metallic and semiconducting nanoparticles to carbon nanomaterials. In: *Anisotropic Nanomaterials: Preparation, Properties, and Applications*, Chapter 6 (ed. Q. Li), 209–240. Heidelberg: Springer-Verlag.
81 Gutierrez-Cuevas, K.G., Wang, L., Zheng, Z. et al. (2016). Frequency-driven self-organized helical superstructures loaded with mesogen-grafted silica nanoparticles. *Angew. Chem. Int. Ed.* 55: 13090.
82 Montbach, E., Venkataraman, N., Doane, J.W. et al. (2008). Novel optically addressable photochiral displays. *Soc. Inf. Disp. – Dig. Tech. Pap.* 39: 919.
83 Venkataraman, N., Magyar, G., Montbach, E. et al. (2009). Thin flexible photosensitive cholesteric displays. *J. Soc. Inf. Disp.* 17: 869.
84 Li, L., Kang, S., Harden, J. et al. (2008). Nature inspired light-harvesting liquid crystalline porphyrins for organic photovoltaics. *Liq. Cryst.* 35: 233.
85 Zheng, Z., Li, Y., Bisoyi, H.K. et al. (2016). Three-dimensional control of the helical axis of a chiral nematic liquid crystal by light. *Nature* 531: 352.
86 Sun, Q., Dai, L., Zhou, X. et al. (2007). Bilayer- and bulk-heeterojunction solar cells using liquid crystalline porphyrins as donors by solution processing. *Appl. Phys. Lett.* 91: 253505.
87 Zhou, X., Kang, S., Kumar, S. et al. (2008). Self-assembly of porphyrin and fullerene supramolecular complex into highly ordered nanostructure by thermal annealing. *Chem. Mater.* 20: 3551.
88 Zhou, X., Kang, S., Kumar, S., and Li, Q. (2009). Self-assembly of discotic liquid crystal porphyrin into more controllable ordered nanostructure mediated by fluorophobic effect. *Liq. Cryst.* 36: 269.
89 Mathews, M., Zola, R., Hurley, S. et al. (2010). Light-driven reversible handedness inversion in self-organized helical superstructures. *J. Am. Chem. Soc.* 132: 18361.
90 Li, Q., Li, Y., Ma, J. et al. (2011). Directing dynamic control of red, green and blue reflection enabled by a light-driven self-organized helical superstructure. *Adv. Mater.* 23: 5069.
91 Li, Y., Urbas, A., and Li, Q. (2012). Reversible light-directed red, green and blue reflections with thermal stability enabled by a self-organized helical superstructure. *J. Am. Chem. Soc.* 134: 9573.
92 Lin, T.-H., Li, Y., Wang, C.-T. et al. (2013). Red, green and blue reflections enabled in optically tunable self-organized 3D cubic nanostructured thin film. *Adv. Mater.* 25: 5050.

93 Wang, L., Dong, H., Li, Y. et al. (2014). Reversible near-infrared light directed reflection in a self-organized helical superstructure loaded with upconversion nanoparticles. *J. Am. Chem. Soc.* 136: 4480.
94 Zhang, L., Wang, L., Hiremath, U.S. et al. (2017). Dynamic orthogonal switching of a thermoresponsive self-organized helical superstructure. *Adv. Mater.* 29: 1700676.
95 Jau, H.-C., Li, Y., Li, C.-C. et al. (2015). Light-driven wide-range non-mechanical beam steering and spectrum scanning based on a self-organized liquid crystal grating enabled by a chiral molecular switch. *Adv. Opt. Mater.* 3: 166.
96 Zheng, Z., Yuan, C., Hu, W. et al. (2017). Light-patterned crystallographic direction of a self-organized three-dimensional soft photonic crystal. *Adv. Mater.* 29: 1703165.
97 Wang, Y., Singh, G., Agra-Kooijman, D.M. et al. (2015). Room temperature heliconical twist-bend nematic liquid crystal. *CrystEngCommun* 17: 2778.
98 Wang, L., Chen, D., Gutierrez-Cuevas, K. et al. (2017). Optically reconfigurable chiral microspheres of self-organized helical superstructures with handedness inversion. *Mater. Horiz.* 4: 1190.
99 Fan, J., Li, Y., Bisoyi, H.K. et al. (2015). Light-directing omnidirectional circularly polarized reflection from liquid crystal droplets. *Angew. Chem. Int. Ed.* 54: 2160.
100 Wang, Y., Zheng, Z., Bisoyi, H.K. et al. (2016). Thermally reversible full color selective reflection in a self-organized helical superstructure enabled by a bent-core oligomesogen exhibiting twist-bend nematic phase. *Mater. Horiz.* 3: 442.
101 Xiang, J., Li, Y., Li, Q. et al. (2015). Electrically tunable selective reflection of light from ultraviolet to visible and infrared by heliconical cholesterics. *Adv. Mater.* 27: 3014.
102 Wang, L., Dong, H., Li, Y. et al. (2015). Photoluminescence-driven reversible handedness inversion of self-organized helical superstructures enabled by unprecedented near-infrared light nanotransducers. *Adv. Mater.* 27: 2065.
103 White, T., Bricker, R.L., Natarajan, L.V. et al. (2010). Electromechanical and light tunable cholesteric liquid crystals. *Opt. Commun.* 283: 3434.
104 Li, Y., Wang, M., White, T.J. et al. (2013). Azoarenes bearing opposite chiral configurations: light-driven dynamic reversible handedness inversion in self-organized helical superstructure. *Angew. Chem. Int. Ed.* 52: 8925.
105 Li, Y., Xue, C., Wang, M. et al. (2013). Photodynamic chiral molecular switches with thermal stability: from reflection wavelength tuning to handedness inversion of self-organized helical superstructures. *Angew. Chem. Int. Ed.* 52: 13703.
106 Salili, S.M., Xiang, J., Wang, H. et al. (2016). Magnetically tunable selective reflection of light by heliconical cholesterics. *Phys. Rev. E* 94: 042705.
107 Wang, H., Bisoyi, H.K., Urbas, A. et al. (2018). Photochemically and thermally driven full color reflection in a self-organized helical superstructure enabled by a halogen bonded chiral molecular switch. *Angew. Chem. Int. Ed.* 57: 1627.
108 Zheng, Z., Zola, R.S., Bisoyi, H.K. et al. (2017). Dynamic reconfigurable zigzag pattern formation in a soft helical superstructure. *Adv. Mater.* 29: 1701903.

109 Sun, Y., Evans, J.S., Lee, T. et al. (2012). Optical manipulation of shape-morphing elastomeric liquid crystal microparticles doped with gold nanocrystals. *Appl. Phys. Lett.* 100: 241901.

110 Fong, W.-K., Hanley, T.L., Thierry, B. et al. (2012). Controlling the nanostructure of gold nanorod–lyotropic liquid crystalline hybrid materials using near infrared laser irradiation. *Langmuir* 28: 14450.

111 Fong, W.-K., Hanley, T.L., Thierry, B. et al. (2010). Plasmonic nanorods provide reversible control over nanostructure of self-assembled drug delivery materials. *Langmuir* 26: 6136.

112 Fong, W.-K., Hanley, T.L., Thierry, B. et al. (2014). Understanding the photothermal heating effect in non-lamellar liquid crystalline systems, and the design of new mixed lipid systems for photothermal on-demand drug delivery. *Phys. Chem. Chem. Phys.* 16: 24936.

113 Pezzi, L., de Sio, L., Veltri, A. et al. (2015). Photo-thermal effects in gold nanoparticles dispersed in thermotropic nematic liquid crystals. *Phys. Chem. Chem. Phys.* 17: 20281.

114 Liu, X., Wei, R., Hoang, P.T. et al. (2015). Reversible and rapid laser actuation of liquid crystalline elastomer micropillars with inclusion of gold nanoparticles. *Adv. Funct. Mater.* 25: 3022.

115 Yang, H., Liu, J., Wang, Z. et al. (2015). Near-infrared-responsive gold nanorod/liquid crystalline elastomer composites prepared by sequential thiol-click chemistry. *Chem. Commun.* 51: 12126.

116 Gutierrez-Cuevas, K.G., Wang, L., Xue, C. et al. (2015). Near infrared light-driven liquid crystal phase transition enabled by hydrophobic mesogen grafted plasmonic gold nanorods. *Chem. Commun.* 51: 9845.

117 Pezzi, L., de Sio, L., Palermo, G. et al. (2016). Nematic liquid crystals used to control photo-thermal effects in gold nanoparticles. *Proc. SPIE* 9769: 97690C.

118 de Sio, L., Placido, T., Comparelli, R. et al. (2016). Plasmonic photoheating of gold nanorods in thermo-responsive chiral liquid crystals. *J. Opt.* 18: 125005.

119 Wang, L., Gutierrez-Cuevas, K.G., Urbas, A., and Li, Q. (2016). Near-infrared light-directed handedness inversion in plasmonic nanorod-embeded helical superstructure. *Adv. Opt. Mater.* 4: 247.

120 Wang, L., Gutierrez-Cuevas, K.G., Bisoyi, H.K. et al. (2015). NIR light-directing self-organized 3D photonic superstructures loaded with anisotropic plamsonic hybrid nanorods. *Chem. Commun.* 51: 15039.

121 Palermo, G., Cataldi, U., de Sio, L. et al. (2016). Optical control of plasmonic heating effects using reversible photo-alignment of nematic liquid crystals. *Appl. Phys. Lett.* 109: 191906.

122 Palermo, G., Cataldi, U., Pezzi, L. et al. (2017). Thermo-plasmonic effects on E7 nematic liquid crystal. *Mol. Cryst. Liq. Cryst.* 649: 45.

123 Yang, L., Setyowati, K., Li, A. et al. (2008). Reversible infrared actuation of carbon nanotube-liquid crystalline elastomer nanocomposites. *Adv. Mater.* 20: 2271.

124 Ji, Y., Huang, Y.Y., Rungsawang, R., and Terentjev, E.M. (2010). Dispersion and alignment of carbon nanotubes in liquid crystalline polymers and elastomers. *Adv. Mater.* 22: 3436.

125 Li, C., Liu, Y., Lo, C., and Jiang, H. (2011). Reversible white-light actuation of carbon nanotube incorporated liquid crystalline elastomer nanocomposites. *Soft Matter* 7: 7511.
126 Marshall, J.E., Ji, Y., Torras, N. et al. (2012). Carbon-nanotube sensitized nematic elastomer composites for IR-visible photo-actuation. *Soft Matter* 8: 1570.
127 Li, C., Liu, Y., Huang, X., and Jiang, H. (2012). Direct sun-driven artificial heliotropism for solar energy harvesting based on a photo-thermomechanical liquid-crystal elastomer nanocomposites. *Adv. Funct. Mater.* 22: 5166.
128 Kohlmeyer, R.R. and Chen, J. (2013). Wavelength-selective, IR light-driven hinges based on liquid crystalline elastomer composites. *Angew. Chem. Int. Ed.* 52: 9234.
129 Li, C., Huang, X., Li, C., and Jiang, H. (2015). Reversible photo actuated bulk nanocomposite with nematic liquid crystalline elastomer matrix. *Mol. Cryst. Liq. Cryst.* 608: 146.
130 Wang, M., Sayed, S.M., Guo, L. et al. (2016). Multi-stimuli responsive carbon nanotube incorporated polysiloxane azobenzene liquid crystalline elastomer composites. *Macromolecules* 49: 663.
131 Yu, L., Cheng, Z., Dong, Z. et al. (2014). Photomechanical response of polymer-dispersed liquid crystals/graphene oxide nanocomposites. *J. Mater. Chem. C* 2: 8501.
132 Yu, L. and Yu, H. (2015). Light-powered tumbler movement of graphene oxide/polymer nanocomposites. *ACS Appl. Mater. Interfaces* 7: 3834.
133 Quinn, M.D.J., Du, J., Boyd, B.J. et al. (2015). Lipid liquid crystal phase change induced through near-infrared irradiation of entrained graphene particles. *Langmuir* 31: 6605.
134 Quinn, M.D.J., Wang, T., Du, J. et al. (2017). Graphene as a photothermal actuator for control of lipid mesophase structure. *Nanoscale* 9: 341.
135 Yang, Y., Zhan, W., Peng, R. et al. (2015). Graphene-enabled superior and tunable photomechanical actuation in liquid crystalline elastomer nanocomposites. *Adv. Mater.* 27: 6376.
136 Li, C., Liu, Y., Huang, X. et al. (2015). Light actuation of graphene-oxide incorporated liquid crystalline elastomer nanocomposites. *Mol. Cryst. Liq. Cryst.* 616: 83.
137 Wang, L., Bisoyi, H.K., Zheng, Z. et al. (2017). Stimuli-directed self-organized chiral superstructures for adaptive windows enabled by mesogen-functionalized graphene. *Mater. Today* 20: 230.
138 Vallooran, J.J., Handschin, S., Bolisetty, S., and Mezzenga, R. (2012). Twofold light and magnetic responsive behavior in nanoparticle-lyotropic liquid crystal systems. *Langmuir* 28: 5589.
139 Ono, H. and Kawatsuki, N. (1997). Photothermal response characterstics of host-guest liquid crystals. *Jpn. J. Appl. Phys.* 36: 761.
140 Ono, H., Harato, Y., and Kawatsuki, N. (1998). Photothermal effects in liquid crystals on dye-doped polymer films. *J. Appl. Phys.* 83: 4957.
141 Guo, L., Liu, M., Sayed, S.M. et al. (2016). A calamitic mesogenic near-infrared absorbing croconaine dye/liquid crystalline elastomer composite. *Chem. Sci.* 7: 4400.

142 Liu, W., Guo, L., Lin, B. et al. (2016). Near-infrared responsive liquid crystalline elastomers containing photothermal conjugated polymers. *Macromolecules* 49: 4023.

143 Gelebart, A.H., Vantomme, G., Meijer, E.W., and Broer, D.J. (2017). Mastering the photothermal effect in liquid crystal networks: a general approach for self-sustained mechanical oscillators. *Adv. Mater.* 29: 1606712.

144 Liu, L., Liu, M., Deng, L. et al. (2017). Near-infrared chromophore functionalized soft actuator with ultrafast photoresponsive speed and superior mechanical property. *J. Am. Chem. Soc.* 139: 11333.

145 Li, Z., Yang, Y., Wang, Z. et al. (2017). Polydopamine nanoparticles doped in liquid crystal elastomers for producing dynamic 3D structures. *J. Mater. Chem. A* 5: 6740.

146 Tian, H., Wang, Z., Chen, Y. et al. (2018). Polydopamine-coated main-chain liquid crystal elastomer as optically driven artificial muscle. *ACS Appl. Mater. Interfaces* 10: 8307.

147 Paul, D.R. and Robeson, L.M. (2008). Polymer nanotechnology: nanocomposites. *Polymer* 49: 3187.

148 Moniruzzaman, M. and Winey, K.I. (2006). Polymer nanocomposites containing carbon nanotubes. *Macromolecules* 39: 5194.

149 Yuan, Z., Xiao, X., Li, J. et al. (2018). Self-assembled graphene-based architectures and their applications. *Adv. Sci.* 5: 1700626.

150 Kumar, S.K., Benicewicz, B.C., Vaia, R.A., and Winey, K.I. (2017). 50th anniversary perspective: are polymer nanocomposites practical for applications? *Macromolecules* 50: 714.

151 Ionov, L. (2015). Polymer actuators. *Langmuir* 31: 5015.

152 Lu, S., Ahir, S., Valesco, V. et al. (2009). Photo-mechanical actuation of carbon nanotubes: mechanisms and applications in micro and nano-devices. *J. Micro - Nano Mechatron.* 5: 29.

153 Koerner, H., Price, G., Pearce, N.A. et al. (2004). Remotely actuated polymer nanocomposites - stress-recovery of carbon-nanotube-filled thermoplastic elastomers. *Nat. Mater.* 3: 115.

154 Ahir, S.V. and Terentjev, E.M. (2005). Photomechanical actuation in polymer-nanotube composites. *Nat. Mater.* 4: 491.

155 Ahir, S.V., Squires, A.M., Tajbakhsh, A.R., and Terentjev, E.M. (2006). Infrared actuation in aligned polymer-nanotube composites. *Phys. Rev. E* 73: 085420.

156 Lu, S. and Panchapakesan, B. (2007). Photomechanical responses of carbon nanotube/polymer actuators. *Nanotechnology* 18: 305502.

157 Miyako, E., Nagata, H., Hirona, K., and Hirotsu, T. (2008). Carbon-nanotube-polymer composite for light-driven microthermal control. *Angew. Chem. Int. Ed.* 47: 3610.

158 Czanikova, K., Krupa, I., Iicikova, M. et al. (2012). Photo-actuating materials based on elastomers and modified carbon nanotubes. *J. Nanophotonics* 6: 063522.

159 Czanikova, K., Torras, N., Esteve, J. et al. (2013). Nanocomposite photoactuators based on an ethylene vinyl acetate copolymer filled with carbon nanotubes. *Sens. Actuators, B* 186: 701.

160 Tang, B., Wang, Y., Qiu, M., and Zhang, S. (2014). A full-band sunlight-driven carbon nanotube/PEG/SiO_2 composites for solar energy storage. *Sol. Energy Mater. Sol. Cells* 123 (7).

161 Ilcikova, M., Mrlik, M., Sedlacek, T. et al. (2014). Synthesis of photoactuating acrylic thermoplastic elastomers containing diblock copolymer-grafted carbon nanotubes. *ACS Macro Lett.* 3: 999.

162 Shen, Y., Skirtach, A.G., Seki, T. et al. (2010). Assembly of fullerene-carbon nanotubes: temperature indicator for photothermal conversion. *J. Am. Chem. Soc.* 132: 8566.

163 Asanuma, H., Subedi, P., Hartmann, J. et al. (2013). Nanoplasmonic modification of the local morphology, shape, and wetting properties of nanoflake microparticles. *Langmuir* 29: 7464.

164 Kim, H., Abdala, A.A., and Macosko, C.W. (2010). Graphene/polymer nanocomposites. *Macromolecules* 43: 6515.

165 Liang, J., Xu, Y., Huang, Y. et al. (2009). Infrared-triggered actuators from graphene-based nanocomposites. *J. Phys. Chem. C* 113: 9921.

166 Loomis, J., King, B., and Panchapakesan, B. (2012). Layer dependent mechanical responses of graphene composites to near-infrared light. *Appl. Phys. Lett.* 100: 073108.

167 Loomis, J., Fan, X., Khosravi, F. et al. (2013). Graphene/elastomer composite-based photothermal nanopositioners. *Sci. Rep.* 3: 1900.

168 Ansari, S., Neelanchery, M.M., and Ushus, D. (2013). Graphene/poly(styrene-*b*-isoprene-*b*-styrene) nanocomposite optical actuators. *J. Appl. Polym. Sci.* 130: 3902.

169 Meng, D., Yang, S., Guo, L. et al. (2014). The enhanced photothermal effect of graphene/conjugated polymer composites: photoinduced energy transfer and applications in photocontrolled switches. *Chem. Commun.* 50: 14345.

170 Muraldharan, M.N., Shinu, K.P., and Seema, A. (2016). Optically triggered actuation in chitosan/reduced graphene oxide nanocomposites. *Carbohydr. Polym.* 144: 115.

171 Leeladhar, P., Raturi, A., and Kumar, J.P.S. (2017). Graphene-polydimethylsiloxane/chromium bilayer-based flexible, reversible, and large bendable photomechanical actuators. *Smart Mater. Struct.* 26: 095030.

172 Robinson, J.T., Tabakman, S.M., Liang, Y. et al. (2011). Ultrasmall reduced graphene oxide with high near-infrared absorbance for photothermal therapy. *J. Am. Chem. Soc.* 133: 6825.

173 Feng, Y., Qin, M., Guo, H. et al. (2013). Infrared actuated recovery of polyurethane filled by reduced graphene oxide/carbon nanotube hybrids with high energy density. *ACS Appl. Mater. Interfaces* 5: 10882.

174 Niu, D., Jiang, W., Liu, H. et al. (2016). Reversible bending behaviors of photomechanical soft actuators based on graphene nanocomposites. *Sci. Rep.* 6: 27366.

175 Yang, J., Choi, J., Bang, D. et al. (2011). Convertible organic nanoparticles for near-infrared photothermal ablation of cancer cells. *Angew. Chem. Int. Ed.* 50: 441.

176 Zhou, J., Lu, Z., Zhu, X. et al. (2013). NIR photothermal therapy using polyaniline nanoparticles. *Biomaterials* 34: 9584.

177 Geng, J., Sun, C., Liu, J. et al. (2015). Biocompatible conjugated polymer nanoparticles for efficient photothermal tumor therapy. *Small* 11: 1603.

178 Lyu, Y., Xie, C., Chechetka, S.A. et al. (2016). Semiconducting polymer nanobioconjugates for targeted photothermal activation of neurons. *J. Am. Chem. Soc.* 138: 9049.

179 Lyu, Y., Fang, Y., Miao, Q. et al. (2016). Intraparticle molecular orbital engineering of semiconducting polymer nanoparticles as amplified theranostics for in vivo photoacoustic imaging and photothermal therapy. *ACS Nano* 10: 4472.

180 Cao, Y., Dou, J.H., Zhao, N. et al. (2017). Highly efficient NIR-II photothermal conversion based on an organic conjugated polymer. *Chem. Mater.* 29: 718.

181 Satarkar, N.S., Biswal, D., and Hilt, J.Z. (2010). Hydrogel nanocomposites: a review of applications as remote controlled biomaterials. *Soft Matter* 6: 2364.

182 Thoniyot, P., Tan, M.J., Karim, A.A. et al. (2015). Nanoparticle-hydrogel composites: concepts, design, and applications of these promising, multi-functional materials. *Adv. Sci.* 2: 1400010.

183 Ionov, L. (2014). Hydrogel-based actuators: possibilities and limitations. *Mater. Today* 17: 494.

184 Cametti, M. and Dzolic, Z. (2014). New frontiers in hybrid materials: nobel metal nanoparticles-supramolecular gel systems. *Chem. Commun.* 50: 8273.

185 Leijten, J., Seo, J., Yue, K. et al. (2017). Spatially and temporally controlled hydrogels for tissue engineering. *Mater. Sci. Eng., R* 119: 1.

186 Mebrouk, K., Debnath, S., Fourmigue, M., and Camerel, F. (2014). Photothermal control of the gelation properties of nickel bis(dithiolene) metallogelators under near-infrared irradiation. *Langmuir* 30: 8592.

187 Wang, C., Wang, X., Dong, K. et al. (2016). Injectable and responsively degradable hydrogel for personalized photothermal therapy. *Biomaterials* 104: 129.

188 Zheng, Z., Hu, J., Wang, H. et al. (2017). Dynamic softening or stiffening a supramolecular gel by ultraviolet or near-infrared light. *ACS Appl. Mater. Interfaces* 9: 24511.

189 Okano, K., Nogami, A., and Asakura, K. (2014). High-strength gel actuator driven by a photothermal effect. *Polym. J.* 46: 827.

190 Mohamed, M.B., Ahmadi, T.S., Link, S. et al. (2001). Hot electron and phonon dynamics of gold nanoparticles embedded in a gel matrix. *Chem. Phys. Lett.* 343: 55.

191 Kim, J.-H. and Lee, T.R. (2004). Thermo- and pH-responsive hydrogel-coated gold nanoparticles. *Chem. Mater.* 16: 3647.

192 Kawano, T., Niidome, Y., Mori, T. et al. (2009). PNIPAM gel-coated gold nanorods for targeted delivery responding to a near-infrared laser. *Bioconjugate Chem.* 20: 209.

193 Gramlich, W.M., Holloway, J.L., Rai, R., and Burdick, J.A. (2014). Transdermal gelation of methacrylated macromers with near-infrared light and gold nanorods. *Nanotechnology* 25: 014004.

194 Stower, R.S., Allen, S.C., and Suggs, L.J. (2015). Dynamic phototuning of 3D hydrogel stiffness. *Proc. Natl. Acad. Sci. U.S.A.* 112: 1953.

195 Collins, S.S.E., Wei, X., McKenzie, T.G. et al. (2016). Single gold nanorod charge modulation in an ion gel device. *Nano Lett.* 16: 6863.

196 Park, H., Yang, S., Kang, J.Y., and Park, M.-H. (2016). On-demand drug delivery system using micro-organogels with gold nanorods. *ACS Med. Chem. Lett.* 7: 1087.

197 Sershen, S.R., Mensing, G.A., Ng, M. et al. (2005). Independent optical control of microfluidic valves formed from optomechanically responsive nanocomposite hydrogels. *Adv. Mater.* 17: 1366.

198 Shiotani, A., Mori, T., Niidome, T. et al. (2007). Stable incorporation of gold nanorods into *N*-isopropylacrylamide hydrogels and their rapid shrinkage induced by near-infrared laser irradiation. *Langmuir* 23: 4012.

199 Erdogan, H., Sakalak, H., Yazuv, M.S., and Demirel, G. (2013). Laser-triggered degelation control of gold nanoparticle embedded peptide organogels. *Langmuir* 29: 6975.

200 Zhou, Y., Hauser, A.W., Bende, N.P. et al. (2016). Waveguiding microactuators based on a photothermally responsive nanocomposite hydrogel. *Adv. Funct. Mater.* 26: 5447.

201 Lo, C.-W., Zhu, D., and Jiang, H. (2011). An infrared-light responsive graphene-oxide incorporated poly(*N*-isopropylacrylamide)hydrogel nanocomposite. *Soft Matter* 7: 5604.

202 Zhang, X., Pint, C.L., Lee, M.H. et al. (2011). Optically- and thermally-responsive programmable materials based on carbon nanotube-hydrogel polymer composites. *Nano Lett.* 11: 3239.

203 Zhang, E., Wang, T., Hong, W. et al. (2014). Infrared-driving actuation based bilayer graphene oxide-poly(*N*-isopropylacrylamide) nanocomposite hydrogels. *J. Mater. Chem. A* 2: 15633.

204 Zhu, C.-H., Lu, Y., Peng, J. et al. (2012). Photothermally sensitive poly(*N*-isopropylacrylamide)/graphene oxide nanocomposite hydrogels as remote light-controlled liquid microvalves. *Adv. Funct. Mater.* 22: 4017.

205 Hsiao, C.-W., Chuang, E.-Y., Chen, H.-L. et al. (2015). Photothermal tumor ablation in mice with repeated therapy sessions using NIR-absorbing micellar hydrogels formed in situ. *Biomaterials* 56: 26.

206 Fujigaya, T., Moromoto, T., Niidome, Y., and Nakashima, N. (2008). NIR lase-driven reversible volume phase transition of single-walled carbon nanotube/poly(*N*-isopropylacrylamide)composite gels. *Adv. Mater.* 20: 3610.

207 Li, W., Wang, J., Ren, J., and Qu, X. (2013). 3D graphene oxide-polymer hydrogel: near-infrared light-triggered active scaffold for reversible cell capture and on-demand release. *Adv. Mater.* 25: 6737.

208 Wang, E., Desai, M.S., and Lee, S.-W. (2013). Light-controlled graphene-elastin composite hydrogel actuators. *Nano Lett.* 13: 2826.

209 Cong, H.-P., Qiu, J.-H., and Yu, S.-H. (2015). Thermoresponsive poly(*N*-isopropylacrylamide)/graphene/Au nanocomposite hydrogel for water treatment by a laser-assisted approach. *Small* 11: 1165.

210 Kim, D., Lee, H.S., and Yoon, J. (2016). Highly bendable bilayer-type photo-actuators comprising of reduced graphene oxide dispersed in hydrogels. *Sci. Rep.* 6: 20921.

211 Satarkar, N.S., Zhang, W., Eitel, R.E., and Hilt, J.Z. (2009). Magnetic hydrogel nanocomposites as remote controlled microfluidic valves. *Lab Chip* 9: 1773.

212 Khoshmanesh, K., Tang, S.-Y., Zhu, J.Y. et al. (2017). Liquid metal enabled microfluidics. *Lab Chip* 17: 974.

213 Dickey, M.D. (2014). Emerging applications of liquid metals featuring surface oxides. *ACS Appl. Mater. Interfaces* 6: 18369.

214 Yoon, J., Bian, P., Kim, J. et al. (2012). Local switching of chemical patterns through light-triggered unfolding of creased hydrogel surfaces. *Angew. Chem. Int. Ed.* 51: 7146.

215 Lee, E., Lee, H., Yoo, S.I., and Yoon, J. (2014). Photothermally triggered fast responding hydrogels incorporating a hydrophobic moiety for light-controlled microvalves. *ACS Appl. Mater. Interfaces* 6: 16949.

216 Zhu, C.H., Lu, Y., Chen, J.-F., and Yu, S.-H. (2014). Photothermal poly(*N*-isopropylacrylamide)/Fe_3O_4 nanocomposite hydrogel as a movable position heating source under remote control. *Small* 10: 2796.

217 Lee, E., Kim, D., Kim, H., and Yoon, J. (2015). Photothermally driven fast responding photo-actuators fabricated with comb-type hydrogels and magnetite nanoparticles. *Sci. Rep.* 5: 15124.

218 Lim, D., Lee, E., Kim, H. et al. (2015). Multi stimuli-responsive hydrogel microfibers containing magnetite nanoparticles prepared using microcapillary devices. *Soft Matter* 11: 1606.

2

Photoresponsive Supramolecular Polymers

Yuichi Kitamoto, Keisuke Aratsu, and Shiki Yagai

Chiba University, Institute for Global Prominent Research (IGPR) and Division of Advanced Science and Engineering, Graduate School of Science and Engineering, 1-33 Yayoi-cho, Inage-ku, Chiba, 263-8522, Japan

2.1 Introduction

According to the definition by Meijer and coworkers in their seminal review [1], supramolecular polymers are polymeric assemblies of small molecular units that are brought together through reversible and directional noncovalent interactions, leading to polymeric properties in solution and bulk states [1–5]. Owing to the reversible nature of their main-chain formation, supramolecular polymers have attracted considerable attention as highly stimuli-responsive polymeric materials compared to conventional polymer counterparts. Needless to say, supramolecular polymers have high responsiveness in terms of "degree of polymerization" to temperature change mainly due to entropic reasons, and to solvent media mainly due to enthalpic reasons. Such high responsiveness to external environment has enabled not only the exploration of temperature- and solvent-dependent supramolecular polymer systems, but also the analyses of the supramolecular polymerization process regulated by these parameters [6]. From a technical viewpoint, furthermore, supramolecular polymers have another advantage over conventional polymers, i.e. quite ambient polymerization conditions without using any chemical reagents. This allows more flexible monomer design and enables introduction of a wide variety of functional molecules into supramolecular monomers.

Another possible advantage coming from the reversible nature of supramolecular polymers can be found when one can directly manipulate the structures of monomer units by external stimuli. Because the overall enthalpic and entropic contributions to stabilizing supramolecular polymers depend strongly on the conformation and flexibility of monomer units, the alteration of these factors for monomer units should establish a new equilibrium for the given supramolecular polymer systems. Although temperature and solvent could induce large changes in the state and the structure of supramolecular polymers, they are not "specific" for molecular components – not "directional" as stimuli, and not "well-controllable" in terms of on–off switching. In this sense, the utilization of

Photoactive Functional Soft Materials: Preparation, Properties, and Applications, First Edition.
Edited by Quan Li.

photochromic molecules is a very reasonable approach to control the structure of monomer units at will [7–10]. When a reversible transformation between two photoisomers having different absorption bands is induced by light of a specific wavelength region, fully regulated two-state photoswitch of the monomer structure can be achieved. Furthermore, in comparison with other chemical reactions using chemical substance, photochromic reactions generally produce no chemical wastes, and their repeatability is a very intriguing property that is reminiscent of biochemical transformations and photoregulation systems omnipresent in living systems. Accordingly, many supramolecular chemists have designed a variety of photochromic monomer units and noncovalently polymerized them to establish diverse photoresponsive supramolecular polymers showing a photocontrollable degree of aggregation and/or supramolecular architectures. However, it has not been overviewed yet how photoinduced structure changes in monomer units can induce a variety of changes in the structures and properties of supramolecular polymers.

This chapter highlights how we can utilize photochromic molecules to control (i) the degree of supramolecular polymerization, (ii) the supramolecular polymerization process, and (iii) topologies of one-dimensional chains. The first topic will be illustrated by focusing on photoresponsive supramolecular polymers based on host–guest and metal-ligand complexation (Section 2.2), complementary hydrogen-bonding interactions (Section 2.3), and π–π stacking interactions between photochromic moieties (Section 2.4). For the second topic, supramolecular polymers based on π–π stacking interactions play important roles to discuss the underpinning principles (Section 2.4). The third topic will address specific design of supramolecular monomers inducing topological changes of one-dimensional supramolecular chains, along with maintaining the degree of polymerization during photoisomerization (Section 2.5). The application of photochromic molecules for other functional materials, such as vesicles, micelles, gels, organic nanotubes, biomolecular systems, and molecular machines, are summarized in other important reviews [11–17], and hence we do not address these issues in this chapter. Photoresponsive liquid crystals, which are one of the important smart soft materials, can also be studied in the other chapters of this book.

2.2 Photoresponsive Supramolecular Polymers by Host–Guest and Coordination Systems

The manipulation of association and dissociation events in supramolecular host–guest complexes by light has been attempted already in the late 1980s by Shinkai et al. [18]. A self-complementary azobenzene derivative **1** bearing an ω-ammoniumalkyl moiety on one end and a crown ether moiety on the other end was designed to form crown–ammonium complexes either intermolecularly or intramolecularly (tail-biting), depending on the isomeric form of the azobenzene core (Figure 2.1a). Although the polymerizable *trans*-isomer affords only cyclic oligomers by the intermolecular complexation, upon *trans-to-cis* photoisomerization it transforms into a "dormant" monomer by the intramolecular complexation of the bent *cis*-isomer. They have also developed photocontrollable polymeric

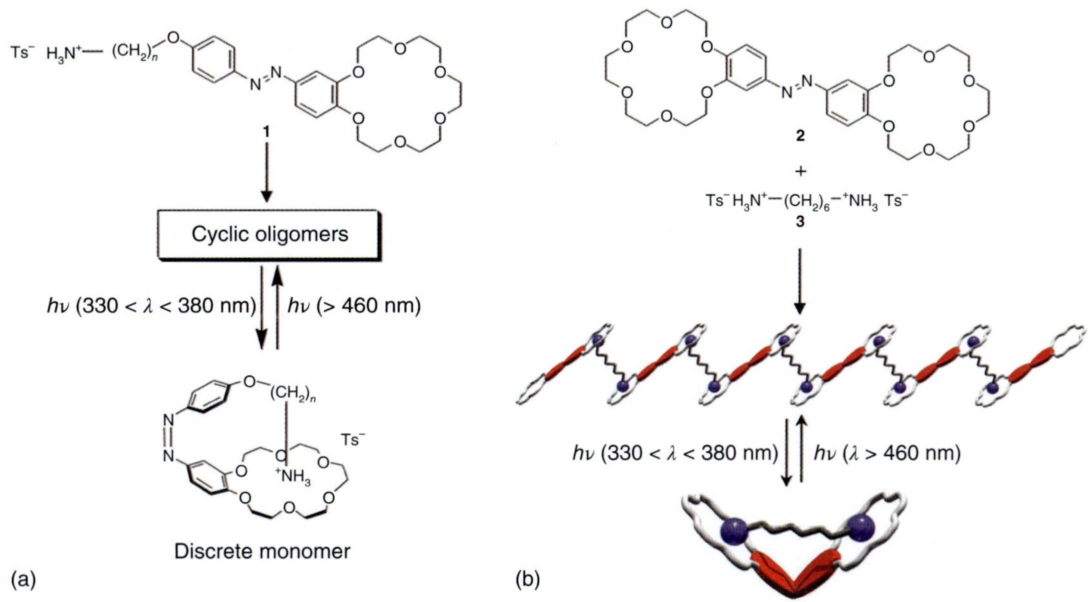

Figure 2.1 Photoresponsive host–guest assemblies designed by Shinkai et al. (a) One-component system achieved by self-complementary azobenzene **1**. (b) Two-component system achieved by complexation of azobis(benzocrown ethers) **2** with α,ω-diammoniumalkane **3**.

supramolecular complexes consisting of a ditopic azobis(benzocrown ether) **2** and an oligomethylene-α,ω-diammonium cation **3** (Figure 2.1b) [19]. This binary system also shows a similar result with that of **1**. Accordingly, these systems already demonstrated a basic idea for creating photoresponsive supramolecular polymers based on host–guest supramolecular chemistry.

Cyclodextrin is one of the most important macrocyclic host molecules applied to numerous research fields [20–25]. Harada and coworkers developed a photoresponsive host–guest supramolecular polymer by using a stilbene-bridged bis(β-cyclodextrin) **4** as a macrocyclic host and a ditopic adamantane guest **5** (Figure 2.2). The *trans-to-cis* and *cis-to-trans* isomerization of the stilbene-bis(β-cyclodextrin) **4** are reversibly controllable in aqueous solution with UV-light irradiation at 365 and 254 nm, respectively. Rotating frame nuclear Overhauser effect spectroscopy (ROESY) analysis revealed that adamantane guest **5** can complex with both the *trans-* and *cis-*isomers of **4** (*trans-* and *cis-***4**). The *trans-***4** formed 1 : 1 supramolecular dimers or small assemblies with **5**, which was suggested from concentration-independent hydrodynamic radius (D_h) observed by pulse field gradient nuclear magnetic resonance (NMR). Upon UV-light irradiation with 365 nm, the D_h value became smaller with increasing concentration of the mixtures, which demonstrated that the photogenerated *cis-***4** formed linear supramolecular polymers upon complexing with **5**. AFM measurements showed that the mixture of *trans-***4** and **5** produced only small objects with the overall lengths of c. 30 nm. Conversely, for the mixture of *cis-***4** and **5**, supramolecular wires with the overall lengths of c. 350 nm were visualized [26].

From the standpoint of versatile macrocyclic host molecules, pillar[*n*]arenes have attracted increasing interest owing to an outstanding applicability to various supramolecular systems based on host–guest chemistry [27, 28]. As the inventor of pillar[*n*]arenes, Ogoshi et al. established photoreversible supramolecular copolymerization of an azobenzene-bridged bis(pillar[5]arene) **6** as a ditopic host with hexamethylenebis(pyridinium) **7** as a ditopic guest (Figure 2.3) [29]. Diffusion-ordered NMR spectroscopy (DOSY) and ^{1}H NMR analyses in dichloromethane-d_2 (CD_2Cl_2) revealed that the host–guest mixture of the *trans-*isomer of **6** (*trans-***6**) and **7** affords supramolecular copolymers in high concentration. Upon irradiation of the mixture with UV light (λ = 365 nm), a significant increase of the diffusion constant (D) was observed, indicating a disassociation of the supramolecular copolymers into relatively low molecular weight oligomers composed of *cis-***6** and **7**. The photoresponsive assembly and disassembly processes have been shown reversibly by alternating irradiation of the mixture with UV and visible (λ = 436 nm) light. Other host–guest supramolecular polymers including calix[*n*]arenes [30–32] and cucurbit[*n*]urils [33] have also been applied to photoresponsive polymerization/depolymerization systems. These photoswitchable host–guest supramolecular polymers can find a wide range of applications, for instance, self-healable supramolecular gels and vesicles, controllable aggregation of particles, and others [34–37].

Metal–ligand interactions are also useful to construct well-defined supramolecular architectures [38–40]. Huang and coworkers successfully prepared

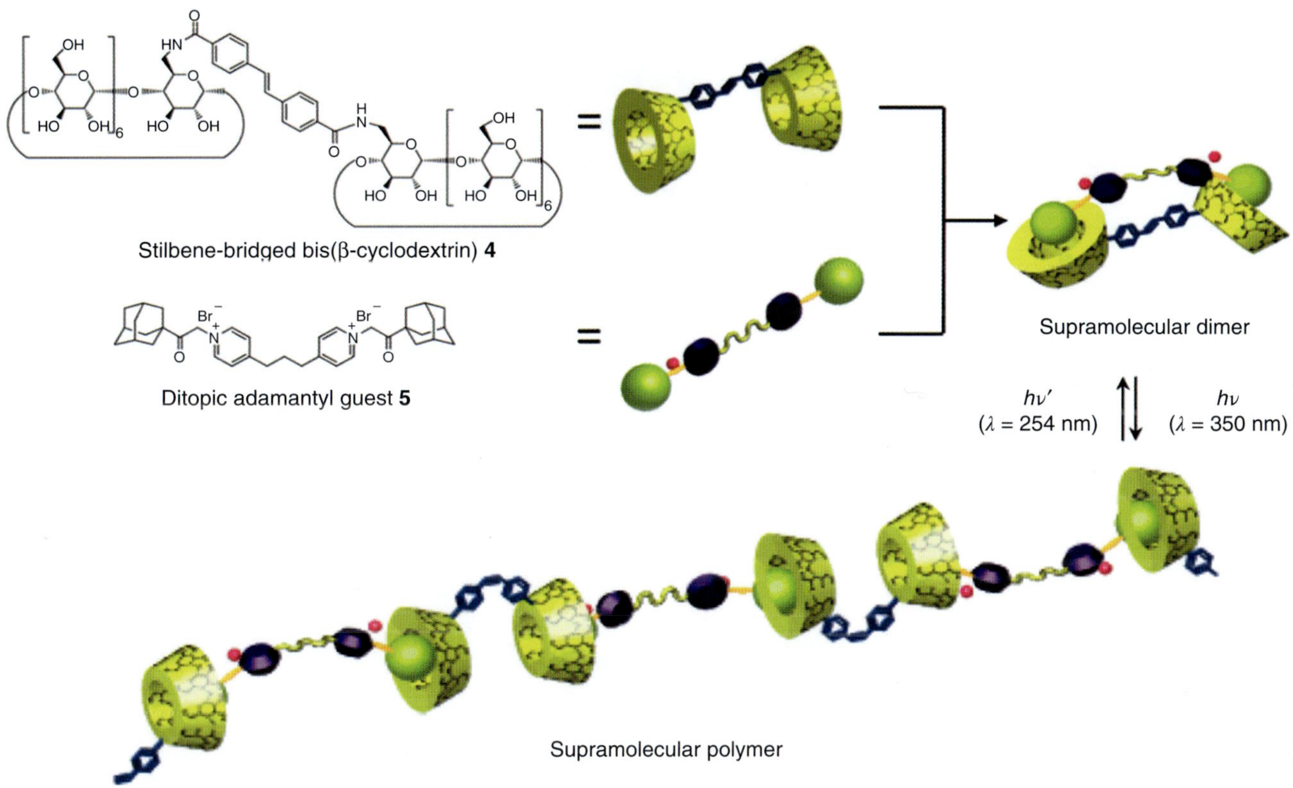

Figure 2.2 Schematic representation of photocontrolled transformation between supramolecular polymer and dimer in the host–guest complex of stilbene-bridged bis(β-cyclodextrin) **4** with ditopic adamantyl guest **5**. Source: Kuad et al. 2007 [26]. Copyright 2007. Reproduced with permission from The American Chemical Society.

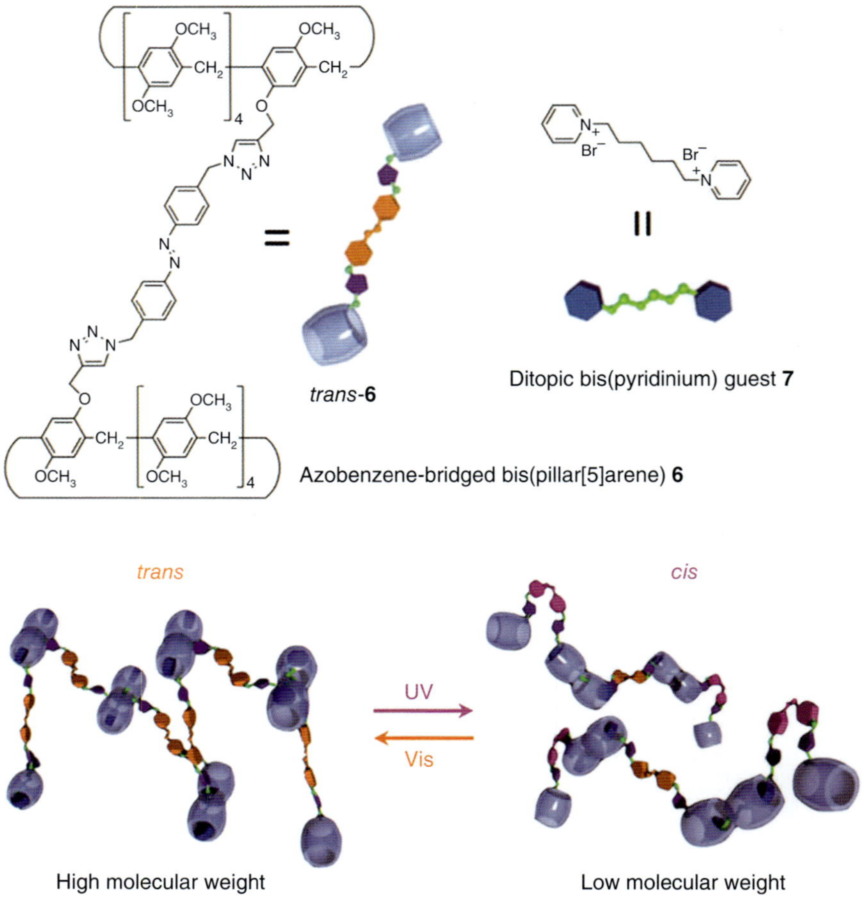

Figure 2.3 Schematic representation of photoresponsive supramolecular polymers composed of azobenzene-bridged bis(pillar[5]arene) **6** and homoditopic hexamethylenebis(pyridinium) guest **7**.

a photoresponsive metallosupramolecular polymer using a stiff stilbene functionalized with pyridine ligands [41]. They designed a stiff stilbene derivative **8** bearing two pyridyl groups as a ligand and a bis(*trans*-Pt(II))benzene complex **9** as a precursor of the metal center (Figure 2.4a). Multinuclear NMR and ESI-MS measurements showed that equimolar mixing of the *cis*-isomer of **8** (*cis*-**8**) with bis-Pt(II) complex **9** in CH_2Cl_2 yielded a discrete metallocyclic tetramer (*cis*-$\mathbf{8}_2 \cdot \mathbf{9}_2$) with high efficiency. Upon UV-light irradiation (λ = 387 nm) of the solution of the metallacycles, *cis-to-trans* photoisomerization of **8** occurred, which gave rise to the transformation from the discrete metallacycles to linear metallosupramolecular polymers (Figure 2.4b), as evidenced by dynamic light scattering (DLS) measurements. Upon *trans-to-cis* back isomerization of **8** with UV-light irradiation (λ = 360 nm), however, quantitative formation of the original discrete metallacycle was not observed as a result of a *cis:trans* ratio of 47 : 53 at a photostationary state (PSS). Instead, the average D_h of the resulting aggregates was larger than that of the original discrete metallacycles, suggesting the

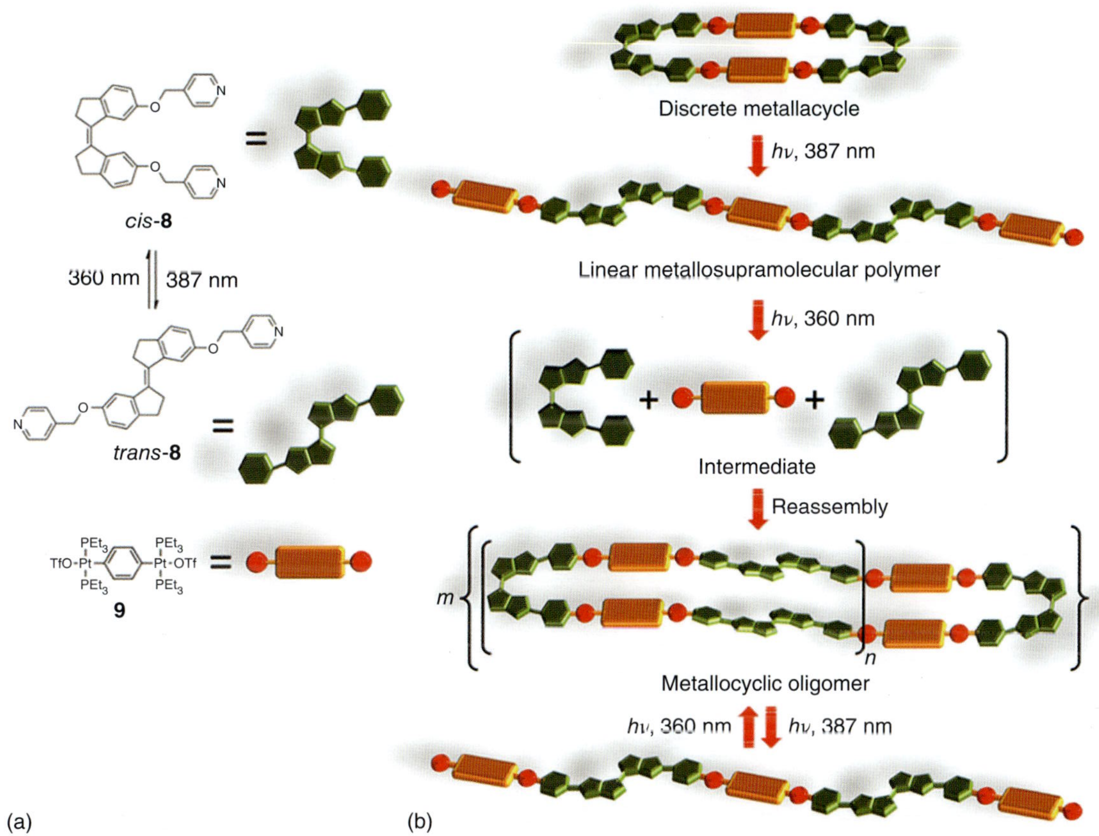

Figure 2.4 (a) Molecular structures of stiff-stilbene derivative **8** functionalized by pyridyl groups and bis(*trans*-Pt(II)) benzene **9**. (b) Schematic representation of photoresponsive coordination supramolecular polymer system composed of **8** and **9**.

formation of metallocyclic oligomers by integrative self-sorting of *cis*- and *trans*-**8** with **9** (Figure 2.4b) [42]. The transformation between the linear metallosupramolecular polymers and the metallocyclic oligomers was photochemically reversible by alternating irradiation at the two wavelengths. The authors presumed that the photoresponsive transformation would proceed in an off-pathway mechanism. Transmission electron microscopic (TEM) analysis revealed that spherical nanoparticles formed by the discrete metallacycles converted upon UV-light irradiation at 387 nm into long nanofibers with a width of c. 50 nm and a length of >1 μm.

2.3 Photoresponsive Supramolecular Polymers by Complementary Hydrogen Bonds

This section focuses on photoresponsive supramolecular polymers, wherein photochromic units are noncovalently polymerized through multiple hydrogen-bonding interactions. The high reversibility of hydrogen bonds plays an essential role in the dynamic properties of many supramolecular polymers. Because photoisomerization of azobenzene and other photochromic molecules generally occurs on subpicosecond to picosecond timescales, monomer units can change their structures upon absorbing light in the monomeric state supplied through aggregate–monomer exchange/reshuffling even though they are very tightly aggregated. The photoisomerized monomers should reassemble to form different assembled structures or even remain monomeric as a result of alteration of conformation or flexibility by photoisomerization.

Ghadiri and coworkers designed a pioneering photoresponsive supramolecular oligomer by using a peptide-functionalized azobenzene **10**, wherein two macrocyclic octapeptides can noncovalently dimerize through multiple hydrogen-bonding interactions (Figure 2.5) [43]. The *trans*-isomer of **10** intermolecularly self-assembled to form linear and polydisperse hydrogen-bonded aggregates in chloroform ($CHCl_3$). NMR studies revealed that upon UV-light irradiation at 366 nm the linear aggregates transform into discrete species due to intramolecular hydrogen bonding. Although these photoswitchable assembled states are reversible, a remarkable resistance of the *cis-to-trans* thermal isomerization was observed as a result of the strong intramolecular hydrogen-bonding interactions. An analogous effect was observed for the host–guest complex of **2** with alkali metal cations [44].

Complementary multiple hydrogen-bonding (CMHB) interactions are essential for the biological recognition process and are one of the most powerful tools to control the organization of small molecular building blocks. High directionality and specificity with reversibility of CMHB interactions have motivated many supramolecular chemists to create various supramolecular assemblies. For example, compounds having complementary centrosymmetric hydrogen-bonding arrays, such as DAD and ADA (D = hydrogen bond donor, A = hydrogen bond acceptor), are capable of assembling into cyclic or polymeric aggregates [45]. Because heteroaromatic compounds generally possess a rigid and geometrically well-defined skeleton, they can be used as highly specific and directional

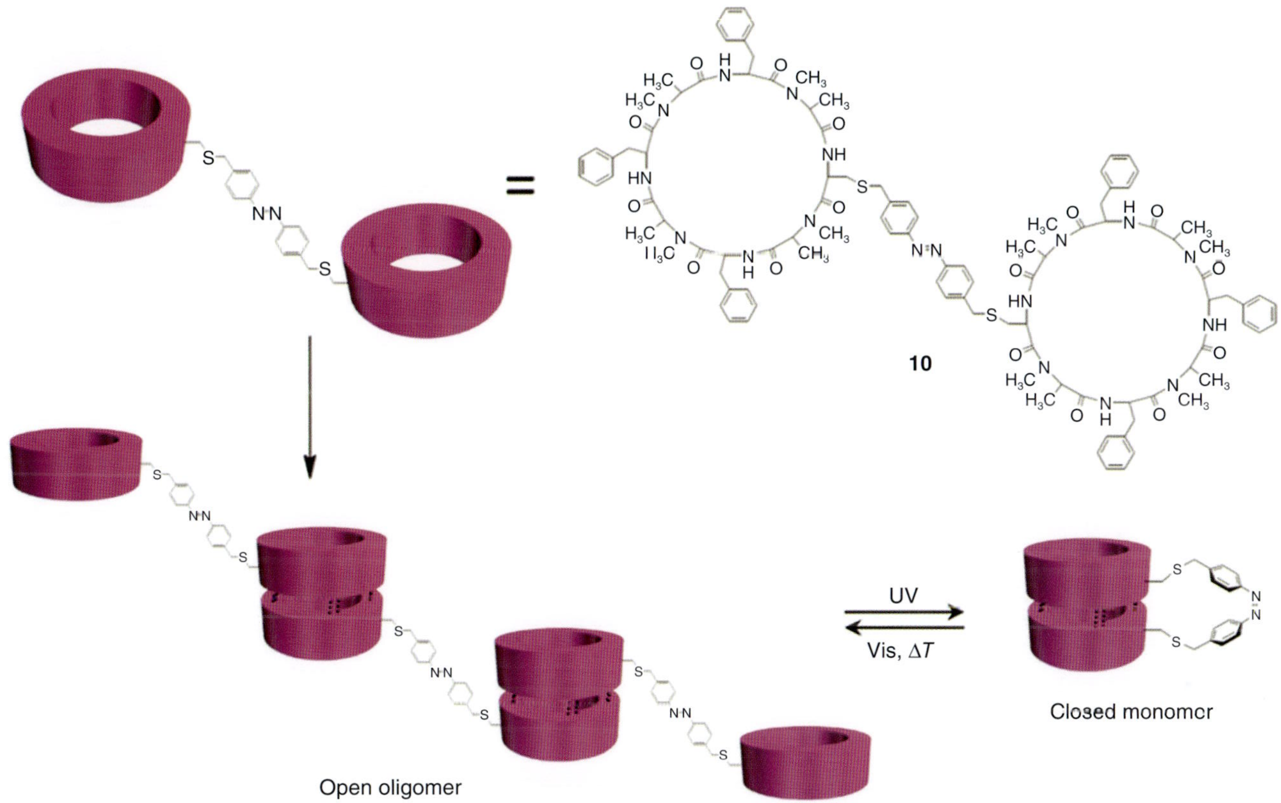

Figure 2.5 Schematic representation of photoreversibly switchable supramolecular oligomers of azobenzene-tethered cyclic octapeptide dimer **10**.

multiple hydrogen-bonding modules. Needless to say, their high fidelity is epitomized by the base-pairing in DNA and RNA strands. The conceptual background and studies of CMHB interactions organizing functional dyes and π-conjugated systems have been addressed in several reviews [46–49].

Ureidopyrimidinone (Upy) is one of the most widely used CMHB units to prepare supramolecular polymers. A mono-Upy derivative **11** has been known to dimerize in even $CHCl_3$ with a high dimerization constant exceeding $10^6\,M^{-1}$ [50]. Meijer and coworkers first demonstrated supramolecular polymers based on CMHB interactions by using a self-complementary bis-Upy derivative **12**. To control the degree of polymerization of **12** by light, they developed a photoresponsive depolymerization system by *in situ* photogeneration of chain stopper **11** (Figure 2.6) [51]. For this purpose, they designed a "caged" stopper **13** protected with *ortho*-nitrobenzyl ether group. Compound **13** was smoothly photodeprotected to give stopper molecule **11** by UV-light irradiation [52], which can be confirmed by ^1H NMR spectroscopy. Viscosity (η_{rel}) measurements of a supramolecular polymer solution of **12** in $CDCl_3$ in the presence of 0.1 equiv. of **13** showed that after two hours of irradiation with UV light, the original η_{rel} value of 10.79 drastically decreased to c. 1.92, comparable to the value observed in the presence of 0.1 equiv. of externally added **11** instead of **13**. The significant drop in η_{rel} value indicates that the formation of hetero-assemblies of *in situ*

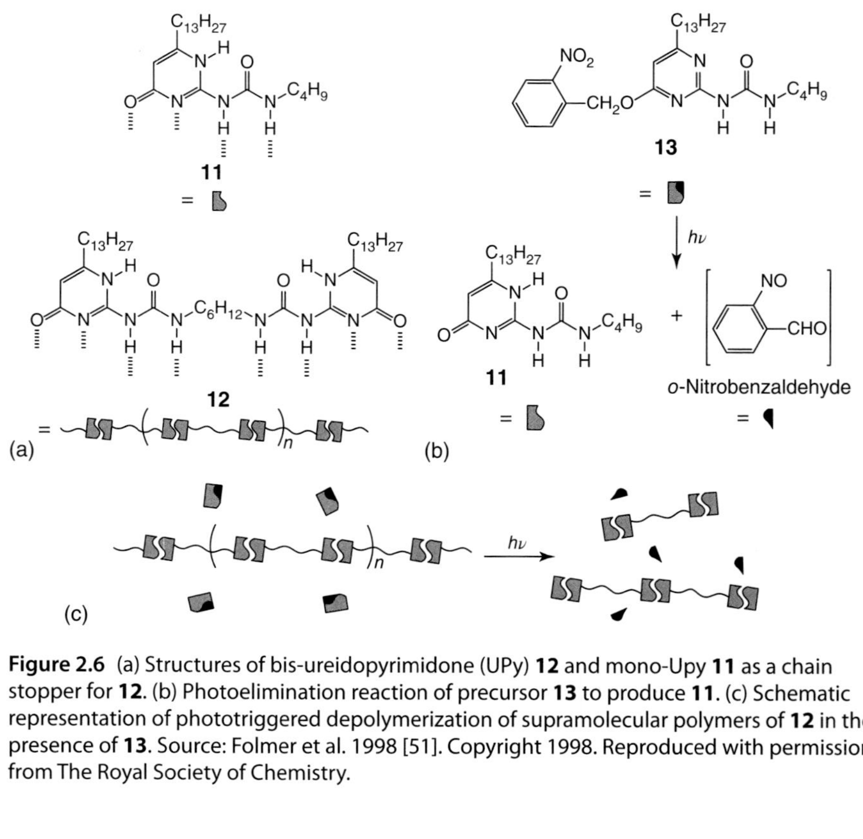

Figure 2.6 (a) Structures of bis-ureidopyrimidone (UPy) **12** and mono-Upy **11** as a chain stopper for **12**. (b) Photoelimination reaction of precursor **13** to produce **11**. (c) Schematic representation of phototriggered depolymerization of supramolecular polymers of **12** in the presence of **13**. Source: Folmer et al. 1998 [51]. Copyright 1998. Reproduced with permission from The Royal Society of Chemistry.

photogenerated **11** with **12** inhibits the self-assembly of 12, leading to phototriggered depolymerization of supramolecular polymers of 12.

The *open-* and *closed*-ring isomers of diarylethenes are generally thermally inactive toward reversible isomerization [53, 54], which enables more stable and sustainable two-state switching in comparison with azobenzene systems [55, 56]. Takeshita et al. prepared a diarylethene derivative **14** functionalized on both ends by Upy quadruple hydrogen-bonding groups (Figure 2.7) [57]. ^{1}H NMR study of the *open*-ring isomer of **14** ($\mathbf{14}_{open}$) in $CDCl_3$ showed no shifts of the NH proton signals, and only small particles less than 10 nm were detected by DLS. The ring-closing reaction of $\mathbf{14}_{open}$ proceeded in $CHCl_3$ with a quantum yield of 0.26, indicating that the major and minor conformers are antiparallel and parallel, respectively [58]. DLS measurements showed that UV-light irradiation ($\lambda = 366$ nm) of the $CHCl_3$–ethanol (97 : 3) solution of $\mathbf{14}_{open}$ resulted in a gradual increase in the original particle size (<10 nm), which became constant at around 650 nm after three hours, indicating the formation of supramolecular polymers by the photogenerated *closed*-ring isomer $\mathbf{14}_{closed}$. Subsequent visible-light irradiation ($\lambda > 540$ nm) decreased the particle size from 650 to 400 nm, whereas the resulting supramolecular aggregates with the size of c. 400 nm did not further collapse even after the ring-opening reaction was accomplished. Conversely, after aging the solution at 60 °C for 30 minutes, the aggregates disassembled into the original small particles (<10 nm). The author concluded that the strong CMHB interactions prevent the large aggregates consisting of antiparallel $\mathbf{14}_{open}$ from disassembling into its own small aggregates.

Intermolecular photoreactions such as photocyclization and photodimerization reactions observed for acenes can also be applied to designing photoresponsive supramolecular polymers. Arai and coworkers employed a [4 + 4] photodimerization reaction to establish a photoreversible supramolecular polymer system [59]. Anthracene **15** equipped with an Upy moiety initially formed hydrogen-bonded dimers in $CDCl_3$ (Figure 2.8a). ^{1}H NMR and UV–vis studies showed that **15** undergoes photodimerization upon UV-light irradiation (>310 nm) in $CDCl_3$ to form the photodimer **16**. The NH proton signals associated with intermolecular hydrogen bonds became broadened upon photodimerization, and the signals eventually became too broad to be observed after reaching a PSS (conversion = c. 80%), indicating the formation of oligomeric and/or polymeric assemblies consisting of **16**. Considering the conversion yield at the PSS, hydrogen-bonded oligomers of **16** should be capped with unreacted **15** (Figure 2.8b). The photoinduced oligomerization also showed thermal reversibility: upon heating the resulting oligomer solution at 125 °C, the retro-reaction quantitatively proceeded with the rate constant of $3 \times 10^{-4}\ s^{-1}$. Since the temperatures at which the half-life of the photodimer is one hour are very different between a "head-to-head" photodimer (38 °C) and a "head-to-tail" photodimer (149 °C) for 9-methylanthracene, the above result also suggests that the photodimer **16** has a "head-to-tail" conformation [60].

Yang and coworkers described a photoresponsive supramolecular polymer based on a stiff stilbene derivative **17** functionalized with Upy moieties (Figure 2.9a) [61]. Unlike azobenzene, *cis-to-trans* thermal isomerization of

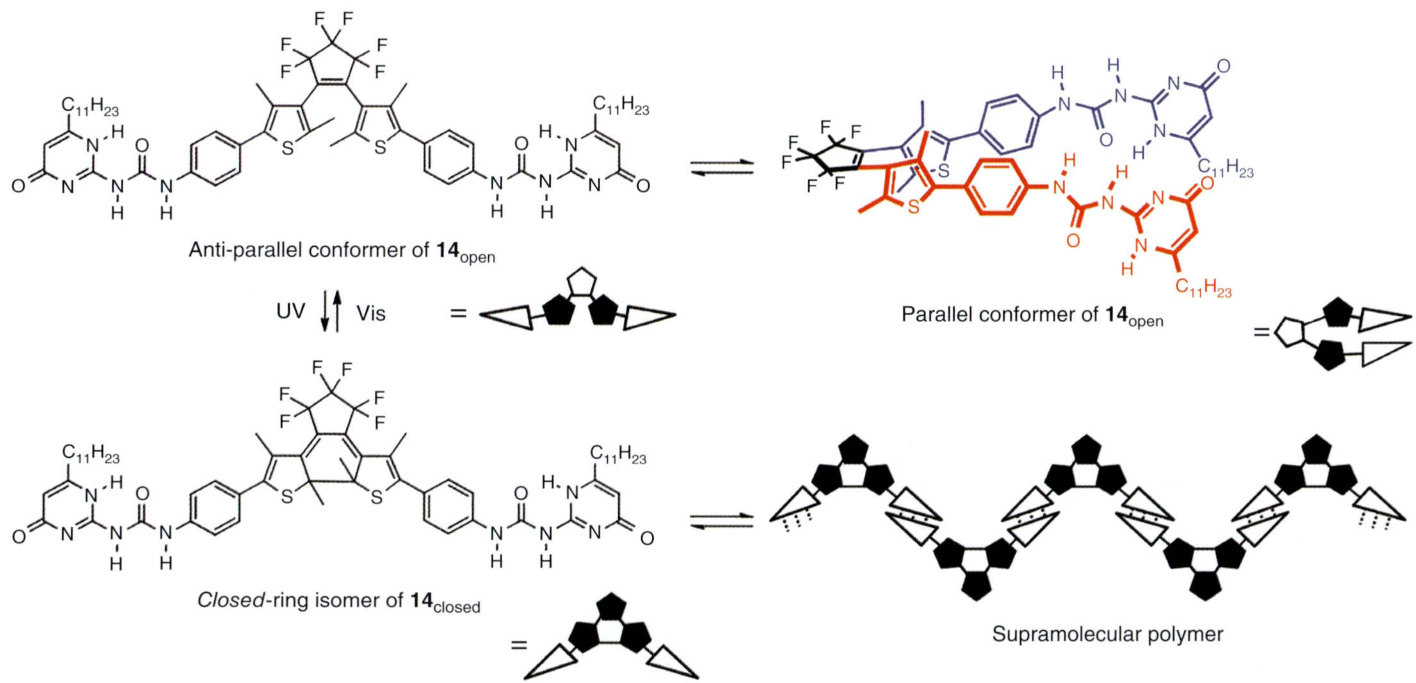

Figure 2.7 Schematic representation of photoresponsive supramolecular polymers of self-complementary diarylethene **14** functionalized on both ends by Upy groups.

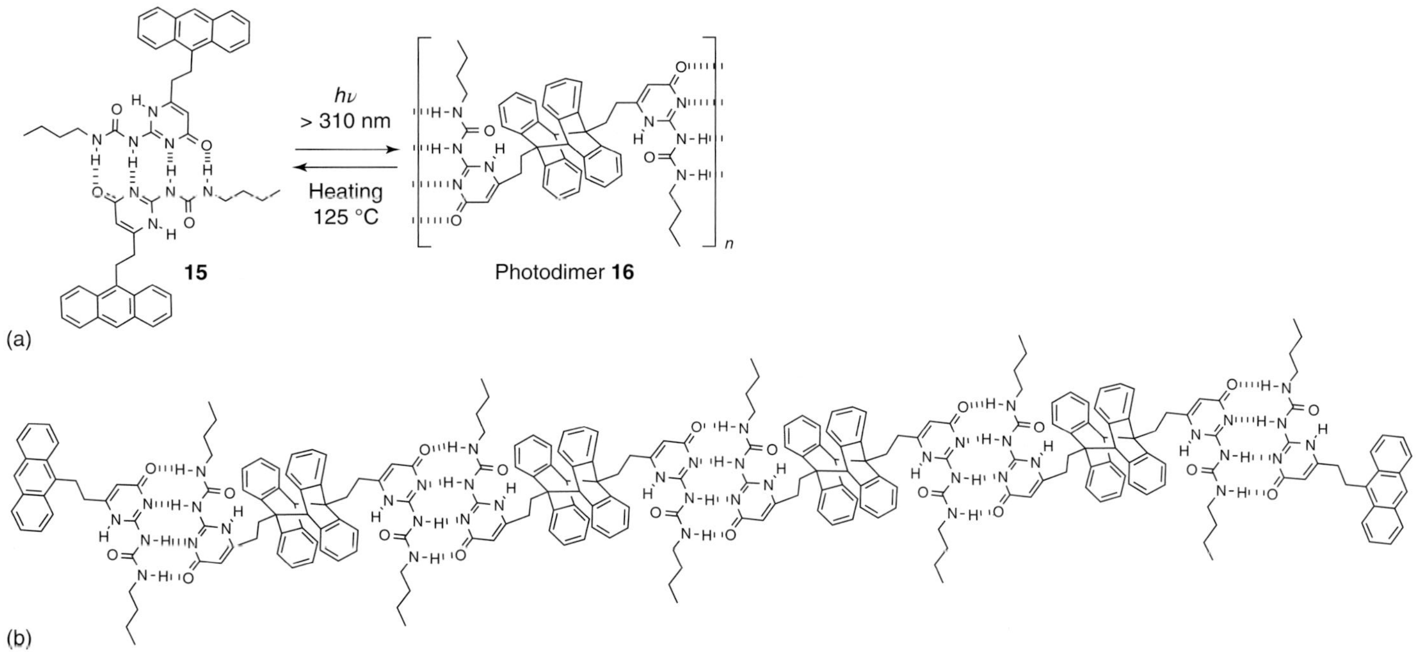

Figure 2.8 (a) Photocyclization of Upy-functionalized anthracene **15** and thermal retro-cyclization reaction of the photodimer **16**. (b) Plausible supramolecular oligomeric form of **16** end-capped with **15** in 25 mM $CDCl_3$ solution. Source: Ikegami et al. 2003 [59]. Copyright 2003. Reproduced with permission from The Royal Society of Chemistry.

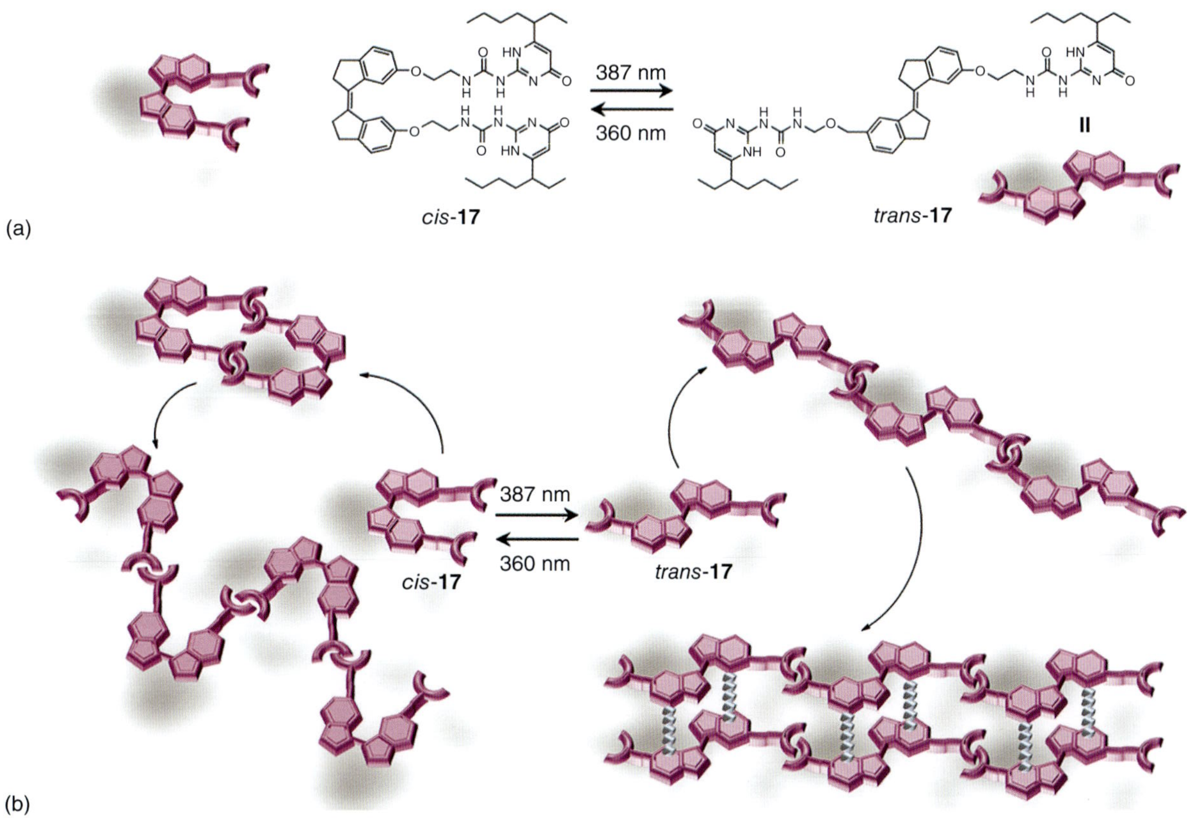

Figure 2.9 (a) Photoisomerization of Upy-functionalized ditopic stiff stilbene **17**. (b) Cartoon representation of photoresponsive supramolecular polymerization of *cis*- and *trans*-**17**.

stilbene is negligibly slow at temperatures below c. 147 °C [62]. ^{1}H NMR and DOSY spectra of **17** ($c = 5 \times 10^{-2}$ M) in $CDCl_3$ revealed that *cis*-**17** (at a PSS, >99%) forms two distinct species corresponding to a cyclic dimer as the major component and a monomer as the minor component. In contrast, *trans*-**17** (at a PSS, >99%) displayed smaller diffusion constants compared with *cis*-**17**, indicating that *trans*-**17** can self-assemble into supramolecular polymers. No critical point was observed for *trans*-**17** in viscosity measurements upon increasing its concentration. For *cis*-**17**, however, the critical polymerization concentration (c. 0.1 M) was observed. On the basis of the viscosity measurements, it was proposed that the supramolecular polymerization of *cis*-**17** proceeds via a ring–chain exchange mechanism in concentrations above 0.1 M, whereas *trans*-**17** follows an isodesmic growth mechanism (Figure 2.9b). Scanning electron microscopy (SEM) observation at high concentration (0.4 M) revealed that *cis*-**17** self-assembles into nanofibers with diameters of 0.3–3 μm and lengths of hundreds of micrometers, whereas *trans*-**17** forms a gel. UV-light irradiation (λ = 365 nm) of the gel of *trans*-**17** caused the *trans-to-cis* photoisomerization to give *cis*-**17**, leading to the dissolution of the gel.

Chi et al. reported a photoreversible transformation of hydrogen-bond-directed helical supramolecular polymers of a diarylethene derivative (*S*,*S*)-**18** to granular aggregates (Figure 2.10) [63]. The supramolecular chirality is governed by (*S*)-phenylalanine moieties of (*S*,*S*)-**18** and detected through excitonic coupling of naphthalimide chromophores cross-linking diarylethene and (*S*)-phenylalanine moieties of (*S*,*S*)-**18** (Figure 2.10a). In tetrachloromethane (CCl_4), the *open*-ring isomer of (*S*,*S*)-**18** ((*S*,*S*)-$\mathbf{18}_{\text{open}}$) showed a sharp circular dichroism (CD) signal with a characteristic Cotton effect corresponding to the absorption regions of the diarylethene and naphthalimide chromophores. The methyl ester derivative of (*S*,*S*)-$\mathbf{18}_{\text{open}}$ showed much weaker CD signals compared with the carboxylic acid (*S*,*S*)-$\mathbf{18}_{\text{open}}$, suggesting that (*S*,*S*)-$\mathbf{18}_{\text{open}}$ forms chiral supramolecular assemblies through intermolecular hydrogen bonding between the carboxylic groups. AFM visualized left-handed helical nanofibers with a height of c. 6 nm and a helical pitch of c. 50 nm (Figure 2.10b). Upon UV-light irradiation (λ = 365 nm) of the CCl_4 solution of (*S*,*S*)-$\mathbf{18}_{\text{open}}$, the CD signal remarkably decreased with the photoconversion of (*S*,*S*)-$\mathbf{18}_{\text{open}}$ to afford (*S*,*S*)-$\mathbf{18}_{\text{closed}}$ in 48% yield at a PSS. AFM showed the transformation of the original helical nanofibers into the granular particles (Figure 2.10c). The authors attributed this transformation to the rigidity of the diarylethene core, which hampers (*S*,*S*)-$\mathbf{18}_{\text{closed}}$ from adopting a helical geometry. This morphology transformation is completely photoreversible, as AFM imaging after the visible-light irradiation (λ > 510 nm) recovered the original helical nanofibers with the active CD signals (Figure 2.10d). The optical isomer (*R*,*R*)-$\mathbf{18}_{\text{open}}$ showed the mirror image CD signal of the one observed for (*S*,*S*)-$\mathbf{18}_{\text{open}}$, and AFM image for (*R*,*R*)-$\mathbf{18}_{\text{open}}$ confirmed right-handed helical nanofibers. Based on density functional theory (DFT) calculations and molecular modeling, the authors proposed that the parallel conformer of (*S*,*S*)-$\mathbf{18}_{\text{open}}$ can self-assemble through intermolecular hydrogen bonds between the carboxylic groups to form primary helical strands (Figure 2.10e,f), which further organize into secondary helical fibers to accommodate the structural strain of the primary helical strands (Figure 2.10g). Considering this proposed assembled model, it can

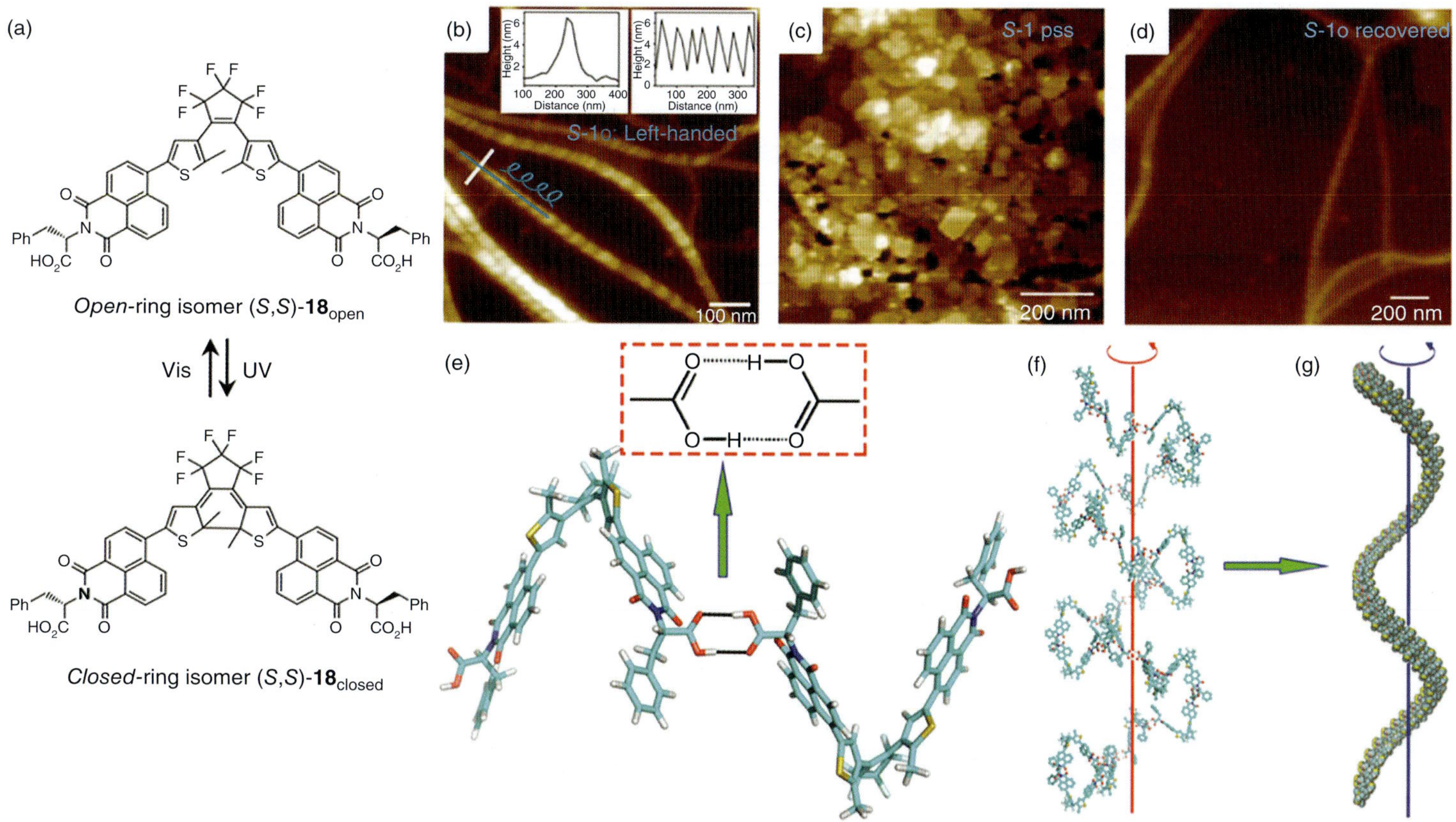

Figure 2.10 (a) Photoisomerization between *open*- and *closed*-ring isomers of diarylethene (*S*,*S*)-**18** bearing (*S*)-phenylalanine-functionalized naphthalimide units. (b–d) AFM images of supramolecular assemblies (b) before irradiation of a (*S*,*S*)-$\mathbf{18}_{open}$ solution in CCl_4 (insert: cross sectional analysis), (c) after irradiation of the (*S*,*S*)-$\mathbf{18}_{open}$ solution with UV light at a PSS ($\lambda = 365$ nm), and (d) after subsequent visible-light irradiation of the resulting solution ($\lambda > 510$ nm). (e–g) Molecular models of helical assembly of (*S*,*S*)-$\mathbf{18}_{open}$ for (e) intermolecular hydrogen-bonding motif between the neighboring two molecules, (f) energy-optimized primary helical strand, and (g) computer-generated representation of secondary helix. Source: Cai et al. 2016 [63]. Copyright 2016. Reproduced with permission from American Chemical Society.

be inferred that the photoinduced transformation could proceed via aggregate–monomer exchange, although the authors did not describe this (see Section 2.1).

Yagai et al. have designed a unique photoresponsive supramolecular polymer wherein the parallel conformer of a diarylethene monomer engages actively in supramolecular polymerization [64]. Based on the previous study on photoresponsive supramolecular complexes composed of a melamine-functionalized diarylethene and a barbiturated merocyanine dye [65], they newly synthesized a tetratopic diarylethene receptor **19** bearing ditopic melamine groups on both ends and investigated the complexation with a ditopic tetraphenoxy perylene bisimide (PBI) guest **20**, which has a propensity to form slipped (J-type) aggregates (Figure 2.11a) [66, 67]. UV–vis titration experiments of **20** with the *open*-ring isomer of **19** ($\mathbf{19}_{open}$) in toluene revealed the formation of 2 : 1 supramolecular co-assemblies $[\mathbf{19}_{open}{\cdot}(\mathbf{20})_2]_n$, wherein the PBI chromophores form a J-type stacking mode. AFM imaging of $[\mathbf{19}_{open}{\cdot}(\mathbf{20})_2]_n$ showed well-defined nanofibers with lengths reaching nearly a micrometer-scale regime (Figure 2.11b). A close-up AFM imaging revealed that the nanofibers have a helically twisted topology with a height of 2 nm and a helical pitch of 20 nm (inset in Figure 2.11b). Based on the spectroscopic and morphological observations, the authors constructed an oligomeric co-assembly composed of the parallel conformer of $\mathbf{19}_{open}$ and the dimeric J-type stack (J-dimer) of **20** by means of force-field molecular mechanics calculation, which could reproduce the helically extended structure due to the twisted geometry of **19** (Figure 2.11d,e). Upon UV-light irradiation (λ = 313 nm) of the toluene solution of $[\mathbf{19}_{open}{\cdot}(\mathbf{20})_2]_n$, photocyclization of $\mathbf{19}_{open}$ proceeded to afford $\mathbf{19}_{closed}$ while maintaining the J-band. AFM observations revealed that the helical nanofibers were converted to granular nanostructures with diameters of c. 10 nm and heights of c. 1.2 nm (Figure 2.11c). Because the conrotatory photocyclization can proceed only from the antiparallel conformer of $\mathbf{19}_{open}$ [68, 69], aggregate–monomer exchange should occur to afford monomeric $\mathbf{19}_{open}$, which quickly reaches parallel–antiparallel conformational equilibrium. The ring-opening reaction of $\mathbf{19}_{closed}$ by visible-light irradiation (λ = 630 nm) reproduced $\mathbf{19}_{open}$, and reconstructed the original fibrous nanostructures. This cycle could be repeated at least three times, thus demonstrating the photoreversibility of this supramolecular system. Notably, the granular aggregates composed of $\mathbf{19}_{closed}$ and the J-dimer $(\mathbf{20})_2$ cannot be obtained if $\mathbf{19}_{closed}$ and **20** were directly mixed. Accordingly, the J-dimer as well as the granular nanostructures could form only through the photoconversion of supramolecular copolymers $[\mathbf{19}_{open}{\cdot}(\mathbf{20})_2]_n$ (Figure 2.11f).

2.4 Photoresponsive Supramolecular Polymers by Stacking of Photochromic Molecules

In Section 2.3, we focused on photoresponsive supramolecular polymers whose main chains are formed by hydrogen bonds. Another fascinating approach to supramolecular polymers is the use of π–π stacking interactions between functional π-conjugated molecules. Because most photochromic molecules have

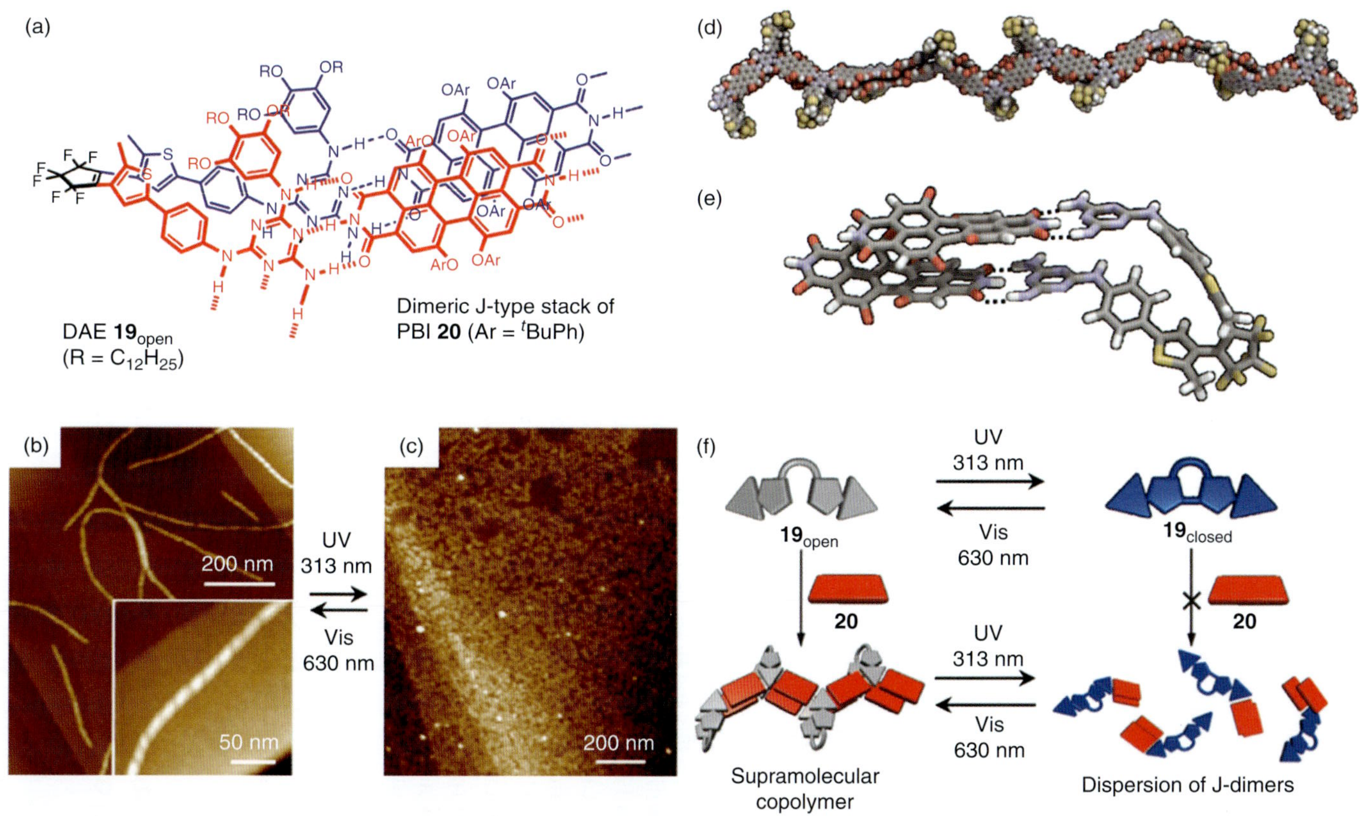

Figure 2.11 (a) Binding motif of local supramolecular structure formed by parallel conformer of diarylethene $\mathbf{19}_{open}$ and a J-type dimer of PBI **20**. (b) AFM images of helical nanofibers of $[\mathbf{19}_{open}{\cdot}(\mathbf{20})_2]_n$ prepared in toluene. (c) AFM image of granular aggregates formed by UV-light irradiation ($\lambda = 313$ nm) of the nanofibers of $[\mathbf{19}_{open}{\cdot}(\mathbf{20})_2]_n$. (d, e) Energy-optimized structures of [parallel-$\mathbf{19}_{open}{\cdot}(\mathbf{20})_2]_{12}$. (e) A local supramolecular structure of $\mathbf{19}_{open}{\cdot}(\mathbf{20})_2$ extracted from the energy-optimized structure. (f) Schematic representation of the photoregulated aggregation pathway of mixture of $\mathbf{19}_{open}/\mathbf{19}_{closed}$ and **20**.

fertile π-conjugated systems, they are expected to self-assemble into one-dimensional aggregates by proper functionalization with auxiliary interacting units. Moreover, photochromic molecules show fruitful changes in their molecular structures and chemical properties upon photoisomerization. This can be exploited to control or regulate the degree of supramolecular polymerization under proper conditions.

In the case of azobenzene, the *trans*-isomer has a planar geometry favorable for π–π stacking, whereas the *cis*-isomer has a bent conformation unfavorable for π–π stacking [70, 71]. Yagai et al. reported an archetypal example of photoresponsive supramolecular polymers consisting of azobenzene-appended hydrogen-bonded supermacrocycles (rosettes) [72–74]. A cyanurate dCA bearing sterically less bulky dodecyl groups aggregated complementarily with a bis(azobenzene)-functionalized melamine **21** to afford rosette $(\mathbf{21}\cdot\mathrm{dCA})_3$ in $CHCl_3$, which further self-assembles in nonpolar media into higher order columnar structures through π–π stacking interactions (Figure 2.12). At millimolar concentration regimes, this hierarchical association of rosettes resulted in the formation of elongated and entangled fibers organizing into organogels. Upon irradiation of cyclohexane organogels with UV light (λ = 350 nm), a gel-to-sol phase transition was observed even though the *trans-to-cis* photoisomerization efficiency was moderate (20–40% *cis*-isomer, Figure 2.12b). The resulting solution phase could revert to the original gel phase with visible-light irradiation (λ = 450 nm) and subsequent aging (Figure 2.12b). Because the planarity of rosette $(\mathbf{21}\cdot\mathrm{dCA})_3$ critically influences its stacking trend, even partial *trans-to-cis* isomerization of azobenzene moieties at the periphery can strongly deteriorate its supramolecular polymerization capability. In other words, the impact of photoinduced structural change of azobenzene on aggregation can be amplified through rosette formation. For a diluted supramolecular polymer solution, a similar photoresponsive behavior was confirmed by DLS measurements.

A photoresponsive behavior in contrast to the above example has been reported by Pavan and coworkers [75] for supramolecular tubular nanoassemblies composed of a V-shaped chiral azobenzene amphiphile **22** (Figure 2.13a). The original design of this amphiphilic molecule has been reported by Lee and coworkers, who demonstrated that a series of the V-shaped amphiphiles organize into tubular supramolecular polymers through the formation of hexameric supermacrocycles [76–79]. For *trans*-azobenzene-appended **22** (*trans,trans*-**22**), the formation of tubular nanoassemblies in aqueous solution has been confirmed by TEM observation. The external diameter of c. 7 nm roughly corresponds to the estimated diameter of the hexameric supermacrocycle of *trans,trans*-**22** from molecular models (Figure 2.13b). A pronounced Cotton effect was observed for the absorption region of the azobenzene chromophore in CD measurements, indicating that supermacrocycles stack with an offset arrangement in a preferred handedness. Interestingly, a nonlinear disassembly process was observed upon irradiation of the nanoassemblies with UV light (λ = 365 nm). The initial five minutes of irradiation resulted in a drastic decrease of the CD signal at 400 nm despite the fact that the fraction of photogenerated *cis*-isomer was only 7%. In the absorption spectra, a bathochromic shift of the absorption maximum was observed. These complex behaviors were ascribed to the annihilation of the

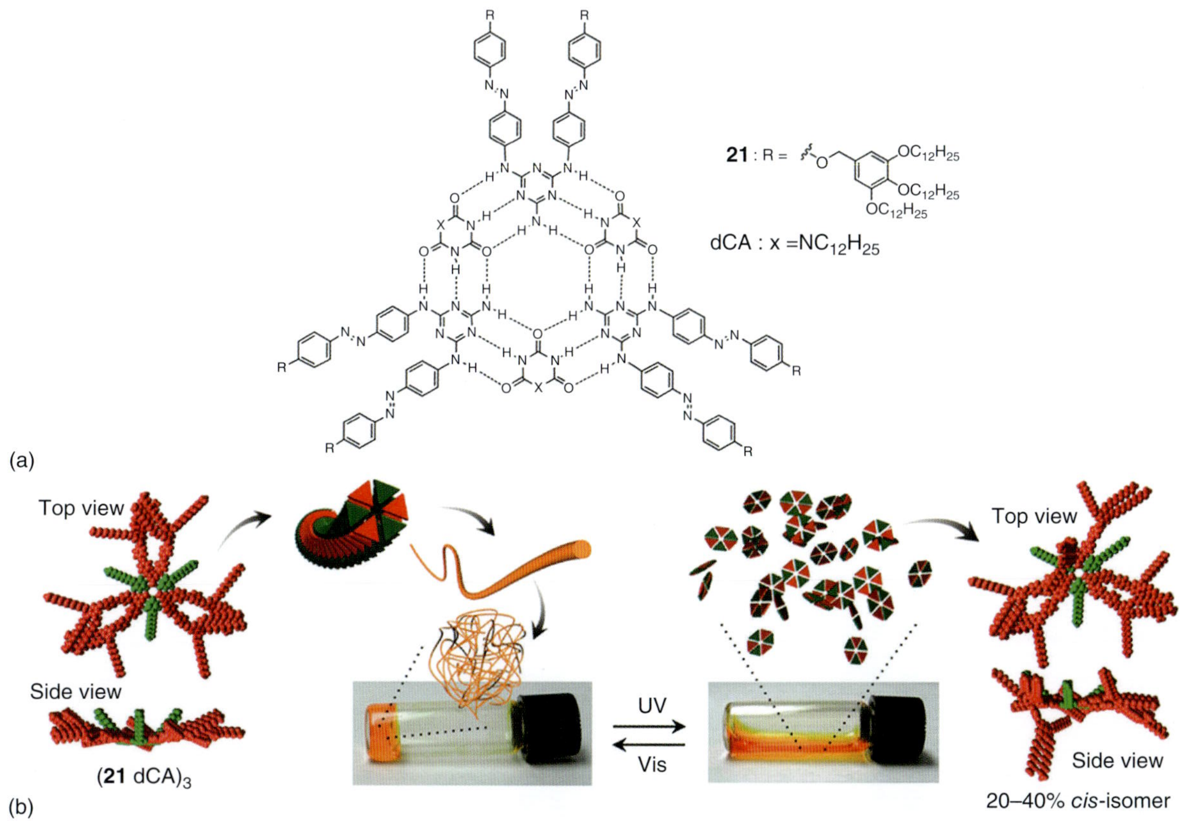

Figure 2.12 (a) Hydrogen-bonded hexameric rosette (**21**·dCA)$_3$ consisting of dodecylcyanurate dCA and azobenzene-functionalized melamine **21**. (b) Schematic representation of photoreversible supramolecular polymerization of rosette (**21**·dCA)$_3$, leading to gel–sol transition.

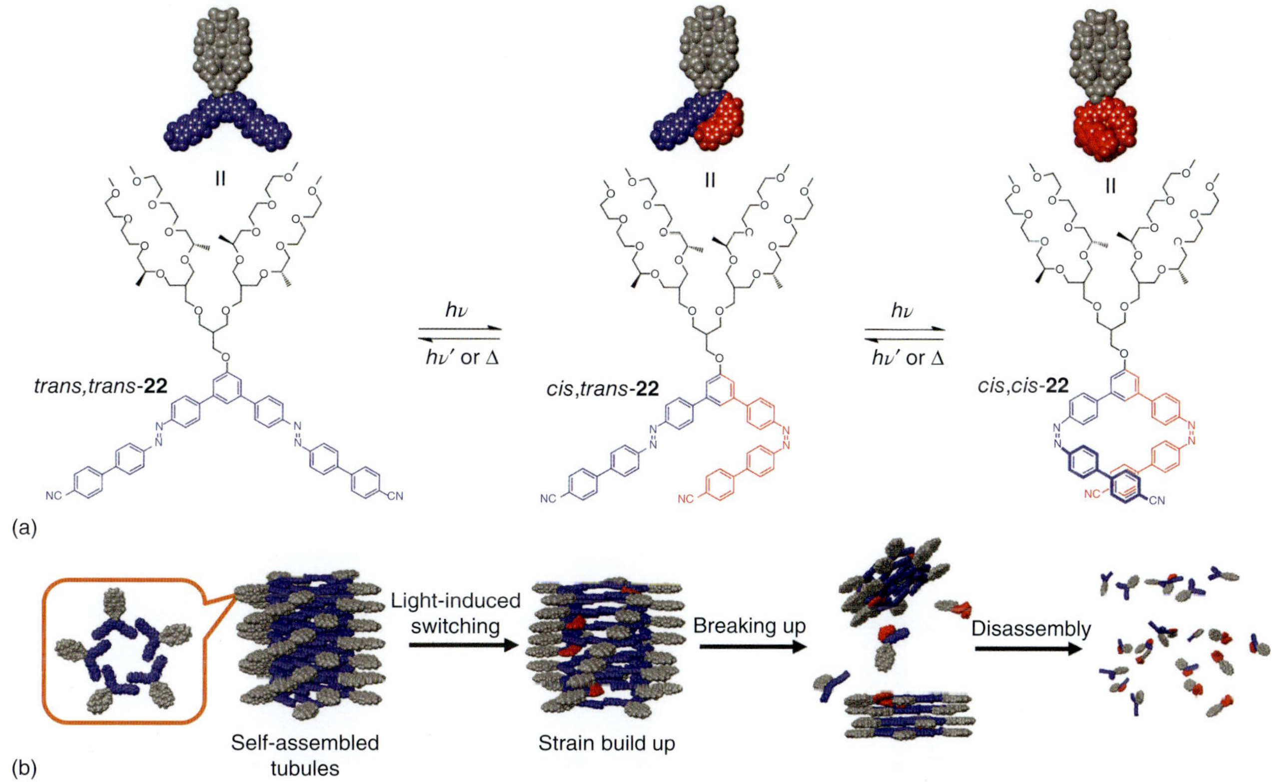

Figure 2.13 (a) Photoisomerization of V-shaped amphiphile **22** bearing chiral hydrophilic chains and hydrophobic azobenzene units. (b) Proposed mechanism for photoresponsive collapse of the supramolecular tubules formed by **22**.

helical twist due to the formation of *cis,trans*-**22** and *cis,cis*-**22**, which were still embedded in the tubular nanoassemblies. The collapse of the tubular nanoassemblies was probed by fluorescent analysis using Nile red as an internal probe that could be encapsulated in the tubules. A gradual release of this lipophilic dye was observed after the UV-light irradiation for c. 10 minutes, as shown by a steady decrease in the fluorescence signal from the dye. The authors proposed that photogenerated *cis,trans*-**22** and *cis,cis*-**22** give birth to strain energy that can trigger the depolymerization of the tubular nanoassemblies, and disassembly of the tubules takes place when the amount of these *cis*-isomers reaches the critical concentration. If water-miscible organic solvents such as acetonitrile were added as a co-solvent, the depolymerization process more linearly responded to the increase of the *cis*-isomers. The observation indicates that a far-from-equilibrated state could be achieved in pure water to show the initial lag regime by producing strain but not breaking the assemblies.

For synthesis of covalent polymers, light is one of the attractive energy sources that can initiate covalent bond formation in a chain reaction, e.g. living radical polymerization, click reaction, and thiol–ene and thiol–yne coupling reactions [80, 81]. With respect to supramolecular polymerization, if nucleation and/or elongation steps could be regulated by photoinduced geometrical change of monomer units, such a photoinitiated process may realize precise supramolecular polymerization. In this regard, poorly aggregative photoisomers, e.g. *cis*-azobenzene, are promising "dormant" monomers that can retard spontaneous nucleation. Recently, Sugiyasu and coworkers have successfully employed phototriggered supramolecular polymerization of an azobenzene monomer to realize photoregulated living supramolecular polymerization [82]. According to the typical design strategy for functional organogelators, they designed and synthesized an amide-functionalized azobenzene derivative **23** (Figure 2.14a). In this study, *cis*-**23** behaves as the dormant monomer that can be activated for supramolecular polymerization through *cis-to-trans* photoisomerization. Upon cooling a hot methylcyclohexane (MCH)-toluene (9 : 1) solution of *trans*-**23** to 10 °C, the azobenzene absorption band bathochromically shifted in a nonsigmoidal manner against temperature change, suggesting the formation of J-type aggregates via the cooperative nucleation–elongation mechanism. AFM observation showed linear supramolecular fibers with lengths of several micrometers (Figure 2.14b). In sharp contrast, *cis*-**23**, which had been prepared by cooling a hot solution of *trans*-**23** with UV-light irradiation (λ = 351 nm), afforded amorphous films, indicating the less aggregative nature of *cis*-**23** due to its bent conformation. Upon irradiation of *cis*-**23** with visible light (λ = 520 nm), spontaneous nucleation of the photogenerated *trans*-**23** occurred when its concentration reached critical value ($\sim 1.0 \times 10^{-5}$ M). Accordingly, careful visible-light irradiation of *cis*-**23** can kinetically supply the active monomer. Based on these preliminary results, they irradiated the solution of *cis*-**23** with visible light ([*cis*-**23**] = 3.75×10^{-5} M) but in the presence of "seeds" of *trans*-**23** ([seed] = 1.25×10^{-5} M). Interestingly, the absorption spectra did not show the formation of free monomer *trans*-**23** upon visible-light irradiation, but the direct growth of the J-aggregates was observed. This observation suggests the successful photoregulated living supramolecular polymerization at the termini of seeds.

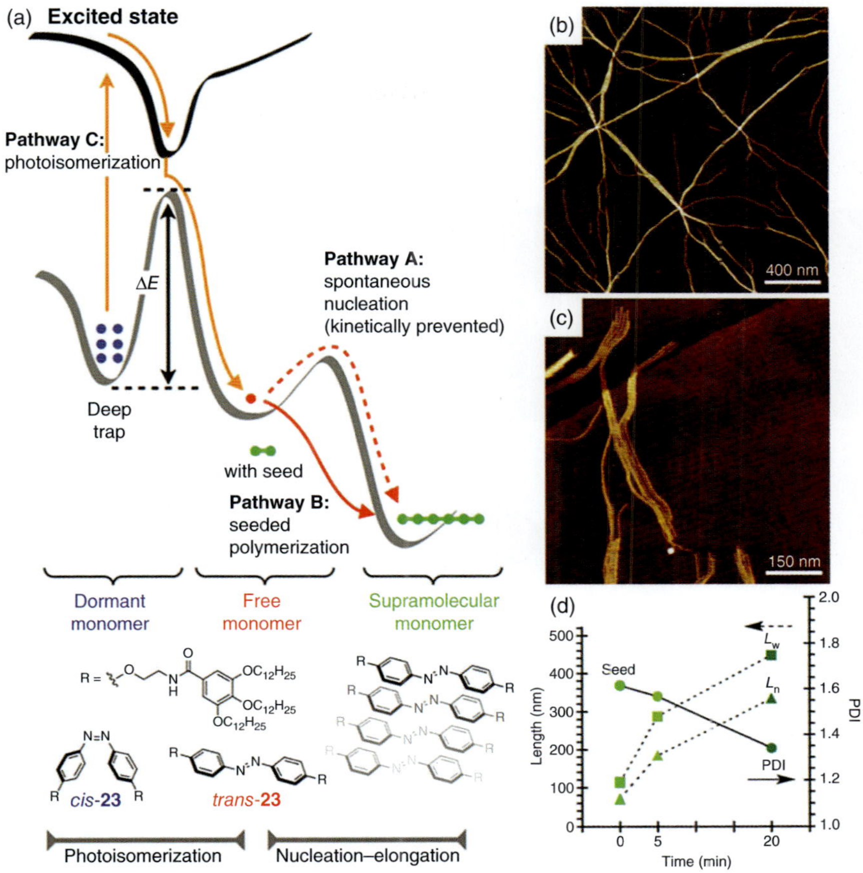

Figure 2.14 (a) Combined energy landscape of supramolecular polymerization and photoisomerization. The photoisomerization pathway C can overcome the large energy barrier (ΔE), supplying free monomers for subsequent seeded supramolecular polymerization pathway B. (b) AFM image of long nanofibers of *trans*-**23** formed through spontaneous supramolecular polymerization. (c) AFM image of nanofibers formed through photoregulated living supramolecular polymerization using *cis*-**23**. (d) Plots of the number-averaged length (L_n), weight-averaged length (L_w), and polydispersity index (PDI = L_w/L_n) of the seed and supramolecular polymers. Source: Endo et al. [82]. Copyright 2016. Reproduced with permission from The American Chemical Society.

Indeed, the formation of nanofibers with controlled lengths and a narrow polydispersity were confirmed by AFM (Figure 2.14c,d).

Yagai et al. have applied phototriggered self-assembly to prepare supramolecular polymers whose lengths are different from those prepared by thermodynamically controlled self-assembly pathways. They previously reported that barbiturated naphthalene **24** can self-assemble into well-defined cylindrical supramolecular polymers through the formation of hydrogen-bonded rosettes (Figure 2.15a) [83, 84]. To control the self-assembly of **24**, they newly designed and synthesized an azobenzene-functionalized derivative **25** [85]. Upon cooling a hot MCH solution of *trans*-**25** to 20 °C, a hypsochromic shift was observed for

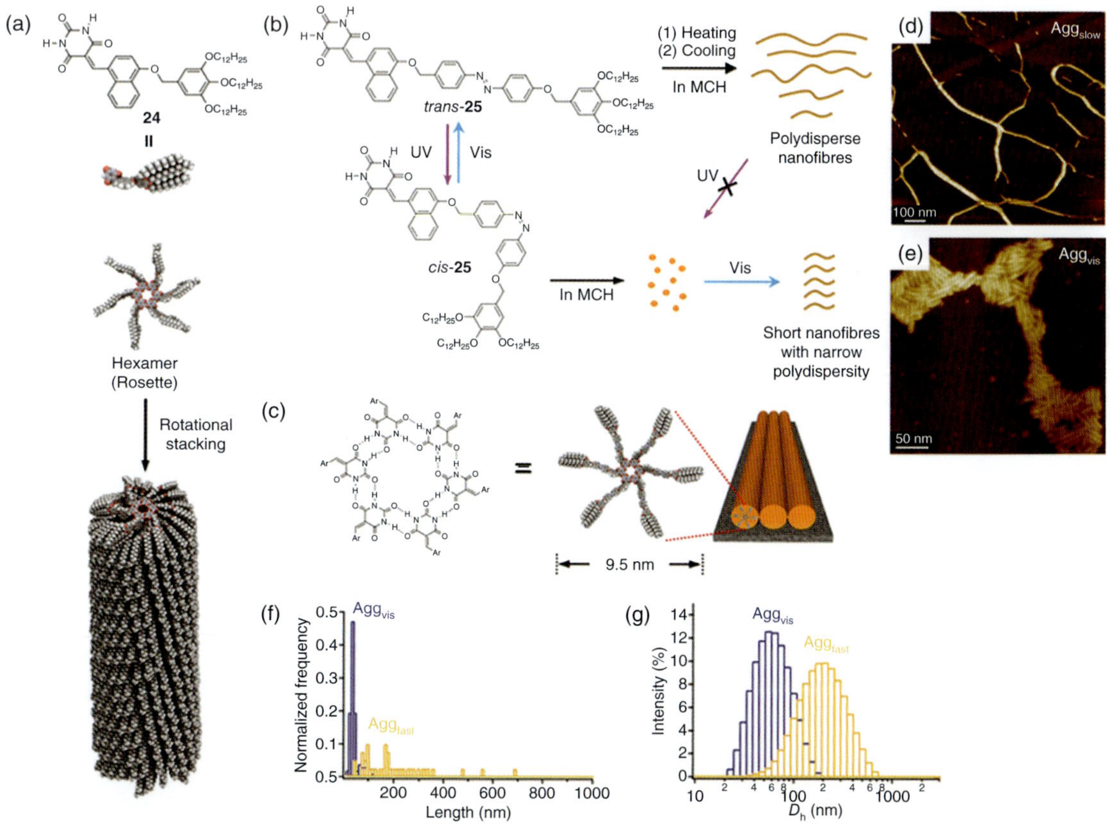

Figure 2.15 (a) Schematic representation of supramolecular polymerization of barbiturated naphthalene **24** through the formation of rosettes. (b) Schematic representation of thermal- and photo-controlled supramolecular polymerization of *trans*- and *cis*-**25** leading to nanofibers with different polydispersity and length regime. (c) Cartoon representation of cylindrical nanofibers from π-stacked rosettes of *trans*-**25**. (d, e) AFM images of (d) $\mathbf{Agg_{slow}}$ and (e) $\mathbf{Agg_{vis}}$. (f) Histogram of the length distribution of $\mathbf{Agg_{vis}}$ and $\mathbf{Agg_{fast}}$ obtained from AFM images. (g) DLS-derived size distribution of hydrodynamic diameter (D_h) of $\mathbf{Agg_{vis}}$ and $\mathbf{Agg_{fast}}$.

the absorption band of the azobenzene chromophores, indicating H-type stacking upon supramolecular polymerization. The plot of this absorption as a function of temperature "(cooling curve)" follows a nucleation–elongation mechanism. Notably, the subsequent "heating curve" corresponding to the depolymerization process shifted toward higher temperatures in comparison with the cooling curve, showing a higher critical temperature T_e than the elongation temperature T_e' in the cooling process. The observed thermal hysteresis suggests that the polymerization proceeds under a kinetically controlled process, which was also supported by a van't Hoff analysis. AFM analysis showed that lengths of aggregates of *trans*-**25** are dependent on the cooling rate. Lengths of slowly prepared aggregates (**Agg$_{slow}$**) with a cooling rate of 1.0 K min^{-1} exceeded 1000 nm, whereas those of kinetically prepared aggregates (**Agg$_{fast}$**) by fast cooling (quenching) were in the range of 50–400 nm (Figure 2.15c,d). Upon UV-light irradiation (λ = 365 nm) of the MCH solution of **Agg$_{slow}$** at 20 °C, *trans*-*to*-*cis* photoisomerization occurred (Figure 2.15b, *trans*:*cis* ratio at a PSS is 79 : 21) but no significant morphology change was observed. Conversely, when a *cis*-**25**-rich $CHCl_3$ solution was injected into MCH, amorphous films were imaged by AFM. Subsequent irradiation of the *cis*-rich solution with visible light (λ = 470 nm) reproduced a *trans*-rich state (*trans*:*cis* = 98 : 2) in a short period of 10 seconds, thus allowing supramolecular polymerization of the photogenerated *trans*-**25** under very kinetic conditions. This phototriggered kinetic pathway enabled the formation of short cylindrical nanofibers (**Agg$_{vis}$**) with a narrower polydispersity index (PDI = L_w/L_n = 1.26) compared with **Agg$_{fast}$** (PDI = 1.50) (Figure 2.15e,f). Different size distributions of the aggregates were also shown by DLS measurements (Figure 2.15g). In the case of the phototriggered kinetic pathway, the more "supercooled" situation might be realized compared to the quenching procedure. In such a state where diffusion of the solute is not sufficient to supply monomers to the nucleus formed at the initial stage, further nucleation of monomers can take place, leading to the formation of a greater amount of supramolecular polymers with smaller aggregation numbers (shorter lengths) compared to those formed by the quenching procedure. These kinetically trapped small aggregates might be useful as "seeds" for living supramolecular polymerization [86, 87].

Unlike azobenzene-based supramolecular monomers, the relative aggregation capabilities of the two isomeric forms of diarylethene vary depending on molecular design (Figure 2.16) [88–91]. For instance, in the case of amide-functionalized diarylethene **26** reported by Feringa and coworkers, the rigid *closed*-ring isomer of **26** (**26**$_{closed}$) can form thermally stable organogels in comparison with the *open*-ring isomer **26**$_{open}$ [88]. A similar trend was observed for an amphiphilic diarylethene **27** reported by Hirose et al., who found that **27**$_{closed}$ can assemble into nanoaggregates in water through a π–π stacking interaction between the expanded π-surfaces whereas **27**$_{open}$ cannot assemble under the same condition [89].

In contrast, an inverse effect of the *open*- and *closed*-ring isomers on the on/off switching of aggregation has been observed for a hydrogen-bonding diarylethene **28** wherein the flexible *open*-ring isomer **28**$_{open}$ displayed higher solution viscosity compared with **28**$_{closed}$ [90]. More prominent impact has been reported

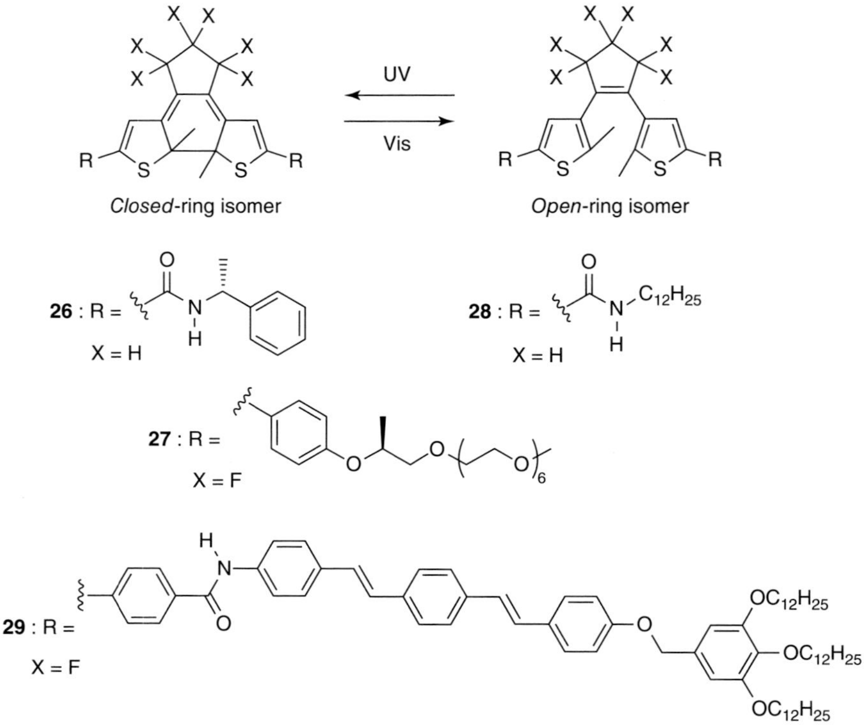

Figure 2.16 Molecular structures of self-assembling diarylethenes **26**–**29** having different substituents that can enhance supramolecular polymerization.

by Yagai et al. They synthesized a diarylethene **29** equipped with highly aggregative oligo(*p*-phenylenevinylene) (OPV) wings and found that $\mathbf{29}_{\text{open}}$ can form π-stacked nanoaggregates in MCH to form a gel (Figure 2.17b), whereas $\mathbf{29}_{\text{closed}}$ does not form the gel owing to the lack of extended assembly (Figure 2.17a) [91]. Based on a molecular modeling calculation, the authors proposed that the central two methyl groups of the diarylethene core of **29** can work as regulators for aggregation, because they are mobile by the rotation of thiophene rings in the *open* form, whereas in the *closed* form they align along the stacking direction (Figure 2.17c–f). This steric effect of the methyl groups should be pronounced if the aggregation is driven by π–π stacking interaction, which requires close contact between the building blocks. While the solution of $\mathbf{29}_{\text{closed}}$ can be converted to the gel of $\mathbf{29}_{\text{open}}$ by visible-light irradiation (λ = 600 nm), the gel of $\mathbf{29}_{\text{open}}$ cannot be reconverted to the solution upon irradiation with UV light (λ = 313 nm). Because the two thiophene rings of $\mathbf{29}_{\text{open}}$ are fixed in a twisted conformation by strong aggregation, ring-closure reaction might not be allowed.

A similar observation has been made by Ma et al. for the supramolecular polymerization of a diarylethene-introduced benzene-1,3,5-tricarboxamide (BTA) that has been extensively cultivated as a scaffold of the supramolecular monomer by Meijer and coworkers [92, 93]. A series of this C_3-symmetrical disk-shaped molecule has been shown to self-assemble into robust one-dimensional supramolecular structures owing to their high degree of functionalizability [94].

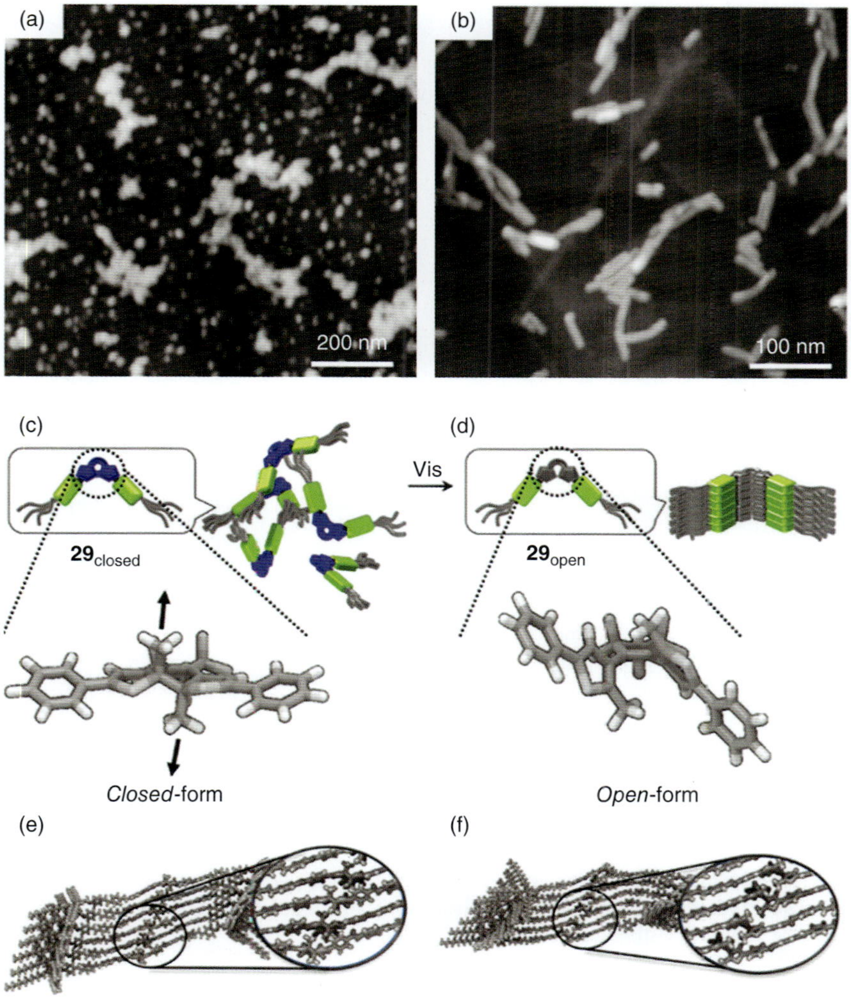

Figure 2.17 (a, b) AFM images of (a) amorphous aggregates of **29**$_{closed}$ and (b) tape-like nanoaggregates of **29**$_{open}$. (c, d) Schematic representation and molecular-modeled structures of (c) **29**$_{closed}$ and (d) **29**$_{open}$. (e, f) Energy-optimized structures of hexameric stack of (e) **29**$_{closed}$ and (f) **29**$_{open}$.

BTA supramolecular monomer **30** bearing three diarylethene units at its peripheral arms has been thus synthesized (Figure 2.18a) [95]. The FT-IR spectrum of a toluene solution of the *open*-ring isomer of **30** (**30**$_{open}$) showed the formation of intermolecular hydrogen bonds between the amide groups. The fluorescence intensity of **30**$_{open}$ in toluene was largely enhanced compared with that in dimethylformamide (DMF) at the same concentration ($c = 1 \times 10^{-4}$ M), suggesting a fixed molecular arrangement that can restrict molecular motion. Hence, BTA **30**$_{open}$ can form a one-dimensional supramolecular structure through synergistic hydrogen bonding and π–π stacking interactions. SEM visualized crystalline one-dimensional fibers. The photocylization of **30**$_{open}$ with UV-light irradiation

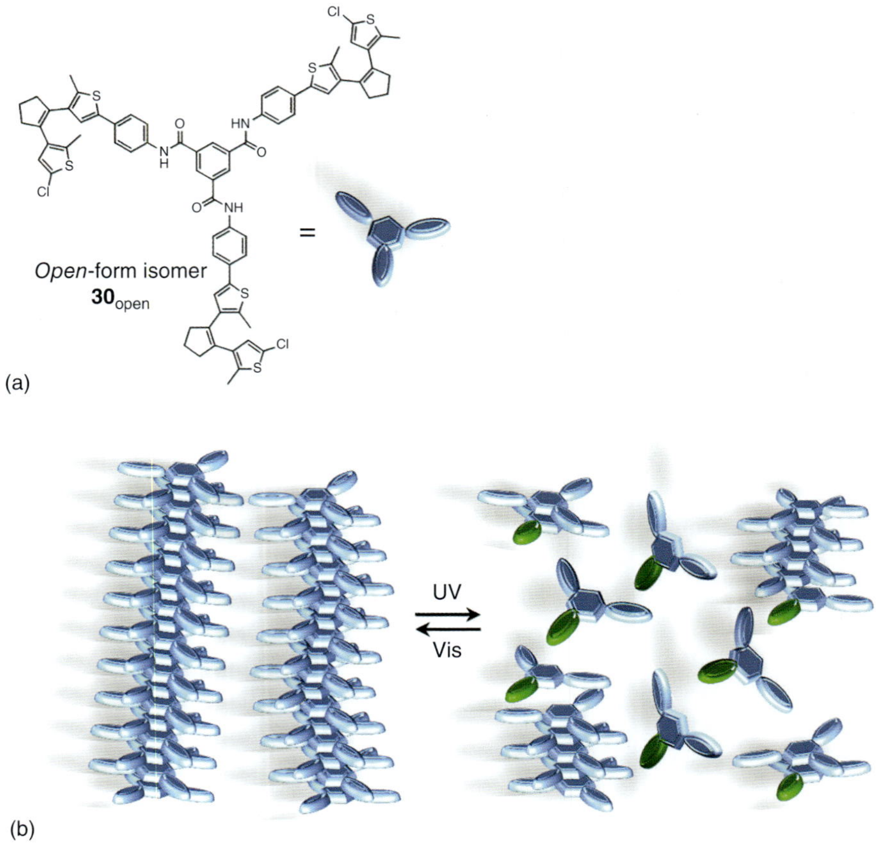

Figure 2.18 (a) Molecular structure of diarylethene-functionalized benzene-1,3,5-trioxamide molecule **30**. (b) Cartoon representation of photoreversible depolymerization/polymerization processes of **30** by UV/visible-light irradiation.

(λ = 365 nm) resulted in a low yield of the *closed*-ring isomer of the diarylethene units according to the intensity of the new visible absorption band, which can be ascribed to the tight molecular packing of $\mathbf{30}_{open}$. TEM observation revealed that the supramolecular fibers of $\mathbf{30}_{open}$ with diameters of 15–20 μm could be depolymerized by UV-light irradiation (λ = 365 nm) for c. 1 hour. This process was photochemically reversible by ring-opening reaction of the diarylethene units with visible-light irradiation (λ > 450 nm) of the resulting solution of dissociated **30** (Figure 2.18b). Molecular dynamics (MD) simulations suggested that the conformation of $\mathbf{30}_{close}$ bearing three ring-closed diarylethene units is unfavorable for aggregation compared with $\mathbf{30}_{open}$ due to the steric hindrance of the central methyl groups of the rigidified diarylethene units upon one-dimensional stacking, which had been originally proposed by Yagai et al. (Figure 2.17). Such an adaptability of diarylethene for extended aggregation is one of the fascinating properties that support utilizing this remarkable photochromic molecule to create photoresponsive supramolecular polymers.

Recently, Ghosh and coworkers reported the phototriggered supramolecular polymerization of naphthalenediimide (NDI) by designing "caged" pro-monomer **31** (Figure 2.19) [96]. Compound **31** is protected on its amide site, responsible for aggregation, with an *ortho*-nitrobenzyl group that has been also applied for the caged stopper **13** by Meijer and coworkers (Figure 2.6) [51]. Supramolecular monomer **32** self-assembled upon cooling its hot MCH solutions to 15 °C to form entangled long fibrils as shown by AFM (Figure 2.19b). During cooling, the π–π^* absorption band of the NDI chromophores underwent a redshift and the temperature dependence of this redshift followed a sigmoidal transition, indicating the formation of J-type aggregates via the isodesmic mechanism. On the other hand, any self-assemblies were not observed for the caged monomer **31**, as shown by UV–vis and FT-IR measurements. When the MCH solution of **31** ($c = 1 \times 10^{-3}$ M) was irradiated with UV light ($\lambda = 254$ nm), the *ortho*-nitrobenzyl group was quantitatively cleaved to yield active monomer **32**, leading to the gelation of the solution. During this phototriggered supramolecular polymerization, the D_h measured by DLS and the solution viscosity gradually increased in line with the conversion of caged **31** to activated **32**. AFM images of the photogenerated supramolecular polymers showed spherulite structures (Figure 2.19c–e), which are different from the one-dimensionally extended fibrillar topology observed for the spontaneous polymerization of **32** (Figure 2.19b). The authors suggested that the phototriggered supramolecular polymerization may occur via an in-built growth mechanism, which is different from the spontaneous process. In the early stage of the photoelimination reaction of the caged **31**, activated **32** could form co-aggregates with **31** through hydrogen bonds, leading to morphology differences in the growth of supramolecular fibers.

Hirose et al. demonstrated an interesting strategy for control over the degree of supramolecular polymerization by photoswitching active/inactive states of monomers through metal–ligand complexation [97]. For this purpose, they employed symmetrically amide-functionalized zinc porphyrin **33** bearing chiral aliphatic chains [98] and phenylazopyridine **34** possessing bulky mesityl groups as a Zn-ligand (Figure 2.20a). The photoresponsive complexation of phenylazopyridine derivatives with metalloporphyrins had been well established by Inoue and coworkers [99] and Otsuki and Narutaki [100]. In a previous paper, Meijer and coworkers disclosed the self-assembly of **33** as follows [98]. Upon cooling a hot MCH solution of **33** to 20 °C, the Soret absorption band hypsochromically shifted in a nonsigmoidal manner against temperature change, indicating a cooperative self-assembly process. The self-assembly of **33** showed a sharp CD signal with a characteristic Cotton effect in the Soret band of porphyrin chromophore, suggesting the helical stacking of porphyrin, which should be supported by amide–amide hydrogen bonding as shown by FT-IR. AFM imaging visualized well-defined fibrillar supramolecular polymers of **33** with micrometer lengths. To simulate the photoswitchability of the cooperative self-assembly of **33** in the presence of **34**, they designed a multiple-equilibrium model using five parameters including equilibrium constants K_2 and K in the cooperative process of **33** obtained by fitting the above temperature-dependent data with a K_2–K (dimerization–elongation) model, binding constants of the *trans*- and *cis*-isomer

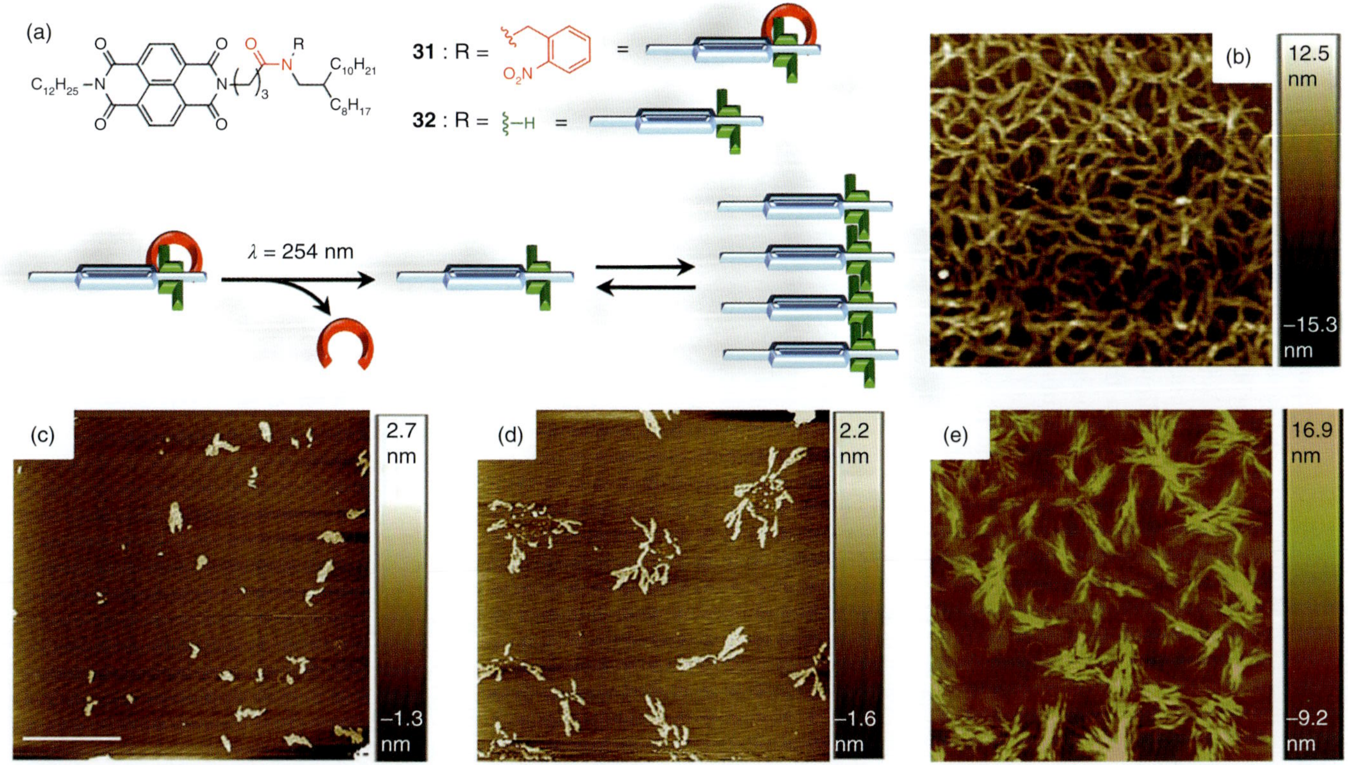

Figure 2.19 (a) Molecular structures of caged pro-monomer **31** and activated **32**, and cartoon representation of photoinitiated supramolecular polymerization of **32**. (b) AFM image of entangled long fibrils obtained by spontaneous polymerization of **32** in MCH. (c–e) AFM images of supramolecular polymers obtained by UV-light irradiation of **31** in MCH for (c) 2 minutes, (d) 6 minutes, and (e) 12 minutes. Source: Pal et al. 2016 [96]. Copyright 2016. Reproduced with permission from Wiley-VCH Verlag GmbH & Co. KGaA.

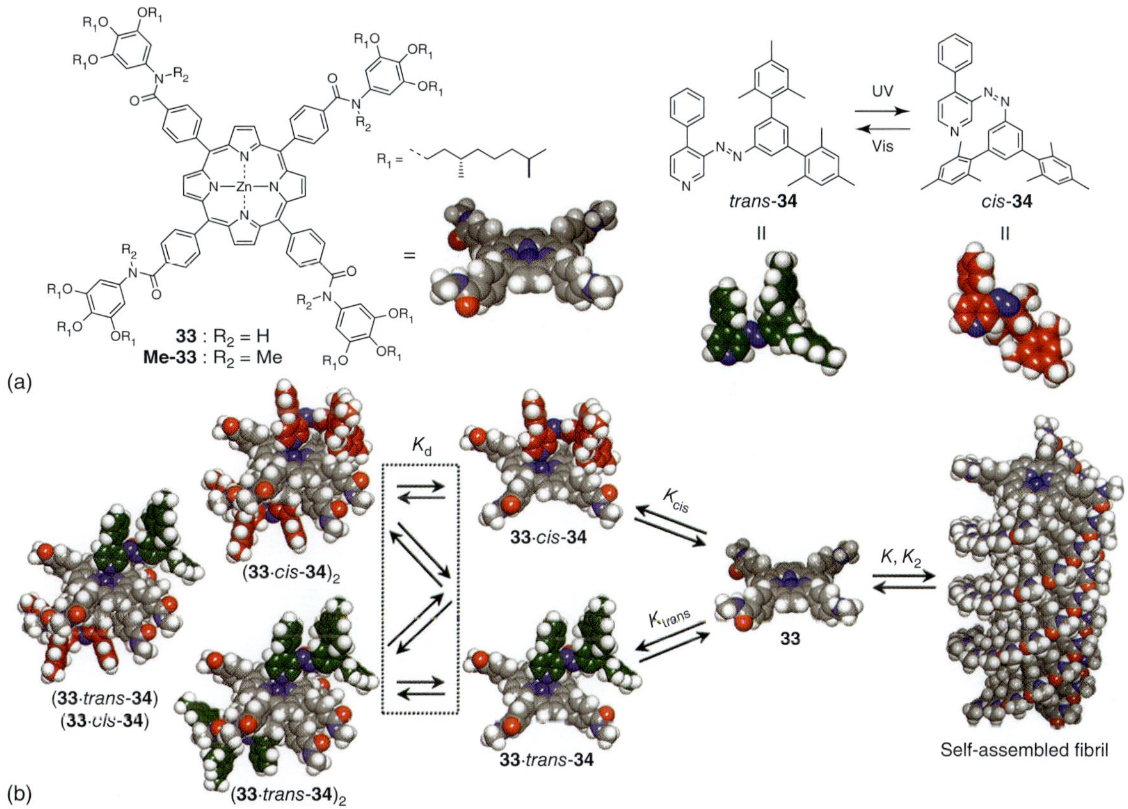

Figure 2.20 (a) Molecular structures of zinc porphyrin **33** functionalized with four chiral amide groups and phenylazapyridine **34**, and photoisomerization of **34** with UV/visible light. (b) Schematic representation of photoswitchable supramolecular polymerization of **33** in the presence of photochromic molecule **34** as Zn-ligand.

of **34** (*trans*- and *cis*-**34**) with **33** (K_{trans}, K_{cis}), and the dimerization constant K_d of a porphyrin–pyridine complex **33·34** (Figure 2.20b). The K_{trans} and K_{cis} were determined to be $3.6 \times 10^4\,M^{-1}$ and $8.1 \times 10^3\,M^{-1}$, respectively, by the titration experiment of *N*-methylated derivative **Me-33** with *trans*-**34** (97% at PSS_{trans}) and successive UV irradiation ($\lambda = 360\,nm$) into *cis*-**34** (95% at PSS_{cis}). For K_d, the value of **33**·pyridine was used because the dimerization process of **33** is independent of the ligand type. The simulation with the cooperative model in the presence of 40 equiv. of **34** predicted larger porphyrin stacks of 3100 monomers at PSS_{cis} while only 2% of short porphyrin stacks was predicted at PSS_{trans}. Conversely, simulation with the isodesmic model resulted in a less photoinduced switchability of degree of polymerization, and porphyrin stacks of 22 monomers were estimated at PSS_{cis}. Taking into consideration the above simulation and the result of the titration experiment of **33** with *trans*-**34** as well as insufficient photoisomerization of **34** in the presence of porphyrins, they prepared an MCH solution of **33** containing 54 equiv. of *trans*-**34**. When this solution was UV irradiated, *trans*-**34**:*cis*-**34** = 10 : 90 was achieved at PSS_{cis}, and the growth of the Soret band and the CD signal suggesting the supramolecular polymerization of **33** was observed. The subsequent irradiation of the *cis*-rich solution with visible light ($\lambda = 420\,nm$) converted *cis*-**34** to *trans*-**34**, which again disassociated supramolecular polymers of **33**. The fraction of stacked porphyrin monomers increased significantly from 1% at PSS_{trans} to 81% at PSS_{cis}, which is in good agreement with the simulated results (Figure 2.20b). These photoinduced depolymerization/repolymerization processes resulted in a change in viscosity from 0.637 to 0.704 cP for 4.0×10^{-5} M MCH solution of **33** upon *trans-to-cis* isomerization of **34** ($c = 1.88\,mM$).

2.5 Photoresponsive Supramolecular Polymers with Photocontrollable 1D Topology

The photoisomerization of supramolecular monomers without breaking noncovalent interactions (without monomer exchange) is one of the challenging but intriguing research topics, because this could realize a topological control of supramolecular main chains. The topological control of one-dimensional polymer chain is a key to induce a specific property beyond the functionality of monomer building blocks, as shown by the diverse topological features of proteins. To realize this, sufficiently strong intermolecular interactions are required between supramolecular monomers because their structural changes upon photoisomerization could cause significant changes in enthalpy and entropy associated with self-assembly.

In this chapter, we focus on supramolecular polymers whose one-dimensional topology can be transformed by light stimuli in such a manner. In general, functions of naturally occurring and synthetic polymers are dominated not only by the property of the monomer units but also by the topology of their one-dimensional chains. Photochromic molecules serve as an essential trigger in many important biological photoreceptors that can exhibit conformational changes. For instance, channel rhodopsins, the light-gated cation channels, show a conformational transformation of one state to the other in response to light, finally leading to physiological responses such as opening/closing ion channels [101, 102]. Therefore, photocontrol over the topology of one-dimensional supramolecular polymer chains is a fascinating and challenging research topic that paves the way for the creation of supramolecular polymers with dynamically controllable functions.

The research group of Stupp and coworkers has introduced photoresponsive units into their peptide amphiphile systems. For instance, they exploited a photoelimination reaction that triggered the untangling of supramolecular quadruple helical fibers composed of a peptide amphiphile **35** bearing a pentadecanyl chain (Figure 2.21) [103]. The photocleavable *ortho*-nitrobenzyl group was introduced as a bulky substituent into the *N*-terminal amide position of **35** (Figure 2.21a). Upon cooling a hot aqueous solution (pH 11) of **35** to 25 °C, the formation of quadruple helical fibers was observed by TEM (Figure 2.21b,c, arrow I). A closer inspection of the TEM image revealed that the entangled helix is formed by two smaller double helical fibers (arrows II), which are further composed of two nonhelical, cylindrical elementary fibrils (arrows III). The untangled double helical fibers and elementary fibrils were intertwined again with the neighboring supramolecular strands (Figure 2.21b, arrow IV). The elementary fibrils had a diameter of 11 nm, which was approximately twice as large as the calculated length of **35** (5.6 nm). The widths and pitches were 33 ± 2 nm and 92 ± 4 nm, respectively, for the quadruple helices, and 24 ± 1 nm and 106 ± 3 nm, respectively, for the double helices. In addition, both helices had right-handed helical structures. FT-IR studies revealed that the self-assembly of **35** is driven by β-sheet formation through hydrogen bonding and hydrophobic interactions between the amphiphilic chains (Figure 2.21c). The self-assembly of **35** resulted in sharp CD signals in the absorption region of the *ortho*-nitrobenzyl moiety, suggesting that the steric repulsion between the bulky substituents may induce higher order helical twisting. Accordingly, the irradiation of the aqueous solution of **35** with UV light (λ = 350 nm) for five minutes converted these higher order helical fibers into the elementary cylindrical fibrils with a diameter of 11 nm, which was demonstrated by TEM observation (Figure 2.21d,e). MALDI-TOF MS and UV–vis analyses revealed that the *ortho*-nitrobenzyl moiety was photoeliminated in c. 97% yield, thus producing deprotected amphiphile **36,** which did not induce higher order helical architecture due to lack of the bulky substituent.

Similar photoinduced untangling of helically intertwined supramolecular polymers into linear fibrils has been reported by Yagai et al. (Figure 2.22) [104]. They designed a diarylethene derivative **37** bearing two monotopic melamine units and a ditopic cyanurate-appended oligothiophene **38** (Figure 2.22a). The aggregation of these complementary components via CMHB interactions could lead to the formation of an ABAB type supramolecular copolymer $(\mathbf{37}\cdot\mathbf{38})_n$ (Figure 2.22b). The complexation between the *open*-form isomer of **37** ($\mathbf{37}_{\mathrm{open}}$) and **38** at 1 : 1 stoichiometry was confirmed by UV–vis titration experiments in cyclohexane, wherein the oligothiophene moiety of **38** quantitatively formed H-type aggregates upon addition of $\mathbf{37}_{\mathrm{open}}$. At millimolar concentrations, gelation of cyclohexane was observed, suggesting the formation of elongated supramolecular polymers of $(\mathbf{37}_{\mathrm{open}}\cdot\mathbf{38})_n$ (Figure 2.22d, left). TEM and AFM observations

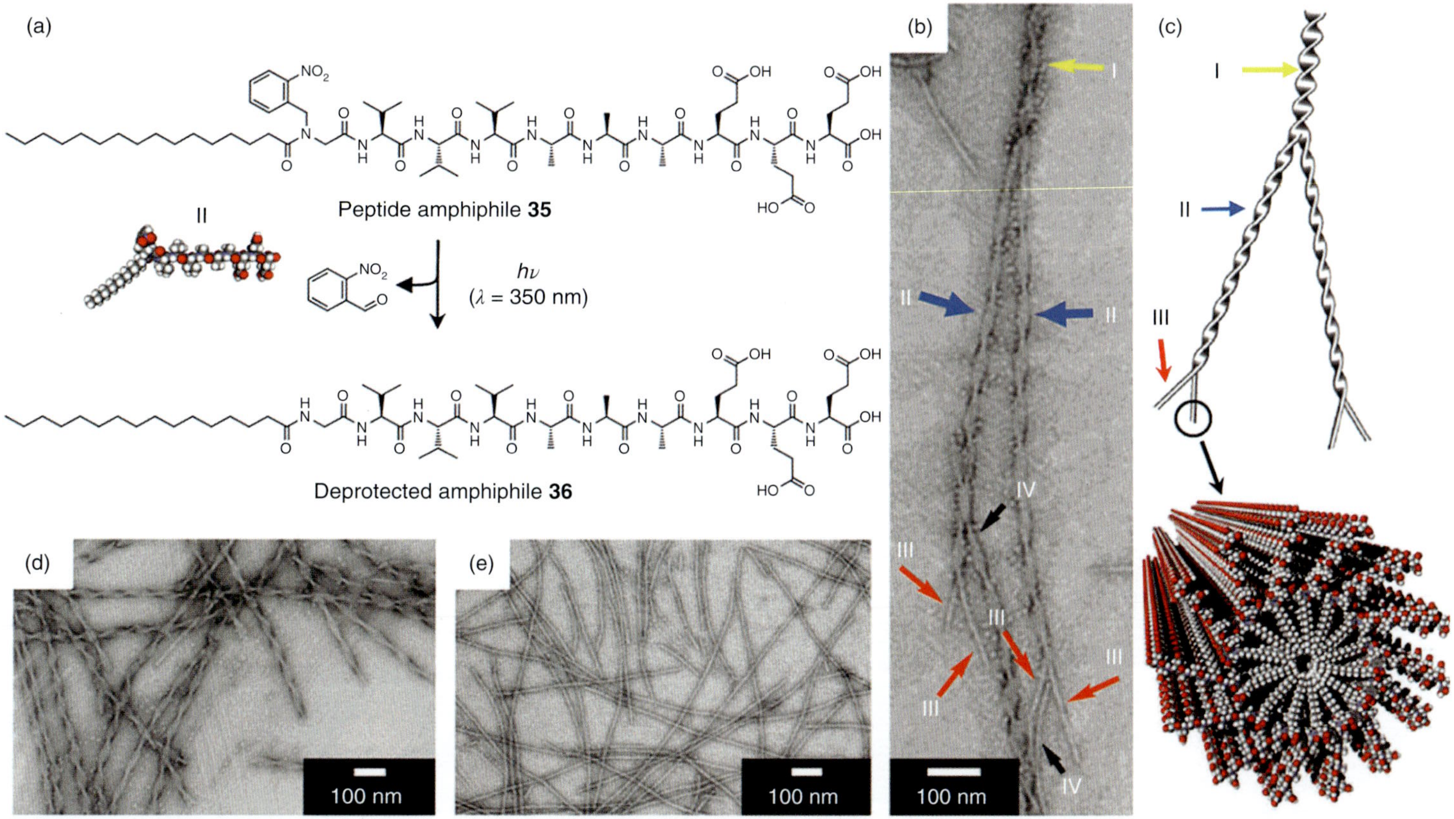

Figure 2.21 (a) Photoelimination reaction of peptide amphiphile **35** to give **36**. (b, d) TEM images of self-assembled **35** in aqueous solution. The quadruple helical fibers (arrow I) are composed of two double helical fibers (arrow II) tangled with cylindrical elementary fibrils (arrow III). Arrow IV points to the regime of the inter-fibrous tangling. (c) Cartoon representation of hierarchical nanostructure of elementary fibrils, double helices, and quadruple helices of **35**. (e) TEM image of the elementary cylindrical fibrils of **36** obtained by UV-light irradiation of self-assembled **35**. Source: Muraoka et al. 2008 [103]. Copyright 2008. Reproduced with permission from The American Chemical Society.

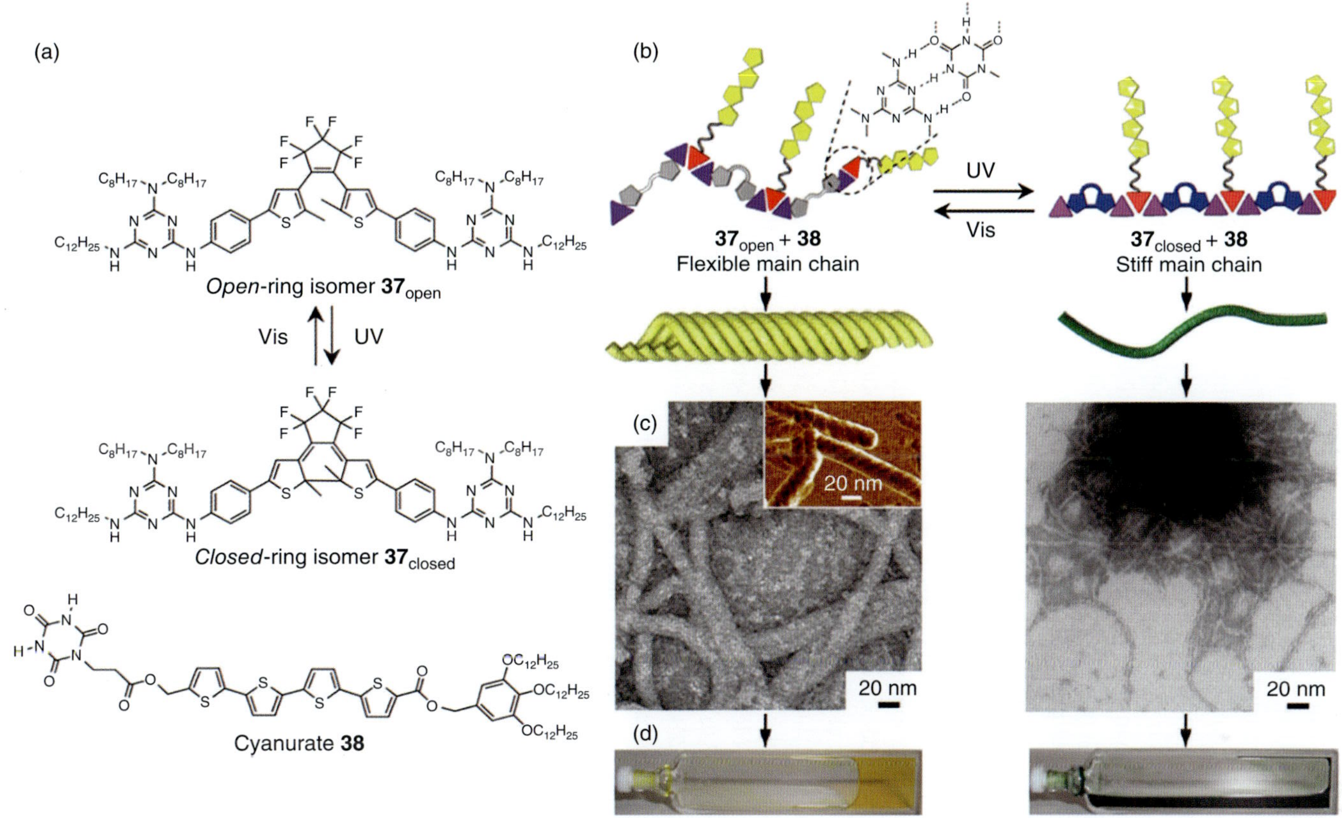

Figure 2.22 (a) Molecular structures of monotopic melamine-functionalized diarylethene **37** and oligothiophene-functionalized cyanurate **38**. (b) Schematic representation of photoreversible topological transformation of supramolecular copolymers ($\mathbf{37}_{open}\cdot\mathbf{38})_n$ and ($\mathbf{37}_{closed}\cdot\mathbf{38})_n$. (c) TEM images of thick helical fibers of ($\mathbf{37}_{open}\cdot\mathbf{38})_n$ and thin fibrils of ($\mathbf{37}_{closed}\cdot\mathbf{38})_n$ in cyclohexane. The inset for ($\mathbf{37}_{open}\cdot\mathbf{38})_n$ shows the corresponding AFM image of thick helical fibers. (d) Photographs showing reversible gel–sol transition achieved by photointerconversion between ($\mathbf{37}_{open}\cdot\mathbf{38})_n$ and ($\mathbf{37}_{closed}\cdot\mathbf{38})_n$.

visualized thick helical fibers with widths of c. 20 nm (Figure 2.22c, left). Upon UV-light irradiation (λ = 313 nm) of the solution of ($\mathbf{37}_{open}$·$\mathbf{38}$)$_n$, ring-closing reaction of $\mathbf{37}_{open}$ occurred and induced a structural transition of thick helical fibers into thin fibrils at a PSS ($\mathbf{37}_{open}$:$\mathbf{37}_{closed}$ = 13 : 87), as shown by TEM (Figure 2.22c, right). For the gel, this nanostructural transition induced a gel-to-sol transition (Figure 2.22d, right). The original helical nanofibers were regenerated by visible-light irradiation (λ = 602 nm) of the UV-irradiated solution. This unique nanostructural transition could be attributed to the change in molecular rigidity of the diarylethene moiety upon the ring-closing reaction. The supramolecular copolymers ($\mathbf{37}_{open}$·$\mathbf{38}$)$_n$ would be flexible due to the twistable conformation of $\mathbf{37}_{open}$, allowing the primary chains to further organize into higher order helical structures with folding and intertwining through the interchain π–π stacking between the pendent oligothiophene units of **38**. On the other hand, due to a conformational stiffness of $\mathbf{37}_{closed}$, supramolecular copolymers ($\mathbf{37}_{closed}$·$\mathbf{38}$)$_n$ are unable to further organize into higher order structures, and exist as the primary structure.

In another example by Stupp and coworkers, photoresponsive supramolecular polymers have been constructed from a tripeptide amphiphile **39** bearing an azobenzene unit at one terminus (Figure 2.23a) [105]. In this system, the helical pitch of the superhelical nanostructure could be controlled by light. Amphiphile **39**, when the azobenzene unit was the *trans*-isomer (*trans*-**39**), self-assembled in cyclohexyl chloride to form superhelical nanofibers that were well visualized by AFM (Figure 2.23b). The superhelical nanofibers had a uniform helical pitch of 78 nm (Figure 2.23c,d). According to the mechanism proposed by the authors, *trans*-**39** first forms a "primary" helix, which further twists due to the torsional strain generated by steric repulsion between the bulky end groups (i.e. *trans*-azobenzene moieties). In other words, the bulky *trans*-azobenzene moieties at the peripheries can be accommodated by helical twisting. Interestingly, UV-light irradiation (λ = 360 nm) of the helical nanofibers in cyclohexyl chloride caused a decrease in helical pitches to 40–70 nm (Figure 2.23e–g). The photogenerated *cis*-**39** whose conformation was less planar (sterically bulky) compared with the *trans*-conformer could induce the torsional strain, resulting in a reduction of the helical pitch. The resulting helical pitches varied from nanofiber to nanofiber, which arises from the inhomogeneity of the *trans-to-cis* isomerization efficiency of *trans*-**39** among nanofibers. Conversely, the helical pitch is uniform within each superhelix, indicating that the photoresponsive decrease of the helical pitches undergoes a relaxation process throughout the entire nanostructure.

Recently, an outstanding topological control of one-dimensional supramolecular polymer chains has been accomplished by Yagai and coworkers [106] using barbiturated naphthalene–azobenzene hybrid molecule **40** (Figure 2.24a). They had previously found that a 2,6-substituted naphthalene molecule **41** equipped with barbituric acid and aliphatic wedge self-assembles via the formation of a hydrogen-bonded rosette into toroidal nanostructures with a uniform diameter of c. 20 nm (Figure 2.24b,d,f) [83, 107]. The size uniformity of the toroidal nanostructures established that naphthalene-functionalized barbituric acid rosettes can continuously stack to generate spontaneous (uniform) curvature due to idiosyncratic internal order along the π-stacked rosettes (Figure 2.24f).

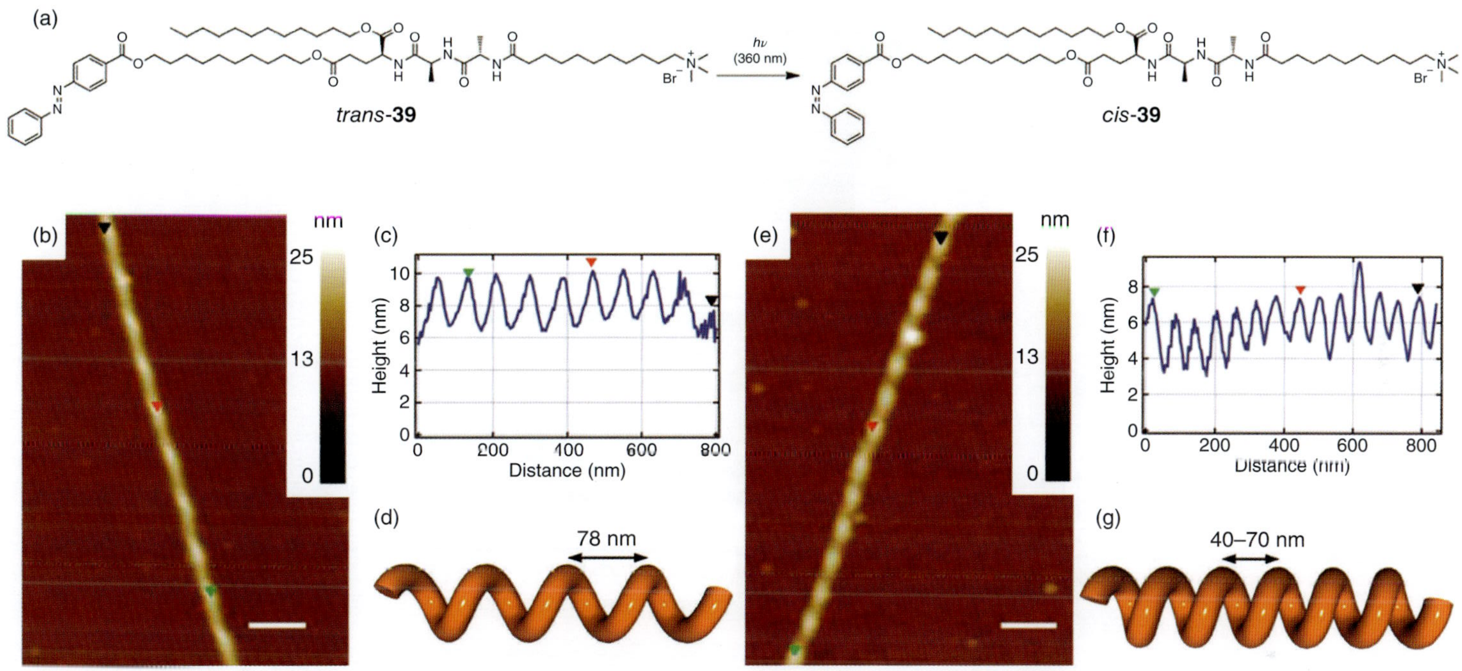

Figure 2.23 (a) Molecular structure and *trans-to-cis* photoisomerization of azobenzene-functionalized peptide amphiphile **39**. (b–g) Photoinduced modulation in helical pitch of self-assembled **39**. (b, e) AFM images and (c, f) corresponding cross-sectional analyses. (d, g) Cartoon representation of helical nanofibers (d) before and (g) after UV-light irradiation. Source: Li et al. 2007 [105]. Copyright 2007. Reproduced with permission from Wiley-VCH Verlag GmbH & Co. KGaA.

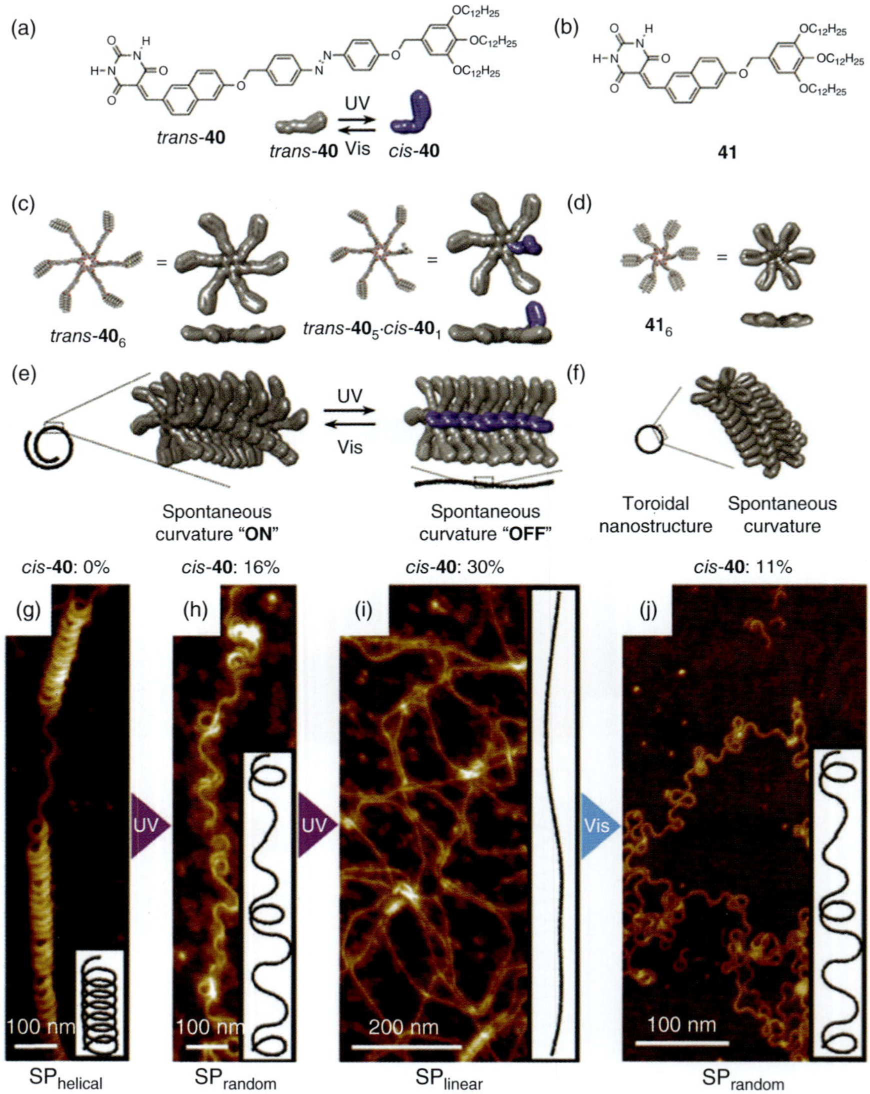

Figure 2.24 (a, b) Molecular structures of (a) naphthalene–azobenzene hybrid molecule **40** and (b) its original naphthalene molecule **41**. The geometrical change of **40** by photoisomerization of the azobenzene moiety is shown by electron cloud model. (c, d) Molecular models of (c) rosettes composed of only *trans*-**40** (denoted as *trans*-$\mathbf{40}_6$), and five *trans*-**40** and one *cis*-**40** (denoted as *trans*-$\mathbf{40}_5$·*cis*-$\mathbf{40}_1$), and (d) rosette of $\mathbf{41}_6$. (e) Schematic representation of photoswitchable curvature in supramolecular polymers of rosettes $\mathbf{40}_6$. (f) Schematic representation of the formation of toroidal supramolecular polymer by rosette $\mathbf{41}_6$. (g–i) AFM images of UV-induced unfolding of $SP_{helical}$ fibers into SP_{linear} through the formation of SP_{random}. (j) AFM image of SP_{random} obtained by subsequent visible-light irradiation of the SP_{linear} in (i).

The additional azobenzene π-surface could strengthen the aggregation capability of the rosette, by which two important outcomes have been induced. First, the increased association force between rosettes prevented the growing supramolecular polymer chains from closing into discrete shorter assemblies, and instead yielded extended curved supramolecular polymers. Second, the strengthened rosette–rosette interactions allowed a direct deformation of columnar aggregates (supramolecular polymer chains) upon photoisomerization of the azobenzene moiety without monomer exchange (Figure 2.24c,e). When a hot MCH or $CHCl_3$–MCH mixed solution of *trans*-**40** was cooled to 20 °C with various cooling rates, well-defined nanofibers that ranged from randomly folded to helically folded supramolecular polymers were confirmed by AFM and TEM (Figure 2.24g). Upon UV-light irradiation (λ = 365 nm) of the solutions of *trans*-**40**, these helical supramolecular polymer fibers ($SP_{helical}$) transformed into extended and unfolded fibers (SP_{linear}) via a randomly folded structure (SP_{random}) (Figure 2.23g–i), indicating that photogenerated *cis*-isomer-containing rosette destroys the spontaneous curvature. Small-angle X-ray scattering (SAXS) analysis also showed a similar result, which indicates the deletion of spontaneous curvature upon UV irradiation. Subsequent visible-light irradiation (λ = 470 nm) that causes the *cis-to-trans* isomerization of *cis*-**40** allowed the extended fibers to refold up to the level of the randomly folded structure (Figure 2.24j). This topological transformation of supramolecular polymer fibers demonstrates that the photoinduced deformation of the folded structure of **40** can regulate the internal order of main chains by the deletion/recovery of spontaneous curvature, leading to control over higher order secondary conformation. Notably, mixing of two solutions exclusively containing folded and extended supramolecular polymers did not show any merging of these fiber morphologies, and each original fiber morphology has been confirmed upon aging for days. Based on this finding, the authors concluded that the photoisomerization of **40** proceeds within the supramolecular polymer fibers, and their unfolding/refolding undoubtedly occurs without aggregate–monomer exchange. As the observed photoinduced unfolding and refolding of supramolecular polymers is reminiscent of the protein unfolding/folding process, the authors expect that this study could advance the frontiers of supramolecular polymers toward applications by mimicking the functionality of biopolymers.

2.6 Summary and Outlook

Without any doubt, supramolecular polymers cannot hold a candle to covalent polymers in terms of bulk physical properties due to the reversible nature of main chains. However, such a weak point can turn into an exceptionally attractive property when one can successfully impart controllability in bond breakage and formation by using external stimuli. Thus, the implementation of host–guest recognition sites, complementary hydrogen-bonding sites, extended π-stacking moieties, and other noncovalent interacting sites to photoreactive units has been addressed by many supramolecular chemists to engineer a variety of supramolecular polymers exhibiting different types of photoresponse. Although the most widely used photoreactive unit is azobenzene, similarly to other photoresponsive research fields, diarylethene as a molecular switch, stilbene as a photodimerizable unit, and *ortho*-nitrobenzyl group as a photoelimination group have been successfully coupled with various supramolecular polymer systems.

From the viewpoint of supramolecular polymer backbones, a relatively unexplored system is one-dimensional columnar structures formed from discotic molecules or supramolecules with extended π-surface. Such π-surface is available through the synthesis of C_3-symmetrical disk-shaped molecules (e.g. BTA) and multicomponent rosette-shaped supermacrocycles (e.g. barbiturated π-conjugated molecule). By introducing photoreactive units at the core or peripheries of these discotic building blocks, we would obtain fascinating photoresponsive supramolecular polymers that can control not only the polymerization/depolymerization but also their one-dimensional topologies without depolymerization. We believe that the latter approach sheds light on new responsiveness and the resulting functionality in supramolecular polymer chemistry.

References

1 Brunsveld, L., Folmer, B.J.B., Meijer, E.W., and Sijbesma, R.P. (2001). Supramolecular polymers. *Chem. Rev.* 101: 4071.

2 De Greef, T.F.A., Smulders, M.M.J., Wolffs, M. et al. (2009). Supramolecular polymerization. *Chem. Rev.* 109: 5687.

3 Yan, X., Wang, F., Zheng, B., and Huang, F. (2012). Stimuli-responsive supramolecular polymeric materials. *Chem. Soc. Rev.* 41: 6042.

4 Yang, L., Tan, X., Wang, Z., and Zhang, X. (2015). Supramolecular polymers: historical development, preparation, characterization, and functions. *Chem. Rev.* 115: 7196.

5 Krieg, E., Bastings, M.M.C., Besenius, P., and Rybtchinski, B. (2016). Supramolecular polymers in aqueous media. *Chem. Rev.* 116: 2414.

6 Fenske, T., Korth, H.-G., Mohr, A., and Schmuck, C. (2012). Advances in switchable supramolecular nanoassemblies. *Chem. Eur. J.* 18: 738.

7 Yagai, S., Karatsu, T., and Kitamura, A. (2005). Photocontrollable self-assembly. *Chem. Eur. J.* 11: 4054.

8 Yagai, S. (2006). Supramolecular complexes of functional chromophores based on multiple hydrogen-bonding interactions. *J. Photochem. Photobiol., C Photochem. Rev.* 7: 164.

9 Yagai, S. and Kitamura, A. (2008). Recent advances in photoresponsive supramolecular self-assemblies. *Chem. Soc. Rev.* 37: 1520.

10 Yao, X., Li, T., Wang, J. et al. (2016). Recent progress in photoswitchable supramolecular self-assembling systems. *Adv. Opt. Mater.* 4: 1322.

11 Dong, R., Zhou, Y., and Zhu, X. (2014). Supramolecular dendritic polymers: from synthesis to applications. *Acc. Chem. Res.* 47: 2006.

12 Xing, P., Sun, T., and Hao, A. (2013). Vesicles from supramolecular amphiphiles. *RSC Adv.* 3: 24776.

13 Draper, E.R. and Adams, D.J. (2016). Photoresponsive gelators. *Chem. Commun.* 52: 8196.

14 Shimizu, T. (2018). Self-assembly of discrete organic nanotubes. *Bull. Chem. Soc. Jpn.* 91: 623.
15 Szymański, W., Beierle, J.M., Kistemaker, H.A.V. et al. (2013). Reversible photocontrol of biological systems by the incorporation of molecular photoswitches. *Chem. Rev.* 113: 6114.
16 Saha, S. and Stoddart, J.F. (2007). Photo-driven molecular devices. *Chem. Soc. Rev.* 36: 77.
17 Collin, J.-P., Heitz, V., and Sauvage, J.-P. (2005). Transition metal complexed catenanes and rotaxanes in motion: towards molecular machines. *Top. Curr. Chem.* 262: 29.
18 Shinkai, S., Yoshida, T., Miyazaki, K., and Manabe, O. (1987). Photoresponsive crown ethers. 19. Photocontrol of reversible association-dissociation phenomena in "tail(ammonium)-biting" crown ethers. *Bull. Chem. Soc. Jpn.* 60: 1819.
19 Shinkai, S., Yoshida, T., Manabe, O., and Fuchita, Y. (1988). Photoresponsive crown ethers. Part 20. Reversible photocontrol of association-dissociation equilibria between azobis(benzo-18-crown-6) and diammonium cations. *J. Chem. Soc., Perkin Trans. 1* 1431.
20 Dodziuk, H. (ed.) (2006). *Cyclodextrins and Their Complexes: Chemistry, Analytical Methods, Applications.* Weinheim: Wiley-VCH.
21 Hedges, A.R. (1998). Industrial applications of cyclodextrins. *Chem. Rev.* 98: 2035.
22 Davis, M.E. and Brewster, M.E. (2004). Cyclodextrin-based pharmaceutics: past, present and future. *Nat. Rev. Drug Discov.* 3: 1023.
23 Mellet, C.O., Fernández, J.M.G., and Benito, J.M. (2011). Cyclodextrin-based gene delivery systems. *Chem. Soc. Rev.* 40: 1586.
24 Chen, G. and Jiang, M. (2011). Cyclodextrin-based inclusion complexation bridging supramolecular chemistry and macromolecular self-assembly. *Chem. Soc. Rev.* 40: 2254.
25 Harada, A., Takashima, Y., and Yamaguchi, H. (2009). Cyclodextrin-based supramolecular polymers. *Chem. Soc. Rev.* 38: 875.
26 Kuad, P., Miyawaki, A., Takashima, Y. et al. (2007). External sitimulus-responsive supramolecular structures formed by a stilbene cyclodextrin dimer. *J. Am. Chem. Soc.* 129: 12630.
27 Xue, M., Yang, Y., Chi, X. et al. (2012). Pillararenes, a new class of macrocycles for supramolecular chemistry. *Acc. Chem. Res.* 45: 1294.
28 Ogoshi, T., Yamagishi, T., and Nakamoto, Y. (2016). Pillar-shaped macrocyclic hosts pillar[n]arenes: new key players for supramolecular chemistry. *Chem. Rev.* 116: 7937.
29 Ogoshi, T., Yoshikoshi, K., Aoki, T., and Yamagishi, T. (2013). Photoreversible switching between assembly and disassembly of a supramolecular polymer involving an azobenzene-bridged pillar[5]arene dimer. *Chem. Commun.* 49: 8785.
30 Asfari, M.-Z., Böhmer, V., Harrowfield, J., and Vicens, J. (eds.) (2001). *Calixarenes 2001.* Dordrecht: Springer Netherlands.
31 Morohashi, N., Narumi, F., Iki, N. et al. (2006). Thiacalixarenes. *Chem. Rev.* 106: 5291.

32 Ikeda, A. and Shinkai, S. (1997). Novel cavity design using calix[n]arene skeletons: toward molecular recognition and metal binding. *Chem. Rev.* 97: 1713.

33 Lagona, J., Mukhopadhyay, P., Chakrabarti, S., and Isaacs, L. (2005). The cucurbit[n]uril family. *Angew. Chem. Int. Ed.* 44: 4844.

34 Ma, X. and Tian, H. (2014). Stimuli-responsive supramolecular polymers in aqueous solution. *Acc. Chem. Res.* 47: 1971.

35 Qu, D.-H., Wang, Q.-C., Zhang, Q.-W. et al. (2015). Photoresponsive host–guest functional systems. *Chem. Rev.* 115: 7543.

36 Harada, A., Takashima, Y., and Nakahata, M. (2014). Supramolecular polymeric materials via cyclodextrin–guest interactions. *Acc. Chem. Res.* 47: 2128.

37 Peng, L., Liu, S., Feng, A., and Yuan, J. (2017). Polymeric nanocarriers based on cyclodextrins for drug delivery: host–guest interaction as stimuli responsive linker. *Mol. Pharmaceutics* 14: 2475.

38 Fujita, M., Tominaga, M., Hori, A., and Therrien, B. (2005). Coordination assemblies from a Pd(II)-cornered square complex. *Acc. Chem. Res.* 38: 369.

39 Mayoral Muñoz, M.J. and Fernández, G. (2012). Metallosupramolecular amphiphilic π-systems. *Chem. Sci.* 3: 1395.

40 Smulders, M.M.J., Riddell, I.A., Browne, C., and Nitschke, J.R. (2013). Building on architectural principles for three-dimensional metallosupramolecular construction. *Chem. Soc. Rev.* 42: 1728.

41 Yan, X., Xu, J.F., Cook, T.R. et al. (2014). Photoinduced transformations of stiff-stilbene-based discrete metallacycles to metallosupramolecular polymers. *Proc. Natl. Acad. Sci. U.S.A.* 111: 8717.

42 Safont-Sempere, M.M., Fernández, G., and Würthner, F. (2011). Self-sorting phenomena in complex supramolecular systems. *Chem. Rev.* 111: 5784.

43 Vollmer, M.S., Clark, T.D., Steinem, C., and Ghadiri, M.R. (1999). Photoswitchable hydrogen-bonding in self-organized cylindrical peptide systems. *Angew. Chem. Int. Ed.* 38: 1598.

44 Shinkai, S., Nakaji, T., Ogawa, T. et al. (1981). Photoresponsive crown ethers. 2. Photocontrol of ion extraction and ion transport by a bis(crown ether) with a butterfly-like motion. *J. Am. Chem. Soc.* 103: 111.

45 Whitesides, G.M., Simanek, E.E., Mathias, J.P. et al. (1995). Noncovalent synthesis: using physical-organic chemistry to make aggregates. *Acc. Chem. Res.* 28: 37.

46 Sánchez, L., Martín, N., and Guldi, D.M. (2005). Hydrogen-bonding motifs in fullerene chemistry. *Angew. Chem. Int. Ed.* 44: 5374.

47 Schenning, A.P.H.J. and Meijer, E.W. (2005). Supramolecular electronics; nanowires from self-assembled π-conjugated systems. *Chem. Commun.* 3245.

48 Würthner, F. (2004). Perylene bisimide dyes as versatile building blocks for functional supramolecular architectures. *Chem. Commun.* 1564.

49 Adhikari, B., Lin, X., Yamauchi, M. et al. (2017). Hydrogen-bonded rosettes comprising π-conjugated systems as building blocks for functional one-dimensional assemblies. *Chem. Commun.* 53: 9663.

50 Sijbesma, R.P., Beijer, F.H., Brunsveld, L. et al. (1997). Reversible polymers formed from self-complementary monomers using quadruple hydrogen bonding. *Science* 278: 1601.

51 Folmer, B.J.B., Cavini, E., Sijbesma, R.P., and Meijer, E.W. (1998). Photo-induced depolymerization of reversible supramolecular polymers. *Chem. Commun.* 1847.
52 Kalbag, S.M. and Roeske, R.W. (1975). A photolabile protecting group for histidine. *J. Am. Chem. Soc.* 97: 440.
53 Nakamura, S. and Irie, M. (1988). Thermally irreversible photochromic systems. A theoretical study. *J. Org. Chem.* 53: 6136.
54 Patel, P.D. and Masunov, A.E. (2011). Theoretical study of photochromic compounds: part 3. Prediction of thermal stability. *J. Phys. Chem. C* 115: 10292.
55 Irie, M. (2000). Diarylethenes for memories and switches. *Chem. Rev.* 100: 1685.
56 Irie, M., Fukaminato, T., Matsuda, K., and Kobatake, S. (2014). Photochromism of diarylethene molecules and crystals: memories, switches, and actuators. *Chem. Rev.* 114: 12174.
57 Takeshita, M., Hayashi, M., Kadota, S. et al. (2005). Photoreversible supramolecular polymer formation. *Chem. Commun.* 761.
58 Takeshita, M., Kato, N., Kawauchi, S. et al. (1998). Photochromism of dithienylethenes included in cyclodextrins. *J. Org. Chem.* 63: 9306.
59 Ikegami, M., Ohshiro, I., and Arai, T. (2003). Hydrogen bonded molecular assembly by reversible cyclization reaction of anthracene. *Chem. Commun.* 1566.
60 Bouas-Laurent, J.-P., Castellan, H., Desvergne, A., and Lapouyade, R. (2001). Photodimerization of anthracenes in fluid solutions: (part 2) mechanistic aspects of the photocycloaddition and of the photochemical and thermal cleavage. *Chem. Soc. Rev.* 30: 248.
61 Xu, J.-F., Chen, Y.-Z., Wu, D. et al. (2013). Photoresponsive hydrogen-bonded supramolecular polymers based on a stiff stilbene unit. *Angew. Chem. Int. Ed.* 52: 9738.
62 Kucharski, T.J. and Boulatov, R. (2011). The physical chemistry of mechanoresponsive polymers. *J. Mater. Chem.* 21: 8237.
63 Cai, Y., Guo, Z., Chen, J. et al. (2016). Enabling light work in helical self-assembly for dynamic amplification of chirality with photoreversibility. *J. Am. Chem. Soc.* 138: 2219.
64 Yagai, S., Iwai, K., Yamauchi, M. et al. (2014). Photocontrol over self-assembled nanostructures of π–π stacked dyes supported by the parallel conformer of diarylethene. *Angew. Chem. Int. Ed.* 53: 2602.
65 Yagai, S., Iwai, K., Karatsu, T., and Kitamura, A. (2012). Photoswitchable exciton coupling in merocyanine-diarylethene multi-chromophore hydrogen-bonded complexes. *Angew. Chem. Int. Ed.* 51: 9679.
66 Kaiser, T.E., Wang, H., Stepanenko, V., and Würthner, F. (2007). Supramolecular construction of fluorescent J-aggregates based on hydrogen-bonded perylene dyes. *Angew. Chem. Int. Ed.* 119: 5637.
67 Schenning, A.P.H.J., von Herrikhuyzen, J., Jonkheijm, P. et al. (2002). Photoinduced electron transfer in hydrogen-bonded oligo(*p*-phenylene vinylene)-perylene bisimide chiral assemblies. *J. Am. Chem. Soc.* 124: 10252.
68 Kobatake, S., Shibata, K., Uchida, K., and Irie, M. (2000). Photochromism of 1,2-bis(2-ethyl-5-phenyl-3-thienyl)perfluorocyclopentene in a single-crystalline phase. Conrotatory thermal cycloreversion of the closed-ring isomer. *J. Am. Chem. Soc.* 122: 12135.

69 Woodward, R.B. and Hoffmann, R. (1965). Stereochemistry of electrocyclic reactions. *J. Am. Chem. Soc.* 87: 395.

70 Bullock, D.J.W., Cumper, C.W.N., and Vogel, A.I. (1965). 989. Physical properties and chemical constitution. Part XLIII. The electric dipole moments of azobenzene, azopyridines, and azopuinolines. *J. Chem. Soc.* 5316.

71 Kawasaki, T., Tokuhiro, M., Kimizuka, N., and Kunitake, T. (2001). Hierarchical self-assembly of chiral complementary hydrogen-bond networks in water: reconstitution of supramolecular membranes. *J. Am. Chem. Soc.* 123: 6792.

72 Yagai, S., Karatsu, T., and Kitamura, A. (2003). Photoresponsive melamine·barbiturate hydrogen-bonded assembly. *Chem. Commun.* 1844.

73 Yagai, S., Nakajima, T., Karatsu, T. et al. (2004). Phototriggered self-assembly of hydrogen-bonded rosette. *J. Am. Chem. Soc.* 126: 11500.

74 Yagai, S., Nakajima, T., Kishikawa, K. et al. (2005). Hierarchical organization of photoresponsive hydrogen-bonded rosettes. *J. Am. Chem. Soc.* 127: 11134.

75 Fredy, J.W., Méndez-Ardoy, A., Kwangmettatam, S. et al. (2017). Molecular photoswitches mediating the strain-driven disassembly of supramolecular tubules. *Proc. Natl. Acad. Sci. U.S.A.* 114: 11850.

76 Lim, Y.-B., Moon, K.-S., and Lee, M. (2009). Recent advances in functional supramolecular nanostructures assembled from bioactive building blocks. *Chem. Soc. Rev.* 38: 925.

77 Huang, Z., Kang, S.K., Banno, M. et al. (2012). Pulsating tubules from noncovalent macrocycles. *Science* 337: 1521.

78 Kim, Y., Li, W., Shin, S., and Lee, M. (2013). Development of toroidal nanostructures by self-assembly: rational designs and applications. *Acc. Chem. Res.* 46: 2888.

79 Kim, Y., Kim, T., and Lee, M. (2013). From self-assembled toroids to dynamic nanotubules. *Polym. Chem.* 4: 1300.

80 Chatani, S., Kloxin, C.J., and Bowman, C.N. (2014). The power of light in polymer science: photochemical processes to manipulate polymer formation, structure, and properties. *Polym. Chem.* 5: 2187.

81 Chen, M., Zhong, M., and Johnson, J.A. (2016). Light-controlled radical polymerization: mechanisms, methods, and applications. *Chem. Rev.* 116: 10167.

82 Endo, M., Fukui, T., Jung, S.H. et al. (2016). Photoregulated living supramolecular polymerization established by combining energy landscapes of photoisomerization and nucleation-elongation processes. *J. Am. Chem. Soc.* 138: 14347.

83 Yagai, S., Goto, Y., Lin, X. et al. (2012). Self-organization of hydrogen-bonding naphthalene chromophores into J-type nanorings and H-type nanorods: impact of regioisomerism. *Angew. Chem. Int. Ed.* 51: 6643.

84 Yamauchi, M., Adhikari, B., Prabhu, D.D. et al. (2017). Supramolecular polymerization of supermacrocycles: effect of molecular conformations on kinetics and morphology. *Chem. Eur. J.* 23: 5270.

85 Yamauchi, M., Kanao, N., Adhikari, B. et al. (2017). Phototriggered supramolecular polymerization of barbituric acid rosette. *Chem. Lett.* 46: 111.

86 Ogi, S., Sugiyasu, K., Manna, S. et al. (2014). Living supramolecular polymerization realized through a biomimetic approach. *Nat. Chem.* 6: 188.

87 Fukui, T., Kawai, S., Fujinuma, S. et al. (2017). Control over differentiation of a metastable supramolecular assembly in one and two dimensions. *Nat. Chem.* 9: 493.

88 de Jong, J.J.D., Lucas, L.N., Kellogg, R.M. et al. (2004). Reversible optical transcription of supramolecular chirality into molecular chirality. *Science* 304: 278.
89 Hirose, T., Matsuda, K., and Irie, M. (2006). Self-assembly of photochromic diarylethenes with amphiphilic side chains: reversible thermal and photochemical control. *J. Org. Chem.* 71: 7499.
90 Lucas, L.N., van Esch, J., Kellogg, R.M., and Feringa, B.L. (2001). Photocontrolled self-assembly of molecular switches. *Chem. Commun.* 759.
91 Yagai, S., Ishiwatari, K., Lin, X. et al. (2013). Rational design of photoresponsive supramolecular assemblies based on diarylethene. *Chem. Eur. J.* 19: 6971.
92 Kulkarni, C., Meijer, E.W., and Palmans, A.R.A. (2017). Cooperativity scale: a structure-mechanism correlation in the self-assembly of benzene-1,3,5-tricarboxamides. *Acc. Chem. Res.* 50: 1928.
93 Palmans, A.R.A. and Meijer, E.W. (2007). Amplification of chirality in dynamic supramolecular aggregates. *Angew. Chem. Int. Ed.* 46: 8948.
94 Cantekin, S., de Greef, T.F.A., and Palmans, A.R.A. (2012). Benzene-1,3,5-tricarboxamide: a versatile ordering moiety for supramolecular chemistry. *Chem. Soc. Rev.* 41: 6125.
95 Li, T., Li, X., Wang, J. et al. (2016). Photoresponsive supramolecular assemblies based on a C_3-symmetric benzene-1,3,5-tricarboxamide-anchored diarylethene. *Adv. Optical Mater.* 4: 840.
96 Pal, D.S., Kar, H., and Ghosh, S. (2016). Phototriggered supramolecular polymerization. *Chem. Eur. J.* 22: 16872.
97 Hirose, T., Helmich, F., and Meijer, E.W. (2013). Photocontrol over cooperative porphyrin self-assembly with phenylazopyridine ligands. *Angew. Chem. Int. Ed.* 52: 304.
98 Helmich, F., Lee, C.C., Nieuwenhuizen, M.M.L. et al. (2010). Dilution-induced self-assembly of porphyrin aggregates: a consequence of coupled equilibria. *Angew. Chem. Int. Ed.* 49: 3939.
99 Iseki, Y., Watanabe, E., Mori, A., and Inoue, S. (1993). Photoresponsive coordination of stilbazoles to metalloporphyrins. *J. Am. Chem. Soc.* 115: 7313.
100 Otsuki, J. and Narutaki, K. (2004). Photochromism of phenylazopyridines and its application to the fluorescence modulation of zinc–porphyrins. *Bull. Chem. Soc. Jpn.* 77: 1537.
101 Pieroni, O., Fissi, A., Angelini, N., and Lenci, F. (2001). Photoresponsive polypeptides. *Acc. Chem. Res.* 34: 9.
102 Kato, H.E., Zhang, F., Yizhar, O. et al. (2012). Crystal structure of the channelrhodopsin light-gated cation channel. *Nature* 482: 369.
103 Muraoka, T., Cui, H., and Stupp, S.I. (2008). Quadruple helix formation of a photoresponsive peptide amphiphile and its light-triggered dissociation into single fibers. *J. Am. Chem. Soc.* 130: 2946.
104 Yagai, S., Ohta, K., Gushiken, M. et al. (2012). Photoreversible supramolecular polymerisation and hierarchical organization of hydrogen-bonded supramolecular co-polymers composed of diarylethenes and oligothiophenes. *Chem. Eur. J.* 18: 2244.
105 Li, L.-S., Jiang, H., Messmore, B.W. et al. (2007). A torsional strain mechanism to tune pitch in supramolecular helices. *Angew. Chem. Int. Ed.* 46: 5873.

106 Adhikari, B., Yamada, Y., Yamauchi, M. et al. (2017). Light-induced unfolding and refolding of supramolecular polymer nanofibers. *Nat. Commun.* 8: 15254.

107 Hollamby, M.J., Aratsu, K., Pauw, B.R. et al. (2016). Simultaneous SAXS and SANS analysis for the detection of toroidal supramolecular polymers composed of noncovalent supermacrocycles in solution. *Angew. Chem. Int. Ed.* 55: 9890.

3

Light-Driven Self-Organized Liquid Crystalline Nanostructures Enabled by Chiral Molecular Switches or Motors: From 1D to 3D Photonic Crystals

Ling Wang and Quan Li

Kent State University, Liquid Crystal Institute and Chemical Physics Interdisciplinary Program, 1425 Lefton Esplanade, Kent, OH 44242, USA

3.1 Introduction

Bottom-up fabrication of functional nanostructures and molecular devices from self-assembling soft materials with tailored photonic properties is currently a burgeoning area of research. The liquid crystals (LCs) capable of forming self-organized, tunable one-dimensional (1D) or three-dimensional (3D) photonic nanostructures represent such an elegant system because of their unique ability to control the propagation of light by various external stimuli [1–8]. Light-driven chiral molecular switches or motors in liquid crystalline media are of fundamental academic interest and technological significance since they could not only bring a new generation of multifunctional supramolecular 1D and 3D photonic materials with unprecedented properties but also exploit the benefits of light stimuli such as the ease of spatial, temporal, and remote control of the irradiation under ambient conditions [9–11]. These photoactive functional soft materials can be facilely fabricated from achiral liquid crystalline hosts by adding appropriate amount of a light-driven chiral molecular switch or motor, and they have found many potential applications in color filters, polarizers, all-optical displays, optical lasers, sensors, energy-saving smart devices, and so on. Fundamentally, the photoresponsiveness of these supramolecular soft materials is a result of both molecular interaction and geometry changes in the chiral molecular switch or motor upon light irradiation, and one of their key features is the existence of a photonic bandgap (PBG) in the range of the wavelength of visible light, which sometimes extents to the near infrared (NIR) region and beyond. This PBG is analogous to the energy bandgap in the semiconductors and can exist in 1D cholesteric liquid crystals (CLCs), or 3D periodic superstructures such as blue phases (BPs), LC microdroplets, and microshells, depending on the number of directions in which the periodicity exists.

The ability of a chiral molecular switch or motor to impart a twist to an achiral LC host is usually quantified in terms of the helical twisting power (HTP), β, which is expressed by the equation $\beta = 1/(pc)$, where c is the molar or weight concentration

Photoactive Functional Soft Materials: Preparation, Properties, and Applications, First Edition.
Edited by Quan Li.

of the chiral switch and p is the helical pitch of the LC. Photoisomerization of the chiral switch results in the changes of the molecular conformation and HTP, which may lead to a change in the pitch or handedness of the chiral LCs [12–17]. One of critical challenges in this endeavor is the design and synthesis of light-driven chiral molecular switches with high HTP values that display large differences in HTP at their different isomeric states. High HTP values reduce the necessary concentrations of the chiral dopants, while large differences in HTP of the chiral dopants enable wide-range pitch tunability of the photoresponsive chiral LCs. Therefore, light-driven chiral molecular switches or motors based on diverse photoresponsive moieties have been developed to fabricate chiral liquid crystalline superstructures with superior photonic properties. Typical photoresponsive moieties include azobenzenes, dithienylethenes, spiropyrans, and their photoisomerization processes are illustrated in Figure 3.1. Azobenzenes could experience *trans–cis* isomerization under UV irradiation, and the *cis* isomer would be driven back to its *trans* form either by visible light or thermal relaxation [18]. Diarylethenes can transform from the colorless ring-open state to the colored ring-closed state under UV irradiation. The reverse process is thermally stable and can be only driven by visible-light irradiations [19]. For spiropyrans, UV irradiation leads to the photochemical cleavage of C—O bond and the transformation from the initial colorless state to the colored merocyanine [20]. When the UV source is removed, they can thermally relax back to colorless initial states quickly. It should be noted that the large change in polarity and charge distribution could be obtained upon photoisomerization between the spiropyran state and the merocyanine state. The aim of this chapter is to highlight the current achievements in the field of light-driven self-organized liquid crystalline nanostructures enabled by chiral molecular switches or motors. Toward this end, we first introduce the recent advances in the development of light-driven CLCs with 1D helical

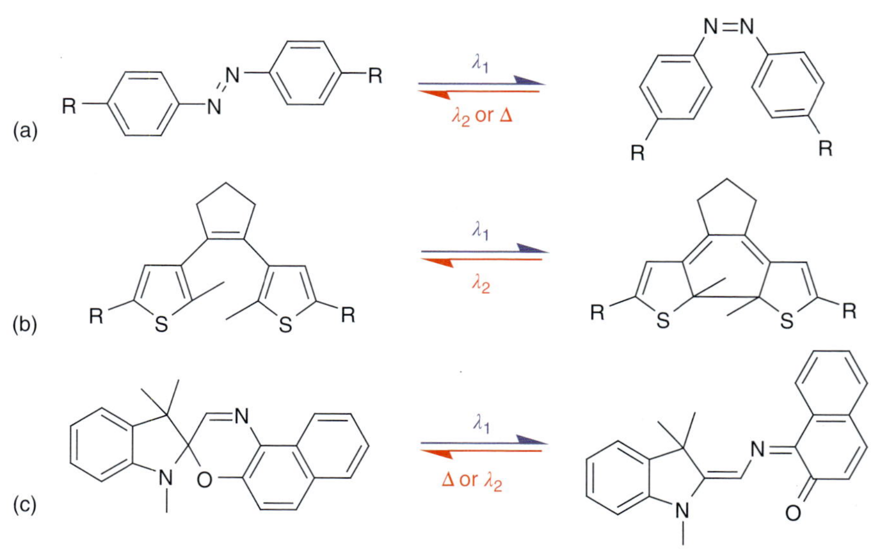

Figure 3.1 Photochromism of (a) azobenzene, (b) dithienylethene, and (c) spiropyran.

superstructures by loading different photoresponsive chiral molecular switches or motors including chiral azobenzene, chiral dithienylethene, chiral spirooxazine, and overcrowded alkenes. Then, we move to a discussion of light-driven BP LCs with 3D cubic nanostructures. We also address the recent endeavor in developing the light-driven chiral liquid crystalline microdroplets and microshells. The last section of this chapter (Section 3.5) provides a perspective for the future scope of these emerging photoactive chiral liquid crystalline nanostructured materials. It should be noted that this chapter does not aim at including all accomplishments in this research field but rather points to important developments by introducing representative examples from different topics.

3.2 Light-Driven Cholesteric Liquid Crystals

Cholesteric phase, i.e. chiral nematic phase, is a variant of the nematic, which spontaneously forms a macroscopic helical structure either when the LC molecules are inherently chiral or when chirality is externally introduced [21–27]. The presence of chirality causes the director profile to assume a twisted configuration throughout the medium. In the cholesteric superstructures, the LC molecules exhibit the spatial orientation in a helical manner along the helical axis, where the helical superstructures are typically characterized by handedness and helical pitch. Handedness describes the direction in which the molecular orientation rotates along the helical axis, while the pitch (p) is the distance over which the director rotates a full 360°. According to Bragg's law, the cholesteric superstructure with homogeneous alignment could selectively reflect circularly polarized light (CPL) with the same handedness as its helix. The reflection wavelength (λ) is typically determined by $\lambda = np$, where n is the average refraction index of LCs. Many efforts have been made to fabricate a light-driven helical superstructure with tunable pitches by doping light-driven chiral molecular switches with superior properties into achiral LC media. For example, the pitch of cholesteric superstructures becomes smaller upon light irradiations at appropriate wavelength, which enables the dynamic reflection color tuning from blue, green, to red and vice versa under light exposures at different wavelengths (Figure 3.2a) [28]. Furthermore, light-driven dynamic handedness inversion of cholesteric superstructures has been obtained by doping unique light-driven chiral molecular switches or motors into achiral LC media. Upon alternating UV- and visible-light irradiations, the cholesteric helix of one handedness unwinds, and disappears at a critical point with infinite pitch, and then rewinds through self-organizing into another helix of the opposite twist sense (Figure 3.2b) [29].

3.2.1 Cholesteric LCs with Chiral Azobenzene Photoswitches

Upon light irradiations at appropriate wavelength, azobenzene-based molecular switches are known to exhibit reversible *trans–cis* isomerization, in which large conformational and polarization changes are often observed. The *trans* form of azobenzene exhibiting a rod-like structure can stabilize the LC superstructure,

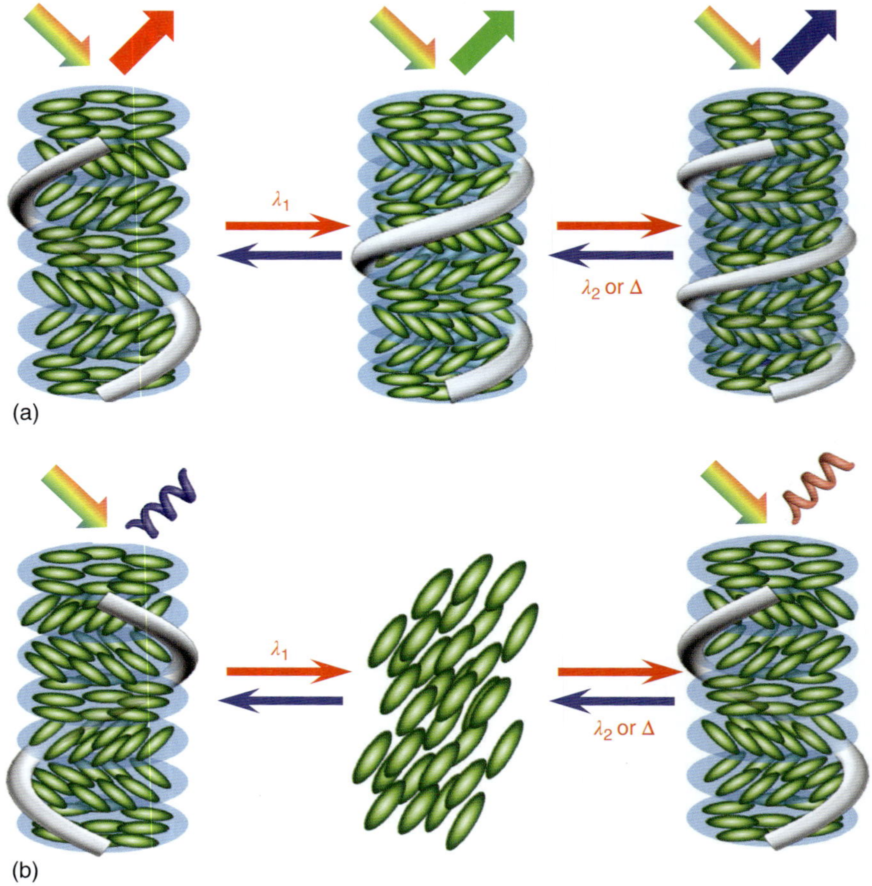

Figure 3.2 (a) Phototunable helical pitch of cholesteric superstructures that induce the RGB photonic reflection. (b) Photoinduced reversible handedness inversion of cholesteric superstructures.

whereas its *cis* form exhibiting a bent-shaped structure destabilizes the LC superstructure. As a result of the dramatic difference in molecular geometry of the *trans* and *cis* forms, the HTPs of these geometric isomers are different, which in turn enables optical tuning of the cholesteric pitch and hence the PBG. In 1971, Sackmann reported the photochemically induced reversible color changes in photoresponsive CLCs prepared by doping azobenzene molecules into a non-halide-based cholesteryl mixture [30]. The PBG reflection of the CLC mixture blue-shifted from 610 to 560 nm upon irradiation with 420 nm light. Although the tuning range is quite narrow, this work is the first demonstration of reversible phototuning of cholesteric reflection colors, providing a pathway to the photoresponsive CLCs. In general, the HTP of the chiral molecular switch in a CLC mixture depends on its structure of the molecules, the nature of the chirality, and the interaction with host molecules. To date, various azobenzene derivatives have been reported for this purpose, such as chiral mesogens doped with achiral azobenzene molecules, pure photoresponsive chiral mesogens, and nematic LC doped with photoresponsive chiral dopants [31–38]. Recently, the studies are mainly focused on the design and synthesis of powerful chiral azobenzenes that exhibit high HTPs as well as large HTP variations upon light irradiation.

Kurihara and coworkers synthesized a series of isosorbide- or isomannide-based chiral photochromic compounds with plural azobenzene groups, and the resulting azobenzene photoswitches exhibited high HTPs as well as large HTP variations upon light irradiation [34]. Gottarelli and coworkers introduced the photoresponsive chiral azobenzene based on the binaphthyl derivatives with axial chirality, where the unique combination of binaphthyl with azobenzene units was expected to generate a family of axially chiral azobenzene photoswitches with high HTPs [39, 40]. Li et al. have developed a series of binaphthyl-derived axially chiral azobenzene compounds [41–46]. These chiral switches have superior solubility in LC hosts and exhibit high HTP values and considerable differences in HTP among their various states, and they were all found to be capable of phototuning the reflection color across the full visible spectrum. Some of these compounds have been employed in the fabrication of chiral photodisplays, which do not require the driving electronics usually used in conventional electro-optic LC displays. Absence of the driving electronics drastically brings down the cost of such displays and makes them lightweight [47, 48].

Some breakthroughs have been made when the binaphthyl-derived axially chiral azobenzene was functionalized with rod-like moieties, where the rod-like units in these chiral dopants resemble the molecular structure of nematic hosts made of rod-shaped molecules. It should be noted that these compounds displayed greater compatibility with the nematic hosts and exhibited exceptionally high HTP values. For example, a light-driven chiral molecular photoswitch as shown in Figure 3.3a was found to exhibit high HTPs and superior solubility in achiral LC hosts, which could be used to develop an optically tunable helical superstructure capable of fast and reversible phototuning of the structural reflection across the entire visible region [49]. To be specific, a mixture of 6.5 wt% chiral photoswitch in nematic LC E7 was capillary filled into a 5-μm thick glass cell with a polyimide planar alignment layer and the cell was painted black on one side. The reflection wavelength of the cell could be tuned starting from the UV region across the entire visible region to the NIR region upon UV irradiation at 365 nm (5.0 mW cm^{-2}) within approximately 50 seconds, whereas its reversible process starting from the NIR region across the entire visible region to the UV region was achieved by visible light at 520 nm (1.5 mW cm^{-2}) or dark thermal relaxation. The reflection colors across the entire visible region were uniform and brilliant as shown in Figure 3.3b,c. The reversible phototuning process was repeated many times without degradation. It is worth noting that the reversible phototuning process across the entire visible region can be achieved in seconds with the increase of light intensity. Furthermore, this chiral switch was used in a photo-addressed colored LC display driven by light and hidden as well as fixed by the application of an electric field from thermal degradation. As illustrated in Figure 3.3d, the reflective image can be hidden in focal conic texture by applying a 30 V pulse and revealed by applying a 60 V pulse. Moreover, by applying a 30 V pulse to an optically written image so as to make the UV irradiated region going to the focal conic texture and the UV un-irradiated region going to the planar

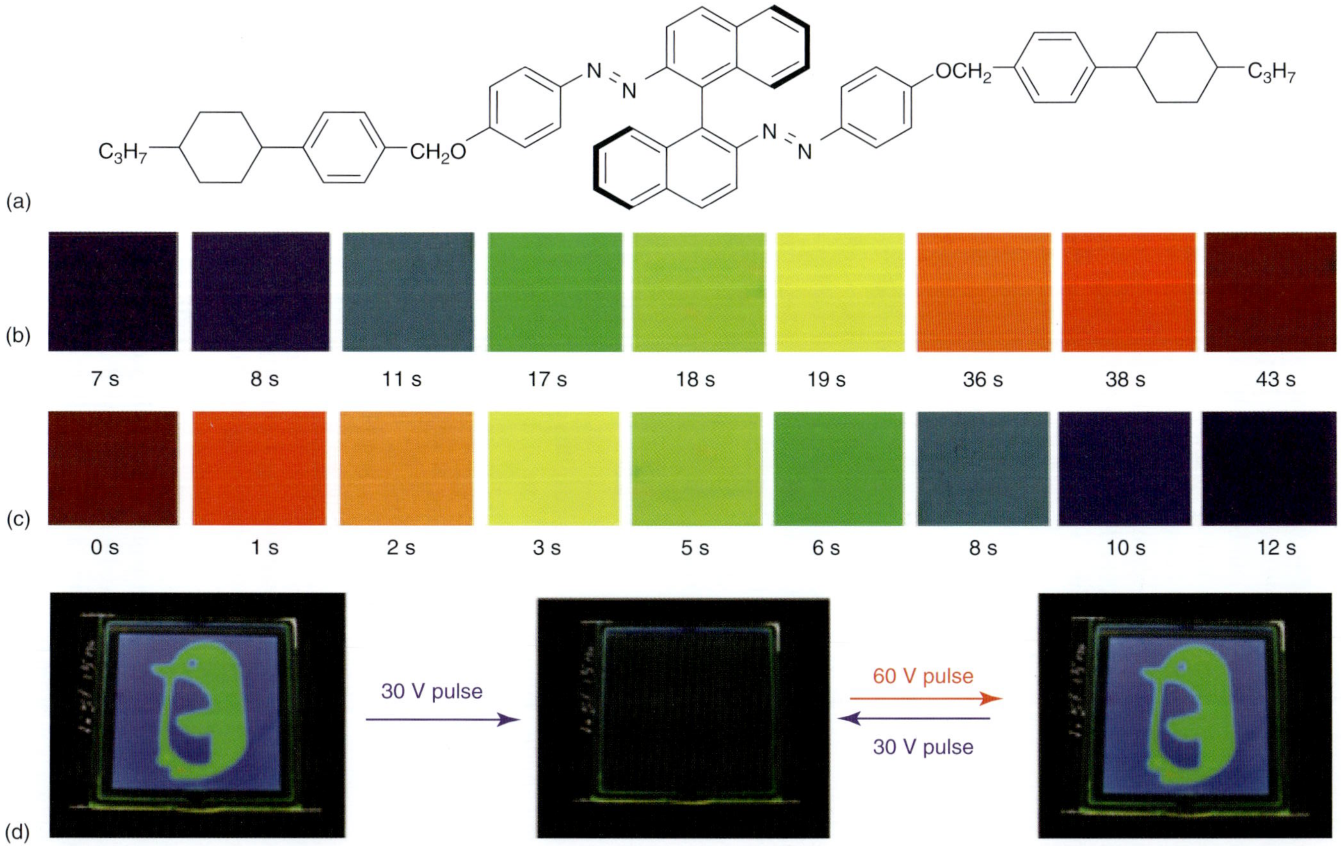

Figure 3.3 (a) Molecular structure of a chiral azobenzene photoswitch with axial chirality. Reflection color images of 6.5 wt% chiral switch in commercially available achiral LC host E7 in 5-μm-thick planar cell; (b) upon UV-light irradiation at 365 nm (5.0 mW cm^{-2}) with different time exposure; (c) reverse back cross the entire visible spectrum upon visible light at 520 nm (1.5 mW cm^{-2}) with different time exposure. The colors were taken from a polarized reflective mode microscope. (d) Images of 5-μm-thick homeotropic alignment cell with 4 wt% chiral switch in E7. The image was recorded in a planar state through a photomask by a UV light (left). The image was hidden by a low voltage pulse in a focal conic state (middle), which was made to reappear by a high voltage pulse (right). Source: Ma et al. 2010 [49]. Copyright 2010. Reproduced with permission from the Royal Society of Chemistry.

texture, an optically written image can be stored indefinitely because the planar and focal conic textures are stable even though the light-driven chiral switch relaxes to the un-irradiated state. It should be noted that wide-range nonmechanical beam steering has been demonstrated by using CLCs with such rod-like molecule-functionalized axially chiral azobenzene [50]. More recently, we reported the first example of a novel halogen-bonded light-driven axially chiral molecular switch, where pyridyl-substituted binaphthyl azobenzene served as a halogen-bond acceptor and the 4-iodotetrafluorophenyl derivative as a halogen-bond donor [51]. This chiral molecular switch was found to exhibit both optical and thermal sensitivity, which enabled red, green, blue (RGB) reflection colors by different light irradiations or by temperature control. This study provides valuable knowledge on the rational design of efficient light-driven switches or motors for amplification and modulation of chirality in dynamic functional soft matter and nanosystems.

Li and coworkers have achieved reversible visible-light-driven dynamic tuning of RGB reflection colors in CLCs by designing a novel light-driven chiral azobenzene molecular switch with extended conjugation [52]. Wang et al. looked toward the possibility of dynamic reflection tuning of cholesteric superstructures by remote NIR irradiation (Figure 3.4) [53]. The upconversion nanoparticles (UCNPs) are known to emit UV and visible light upon NIR exposures [54, 55]. A unique core–shell UCNPs was designed and synthesized, where the intensity ratio of its emissions at different wavelengths was highly dependent on the excitation power density of the NIR irradiation. Then they developed a novel chiral azobenzene photoswitch whose UV and visible absorptions exhibit significant

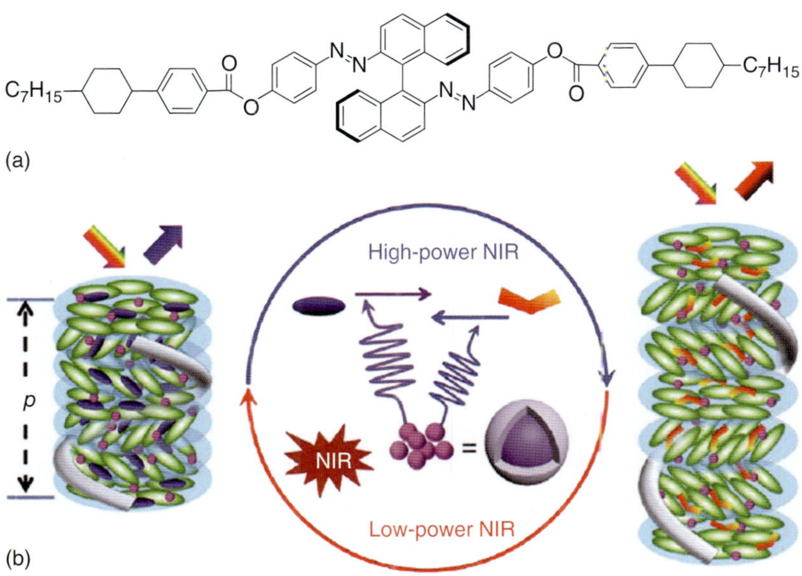

Figure 3.4 (a) Molecular structure of a light-driven chiral azobenzene. (b) Schematic mechanism of the reversible tuning of self-organized helical superstructures with chiral switch and UCNPs upon irradiation with NIR laser at different power densities. Source: Wang et al. 2014 [53]. Copyright 2014. Reproduced from the American Chemical Society.

overlap with the emissions of the UCNPs. The photoresponsive CLCs were formulated by mixing the azobenzene photoswitch and the UCNPs with an achiral liquid crystalline host composed of commercially available components. The resulting nanoparticle impregnated cholesteric superstructure exhibited reversibly tunable RGB reflections by simply changing the power density of the NIR irradiation at 980 nm. The reflection wavelength of such a cholesteric hybrid superstructure redshifted upon NIR irradiation at high power density, whereas the reverse process occurred upon irradiation at lower power density. Such UCNP-assisted photochemistry shall provide important insights into the future fabrication of NIR-photoactive advanced soft materials.

Yu and coworkers demonstrated piecewise RGB reflection tuning of CLCs by developing a photoresponsive tristable chiral molecular switch [56]. The tristable chiral switch was constructed by incorporating two different photoresponsive azobenzene switches, i.e. parent azobenzene (Azo) and *o*-fluoroazobenzene (F-Azo), into one binaphthyl-derived chiral structure with high HTP (Figure 3.5a). It should be noted that *trans–cis* isomerization of F-Azo can be achieved by the irradiations with both UV light and visible light due to the good separation of the n–π^* bands of *trans* and *cis* isomers. The resulting light-driven tristable chiral switch was found to exhibit three different photostationary configurations, i.e. configuration I (*trans, trans, trans, trans*) upon irradiation with 470 nm light, configuration II (*cis, trans, trans, cis*) upon irradiation with 530 nm light, and configuration III (*cis, cis, cis, cis*) upon irradiation with 365 nm light as shown in Figure 3.5b. The CLCs loaded with the tristable chiral switch showed two tuning periods of cholesteric reflection upon appropriate light irradiations, i.e. one period tuned in the visible spectrum and another period in the NIR region (Figure 3.5c). To be specific, upon light irradiation at 470 and 530 nm, the CLCs show dynamic selective reflection in the visible spectrum covering the three primary RGB colors, whereas wide reflection tuning in the NIR region was achieved under irradiation with 530 and 365 nm light. Moreover, photoaddressable images of dynamic RGB colors with black background were further demonstrated based on this mechanism (Figure 3.5d).

Light-driven dynamic handedness inversion of CLCs has attracted increasing attention due to its fundamental significance and the advantages of remote, spatial, and temporal controllability. One can achieve light-driven dynamic handedness inversion through opposite-handed doping of powerful photoresponsive chiral switch and photoinsensitive chiral molecules into achiral LCs [57–60]. This method can eliminate the need for complicated molecular design of chiral switches, but handedness inversion often occurs in infrared regions due to the limited solubility of photoresponsive chiral switch and poor miscibility between two dopants. Many strategies have been developed for the rational molecular design of single chiral molecular photoswitches that could confer handedness inversion capability to CLCs, but the event of light-driven handedness inversion using such methodology is often encountered by chance. Mathews et al. found that light-driven reversible handedness inversion of CLCs could be achieved by judiciously designing and synthesizing a chiral cyclic azobenzene photoswitch based on axially chiral binaphthyl derivatives [61]. Recently, Li et al. further reported a novel design strategy based on the chiral conflict and equilibrium

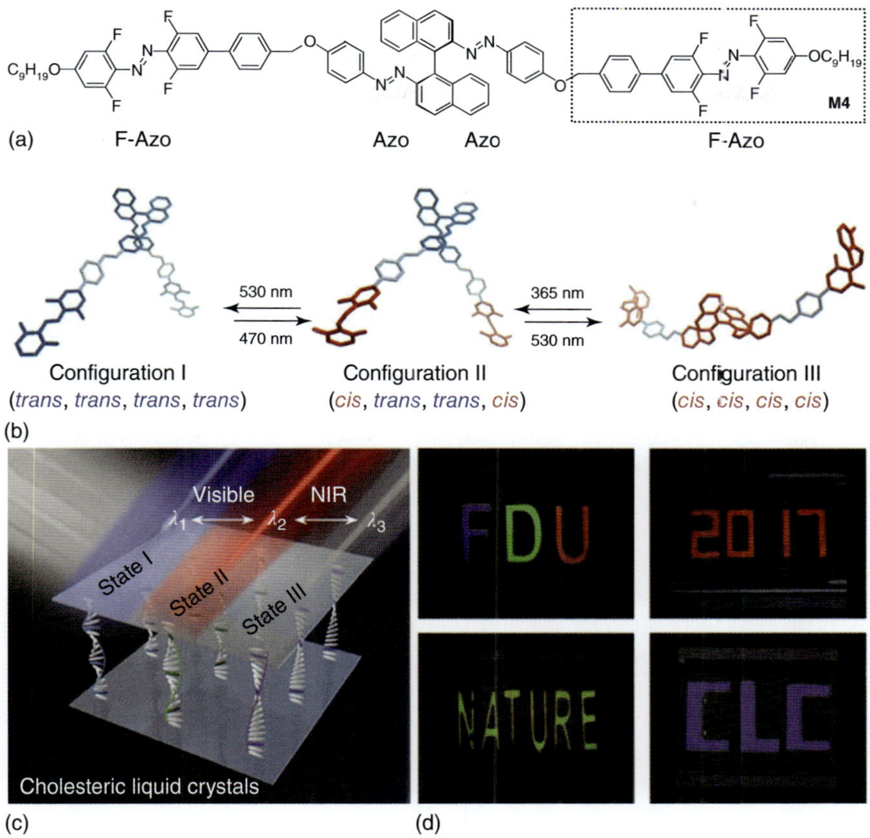

Figure 3.5 (a) Chemical structure of the photoresponsive tristable chiral switch containing Azo near the chiral center and F-Azo far away from the chiral center. (b) Schematic illustration of optimized molecular three configurations in the tristable chiral switch. (c) Schematic illustration of piecewise control over the photoresponsive cholesteric LCs. (d) Real cell images created by piecewise control of the helices in the CLC to show different patterns of RGB colors with a black background in 5-μm-thick antiparallel aligned cells (2.0 cm × 2.5 cm). Source: Qin et al. 2017 [56]. Copyright 2018. Reprinted with permission from John Wiley & Sons.

shifting between multiple chiral moieties in a single molecule [62]. These molecular engineered chiral switches were found to confer handedness inversion in different LC hosts upon light irradiation with moderate to high HTPs, where the chirality of each chiral moiety does not need to be reversed. Since the *R* and *S* configuration of unbridged binaphthyls normally will induce right-handed and left-handed CLCs, the design strategy is to incorporate multiple chiral binaphthyls with opposite chirality into a single molecule functionalized with photoresponsive groups, and helix inversion was thus achieved based on the chiral conflict and equilibrium shift between these chiral moieties. Using the same principle, Guo and coworkers demonstrated light-directed dynamic handedness inversion of CLCs by developing intermolecular hydrogen-bonded chiral azoarenes containing binaphthyl units of opposite chiral configuration [63]. This principle should be applied to other photoresponsive systems toward dynamic chirality inversion capability.

3.2.2 Cholesteric LCs with Chiral Diarylethene Photoswitches

Compared with chiral azobenzene photoswitches, chiral diarylethenes possess the advantage of thermal irreversibility in both the open- and closed-ring isomeric states. However, only a few derivatives of diarylethenes have been reported as chiral molecular dopants with applications as liquid crystalline phase switches [64–66]. Most of the reported chiral diarylethenes exhibit low to moderate HTPs and are not suitable for reflection wavelength tuning. These relatively low HTPs also give rise to the requirement of higher doping concentrations, which often leads to phase separation, coloration, and physical property changes of the LC host. Therefore, it would be of great practical interest to develop chiral diarylethenes with high HTPs as well as large HTP variations upon light irradiation. It was not until recently that we reported some chiral diarylethene switches for wide range reflection color tuning. Li et al. reported the first diarylethene switch with axial chirality, high HTPs, and large HTP variations upon light irradiation for wide-range reflection color tuning (Figure 3.6a) [67]. When doping 0.4 wt% of the dithienylethene photoswitch into the achiral LC host, an achiral LC phase was initially obtained but would transform into the cholesteric upon UV irradiation and vice versa. A higher doping concentration of 7.7 wt% was used for phototuning of the RGB reflection colors of CLC mixtures (Figure 3.6b). The reflection central wavelength of this mixture was around 630 nm at the initial state. Upon UV irradiation, its reflection wavelength was tuned to 530 nm within 10 seconds and further reached a photostationary state in 25 seconds with a reflection central wavelength at 440 nm (Figure 3.6c). This photoinduced state was thermally stable and could be photochemically switched back to a nearly initial state by visible-light irradiation at 550 nm within two minutes. Furthermore, the three primary RGB colors were simultaneously demonstrated in a single thin film by varying UV irradiation times and photomasking at different areas.

Binaphthyl-derived chiral diarylethene switches, i.e. dithienylcyclopentenes, were also demonstrated to enable light-driven dynamic handedness inversion of CLCs with excellent thermal bistability and excellent fatigue resistance [68, 69]. It is believed that the reversible handedness inversion in these systems originated from the conformation change of binaphthyl moieties as well as the stereospecific intermolecular interactions between the binaphthyl moieties and the surrounding nematic LC molecules. Recently, Wang et al. achieved NIR-light-driven dynamic handedness inversion of CLCs by designing and synthesizing a novel category of core–multishell UCNPs and a binaphthyl-derived diarylethene chiral photoswitch (Figure 3.7) [70]. The UCNPs acting as the NIR nanotransducers are composed of dual lanthanide upconverting activators, i.e. Tm^{3+} and Er^{3+} in separate layers, where the Tm^{3+} and Nd^{3+} ions were bridged by an array of Yb^{3+} ions, and the Er^{3+} ions were separated by an optically inert interlayer. The core–multishell nanotransducers were found to exhibit UV and visible luminescence on demand from the individual UCNPs by selective excitation with NIR light at two different wavelengths of 808 and 980 nm. Interestingly, the UV and visible

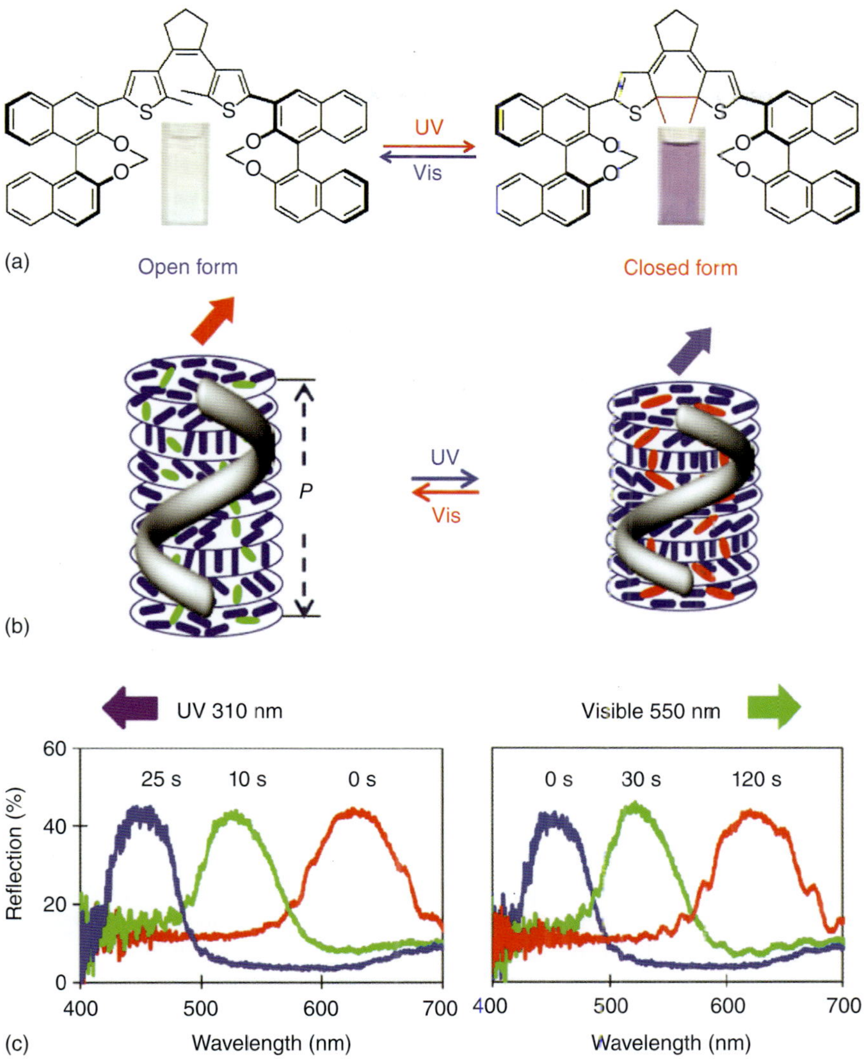

Figure 3.6 (a) Molecular structure of dithienylcyclopentene chiral switch and its photoisomerization. (b) Schematic mechanism of phototunable cholesteric superstructures and corresponding dynamic reflections. (c) Reflective spectra of 7.7 wt% chiral switch in E7 in a 5-μm-thick planar cell at room temperature. (Left) Upon UV irradiation at 310 nm ($30\,mW\,cm^{-2}$) for different times. (Right) Upon visible-light irradiation at 550 nm ($30\,mW\,cm^{-2}$) for different times. Source: Li et al. 2012 [67]. Copyright 2012. Reproduced from American Chemical Society.

luminescences from the nanotransducers were able to drive reversible photocyclization of the chiral diarylethene molecular switch due to the synergetic interaction between the switch and the nanotransducers. The CLCs loaded with such chiral diarylethene photoswitch and UCNP nanotransducers exhibited reversible handedness inversion upon irradiations with NIR light of different wavelengths. Furthermore, Wang et al. also reported NIR-light-directed dynamic

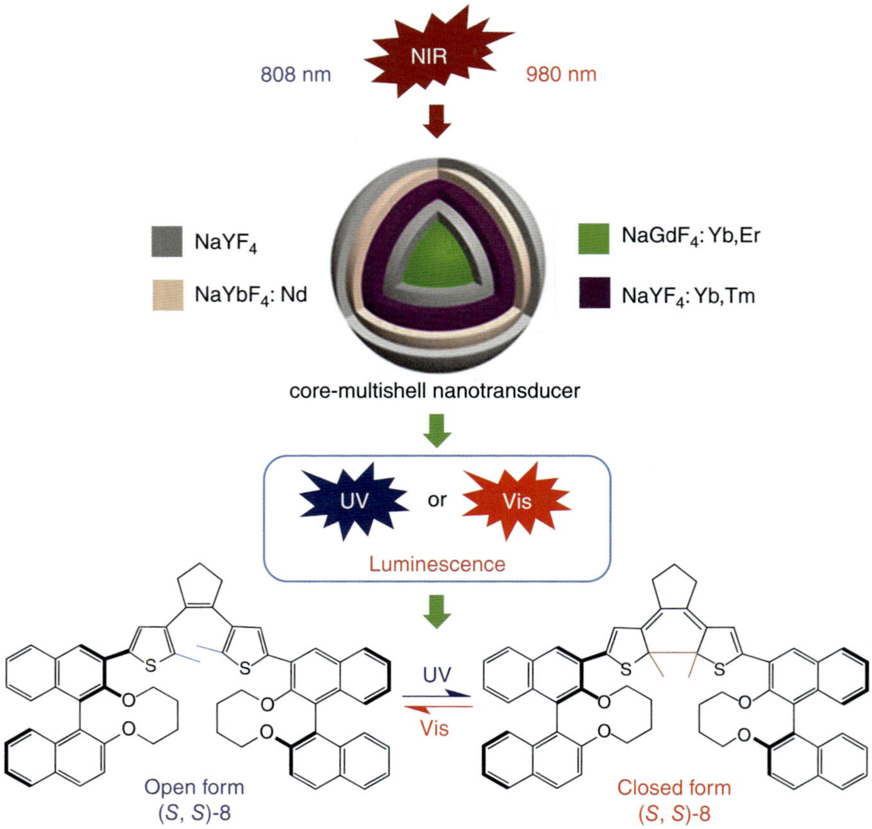

Figure 3.7 Schematic illustration of the core–multishell nanotransducers (UCNPs) and the NIR-light-driven reversible switching process of photochromic chiral diarylethene switch with handedness inversion. Source: Wang et al. 2015 [70]. Copyright 2015. Reproduced with permission from John Wiley & Sons.

reflection of both left- and right-handed CPL through the visible range spanning the RGB wavelengths by developing novel CLCs loaded with mesogen-modified gold nanorods (GNRs) [71]. These studies provide significant impetus toward the development and realization of light-driven functional soft materials for promising photonic applications where CPL is involved.

Recently, Zheng et al. reported the breakthrough of light-driven 3D control of the helical axis in cholesteric superstructures by using the same chiral diarylethene photoswitch [72]. Upon continuous irradiation with UV light, the standing helix transformed to a lying state and the lying helix rotated in the plane of the LC cell until the system reached its photostationary state (Figure 3.8). The system can be driven back in the opposite direction by visible-light irradiation. Facile 3D control over the helical axis of the cholesteric superstructures has also been utilized in a specially engineered bilayer LC cell to obtain 2D diffraction gratings. Interestingly, reversible transformation between 2D to 1D and 1D to zero-dimensional (off state) diffraction states has been achieved under alternating UV- and visible-light irradiations. The helical

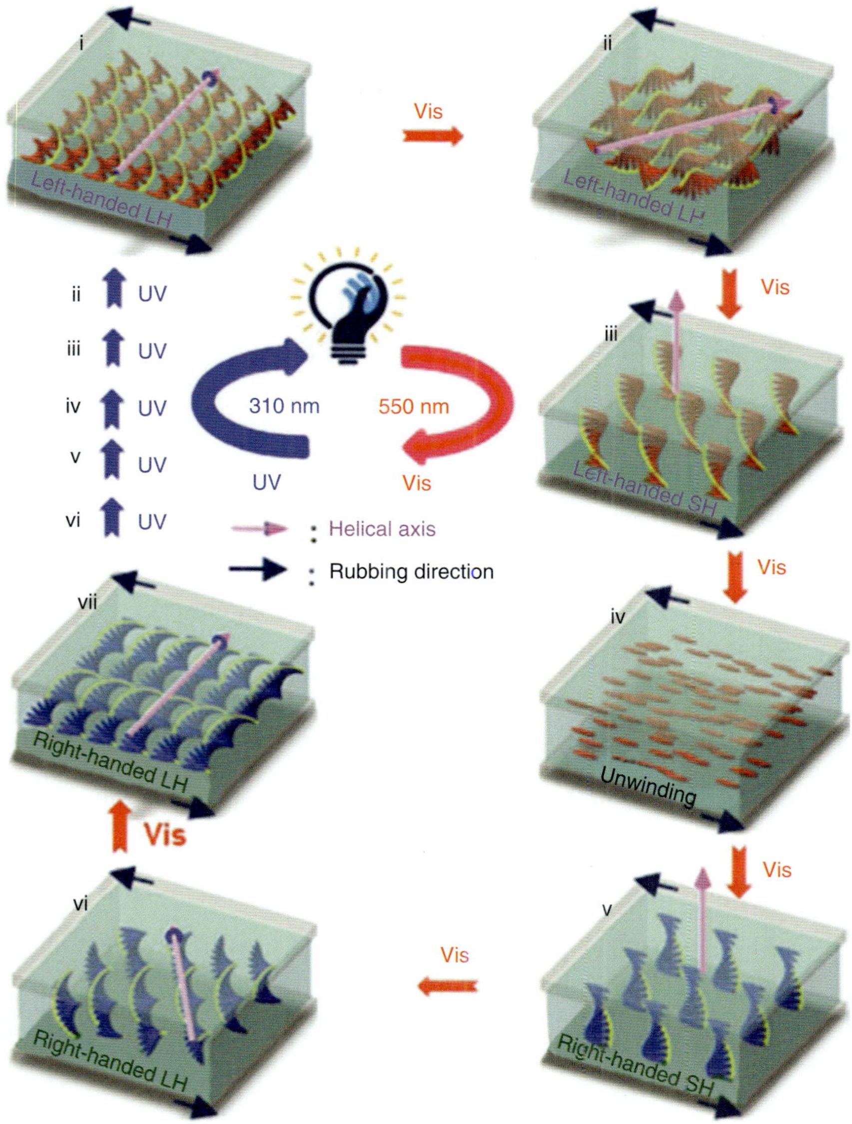

Figure 3.8 Schematic illustration of the light-driven 3D control over the helical axis in cholesteric superstructures loaded with photochromic chiral diarylethene. Source: Zheng et al. 2016 [72]. Copyright 2016. Adapted with permission from Macmillan Publishers Limited.

axis of cholesteric superstructures could be controlled by surface alignments or by applying external voltage, which builds the foundation for their practical application such as reflective displays, thermography, mirrorless lasing, and tunable color filters [73–79]. This work is a step toward the realization of complex, light-activated smart systems and dynamic, reconfigurable 3D architectures through dynamic, remote, and continuous control over the helical axis in 3D space.

3.2.3 Cholesteric LCs with Chiral Spirooxazine and Overcrowded Alkenes

Spirooxazines are an interesting family of photochromic materials due to their unique properties such as excellent photofatigue resistance, strong photocoloration, and fast thermal relaxation. The fatigue resistance of spirooxazines has led to their successful use in various applications including commercialized eyewear. Since the spiro-carbon of a spirooxazine molecule has potential as a chiral center, spirooxazines could be used as chiroptical molecular switches. However, spirooxazines usually exist as racemates. Even if the enantiomers are separated, each enantiomer is racemized by thermal and optical interconversions. Therefore, if the spirooxazines are utilized as chiroptical molecules in achiral nematic LC systems, modification of the spirooxazine with a chiral group is required. Accordingly, Jin et al. synthesized a series of binaphthyl-derived axially chiral spirooxazines [80, 81]. Light-driven cholesteric superstructures with fast thermal relaxation have been demonstrated by using such chiral spirooxazines. However, much effort is still needed to achieve the chiral spirooxazine photoswitches that exhibit high HTPs as well as large HTP variations upon light irradiation.

In addition to chiral azobenzene, diarylethene, and spirooxazine, chiral overcrowded alkenes have also been intensively studied. Feringa et al. pioneered the sterically overcrowded alkenes with intrinsic helical conformations, which are able to induce handedness inversion of the cholesteric helix when used as chiral dopants [82–87]. These types of molecules are also found to be capable of tuning the reflection colors in CLCs. Take the light-driven chiral motor in Figure 3.9a as an example [90]. Its initial HTP at (*P*, *P*)-*trans* form in nematic E7 is +99 μm^{-1}, but generation of a cholesteric helix with an opposite sign of similar pitch is impossible, as the (*M*, *M*)-*trans* form possesses a minor negative HTP ($\beta_M = -7\,\mu m^{-1}$, E7). As a result of the high HTP at (*P*, *P*)-*trans* form, colored LC films were easily generated using this dopant. Photochemical and thermal isomerizations of the motor lead to irreversible color change in the LC film as shown in Figure 3.9b [88]. Interestingly, remarkable stripe rotations were observed in an open film of a photoresponsive CLC with this chiral overcrowded alkene placed on a glass substrate treated with a unidirectional alignment layer, and the collective action of overcrowded-alkene-based molecular motors embedded in the cholesteric matrix has been harnessed to controllably rotate microparticles with remote light

Figure 3.9 (a) Unidirectional rotation of molecular overcrowded alkene motor in a liquid crystalline host, and associated HTPs. (b) Colors of overcrowded-alkene-doped LC phase (6.16 wt% in E7) in time, starting from pure (*P*, *P*)-*trans*-overcrowded alkene upon irradiation with >280-nm light at room temperature, as taken from actual photographs of the sample. The colors shown from left to right correspond to 0, 10, 20, 30, 40, and 80 seconds of irradiation time, respectively. Source: van Delden et al. 2002 [88]. Copyright 2002. Reprinted with permission from National Academy of Sciences. (c) Light-driven rotational reorganization of a CLC film that causes rotation of a microscale object on its surface. Source: Eelkema et al. 2006 [89]. Copyright 2006. Reproduced from with permission from Nature Publishing Group.

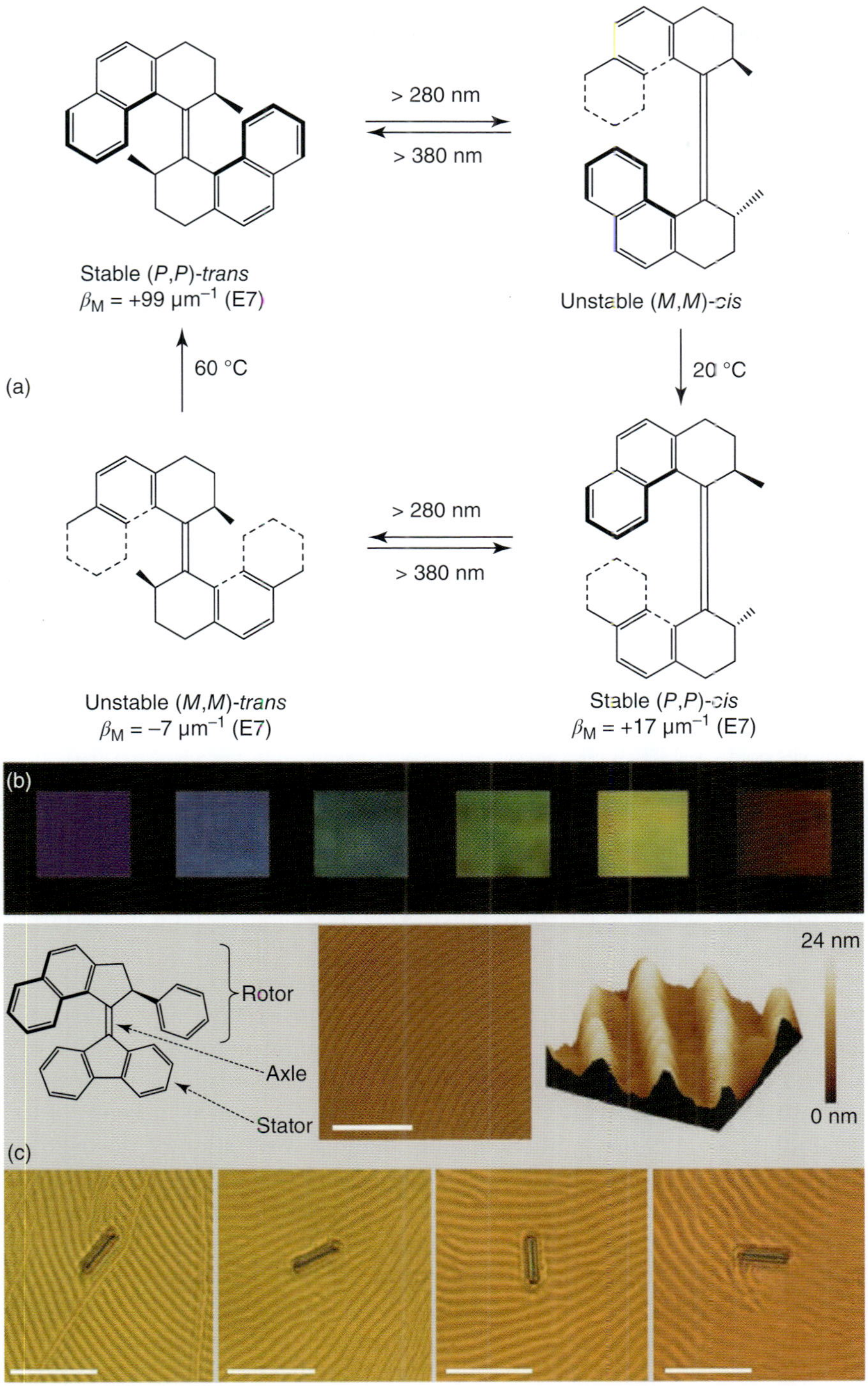

> 280 nm
> 380 nm
Stable (*P*,*P*)-*trans*
β_M = +99 μm^{-1} (E7)
Unstable (*M*,*M*)-*cis*
60 °C
(a)
20 °C
> 280 nm
> 380 nm
Unstable (*M*,*M*)-*trans*
β_M = −7 μm^{-1} (E7)
Stable (*P*,*P*)-*cis*
β_M = +17 μm^{-1} (E7)
(b)
Rotor
Axle
Stator
24 nm
0 nm
(c)

as the fuel (Figure 3.9c) [89, 91, 92]. Moreover, light-driven dynamic reflections with handedness inversion in both visible and infrared regions have been achieved in CLCs loaded with chiral overcrowded alkenes [93, 94].

3.3 Light-Driven Blue Phase Liquid Crystals

Liquid crystalline BPs are among the most interesting self-organized superstructures in the field of LCs, which are known to self-organize into 3D periodic cubic nanostructures [95–97]. This phase usually exists only over a narrow temperature range before the CLCs clear to the isotropic liquid. Similar to CLCs, BPs can be observed either in pure chiral mesogens or can be induced from achiral LCs by adding chiral dopants. The cubic nanostructures are characterized by the double twist cylinders (DTCs) that are stabilized by disclinations. Normally, when chirality in the LC system is high enough, DTCs are more stable than a simple twist structure. However, DTCs cannot be densely packed to fill the entire space and disclination lines appear at the interstices among the cylinders to form a stabilized 3D cubic structure [98–109]. There are three types of BPs reported, namely, BPIII, BPII, and BPI, which are usually observed during cooling from the isotropic to the cholesteric phase in the order of decreasing temperature. BPI and BPII have body-centered cubic and simple cubic structures, respectively, while BPIII is amorphous in nature. BPI and BPII are periodic in three dimensions and the periodicity is typically on the order of the wavelength of visible light. These nanostructures enable BPs to reflect light in the visible region and have been recognized as promising soft self-assembled 3D photonic crystals in recent years [110–112].

Light-driven self-organized BP 3D cubic nanostructures that display distinctive photonic reflection in the three orthogonal directions have received increasing attention, especially in recent years, but the reflection wavelength tuning is usually quite narrow [113–115]. The ability to dynamically tune the PBG in cubic BPs across a wide wavelength range is highly desirable, but it has not been realized due to various obstacles including the instability and irreversibility of BPs under the light irradiations. Li and coworkers made a breakthrough of phototuning full visible range reflection in BPs by using an axially chiral azobenzene photoswitch as shown in Figure 3.10a,b [116]. The initial phase of the doped BPs was the BPII with (100) lattice confirmed by the Kossel diagram as shown in Figure 3.10d (0 second) with a blue reflection color as shown in Figure 3.10c (0 second). Upon 408-nm light irradiation, photoisomerization of the azobenzene dopant enlarged the size of the BP lattice and shifted the corresponding reflection toward a longer wavelength. Increasing the irradiation time resulted in occurrence of the phase transition from BPII to BPI, which might be attributed to the chirality decrease with dopant isomerization. After the photoinduced phase transition took place, the lattice direction changed from (100) of BPII to (110) of BPI with the corresponding change of reflection wavelength from 520 to 575 nm. Subsequently, a redshift in reflection upon further irradiation at 408 nm was observed in BPI. If the irradiation was stopped at any time, the wavelength of the selective reflection does not change, except by the very slow thermal

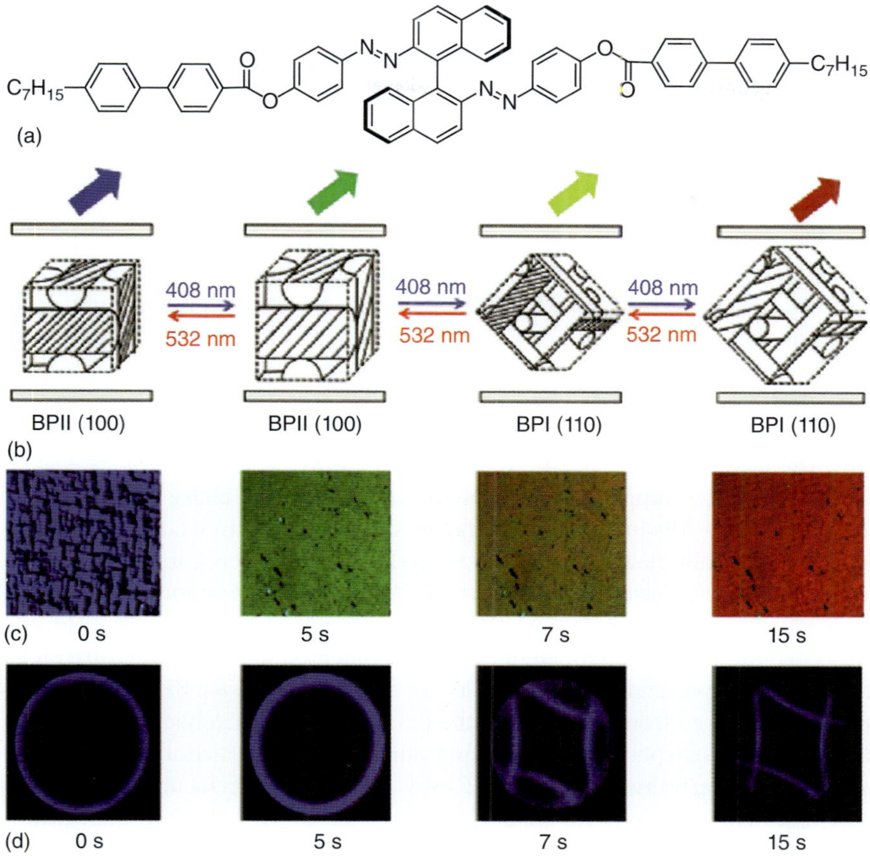

Figure 3.10 (a) Molecular structure of chiral azobenzene photoswitch with axial chirality. (b) Schematic illustration of BP cubic nanostructures. (c) Reflection color images of BPs upon 408-nm light irradiation. (d) The corresponding optical Kossel diffraction diagrams. Source: Lin et al. 2013 [116]. Copyright 2013. Reproduced with permission from John Wiley & Sons.

relaxation effect of the dopant toward its initial *trans* form. The rapid reverse process can occur photochemically with 532-nm light irradiation. Figure 3.10c shows BP images with various durations of pumping observed under a crossed polarized optical microscope at reflection mode. Upon 408-nm light irradiation, the reflection color of the BP thin film changed from blue (BPII) to red (BPI) within 15 seconds. The BP thin films were also indexed using a Kossel diagram as shown in Figure 3.10d. Both blue and green images in Figure 3.10c corresponded to (100) lattices of BPII, while both yellow–green and red images represented (110) lattices of BPI. Figure 3.10d also exhibited the variation of the Kossel diagram with different exposure times at 408 nm. As the irradiation time increased from 0 to 15 seconds, the circle patterns of the (100) lattice of BPII enlarged, switching to a diamond-shaped pattern corresponding to the (110) lattice of BPI, and gradually shrinking. It is noteworthy that increasing the dopant concentration shifted the initial BP reflection to the UV region, which was tuned across the entire visible region with a broad range upon 408-nm light irradiation, and the reflection bands of BPII redshifted from the UV to 520 nm. Once BPII transformed into BPI, the corresponding reflection bands discontinuously jumped to 560 nm, which was further redshifted from 560 to 710 nm within the BPI state.

Wang et al. demonstrated NIR-light-driven dynamic tuning of 3D BP nanostructures by incorporating hydrophobic mesogen-functionalized GNRs (M-GNRs) into a temperature-sensitive BP medium composed of commercially available components [117]. Owing to the excellent thermoresponsive properties of BPs, their 3D nanostructures could be manipulated with the "photothermal effect" of M-GNRs, i.e. converting NIR light into heat through a non-radiative relaxation process of longitudinal surface plasmon resonances (LSPRs) [118–120]. In the experiment, the liquid crystalline composites doped with 0.07 wt% concentrations of M-GNRs were prepared by dispersing the M-GNRs into BP LCs. On cooling from isotropic phase, BPII firstly appeared at 46.4 °C and then transformed into BPI at 44.1 °C. As the temperature was further decreased to 37.2 °C, the phase completely changed into the cholesteric phase (Figure 3.11a). Additionally, the BPs formed the characteristic platelet textures, and the reflection color gradually changed to a longer wavelength with decreasing temperature. Kossel diagram was used to verify the BP lattice structures upon changing the temperature as shown in Figure 3.11b. The isotropic phase did not yield any Kossel diagram. A prominent circular pattern appeared in the temperature range of BPII, which resulted from the (100) lattice of the simple cubic BPII. As the temperature decreased from BPII to BPI, the diffraction ring switched to the characteristic diamond-shaped pattern corresponding to the (110) lattice of body-centered cubic BPI. The introduction of M-GNRs was not only beneficial in stabilizing the cubic nanostructure, but also enabled the dynamic control of 3D BP nanostructures between body-centered cubic and simple cubic symmetry under irradiation using an 808-nm NIR laser due to the significant photothermal effect of M-GNRs (Figure 3.11c). The reverse process occurs upon removal of the NIR laser irradiation. Furthermore, dynamic NIR-light-directed RGB reflections were for the first time demonstrated by tuning the lattice constant of the light-driven 3D soft photonic crystals.

More recently, Zheng et al. demonstrated light-directed crystallographic patterns of self-organized 3D BP nanostructures by using a photoalignment technique (Figure 3.12) [121, 122]. The basic idea is that the cubic nanostructures of BPs exhibit a uniform crystallographic orientation on both microscopic and macroscopic scales at the UV-light-exposed areas, and a random orientation at the unexposed areas. As a result, various desired patterns of uniform crystallographic orientation could be obtained in a BP cell with appropriate patterned photoalignment treatment on one substrate. Interestingly, such an alternate uniform and random orientation pattern formed by the BP lattice could exhibit interesting optical diffraction and responsiveness to different external stimuli. The resultant crystallographic patterns could be reversibly erased and rewritten on demand by light thanks to the rewritable trait of the photoalignment thin film. This work could provide important insights on the controllability of crystallographic orientation of functional soft materials and their diverse photonic applications.

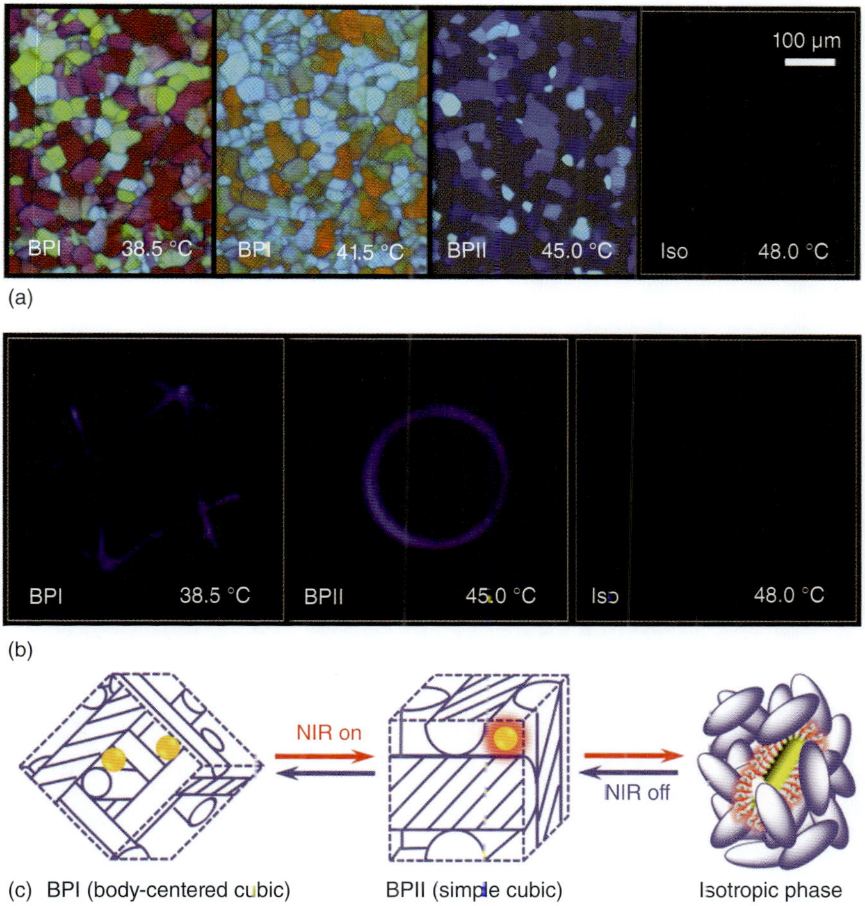

Figure 3.11 NIR-light-directed self-organized BP 3D photonic superstructures loaded with gold nanorods. (a) Typical textures of BPI and BPII nanostructures at different temperatures and (b) corresponding Kossel diagrams, which indicate phase transitions from isotropic to BPII followed by BPII to BPI transition. (c) Schematic illustration of the structural transformations under the NIR-light irradiation. Source: Wang et al. 2015 [117]. Copyright 2015. Reproduced with permission from Royal Society of Chemistry.

3.4 Light-Driven Chiral Liquid Crystal Microdroplets and Microshells

Self-organized 3D superstructures with spherical configurations have recently received considerable attention in the forefront of multitudinous research areas, such as optical sensors or manipulations, soft photonics, and microfluidics [123]. Liquid crystalline microdroplets and microshells, i.e. confining LCs in micrometer- or nanometer-sized superstructures, represent an elegant example of the topic. Recent advances of microfluidic techniques have enabled the fast, simple, and convenient fabrication of stable 3D monodisperse microdroplets with controllable symmetry, shapes, and sizes [124]. Fan et al. demonstrated

Figure 3.12 Photoalignment enabled crystallographic orientation pattern of BP soft lattice, their stimuli-responsive behaviors, and photonic applications. Source: Zheng et al. 2017 [121]. Copyright 2018. Reproduced with permission from John Wiley & Sons.

the photonic cross-communication between monodisperse CLC microdroplets, where their self-assembly was precisely manipulated by using the microfluidic method [125–127]. Only a central reflection spot was observed in the isolated single microdroplet due to the absence of any lateral communication (Figure 3.13a). In the groups of two or more microdroplets, it was surprising that the reflection-related interactions were found between neighboring microdroplets (Figure 3.13b–e). When the microdroplets are densely packed into the hexagonally symmetric monolayer, the circular "lit fire-cracker" patterns were formed because of the photonic cross-communication in the neighboring microdroplets (Figure 3.13f). Moreover, the intensity of photonic cross-interaction was found to depend strongly on the distance between microdroplets, and the photonic cross-communication completely disappeared when the distance was five times larger than the diameter of the microdroplets. It is worth mentioning that omnidirectional RGB reflections in the microdroplets have been

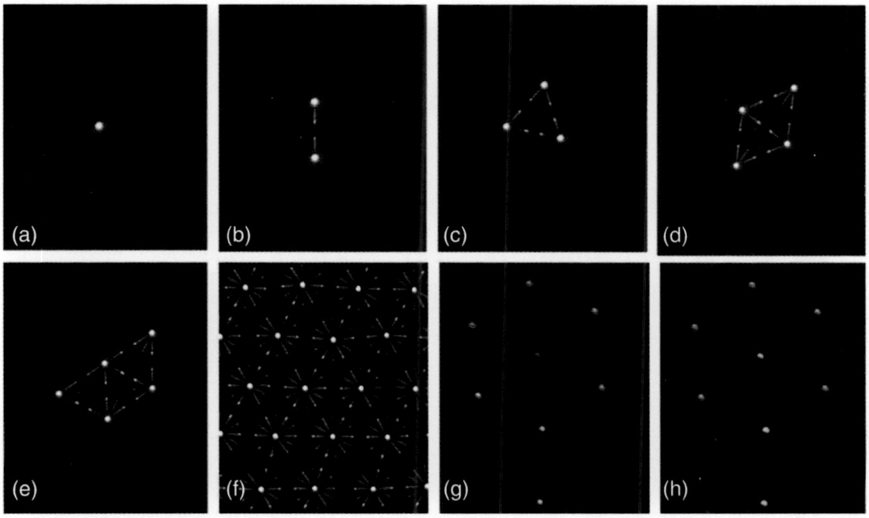

Figure 3.13 (a–f) Polarizing optical microscopy images and photonic cross-communication of linear-, triangular-, and diamond-shaped patterns in the cholesteric microdroplet arrays. (g, h) The "flower-opening" patterns of microdroplets with light-driven iridescent colors. Source: Fan et al. 2015 [125]. Copyright 2015. Reprinted with permission from John Wiley & Sons.

achieved by using a novel thermally stable photochromic chiral molecular switch. They could also readily modulate the reflection color of every single microdroplet independently, and Figure 3.13g,h illustrates the "flower-opening" patterns with light-directed iridescent colors in eight monodisperse microdroplets. Such monodisperse microdroplets and their periodic arrays would not only provide a rich and fascinating platform for fundamental theoretical studies of micro/nanophotonics with geometric confinements but also hold great potential for applications in the photonic and opto-electronic devices.

Recently, Wang et al. reported optically reconfigurable monodisperse cholesteric microspheres with dynamic chirality using the capillary-based microfluidic technique [128]. They judiciously designed and developed a cholesteric system with phototunable handedness inversion capability by the opposite-handedness doping of a powerful left-handed chiral molecular switch and a photoinsensitive right-handed chiral molecule into an achiral liquid crystalline host (Figure 3.14). The selective reflection of the handedness-invertible CLCs was facilely tuned from infrared to visible and ultraviolet regions by appropriately adjusting the relative concentrations of both chiral molecules. The monodisperse chiral microspheres of photoresponsive CLCs with different initial chirality were fabricated using a capillary-based microfluidic technique. Interestingly, they observed the light-driven structural transformations accompanied by dynamic handedness inversion in the cholesteric microdroplets with low chirality. Reducing the concentration of both chiral dopants in the LC host, i.e. low chirality, enabled the direct observation of microstructure evolution upon light irradiations due to the pitch of cholesterics in the micrometer range. As shown in Figure 3.15a, left-handed double spiralized patterns without apparent disclination were initially observed within the microdroplets under polarized optical

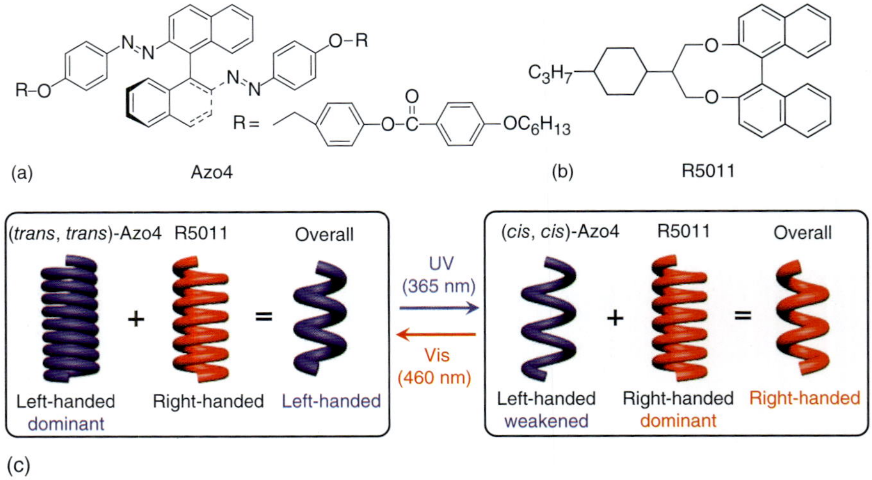

Figure 3.14 Chemical structures of light-driven axially chiral molecular switch **Azo4** (a) and photoinsensitive chiral dopant **R5011** (b). (c) Schematic illustrations of the mechanism of light-driven dynamic handedness inversion in self-organized helical superstructures induced through the doping of **Azo4** and **R5011** with opposite-handedness. At the initial state, the combined doping of two chiral molecules in an achiral nematic LC host yields an overall left-handedness which is dominated by that of (*trans*, *trans*)-**Azo4**. Upon UV irradiation, the left-handed contribution from (*trans*, *trans*)-**Azo4** decreases significantly due to its *trans–cis* isomerization, while the **R5011** maintains a comparable right-handed contribution with the initial state. When the right-handedness becomes dominant, the overall right-handed helical superstructures are achieved. Source: Wang et al. 2017 [128]. Copyright 2017. Reprinted with permission from The Royal Society of Chemistry.

microscope. Upon UV irradiation at 365 nm, the period of spiraling structures gradually increased due to the decrease in the chirality of CLCs (Figure 3.15b–d), followed by the appearance of a bipolar nematic configuration with point defects at the two poles of microdroplets (Figure 3.15e). Continued UV irradiation led to the formation of right-handed double spiralized patterns with decreasing periods (Figure 3.15f–j). The concentric ring pattern with line defects across the droplets (Figure 3.15k) and diametrical configuration (Figure 3.15l) with ring defects were subsequently observed until the photostationary state. The simultaneous observation of diverse characteristic patterns in this single dynamic system is unprecedented although these topological configurations have been theoretically predicted and experimentally investigated [129–132]. It is worth noting that phototunable RGB photonic cross-communication patterns were observed in the handedness invertible CLCs with high chirality. A multicolored complex pattern with a bright red central spot surrounded by multiple radial blue lines was observed at the initial state due to the photonic cross-communications among different cholesteric microdroplets. UV irradiations at 365 nm led to the elongation of the helical pitch of CLCs and subsequent redshift of central and peripheral reflections of microdroplets. Continued UV exposure enabled the vivid observation of handedness inversion in cholesteric microdroplets accompanied by the disappearance of cross-communicated reflections, and the photonic cross-communication reflections with opposite chirality would be

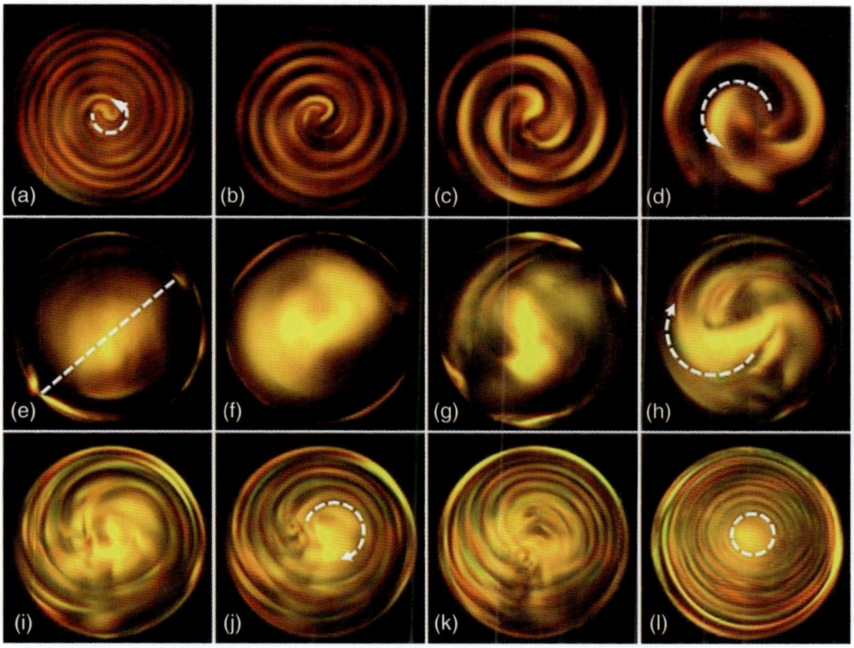

Figure 3.15 Real-time changes in microscopic optical textures of the cholesteric microdroplets with low chirality under the UV illumination (365 nm, 15 mW cm^2). Illumination time is 0 second (a), 5 seconds (b), 15 seconds (c), 30 seconds (d), 45 seconds (e), 60 seconds (f), 70 seconds (g), 85 seconds (h), 90 seconds (i), 100 seconds (j), 120 seconds (k), and 300 seconds (l), respectively. Notes: counterclockwise (a, d) and clockwise (h j) arrows indicate the opposite spirals; dashed line (e) connects two point defects at the poles of the droplets and dashed ring (l) indicates a ring defect. Source: Wang et al. 2017 [128]. Copyright 2017. Reprinted with permission from The Royal Society of Chemistry.

observed if keeping the sample at the photostationary state for around five minutes. These light-driven dynamic processes were reversible and repeated many times without noticeable degradation. The research is expected to provide a new platform for fundamental studies of micro/nanophotonics with geometric confinements, which hold potential applications in the photonic and optoelectronic devices.

Microfluidic techniques have also been demonstrated in fabricating the CLC microshells with spherical configurations that can act as 3D omnidirectional lasers due to the radial molecular arrangement within the cholesteric microshells. Uchida et al. first reported the fabrication of cholesteric microshells with water–oil–water double phase by using the microfluidic method [133], in which the oil phase is composed of dye-doped CLCs. By combining different laser dyes with CLC hosts, three types of laser modes, i.e. distributed feedback (DFB), distributed Bragg reflection (DBR) and whispering-gallery-mode (WGM) resonators, are successfully obtained on the basis of the CLC microshells. Subsequently, Chen et al. demonstrated the fabrication and lasing properties of photochromic monodisperse cholesteric liquid crystalline microshells through capillary-based microfluidic techniques (Figure 3.16) [134], where the photoresponsive CLCs

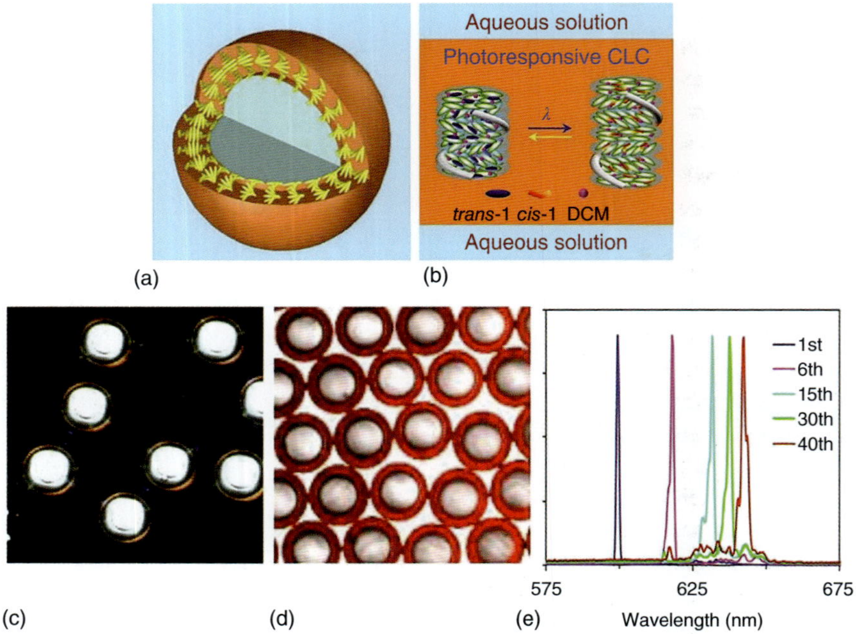

Figure 3.16 Schematic illustrations of photoresponsive monodisperse cholesteric microshell as a water–oil–water double emulsion (a) and the mechanism of phototunable lasing enabled by a light-driven chiral molecular switch (b). (c) Cross-polarized and (d) confocal images of the CLC microshells, and (e) phototuning of the laser emission. Source: Chen et al. 2014 [134]. Copyright 2014. Reprinted with permission from Wiley-VCH.

were developed by loading a chiral azobenzene photoswitch responsive to visible light into a nematic LC host. When the cholesteric microshells loaded with DCM (4-dicyanomethylene-2-methyl-6-p-dimethylamino-styrl-4H-pyran) laser dye were pumped under an external pulsed laser, laser emissions in all optical directions were achieved, where the emission wavelength was found to be able to self-tune because of the photoisomerization of the azobenzene chiral switch.

3.5 Summary and Perspective

Liquid crystalline materials represent a fascinating state of matter that combines crystalline order and fluidic mobility on a molecular and supramolecular level. Such unique combination results in the LC being typically “soft” and responding easily to external stimuli. In this chapter, we provided a glimpse of the research advances in the development of light-driven self-organized liquid crystalline nanostructures enabled by chiral molecular switches or motors. Toward the fabrication of photoactive liquid crystalline nanostructures, a variety of new ideas and strategies have been developed for the design and synthesis of diverse chiral molecular photoswitches with exceptionally high HTPs and handedness inversion capabilities. The high HTP values have enabled the photoresponsive chiral LCs with superior properties by the addition of small quantities of the chiral photoswitches, and the wide tunability of the HTP furnishes reflection colors encompassing the whole visible spectrum and beyond in a reversible manner. Photomodulation of chiral LCs has been achieved by UV-, visible-, and NIR-light irradiation. Through rational molecular design, some chiral photoswitches could confer helix inversion to the cholesteric systems for photonic applications involving CPL. Photoresponsive chiral LCs can be considered as a model system in soft matter research for induction, amplification, and inversion of chirality with dynamic and multifunctional qualities. Photoactive LCs with adaptive PBGs could find many new applications such as tunable mirrorless lasing, optical polarizers, and smart switchable reflective windows to control solar light.

It is anticipated that LC-based photonic crystals could open a new area of research about soft matter photonics. Owing to the unique self-organized properties and high stimuli sensitivity, these liquid crystalline photonic nanostructures can efficiently and selectively transmit and amplify local molecular information in a controllable manner, which undoubtedly offers an attractive route for versatile advanced photonic applications. For example, CLCs can be considered as 1D photonic crystals, while cubic superstructures of BP LCs and spherically symmetric superstructures of LC microdroplets are fascinating for tunable "soft" 3D photonic crystals. These liquid crystalline nanostructures could also function as powerful soft templates to direct the dynamic self-assembly of nanomaterials into various well-defined functional architectures via different noncovalent molecular interactions over multiple length scales. Although significant progress has been made in recent years, looking forward, the marriage of LC science with the emerging capabilities in nanoscience and nanotechnology could provide significant insights for developing programmable and reconfigurable advanced functional materials with myriad possibilities. Taking lessons from nature, researchers could develop novel soft photonic superstructures with extraordinary complexity, where the discrete building blocks can be self-assembled through complex and multiple interactions of liquid crystalline matter. Future investigations in such a promising field with multidisciplinary cooperation of scientists and engineers from different backgrounds would not only broaden our knowledge of fundamental soft matter science, but also promote their diverse technological applications in areas of dynamic photonics, optoelectronics, and beyond [135].

Acknowledgments

The preparation of this chapter benefited from the support to Quan Li by US Air Force Office of Scientific Research, US Air Force Research Laboratory, US National Aeronautics and Space Administration, US National Science Foundation, US Army Research Office, US Department of Defense Multidisciplinary University Research Initiative (DoD MURI), US Department of Energy, and Ohio Third Frontier. We thank all the current and former members of Li's lab as well as his collaborators, whose names are found in the references, for their significant contributions in this project.

References

1 Li, Q. (ed.) (2014). *Nanoscience with Liquid Crystals: From Self-Organized Nanostructures to Applications*. Heidelberg: Springer-Verlag.

2 Wang, L. and Li, Q. (2016). Stimuli-directing self-organized 3D liquid-crystalline nanostructures: from materials design to photonic applications. *Adv. Funct. Mater.* 26: 10.

3 Bisoyi, H.K. and Li, Q. (2014). Liquid crystals. In: *Kirk-Othmer Encyclopedia of Chemical Technology*. New York: Wiley.

4 Wang, L. and Li, Q. (2015). Stimuli-responsive self-organized liquid crystalline nanostructures: from 1D to 3D photonic crystals. In: *Organic and Hybrid Photonic Crystals* (ed. D. Comoretto). New York: Springer.

5 Li, Q. (ed.) (2012). *Liquid Crystals beyond Displays: Chemistry, Physics, and Applications*. Hoboken, NJ: Wiley.

6 Li, Q. (ed.) (2013). *Intelligent Stimuli Responsive Materials: From Well-Defined Nanostructures to Applications*. Hoboken, NJ: Wiley.

7 Li, Q. (ed.) (2011). *Self-Organized Organic Semiconductors: From Materials to Device Applications*. Hoboken, NJ: Wiley.

8 Li, Q. (ed.) (2018). *Functional Organic and Hybrid Nanostructured Materials: Fabrication, Properties, and Applications*. Weinheim: Wiley-VCH.

9 Bisoyi, H.K. and Li, Q. (2014). Light-directing chiral liquid crystal nanostructures: from 1D to 3D. *Acc. Chem. Res.* 47: 3184.

10 Wang, L. and Li, Q. (2018). Photochromism into nanosystems: towards lighting up the future nanoworld. *Chem. Soc. Rev.* 47: 1044.

11 Bisoyi, H.K. and Li, Q. (2016). Light-driven liquid crystalline materials: from photo-induced phase transitions and property modulations to applications. *Chem. Rev.* 116: 15089.

12 Wang, Y. and Li, Q. (2012). Light-driven chiral molecular switches or motors in liquid crystals. *Adv. Mater.* 24: 1926.

13 Bahr, C. and Kitzerow, H.S. (2001). *Chirality in Liquid Crystals*. Heidelberg: Springer-Verlag.

14 Wang, L., Bisoyi, H.K., Zheng, Z. et al. (2017). Stimuli-directed self-organized chiral superstructures for adaptive windows enabled by mesogen-functionalized graphene. *Mater. Today* 20: 230.

15 Wang, Y., Zheng, Z.G., Bisoyi, H.K. et al. (2016). Thermally reversible full color selective reflection in a self-organized helical superstructure enabled by a bent-core oligomesogen exhibiting a twist-bend nematic phase. *Mater. Horiz.* 3: 442.

16 Dierking, I. (2014). Chiral liquid crystals: structures, phases, effects. *Symmetry* 6: 444.

17 Wang, L., Huang, D., Lam, L., and Cheng, Z. (2017). Bowlics: history, advances and applications. *Liq. Cryst. Today* 26: 85.

18 Bandara, H.M.D. and Burdette, S.C. (2012). Photoisomerization in different classes of azobenzene. *Chem. Soc. Rev.* 41: 1809.

19 Irie, M. (2000). Diarylethenes for memories and switches. *Chem. Rev.* 100: 1685.

20 Berkovic, G., Krongauz, V., and Weiss, V. (2000). Spiropyrans and spirooxazines for memories and switches. *Chem. Rev.* 100: 1741.

21 Pieraccini, S., Masiero, S., Ferrarini, A., and Spada, G.P. (2011). Chirality transfer across length-scales in nematic liquid crystals: fundamentals and applications. *Chem. Soc. Rev.* 40: 258.
22 Gutierrez-Cuevas, K.G., Wang, L., Zheng, Z. et al. Frequency-driven self-organized helical superstructures loaded with mesogen-grafted silica nanoparticles. *Angew. Chem. Int. Ed.* 2016, 55: 13090; (2016). *Angew. Chem.* 128, 13284.
23 Zhang, L., Wang, L., Hiremath, U.S. et al. (2017). Dynamic orthogonal switching of a thermoresponsive self-organized helical superstructure. *Adv. Mater.* 29: 1700676.
24 Wang, L. (2016). Self-activating liquid crystal devices for smart laser protection. *Liq. Cryst.* 43: 2062.
25 Li, Y., Wang, M., Urbas, A., and Li, Q. (2013). Photoswitchable but thermally stable axially chiral dithienylperfluorocyclopentene dopant with high helical twisting power. *J. Mater. Chem. C* 1: 3889.
26 Yu, M., Wang, L., Nemati, H. et al. (2017). Effects of polymer network on electrically induced reflection band broadening of cholesteric liquid crystals. *J. Polym. Sci., Part B: Polym. Phys.* 55: 835.
27 Zhang, L., Wang, M., Wang, L. et al. (2016). Polymeric infrared reflective thin films with ultra-broad bandwidth. *Liq. Cryst.* 43: 750.
28 Wang, L., Gutierrez-Cuevas, K.G., and Li, Q. (2017). Photochromic chiral liquid crystal sensors. In: *Liquid Crystalline Sensors*, Chapter 2 (ed. A.P.H.J. Schenning, G. Crawford and D.J. Broer). CRC Press, Taylor and Francis Group.
29 Bisoyi, H.K. and Li, Q. (2016). Light-directed dynamic chirality inversion in functional self-organized helical superstructures. *Angew. Chem. Int. Ed.* 55: 2994.
30 Sackmann, E. (1972). Photochemically induced reversible color changes in cholesteric liquid crystals. *J. Am. Chem. Soc.* 93: 7088.
31 Hrozhyk, U.A., Serak, S.V., Tabiryan, N.V., and Bunning, T.J. (2007). Optical tuning of the reflection of cholesterics doped with azobenzene liquid crystals. *Adv. Funct. Mater.* 17: 1735.
32 Akiyama, H., Mallia, V.A., and Tamaoki, N. (2006). Synthesis, liquid-crystalline properties, and photo-optical studies of photoresponsive oligomeric mesogens as dopants in a chiral glassy liquid crystal. *Adv. Funct. Mater.* 16: 477.
33 Mathews, M., Zola, R.S., Yang, D.-K., and Li, Q. (2011). Thermally, photochemically and electrically switchable reflection colors from self-organized chiral bent-core liquid crystals. *J. Mater. Chem.* 21: 2098.
34 Yoshioka, T., Ogata, T., Nonaka, T. et al. (2005). Reversible-photon-mode full-color display by means of photochemical modulation of a helically cholesteric structure. *Adv. Mater.* 17: 1226.
35 Ruslim, C. and Ichimura, K. (2000). Conformational effect on macroscopic chirality modification of cholesteric mesophases by photochromic azobenzene dopants. *J. Phys. Chem. B* 104: 6529.
36 Kurihara, S., Nomiyama, S., and Nonaka, T. (2000). Photochemical switching between a compensated nematic phase and a twisted nematic phase by photoisomerization of chiral azobenzene molecules. *Chem. Mater.* 12: 9.
37 Ruslim, C. and Ichimura, K. (2002). Photoswitching in chiral nematic liquid crystals: interaction-selective helical twist inversion by a single chiral dopant. *J. Mater. Chem.* 12: 3377.

38 Tamaoki, N., Song, S., Moriyama, M., and Matsuda, H. (2000). Rewritable full-color recording in a photon mode. *Adv. Mater.* 12: 94.
39 Pieraccini, S., Masiero, S., Spada, G.P., and Gottarelli, G. (2003). A new axially-chiral photochemical switch. *Chem. Commun.* 598.
40 Pieraccini, S., Gottarelli, G., Labruto, R. et al. (2004). The control of the cholesteric pitch by some azo photochemical chiral switches. *Chem. Eur. J.* 10: 5632.
41 Li, Q., Green, L., Venkataraman, N. et al. (2007). Reversible photoswitchable axially chiral dopants with high helical twisting power. *J. Am. Chem. Soc.* 129: 12908.
42 Li, Q., Li, Y., Ma, J. et al. (2011). Directing dynamic control of red, green, and blue reflection enabled by a light-driven self-organized helical superstructure. *Adv. Mater.* 23: 5069.
43 Green, L., Li, Y., White, T.J. et al. (2009). Light-driven molecular switches with tetrahedral and axial chirality. *Org. Biomol. Chem.* 7: 3930.
44 White, T.J., Bricker, R.L., Natarajan, L.V. et al. (2009). Phototunable azobenzene cholesteric liquid crystals with 2000 nm range. *Adv. Funct. Mater.* 19: 3484.
45 Chen, X., Wang, L., Chen, Y. et al. (2013). Broadband reflection of polymer-stabilized chiral nematic liquid crystals induced by a chiral azobenzene compound. *Chem. Commun.* 50: 691.
46 Chen, G., Wang, L., Wang, Q. et al. (2014). Photoinduced hyper-reflective laminated liquid crystal film with simultaneous multicolor reflection. *ACS Appl. Mater. Interfaces* 6: 1380.
47 Montbach, E., Venkataraman, N., Doane, J.W. et al. (2008). Novel optically addressable photochiral displays. *Dig. Tech. Pap. - Soc. Inf. Disp. Int. Symp.* 39: 919.
48 Venkataraman, N., Magyar, G., Lightfoot, M. et al. (2009). Thin flexible photosensitive cholesteric displays. *J. Soc. Inf. Disp.* 17: 869.
49 Ma, J., Li, Y., White, T. et al. (2010). Light-driven nanoscale chiral molecular switch: reversible dynamic full range color phototuning. *Chem. Commun.* 46: 3463.
50 Jau, H.C., Li, Y., Li, C.C. et al. (2015). Light-driven wide-range nonmechanical beam steering and spectrum scanning based on a self-organized liquid crystal grating enabled by a chiral molecular switch. *Adv. Optical Mater.* 3: 166.
51 Wang, H., Bisoyi, H.K., Wang, L. et al. (2018). Photochemically and thermally driven full-color reflection in a self-organized helical superstructure enabled by a halogen-bonded chiral molecular switch. *Angew. Chem. Int. Ed.* 57: 1627.
52 Wang, Y., Urbas, A., and Li, Q. (2012). Reversible visible-light tuning of self-organized helical superstructures enabled by unprecedented light-driven axially chiral molecular switches. *J. Am. Chem. Soc.* 134: 3342.
53 Wang, L., Dong, H., Li, Y. et al. (2014). Reversible near-infrared light directed reflection in a self-organized helical superstructure loaded with upconversion nanoparticles. *J. Am. Chem. Soc.* 136: 4480.
54 Haase, M. and Schäfer, H. (2011). Upconverting nanoparticles. *Angew. Chem. Int. Ed.* 50: 5808.
55 Dong, H., Du, S.R., Zheng, X.Y. et al. (2015). Lanthanide nanoparticles: from design toward bioimaging and therapy. *Chem. Rev.* 115: 10725.

56 Qin, L., Gu, W., Wei, J., and Yu, Y. (2017). Piecewise phototuning of self-organized helical superstructures. *Adv. Mater.* 30: 1704941.
57 Gvozdovskyy, I., Yaroshchuk, O., Serbina, M., and Yamaguchi, R. (2012). Photoinduced helical inversion in cholesteric liquid crystal cells with homeotropic anchoring. *Opt. Express* 20: 3499.
58 Lee, Y.H., Wang, L., Yang, H., and Wu, S.T. (2015). Photo-induced handedness inversion with opposite-handed cholesteric liquid crystal. *Opt. Express* 23: 22658.
59 Kurihara, S., Nomiyama, S., and Nonaka, T. (2001). Photochemical control of the macrostructure of cholesteric liquid crystals by means of photoisomerization of chiral azobenzene molecules. *Chem. Mater.* 13: 1992.
60 Ryabchun, A., Bobrovsky, A., Stumpe, J., and Shibaev, V. (2015). Rotatable diffraction gratings based on cholesteric liquid crystals with phototunable helix pitch. *Adv. Optical Mater.* 3: 1273.
61 Mathews, M., Zola, R.S., Hurley, S. et al. (2010). Light-driven reversible handedness inversion in self-organized helical superstructures. *J. Am. Chem. Soc.* 132: 18361.
62 Li, Y., Wang, M., White, T.J. et al. (2013). Azoarenes with opposite chiral configurations: light-driven reversible handedness inversion in self-organized helical superstructures. *Angew. Chem. Int. Ed.* 52: 8925.
63 Fu, D., Li, J., Wei, J., and Guo, J. (2015). Effects of terminal chain length in hydrogen-bonded chiral switches on phototunable behavior of chiral nematic liquid crystals: helicity inversion and phase transition. *Soft Matter* 11: 3034.
64 Denekamp, C. and Feringa, B.L. (1998). Optically active diarylethenes for multimode photoswitching between liquid-crystalline phases. *Adv. Mater.* 10: 1080.
65 Yamaguchi, T., Inagawa, T., Nakazumi, H. et al. (2000). Photoswitching of helical twisting power of a chiral diarylethene dopant: pitch change in a chiral nematic liquid crystal. *Chem. Mater.* 12: 869.
66 Li, Y., Urbas, A., and Li, Q. (2011). Synthesis and characterization of light-driven dithienylcyclopentene switches with axial chirality. *J. Org. Chem.* 76: 7148.
67 Li, Y., Urbas, A., and Li, Q. (2012). Reversible light-directed red, green, and blue reflection with thermal stability enabled by a self-organized helical superstructure. *J. Am. Chem. Soc.* 134: 9573.
68 Hayasaka, H., Miyashita, T., Nakayama, M. et al. (2012). Dynamic photoswitching of helical inversion in liquid crystals containing photoresponsive axially chiral dopants. *J. Am. Chem. Soc.* 134: 3758.
69 Li, Y., Xue, C., Wang, M. et al. (2013). Photodynamic chiral molecular switches with thermal stability: from reflection wavelength tuning to handedness inversion of self-organized helical superstructures. *Angew. Chem. Int. Ed.* 52: 13703.
70 Wang, L., Dong, H., Li, Y. et al. (2015). Luminescence-driven reversible handedness inversion of self-organized helical superstructures enabled by a novel near-infrared light nanotransducer. *Adv. Mater.* 27: 2065.
71 Wang, L., Gutierrez-Cuevas, K.G., Urbas, A., and Li, Q. (2016). Near-infrared light-directed handedness inversion in plasmonic nanorod-embedded helical superstructure. *Adv. Optical Mater.* 4: 247.

72 Zheng, Z., Li, Y., Bisoyi, H.K. et al. (2016). Three-dimensional control of the helical axis of a chiral nematic liquid crystal by light. *Nature* 531: 352.

73 Zheng, Z., Zola, R.S., Bisoyi, H.K. et al. (2017). Controllable dynamic zigzag pattern formation in a soft helical superstructure. *Adv. Mater.* 29: 1701903.

74 Zola, R.S. and Li, Q. (2018). Stimuli-directed helical axis switching in chiral liquid crystalline nanostructure. In: *Functional Organic and Hybrid Nanostructured Materials: Fabrication, Properties, and Applications*, Chapter 8 (ed. Q. Li). Weinheim: Wiley-VCH.

75 Bisoyi, H.K., Bunning, T.J., and Li, Q. (2018). Stimuli-driven control of the helical axis of self-organized soft helical superstructures. *Adv. Mater.* 30: 1706512.

76 Wang, H., Wang, L., Xie, H. et al. (2015). Electrically controllable microstructures and dynamic light scattering properties of liquid crystals with negative dielectric anisotropy. *RSC Adv.* 5: 33489.

77 Wang, H., Wang, L., Chen, M. et al. (2015). Bistable polymer-dispersed cholesteric liquid crystal thin film enabled by a stepwise polymerization. *RSC Adv.* 5: 58959.

78 Xie, H., Wang, L., Wang, H. et al. (2015). Electrically tunable properties of wideband-absorptive and reflection-selective films based on multi-dichroic dye-doped cholesteric liquid crystals. *Liq. Cryst.* 42: 1698.

79 He, W.L., Wang, L., Yang, Z. et al. (2011). Synthesis and optical behaviour of hydrogen-bonded liquid crystals based on a chiral pyridine derivative. *Liq. Cryst.* 38: 1217.

80 Hattori, H. and Uryu, T. (2001). Photochromic chiral liquid crystalline systems containing spiro-oxazine with a chiral substituent II. Photoinduced behaviour. *Liq. Cryst.* 28: 1099.

81 Jin, L., Li, Y., Ma, J., and Li, Q. (2010). Synthesis of novel thermally reversible photochromic axially chiral spirooxazines. *Org. Lett.* 12: 3552.

82 Feringa, B.L. (2017). The art of building small: from molecular switches to motors (nobel lecture). *Angew. Chem. Int. Ed.* 56: 11060.

83 Browne, W.R. and Feringa, B.L. (2006). Making molecular machines work. *Nat. Nanotechnol.* 1: 25.

84 Feringa, B.L., Huck, N.P.M., and van Doren, H.A. (1995). Chiroptical switching between liquid crystalline phases. *J. Am. Chem. Soc.* 117: 9929.

85 Huck, N.P.M., Jager, W.F., de Lange, B., and Feringa, B.L. (1996). Dynamic control and amplification of molecular chirality by circular polarized light. *Science* 273: 1686.

86 Sun, J., Lan, R., Gao, Y. et al. (2017). Stimuli-directed dynamic reconfiguration in self-organized helical superstructures enabled by chemical kinetics of chiral molecular motors. *Adv. Sci.* 5: 1700613.

87 Sun, J., Yu, L., Wang, L. et al. (2017). Optical intensity-driven reversible photonic bandgaps in self-organized helical superstructures with handedness inversion. *J. Mater. Chem. C* 5: 3678.

88 van Delden, R.A., Koumura, N., Harada, N., and Feringa, B.L. (2002). Supramolecular chemistry and self-assembly special feature: unidirectional rotary motion in a liquid crystalline environment: color tuning by a molecular motor. *Proc. Natl. Acad. Sci. U.S.A.* 99: 4945.

89 Eelkema, R., Pollard, M.M., Vicario, J. et al. (2006). Molecular machines: nanomotor rotates microscale objects. *Nature* 440: 163.
90 van Delden, R.A., van Gelder, M.B., Huck, N.P.M., and Feringa, B.L. (2003). Controlling the color of cholesteric liquid-crystalline films by photoirradiation of a chiroptical molecular switch used as dopant. *Adv. Funct. Mater.* 13: 319.
91 Eelkema, R., Pollard, M.M., Katsonis, N. et al. (2006). Rotational reorganization of doped cholesteric liquid crystalline films. *J. Am. Chem. Soc.* 128: 14397.
92 Pijper, D., Jongejan, M.G.M., Meetsma, A., and Feringa, B.L. (2008). Light-controlled supramolecular helicity of a liquid crystalline phase using a helical polymer functionalized with a single chiroptical molecular switch. *J. Am. Chem. Soc.* 130: 4541.
93 Aboff, S.J., Iamsaard, S., Bosco, A. et al. (2013). Time-programmed helix inversion in phototunable liquid crystals. *Chem. Commun.* 49: 4256.
94 White, T.J., Cazella, S.A., Freer, A.S. et al. (2011). Widely tunable, photoinvertible cholesteric liquid crystals. *Adv. Mater.* 23: 1389.
95 Kikuchi, H. (2007). Liquid crystalline blue phases. In: *Liquid Crystalline Functional Assemblies and their Supramolecular Structures*. Heidelberg: Springer-Verlag.
96 Wright, D.C. and Mermin, N.D. (1989). Crystalline liquids: the blue phases. *Rev. Mod. Phys.* 61: 385.
97 Lin, T.-H., Chen, C.-W., and Li, Q. (2015). Self-organized 3D photonic superstructures: blue phase liquid crystal. In: *Anisotropic Nanomaterials: Preparation, Properties, and Applications*, Chapter 9 (ed. Q. Li). Heidelberg: Springer-Verlag.
98 Kikuchi, H., Yokota, M., Hisakado, Y. et al. (2002). Polymer-stabilized liquid crystal blue phases. *Nat. Mater.* 1: 64.
99 Coles, H.J. and Pivnenko, M.N. (2005). Liquid crystal 'blue phases' with a wide temperature range. *Nature* 436: 997.
100 Wang, L., He, W., Xiao, X. et al. (2012). Hysteresis-free blue phase liquid-crystal-stabilized by ZnS nanoparticles. *Small* 8: 2189.
101 Wang, L., He, W., Xiao, X. et al. (2012). Wide blue phase range and electro-optical performances of liquid crystalline composites doped with thiophene-based mesogens. *J. Mater. Chem.* 22: 2383.
102 He, W., Pan, G., Yang, Z. et al. (2009). Wide blue phase range in a hydrogen-bonded self-assembled complex of chiral fluoro-substituted benzoic acid and pyridine derivative. *Adv. Mater.* 21: 2050.
103 Wang, L., He, W., Wang, M. et al. (2013). Effects of symmetrically 2,5-disubstituted 1,3,4-oxadiazoles on the temperature range of liquid crystalline blue phases: a systematic study. *Liq. Cryst.* 40: 354.
104 Li, B., He, W., Wang, L. et al. (2013). Effect of lateral fluoro substituents of rodlike tolane cyano mesogens on blue phase temperature ranges. *Soft Matter* 9: 1172.
105 Wang, L., He, W., Xiao, X. et al. (2012). Low voltage and hysteresis-free blue phase liquid crystal dispersed by ferroelectric nanoparticles. *J Mater. Chem.* 22: 19629.
106 Wang, L., He, W., Wang, Q. et al. (2013). Polymer-stabilized nanoparticle-enriched blue phase liquid crystals. *J. Mater. Chem. C* 1: 6526.

107 He, W.L., Wang, L., Cui, X.P. et al. (2012). Wide temperature range blue phase liquid crystalline materials. *Prog. Chem.* 24: 182.
108 Wang, L., Yu, L., Xiao, X. et al. (2012). Effects of 1,3,4-oxadiazoles with different rigid cores on the thermal and electro-optical performances of liquid crystalline blue phases. *Liq. Cryst.* 39: 629.
109 Castles, F., Day, F.V., Morris, S.M. et al. (2012). Blue-phase templated fabrication of three-dimensional nanostructures for photonic applications. *Nat. Mater.* 11: 599.
110 Cao, W., Muñoz, A., Palffy-Muhoray, P., and Taheri, B. (2002). Lasing in a three-dimensional photonic crystal of the liquid crystal blue phase II. *Nat. Mater.* 1: 111.
111 Coles, H. and Morris, S. (2010). Liquid-crystal lasers. *Nat. Photonics* 4: 676.
112 Wang, M., Zou, C., Sun, J. et al. (2017). Asymmetric tunable photonic bandgaps in self-organized 3D nanostructure of polymer-stabilized blue phase I modulated by voltage polarity. *Adv. Funct. Mater.* 27: 1702261.
113 Chen, X., Wang, L., Li, C. et al. (2013). Light-controllable reflection wavelength of blue phase liquid crystals doped with azobenzene-dimers. *Chem. Commun.* 49: 10097.
114 Chanishvili, A., Chilaya, G., Petriashvili, G., and Collings, P.J. (2005). *Trans–cis* isomerization and the blue phases. *Phys. Rev. E* 71: 051705.
115 Liu, H.Y., Wang, C.T., Hsu, C.Y. et al. (2010). Optically tuneable blue phase photonic band gaps. *Appl. Phys. Lett.* 96: 121103.
116 Lin, T.H., Li, Y., Wang, C.T. et al. (2013). Red, green and blue reflections enabled in an optically tunable self-organized 3D cubic nanostructured thin film. *Adv. Mater.* 25: 5050.
117 Wang, L., Gutierrez-Cuevas, K.G., Bisoyi, H.K. et al. (2015). NIR light-directing self-organized 3D photonic superstructures loaded with anisotropic plasmonic hybrid nanorods. *Chem. Comm.* 51: 15039.
118 Chen, H., Shao, L., Li, Q., and Wang, J. (2013). Gold nanorods and their plasmonic properties. *Chem. Soc. Rev.* 42: 2679.
119 Xue, C., Xiang, J., Nemati, H. et al. (2015). Light-driven reversible alignment switching of liquid crystals enabled by azo thiol grafted gold nanoparticles. *ChemPhysChem* 16: 1852.
120 Gutierrez-Cuevas, K.G., Wang, L., Xue, C. et al. (2015). Near infrared light-driven liquid crystal phase transition enabled by hydrophobic mesogen grafted plasmonic gold nanorods. *Chem. Comm.* 51: 9845.
121 Zheng, Z.G., Yuan, C.L., Hu, W. et al. (2017). Light-patterned crystallographic direction of a self-organized 3D soft photonic crystal. *Adv. Mater.* 29: 1703165.
122 Zhou, K., Bisoyi, H., Jin, J. et al. (2018). Light-driven reversible transformation between self-organized simple cubic lattice and helical superstructure enabled by a molecular switch functionalized nanocage. *Adv. Mater.* 30: 1800237.
123 Cipparrone, G., Mazzulla, A., Pane, A. et al. (2011). Chiral self-assembled solid microspheres: a novel multifunctional microphotonic device. *Adv. Mater.* 23: 5773.
124 Utada, A.S., Lorenceau, E., Link, D.R. et al. (2005). Monodisperse double emulsions generated from a microcapillary device. *Science* 308: 537.

125 Fan, J., Li, Y., Bisoyi, H.K. et al. (2015). Light-directing omnidirectional circularly polarized reflection from liquid-crystal droplets. *Angew. Chem. Int. Ed.* 127: 2188.

126 Noh, J.H., Liang, H.L., Drevensek-Olenik, I., and Lagerwall, J.P.F. (2014). Tuneable multicoloured patterns from photonic cross-communication between cholesteric liquid crystal droplets. *J. Mater. Chem. C* 2: 806.

127 Lee, S.S., Kim, B., Kim, S.K. et al. (2015). Robust microfluidic encapsulation of cholesteric liquid crystals toward photonic ink capsules. *Adv. Mater.* 27: 627.

128 Wang, L., Chen, D., Gutierrez-Cuevas, K.G. et al. (2017). Optically reconfigurable chiral microspheres of self-organized helical superstructures with handedness inversion. *Mater. Horiz.* 4: 1190.

129 Orlova, T., Abhooff, S.J., Yamaguchi, T. et al. (2015). Creation and manipulation of topological states in chiral nematic microspheres. *Nat. Commun.* 6: 7603.

130 Zhou, Y., Bukusoglu, E., Martınez-Gonzalez, J.A. et al. (2016). Structural transitions in cholesteric liquid crystal droplets. *ACS Nano* 10: 6484.

131 Sec, D., Porenta, T., Ravnik, M., and Zumer, S. (2012). Geometrical frustration of chiral ordering in cholesteric droplets. *Soft Matter* 8: 11982.

132 Xu, F. and Crooker, P.P. (1997). Chiral nematic droplets with parallel surface anchoring. *Phys. Rev. E* 56: 6853.

133 Uchida, Y., Takanishi, Y., and Yamamoto, J. (2013). Controlled fabrication and photonic structure of cholesteric liquid crystalline shells. *Adv. Mater.* 25: 3234.

134 Chen, L., Li, Y., Fan, J. et al. (2014). Photoresponsive monodisperse cholesteric liquid crystalline microshells for tunable omnidirectional lasing enabled by a visible light-driven chiral molecular switch. *Adv. Optical Mater.* 2: 845.

135 Wang, L. and Li, Q. (2018). Photochromic organic & hybrid self-organized nanostructures: from molecular design to applications. In: *Functional Organic and Hybrid Nanostructured Materials: Fabrication, Properties, and Applications*, Chapter 3 (ed. Q. Li). Weinheim: Wiley-VCH.

4

Photochemical Chirality Induction and Inversion in Soft Materials

Yuna Kim, Noushaba N. Mafy, and Nobuyuki Tamaoki

Hokkaido University, Research Institute for Electronic Science, Laboratory of Smart Molecules, N20, W10, Kita-Ku, Sapporo, 001-0020, Japan

4.1 Introduction

Chirality or "handedness" is a prerequisite of the biological world, where life exists in a specific chirality. In living organisms, almost all the sugars are exclusively right-handed (D-isomer), amino acids are left-handed (L-isomer), and DNA coils assemble into right-handed helices [1]. The reason behind this homochirality is still an unresolved issue. Inspired by the natural homochirality, the scientific community has been interested in exploring the mechanism of transmission of chirality from the molecular level of monomeric units to the macro- or supramolecular level, as well as the origin of chirality [2, 3]. The supramolecular chirality in nature from self-assembled achiral molecules and its specific handedness play a crucial role in the evolution of life and the development of supramolecular materials. While the chirality of molecules is of configurational and conformational origin, supramolecular level chirality arises from the organization of molecules, with formation of a chiral superstructure by means of long-range positional and orientational orders of molecules [4].

The use of external stimuli to generate and control the asymmetry at various scales has been an important issue to mimic the homochirality of nature and to create smart supramolecular systems. Different types of stimulus such as chiral solvents [5, 6] and chiral additives [7] have been utilized to obtain enantiomeric imbalance from the achiral (racemic) system. Among the possible external stimuli, light is a versatile external stimulus with a tunable circular polarization of right- and left-handedness in addition to light intensity and wavelength, which allow control of the mesomorphic reorganization. Soft materials exhibiting photoresponsive chiroptical functions provide genuine applications that include advanced displays, reflectors, information storages, chiral sensors with amplified responses, chiral recognition processes, and actuators [8, 9]. Moreover, they can offer photoresponsive chiral templates for enantioselective photochemical synthesis [10, 11], and circularly polarized light (CPL)-triggered chiral photopolymerization with visible light [12] or near-infrared (NIR) light source [13]. As a

Photoactive Functional Soft Materials: Preparation, Properties, and Applications, First Edition.
Edited by Quan Li.

photo-trigger/transmitter of chiral information from light to the soft materials, photochromic molecular switches (dopants) such as azobenzene [14–16], spirooxazine [17], fulgide [18, 19], naphthopyran [20], overcrowded alkene [21], and diarylethene [22] have proved their ability [23, 24]. Moreover, the elastic property of soft materials including liquid crystals, polymers, and gels has shown its capability: expression of chiral information to the entire system through amplification [25, 26].

In this chapter, we report on the recent progress harnessing chirality induction in achiral molecules by utilizing light as a chiral stimulus to achieve chiral assemblies based on photochromic switches in soft materials such as liquid crystals and polymeric organizations. In addition, we also discuss the modulation of their induced supramolecular chirality such as helix inversion between opposite handedness and helical twisting power (HTP) control through photochemical reactions. Our chapter would provide an opportunity to better understand molecular chirality control over a mesoscale chiral organization.

4.2 Chirality Induction from Achiral Soft Materials by CPL

Chirality induction is a process in which chiral bias results in a significantly enhanced chiral effect, and CPL is considered as a natural chiral source that can transfer the chiral information of light to the single molecule as well as the supramolecules (Scheme 4.1) [27, 28]. The photochromic molecule exhibiting generation

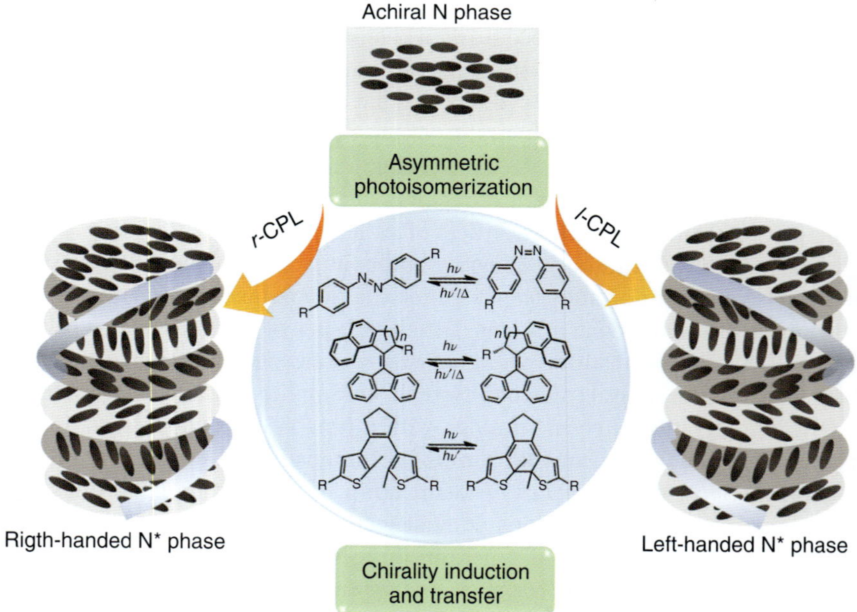

Scheme 4.1 Schematic illustration of the CPL induced chirality: transition from achiral nematic (N) to chiral nematic (N*) organization based on the photoisomerization of photoresponsive molecular switches.

of an enantiomeric excess and dynamic control of racemization has been investigated as a photo-trigger (Figure 4.1) [30]. In these photo-triggers, photochemical interconversion is possible directly between a pair of thermally stable enantiomers, or via non-chiral intermediates. Feringa and coworkers [31] introduced the CPL-induced photoresolution of sterically overcrowded alkene system **1**, but the process required a relatively long irradiation time to get the enantiomeric excess. Tamaoki and coworkers accomplished the fast CPL-induced chirality and molecular-level enantiomeric excess for different types of cyclic and noncyclic azobenzenes exhibiting reversible *trans* (*E*)–*cis* (*Z*) isomerization upon light irradiation (Figure 4.1). The bicyclic azobenzene dimer **2** has a complicated enantio-differentiating photoisomerization path with three isomers (*EE, EZ,* and *ZZ*) associated with planar chirality [32]. The monocyclic planar chiral azo-molecule **3** also exhibits an enantiomeric excess that is induced by CPL irradiation where the *r* (right-handed)- or *l* (left-handed)-CPL at 488 nm repeatedly induces the partial enrichment of the *S* or *R* enantiomers, respectively (Figure 4.1e,f) [29]. Noncyclic achiral azobenzene dimer **4** with two symmetrically arranged azobenzene units generates the chirality and enantiomeric excess at its *EZ* isomer state under CPL irradiation at a single wavelength [33].

Although the photochemical generation of molecular level asymmetry has been achieved as described above, its amplification to the supramolecular system is considered as a challenging issue. The induced chirality plays an important role in the supramolecular system where chirality is generated by the molecular alignment of a large number of achiral molecules, through the asymmetry information transfer process from a small number of chiral molecules.

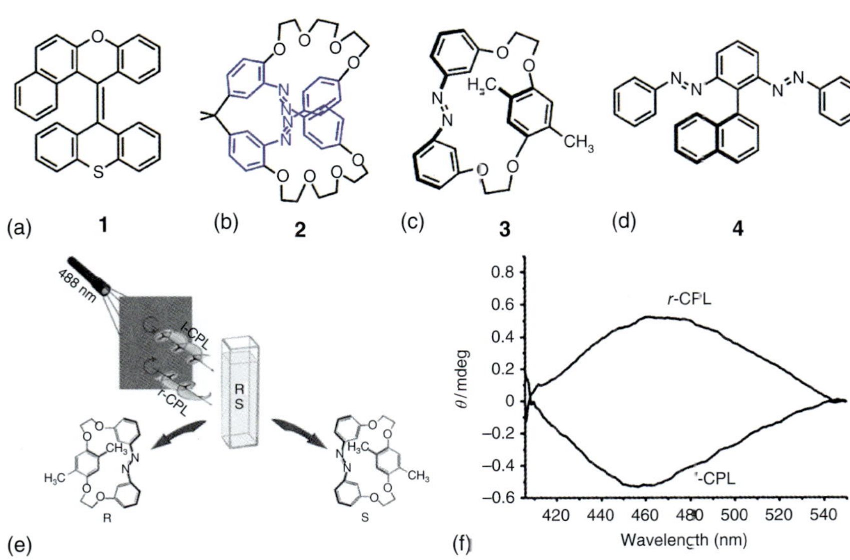

Figure 4.1 Photoresponsive molecules (**1**–**4**) exhibiting photochemical induction of molecular chirality (a-d). Schematic illustration of chirality induction of **3** by *r*- and *l*-CPL irradiation (e), and resultant mirror image CD spectra (f). Source: (e) and (f) Hassim et al. 2011 [29]. Copyright 2011. Reproduced with permission from Wiley-VCH.

Usually in the supramolecular system, asymmetry information transfer occurs via noncovalent weak interactions, such as van der Waals interaction, π–π stacking, hydrogen bonding, electrostatic interaction, and host–guest interaction [8]. In this section, we introduce CPL- or circularly polarized UV light (CPUL)-triggered chirality induction from achiral soft materials doped or substituted with photoswitches including low molecular weight (LMW) liquid crystals, self-assembled supramolecules, and achiral polymers. Facile control of chirality upon CPL irradiation – not only the chiral induction with specific handedness, but also the erasing and changing of the handedness direction – is demonstrated.

4.2.1 Achiral LMW Liquid Crystals

At the LMW supramolecular level, helical organization usually occurs in chiral smectic and chiral nematic (cholesteric, N*) liquid crystals. In these liquid crystals, the director rotates in a helical manner around an axis with a pitch. The pitch (*P*) is the distance along the helical axis when the molecules complete a full turn, and a positive handedness represents a right-handed helix, whereas a negative handedness represents a left-handed helix (Scheme 4.1). The helical pitch length is generally longer than a few hundreds of nanometers in the case of cholesteric liquid crystal (CLC). The periodic helical structures can selectively reflect the CPL of a certain wavelength matching its helical pitch length. The photoinduction of helical organization has been accomplished by irradiating CPL from achiral LMW liquid crystals (LCs).

In 1995, Schuster et al. reported on the first chiroptical trigger for a liquid crystal based on a series of axially chiral bicyclic ketones [28]. The racemic axially chiral bicyclic ketone exhibited a partial photoresolution (*ee* = 0.4%) by CPL irradiation although the enantiomeric enrichment hardly led to the efficient nematic to cholesteric phase transition. In 1996, Feringa et al. described a photoswitchable compound based on overcrowded alkene (**1**) capable of undergoing partial photoresolution within a nematic liquid crystal (NLC), a process that was able to induce the formation of a cholesteric phase [31]. Its two enantiomeric forms – *P* and *M*, due to its inherent helical shape – can interconvert by CPL irradiation at 313 nm (Figure 4.2). When photochemical resolution was undertaken in a nematic host, a cholesteric texture was observed from the racemic compound enriched with one of the enantiomers.

Choi et al. [34] utilized CPL as the irradiation source, which led to enantiomeric excess values in two chiral domains in the B_x phase of bent-core achiral molecules (**5**) as shown in Figure 4.3. The helical direction was arbitrarily regulated by irradiating with *l*- and *r*-handed CPL as observed from circular dichroism (CD) spectra of the B_X phase (Figure 4.3c) of **5** before and after CPL irradiation (Figure 4.3b,c).

Moreover, columnar liquid crystals are also known to be suitable candidates to build chiral supramolecular architectures based on helical organizations. Sierra and coworkers exploited unique columnar liquid crystals such that CPL can be utilized to induce a chiral phase [35]. The authors combined the two systems by combining columnar LCs and supramolecular macrocycles to achieve transferring of the chirality of CPL to the achiral systems. Previous results of their

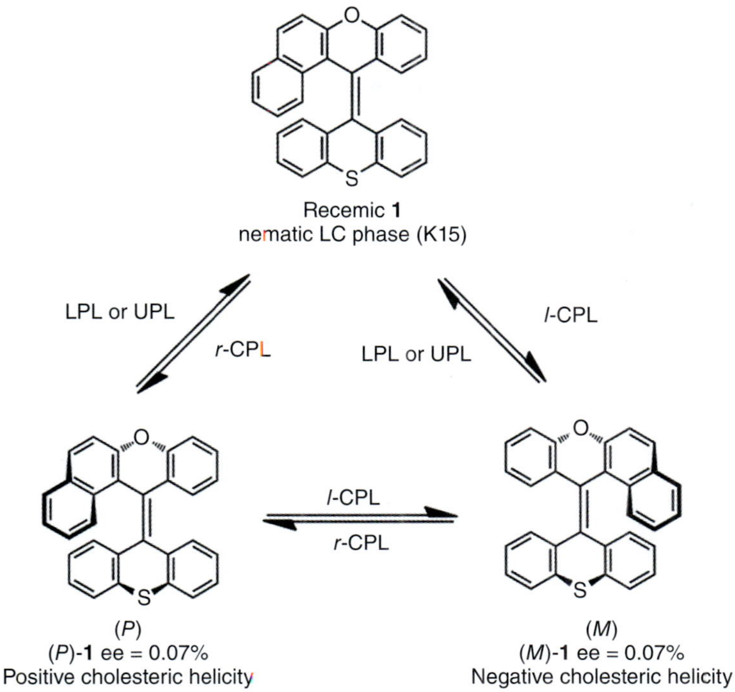

Figure 4.2 CPL-induced enantiomeric excess of overcrowded-alkene-based molecule **1** in a K15 liquid crystalline environment resulting in 3-stage switching. Source: Eelkema and Feringa 2006 [16]. Copyright 2006. Reproduced with permission from The Royal Society of Chemistry.

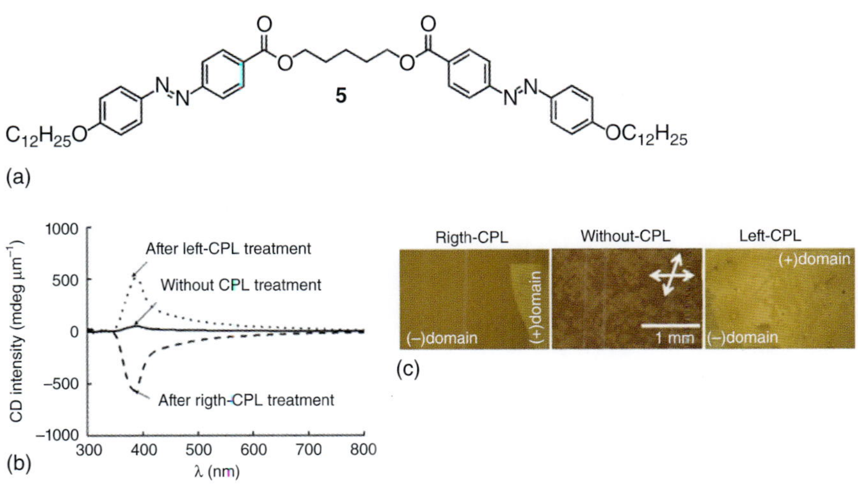

Figure 4.3 (a) Chemical structure of achiral bent-core dimer **5**. (b) CD spectra of the B_x phase of **5** before and after CPL irradiation. (c) Change of the LC texture before and after *r*- and *l*-CPL irradiation. Source: Choi et al. 2006 [34]. Copyright 2006. Reproduced with permission from Wiley-VCH.

research showed that non-mesogenic carboxylic acids endowed with a rigid V shape form H-bonded complexes with a melamine derivative (**6**). These complexes self-organize into helical columnar mesomorphic systems, the chirality of which is controlled by chiral building blocks and/or CPL (Figure 4.4).

Furthermore, by controlling the orthogonal action of H-bonding interaction such as melamine–melamine and melamine–acid (Figure 4.5a, T : A [1 : 1]

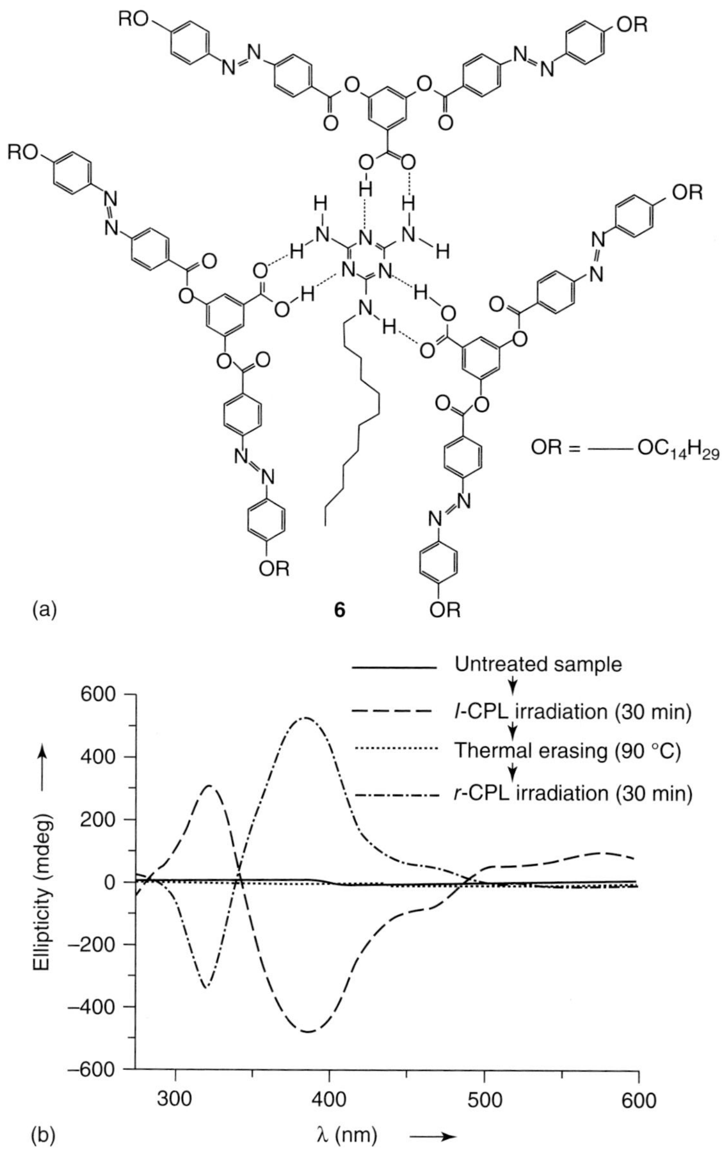

Figure 4.4 (a) Chemical structure of the hydrogen-bonded propeller-like tetrameric complex **6**. (b) CD spectra corresponding to *l*-CPL irradiation, thermal erasing, and *r*-CPL irradiation cycle for a cast film of complex **6**. Source: Vera et al. 2007 [35]. Copyright 2007. Reproduced with permission from Wiley-VCH.

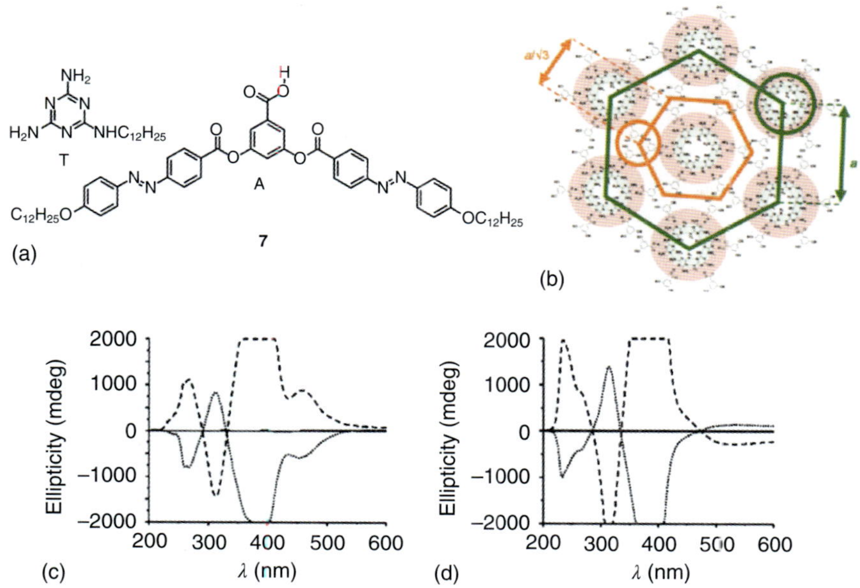

Figure 4.5 (a) Chemical structure of the melanine derivative **T** and V-shaped acid **A**, used to form the supramolecular complexes. (b) Hierarchically self-assembled mesophase: hexagonal two-dimensional packing of the **7** (**T**–**A** [1 : 1]) complexes, and each node of the lattice is formed by a rosette. CD spectra of cast films of **7** complexes (c) and **7**-LiTf (lithium triflate) complexes (d) before (solid line) and after irradiation with 488 nm Ar^+ laser of *l*-CPL (dashed line) and *r*-CPL (dotted line). Source: Reproduced from Ref. [36] with permission. Copyright 2010 Wiley-VCH.

complex), their hierarchical self-assembly was integrated into rosette-like two-dimensional hexagonal columnar (Colh) mesophases (Figure 4.5b) [36]. The modulation of the chirality of Colh mesophases was achieved with CPL. The rosette-type association from 7 exhibits intense CD bands (Figure 4.5c) upon irradiation with CPL, indicating the induction of chirality into the supramolecular systems. On irradiation with light of the opposite handedness, the CD shows the opposite sign, which indicates that the supramolecular chirality of the mesophase can be inverted by the external chiral radiation. In particular, these columns could accommodate the ions of a salt such as lithium triflate (7-LiTf complex), exhibiting comparable CPL-induced chirality control of the mesophase as shown in Figure 4.5d.

A compensated nematic phase is an achiral nematic phase arising from a racemic mixture of equal amounts of chiral dopants with opposite twist sense in host NLC. It is generally observed during the switching between two chiral nematic phases of opposite handedness. Kurihara et al. [37] reported on the compensated nematic (racemic) phase at the nonirradiated initial state. By harnessing the mixture of chiral dopant (**8**) and the chiral azobenzene derivative (**9**) in nematic E44 (20 wt%), a reversible switching between the racemic and the cholesteric phases was obtained by *trans*–*cis* photoisomerization through UV and visible light irradiation, respectively (Figure 4.6). Li et al. [38] demonstrated that axially chiral dithienylperfluorocyclopentene (**10**) is able to induce a nematic-like phase at a

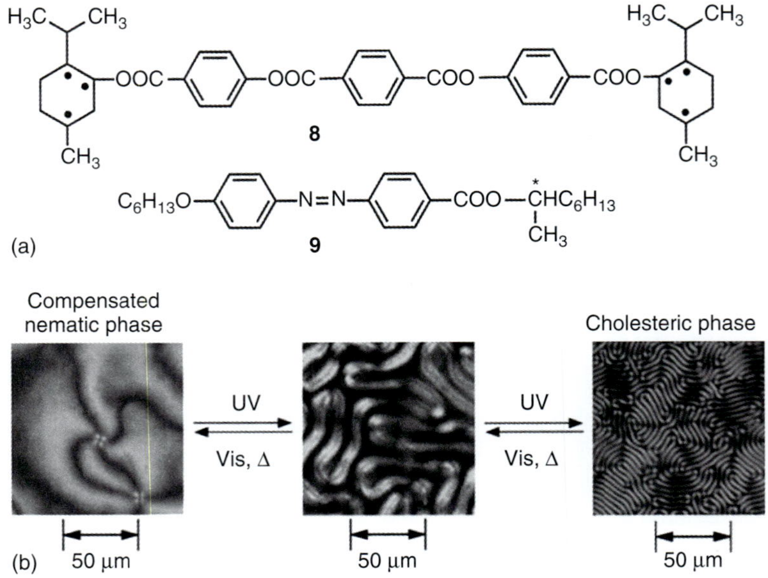

Figure 4.6 (a) Molecular structures of chiral dopants (**8** and **9**). (b) Reversible photochemical phase transition between a compensated nematic phase and a cholesteric phase (doping 20 wt% in E44) on UV and visible light irradiation. Source: Reproduced from Ref. [37] with permission. Copyright 2007 Wiley-VCH.

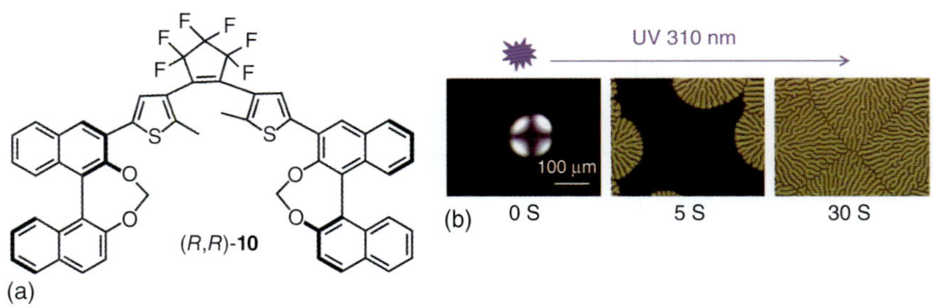

Figure 4.7 (a) Structure of dithienylperfuorocyclopentene (*R,R*)-**10** compound with two bridged binaphthyl units. (b) Cross-polarized optical textures of **10** in 5CB (0.1 mol%) upon UV irradiation from 0 (conoscopic observation) to 30 seconds. Source: Li et al. 2013 [38]. Copyright 2013. Reproduced with permission from The Royal Society of Chemistry.

very low doping concentration (0.1 mol% in 5CB), and it exhibits reversible isothermal phase transition between nematic and cholesteric phases upon light irradiation (Figure 4.7).

4.2.2 Achiral Polymers

In general, CPL or CPUL irradiation is introduced as a light stimulus for the induction and modulation of the chiral helical structures from achiral polymers bearing photochromophores (Figure 4.8). The dynamic control of polymer chiral functions can exploit not only the optical properties but also the dynamic micro

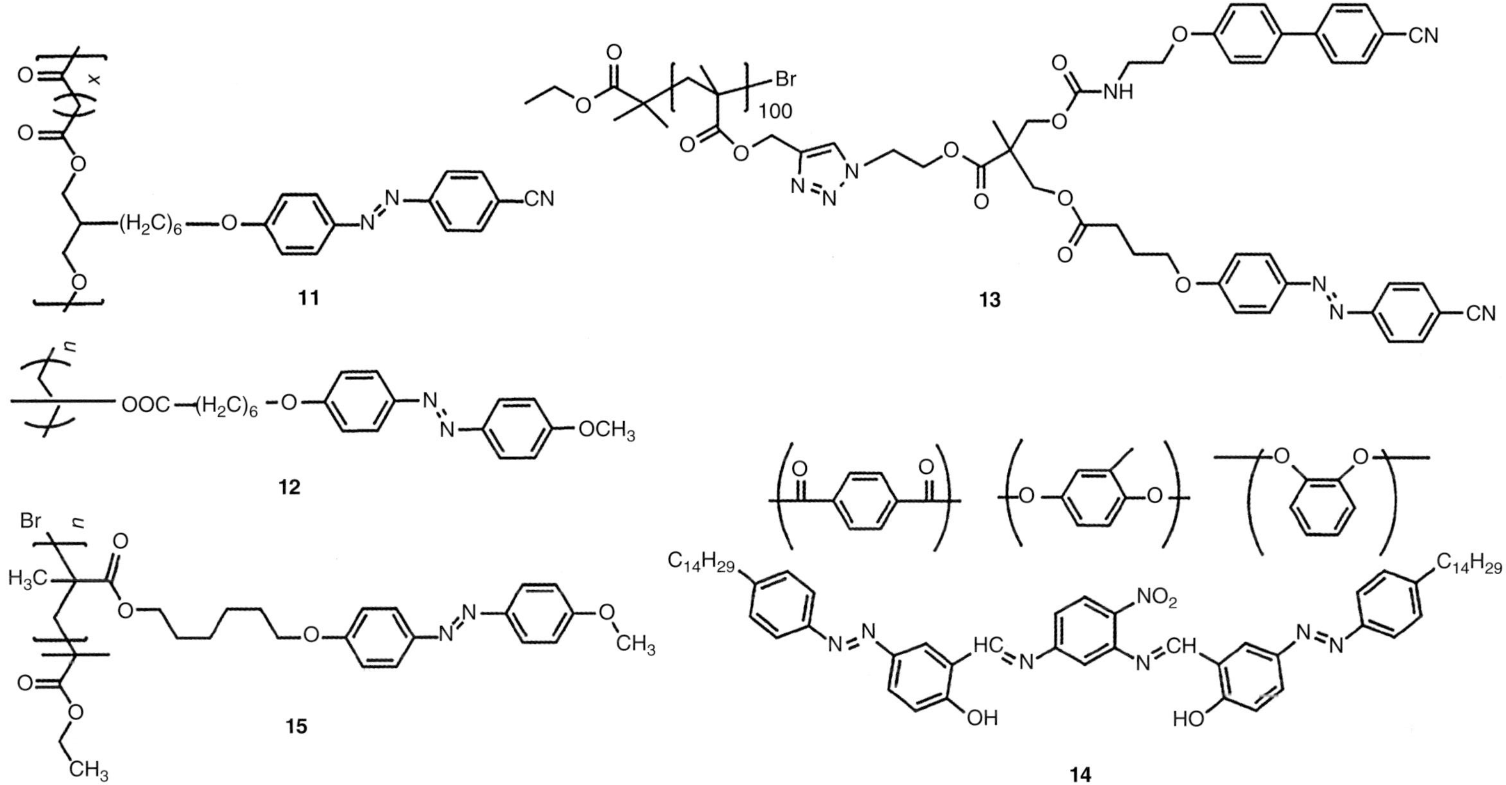

Figure 4.8 Achiral azopolymers exhibiting photoinduced chiral induction.

or macroscopic morphology and shape transformation of the soft matter system. The first demonstration on CD achieved from an achiral polymer was an induced chirality by irradiating it with CPL to **11** bearing azobenzene moieties through its chain alignment [39]. This phenomenon can be explained by the Weigert effect [9, 40]: the photoreaction efficiency of a molecule undergoing photoisomerization largely enhances when the direction of the transition moment of the molecule is parallel to the axis of the polarized light. Azobenzenes undergo repetitive *trans-to-cis* and *cis-to-trans* isomerizations by light irradiation so that the long axes of the molecules lie in a direction perpendicular to the optical field direction. In the case of CPL irradiation, the molecules align in the direction perpendicular to the optical field rotating in the thickness direction; then, the long axes of the molecules become gyratory in the thickness direction [41].

Tejedor et al. [42] studied CD responses of a series of liquid crystalline homopolymers bearing a methoxyazobenzene group in the side chain, (**12**), by illuminating with 488-nm CPL. The illumination of a nematic polymer film produced a helical organization as confirmed from CD spectra (Figure 4.9a), which resembles a chiral nematic phase as observed in the wedge cell image (Figure 4.9c) and under polarized optical microscope (POM) (Figure 4.9d). The CPL-irradiated region with a planar orientation exhibits an iridescent green color that corresponds to a selective reflection of a helicoidal organization. Moreover, vibrational circular dichroism (VCD) spectra after CPL irradiation (Figure 4.9b) indicate that only the mesogenic azobenzene units are involved in the chiral supramolecular organization. Consequently, using the chiral exciting light, supramolecular chirality is generated at the molecular level with an asymmetric photoisomerization, which possibly results in the conversion of a glassy nematic phase into a cholesteric phase at room temperature. Recently, the authors have achieved chiral photoinduction by CPL even in achiral amorphous azopolymers based on a constitutional repeating unit containing both a 4-cyanoazobenzene moiety and a promesogenic 4-cyanobiphenyl moiety (**13**) [43].

Choi et al. reported on main-chain polymeric liquid crystal systems doped with azobenzene-containing W-shaped molecules (**14**) [44]. The nucleated W-shaped molecules underwent preferential twisted conformation by CPL irradiation, and could trigger the local formation of a chiral structure in the doped main-chain polymer LC medium.

Recently, Wang et al. have verified the CPL wavelength dependence on the photogeneration and photoswitching modes and regulated azobenzene supramolecular chirality in an optofluidic medium [45]. The authors chose azobenzene with flexible linkers attached to chain-like polymethacrylate (**15**) dispersed as micrometer-sized aggregates in an optofluidic medium with a tuned refractive index (n_D). According to their careful study, it was revealed that the choice of CPL wavelength (313 nm and 365/405/436 nm) is the determining factor for photogeneration, photoresolution, and photoswitching of the azobenzene polymer aggregates, regardless of the sense of the CPL source (Figure 4.10a–c). Moreover, irradiating the aggregates with a CPL source at 365 nm induces almost optically pure chiral azobenzene aggregates with a maximum dissymmetry ratio of $g_{CD} \approx \pm 1.0 \times 10^{-2}$ at 313 nm. In addition, a distinctive chiroptical switching of **15** aggregates in dichloroethane/methylcyclohexane (0.5/2.5, v/v)

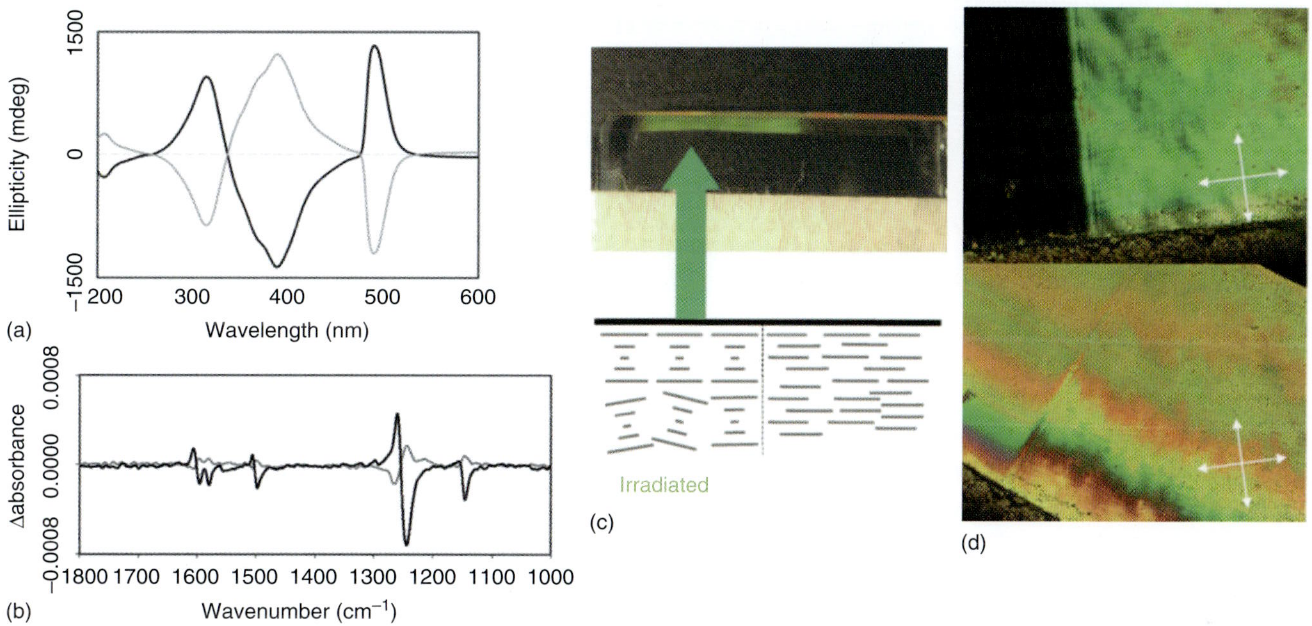

Figure 4.9 CD spectra (a) and VCD spectra (b) of **12** under irradiation with *r*-CPL (gray line) and *l*-CPL (black line) for 30 minutes (20 mW cm^{-2}). **12** in a wedge cell with an aligning layer for planar orientation with partial irradiation by CPL: photograph taken at room temperature of photoinduced iridescent green light (c), and crossed polarized optical textures micrographs (d). Source: Tejedor et al. 2007. [42]. Copyright 2007. Reproduced with permission from Wiley-VCH.

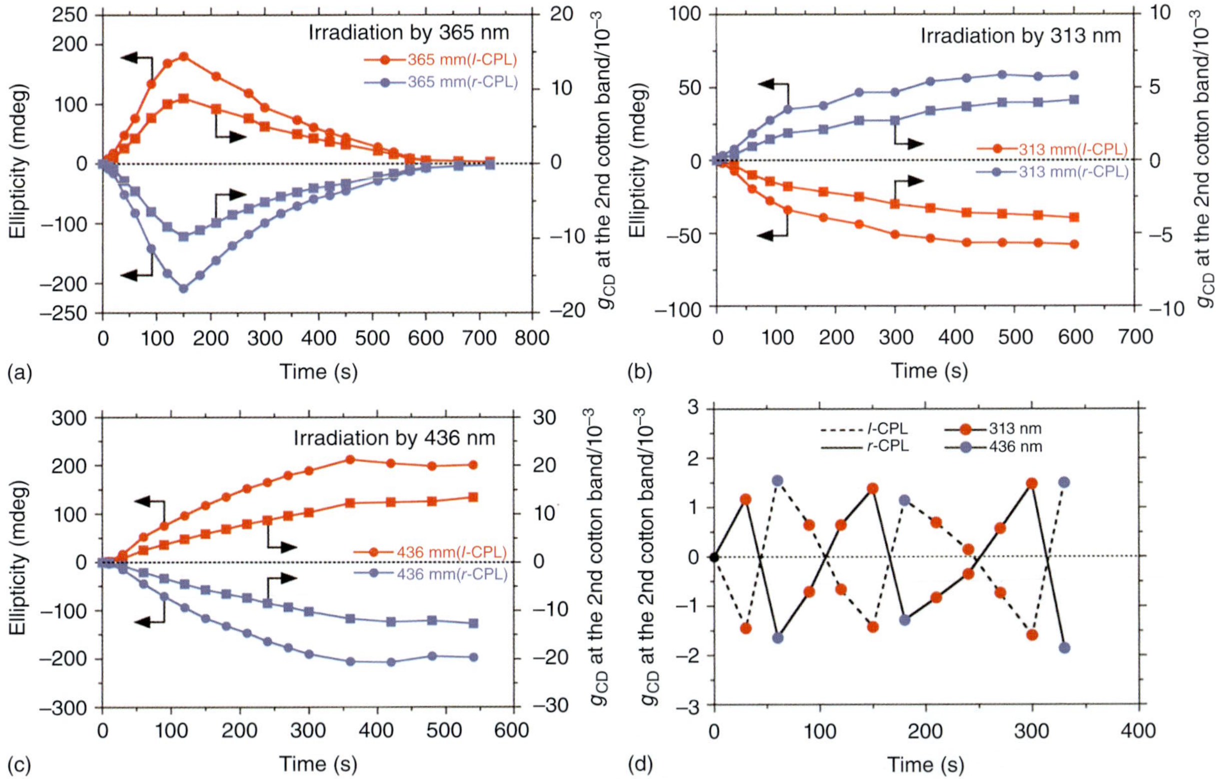

Figure 4.10 g_{CD} values (313 nm) and ellipticity (313 nm) of **15** aggregates in dichloroethane/methylcyclohexane (0.5/2.5, v/v) under (a) 365 nm, (b) 313 nm, and (c) 436 nm CPL irradiation with irradiation time. (d) Chiroptical switching of g_{CD} values (at 313 nm) of **15** aggregates in dichloroethane/methylcyclohexane (0.5/2.5, v/v) under alternating 313 and 436 nm CPL irradiation, using CPL sources with the same sense. Source: Wang et al. 2017 [45]. Copyright 2017. Reproduced with permission from American Chemical Society.

was attainable under alternating 313 and 436 nm CPL irradiation with the same sense (Figure 4.10d).

4.2.3 Self-Assembled Supramolecules

As described above, CPL-triggered chirality induction and switching in achiral vinyl polymers with azobenzene pendant units have been achieved from various studies in the film state, whose chiroptical properties are mostly determined from supramolecular assembly of the pendant groups. On the other hand, diacetylenes are known to undergo topochemical polymerization during UV and γ-irradiation in the solid state. Taking advantage of CPL-triggered rearrangement of azobenzene chromophores in the side chains, Iwamoto and coworkers reported that chiroptically active polydiacetylenes (PDAs) bearing azobenzene residues could be obtained from achiral **16**, by irradiation with either *r*- or *l*-CPUL or CP laser treatment (Figure 4.11) [46]. Upon unpolarized UV polymerization, a chiral structure of the PDA backbone was accidentally obtained and the sign of the chirality was not controlled. In contrast, the designed chiral structure of the PDA backbone could be realized by using CPUL, and the handedness direction could be modulated to opposite enantiomeric structures by irradiation with opposite handedness CPL (Figure 4.11b). The authors inferred that the stereoregular packing of azobenzene chromophores in the side chains plays a dominating role in the formation of enantiomeric helical PDA chains, and the CPL treatment is effective in adjusting the stereoregular packing of azobenzene units.

Meanwhile, Kim and coworkers have reported that helicity induction, modulation, and memory of supramolecular helical assemblies were achieved from an achiral triphenylamine derivative bearing photopolymerizable diacetylene units (**17**) by CPL and CPUL (Figure 4.12) [47]. Irradiation of **17** in a chlorinated solvent by *r*- or *l*-handed CPL resulted in the formation of a preferred-handed

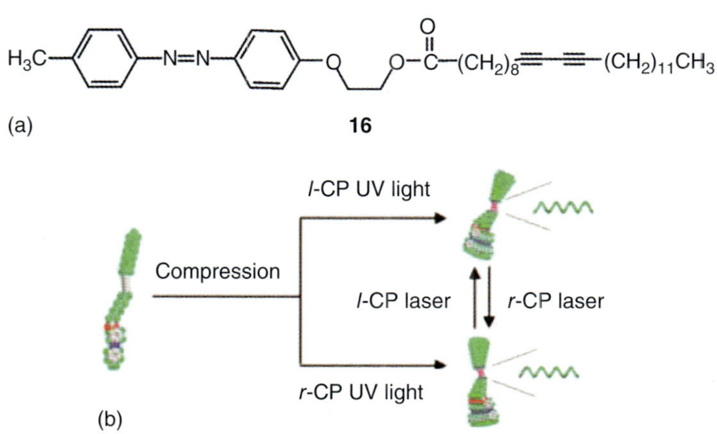

Figure 4.11 (a) Molecular structure of *p*-methyl azobenzene moiety-substituted diacetylene. **16**. (b) Schematic illustration of the formation and modulation of chirality for the film of **16** with CPUL and CP laser, respectively. Source: Zou et al. 2009 [46]. Copyright 2009. Reproduced with permission from The Royal Society of Chemistry.

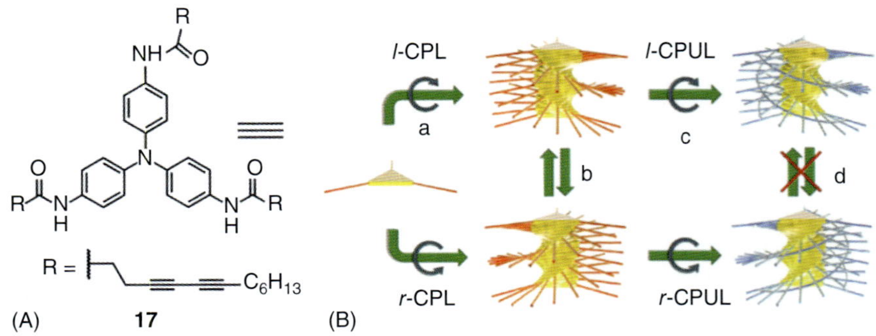

Figure 4.12 Chemical structure of achiral **17**(A) exhibiting manipulation of supramolecular chirality – induction, control and locking – by CPL and CPUL. Irradiation with CPL(B) induces self-assembly of **17** with control over handedness (a) and the reversible handedness switching by counter CPL irradiation (b). Photopolymerization of the diacetylene moieties by CPUL irradiation to produce a covalently joined nano-object (c) and the permanent locking of supramolecular chirality upon CPL irradiation (d). Source: Kim et al. 2015 [47]. Copyright 2015. Reproduced with permission from Nature Publishing Group.

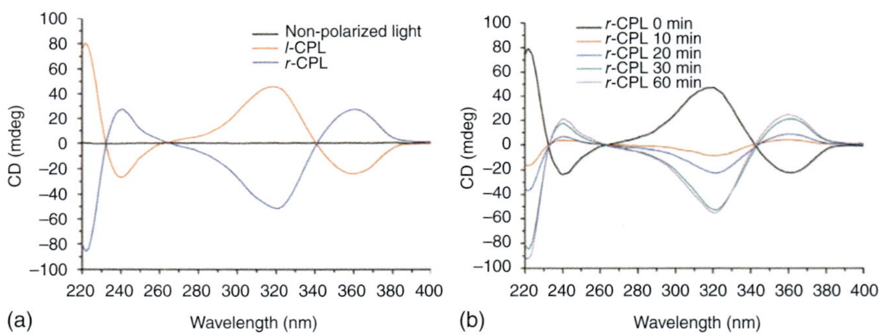

Figure 4.13 Induction and control of supramolecular chirality by CPL. CD spectra of **17** in dichloroethane solution after irradiation with CPL. (a) Spectra of **17** after irradiation with non-polarized light (black), *l*-CPL (red), and *r*-CPL (blue) for 10 minutes. (b) Inversion of CD activity by irradiation with *r*-CPL of the solution that had been pre-exposed to *l*-CPL. Source: Kim et al. 2015 [47]. Copyright 2015. Reproduced with permission from Nature Publishing Group.

supramolecular helical assembly, and the resultant helix sense could be reversibly switched by CPL irradiation with the opposite rotational direction as depicted in Figure 4.13. Interestingly, such switching did not occur after photopolymerization of the diacetylene unit helical assembly of **17** in the preferred handedness by irradiation using CPUL with the same rotational direction as the CPL used for the helicity induction of **17**, due to the covalent fixation of the supramolecular helical structure.

Royes et al. [48] showcased an achiral azobenzene-based gelator exhibiting photochemical chiral induction in supramolecular organization. The organogel based on the achiral crown ether azobenzene gelator (**18**) shown in Figure 4.14a was supported by a 2D layer nanostructure (Figure 4.14b), and a new class of chiral material has been obtained by the 488-nm *r*- and *l*-CPL irradiation of the

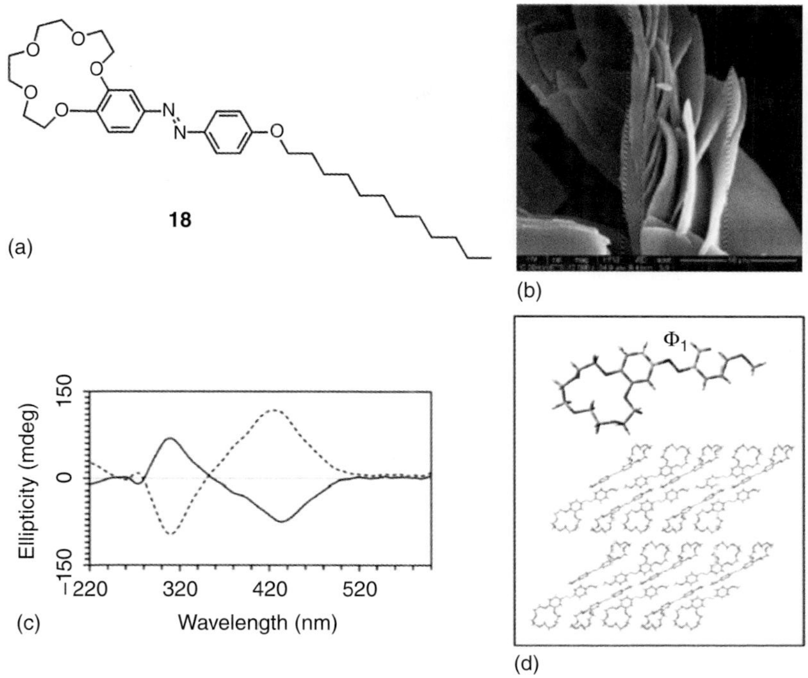

Figure 4.14 (a) Structure of gelator 4′-(*p*-dodecyloxyphenylazo)benzo[15]crown-5 ether **18**. (b) Scanning electron microscopy (SEM) image of the xerogel formed from **18**. (c) ECD spectra of *r*-CPL (dashed line)- and *l*-CPL (solid line)-irradiated samples of **18** for 80 minutes. (d) The crystalline structure of **18**. Source: Royes et al. 2017 [48]. Copyright 2017. Reproduced with permission from The Royal Society of Chemistry.

achiral gel, as confirmed from electronic circular dichroism (ECD) spectra (Figure 4.14c). This chiral photoinduction is possibly attributed to the preferential excitation of nonplanar conformers of the azo-gelator predicted from the crystal structure (Figure 4.14d).

4.3 Photochemical Chirality Inversion from Chiral Soft Materials

Light-induced helicity inversion has drawn profound attention for its importance in mimicking biological systems and fabricating three-dimensional materials and optical devices with tunable chirality [49]. Furthermore, the conversion of light input into the macroscopic motional response of materials offers a significant advantage in the development of artificial molecular machines for soft robotics, micromechanical systems, and artificial muscles [50, 51].

In this section, we introduce different classes of photoresponsive chiral switches doped or tethered to chiral soft materials such as CLCs, helical polymers, and self-assembled supramolecular systems exhibiting light-driven helical sense inversion (Scheme 4.2). Especially the helix inversion of intrinsically chiral

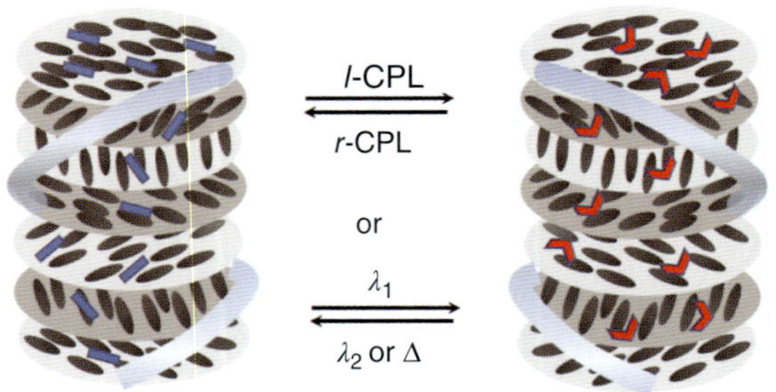

Scheme 4.2 Reversible handedness inversion of a helical superstructure based on the photochemical reaction of chiral dopants by irradiation with *l*- and *r*-CPL, or unpolarized light of two different wavelengths of λ_1 and λ_2 (e.g. UV or visible light) or heat.

materials is introduced. The achiral molecules processed with chiral induction by CPL are excluded from the discussion here to avoid overlap of contents, because their chiroptical intensity control such as inversion to an opposite handedness using unpolarized light or CPL has already been discussed in Section 4.2.

4.3.1 Photoresponsive Chiral Dopants for Cholesteric Liquid Crystals

When a photoresponsive chiral dopant is dissolved in an achiral NLC, its molecular chirality can be transferred to the NLC medium, resulting in the photoresponsive chiral nematic (N* or cholesteric) liquid crystal phase. The CLC superstructure is endowed with a helical pitch (*P*) and handedness. The pitch (*P*) is the distance along the helical axis when the molecules complete a full turn, and a positive sign represents a right-handed CLC while a negative sign represents a left-handed CLC. The right-handed CLC can selectively reflect *r*-CPL and transmit *l*-CPL (those of opposite for left-handed CLC) with the reflection wavelength corresponding to the pitch length of the helical superstructure. The ability of a chiral molecule to induce a helical structure in an achiral NLC can be quantified in terms of the HTP (β), as expressed in Eq. (4.1).

$$\beta = (PC)^{-1} \tag{4.1}$$

where *C* is the concentration of the chiral dopant and *P* is the pitch [52].

At the helix inversion point, the net HTP value of the chiral dopant becomes zero due to the presence of equal amounts of the isomers with opposite signs. Consequently, the pitch approaches infinite value, resulting in an achiral nematic phase. This nematic phase can also be referred as a racemic mixture of left-handed and right-handed CLCs [53].

In the CLC system, it is noteworthy that a high initial HTP value and its large photoinduced variation are advantageous in terms of the amount required (low doping concentration) to induce and modulate the cholesteric phase, which contributes to the selective reflection color to be tuned over a wide range of wavelengths and the large amplitude of mechanical motion control [20, 54, 55]. Thus, numerous investigations have demonstrated a significantly wide range of pitch modulations based on chiral nematic phases doped with chiral photoresponsive switch molecules [54–57]. On the other hand, reports are relatively rare on the inversion of the handedness in CLCs despite its merits in switching the chirality of the CPL that they selectively reflect, the so-called "wavelength-tunable handedness-specific CPL" [58] for advanced photonics.

4.3.1.1 Azobenzenes

Azobenzene undergoing *trans–cis* isomerization appears to be a very attractive photo-trigger in the study of the structure–property relation in cholesteric media, because of its photo fatigue resistance and the simplicity of the molecule. Among the chiral azobenzene dopants undergoing *trans–cis* photoisomerization, some planar chiral and axially chiral azobenzene derivatives have been reported to exhibit chiral inversion by UV and visible light irradiation. Planar chiral azobenzenes [54, 59, 60] are known to exhibit large HTP photoswitching, which is ascribed not only to the close distance between the chiral center and the mesogen unit but also to unifying the photoresponsive moieties and chiral moieties in a single moiety by cyclization. Tamaoki et al. demonstrated an interesting phenomenon that the bicyclic azophanes with planar chirality (**19**) shown in Figure 4.15A could reversibly change not only the HTP but also the handedness of a CLC by light irradiation without any molecular racemization of enantiomers during the photoisomerization (Figure 4.15B) [61]. (*R*) *E,E*-**19** of 1 wt% was dissolved in ZLI-1132 and introduced into a wedge-type cell at room temperature. As depicted in Figure 4.15C, a reversible change occurred in pitch length by irradiation of UV and visible light. N phase was obtained in between the transition of cholesteric phases with opposite handedness upon continuous exposure to UV light, and the photoinduced maximum HTP switching ratio was over 600% (Table 4.1).

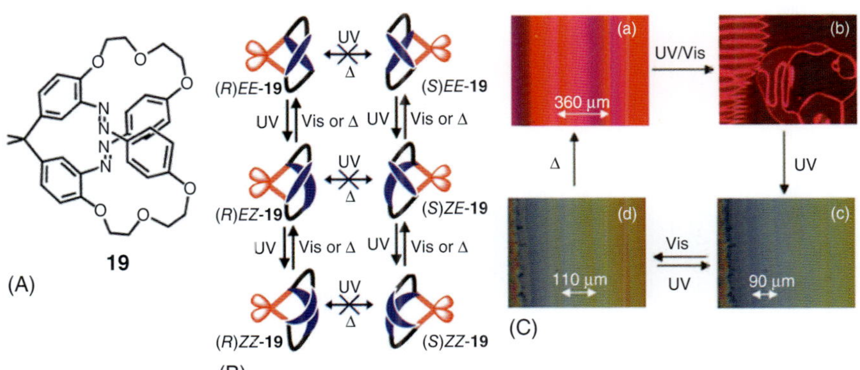

Figure 4.15 (A) Structure of bicyclic azophanes **19**. (B) Schematic representation of photochemical and thermal processes taking place in its *R* and *S* enantiomers. (C) POM images of Cano wedge cell filled with **19** in ZLI-1132 at room temperature: (a) before irradiation, (b) N phase obtained by exposure of sample to UV or vis irradiation, (c) PSS_{UV}, (d) PSS_{vis}. Source: Mathews and Tamaoki 2009 [61]. Copyright 2009. Reproduced with permission from The Royal Society of Chemistry.

Table 4.1 Photoresponsive chiral dopants exhibiting chirality inversion and large HTP photoswitching ratio.

Photoresponsive chiral dopant structure	Number	Host NLC	β (μm^{-1}) Initial	PSS$_{UV}$	PSS$_{Vis}$	$\lvert\Delta\beta/\beta_{ini}\rvert$ (%)[a)]	References
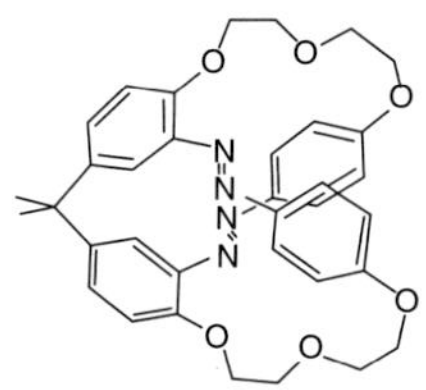	(*R*)-19	5CB	−6	+32	+26	633	[61]
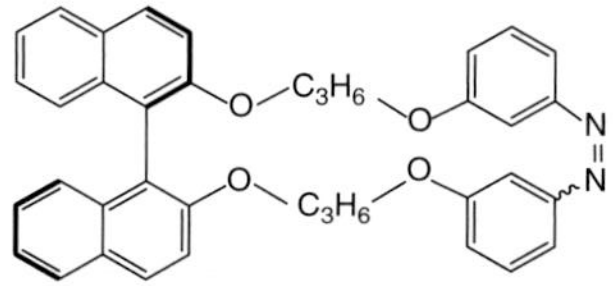	(*R*)-20	ZLI-1132	+8	−26	+6	425	[62]
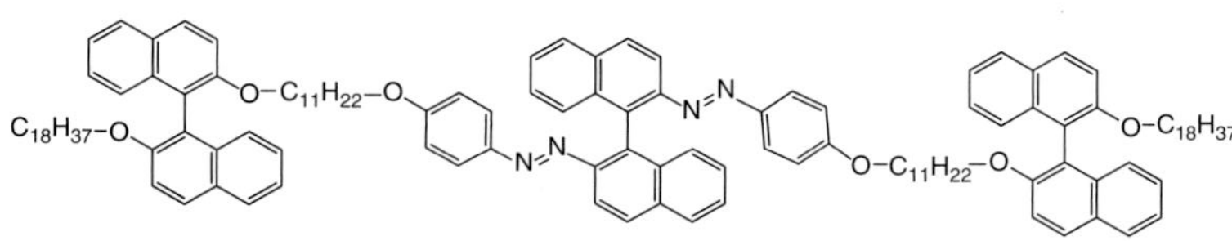	(*S,R,S*)-21	5CB	+59	−65	+54	210	[63]
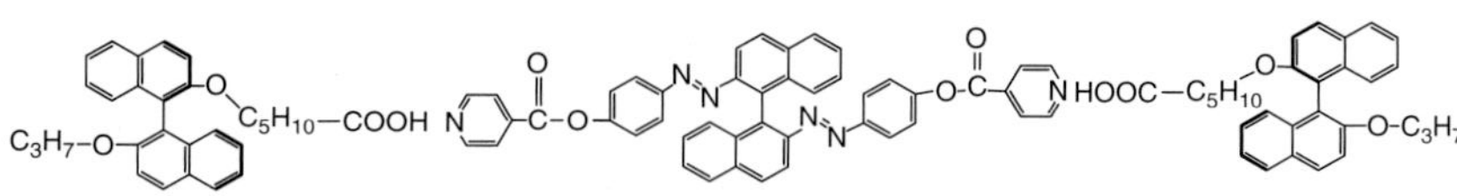	(*R,S,R*)-22	5CB	−0.39	+2.80	–	818	[64]

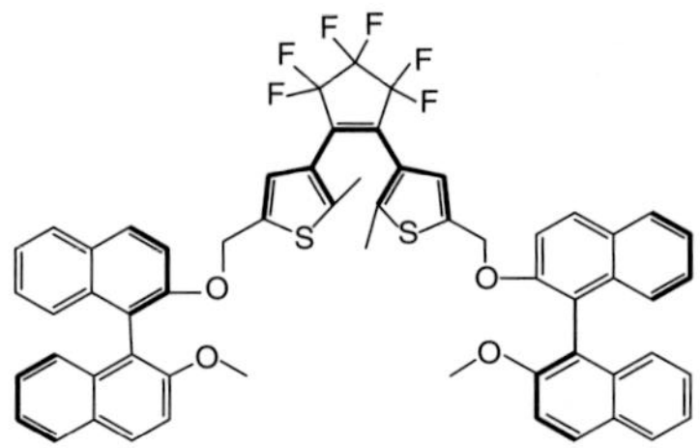	(*R*)-23	PCH	+6.6	−8.3	–	226	[65]
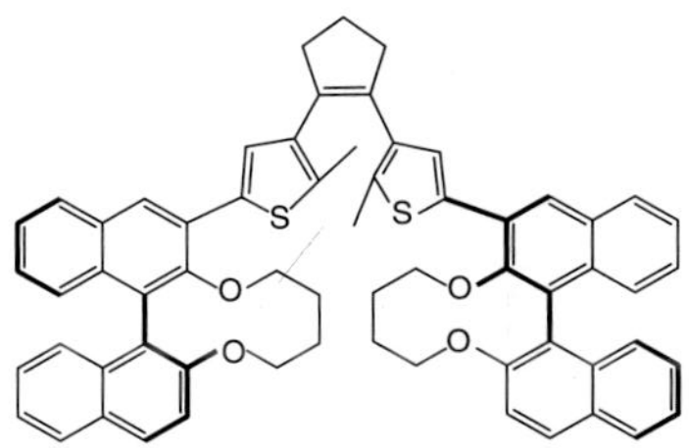	(*S*)-24	ZLI-1132	+6	−25	–	517	[66]
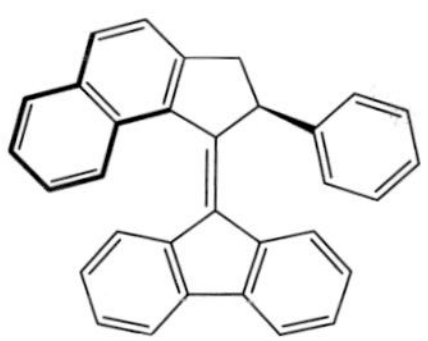	(*P*)-25	E7	+90	−59	–	166	[67]
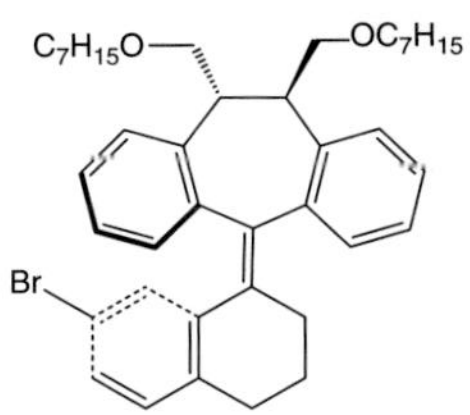	(*R*,*R*,*P*)-26	E7	+7.2	−12.1[b)]	+6.1[c)]	268	[68]

a) Change ratio in HTP between the initial (β_{ini}) and PSS_{UV} states.
b) Irradiation wavelength of 290 nm.
c) Irradiation wavelength of 330 nm.

Axially chiral binaphthyl moieties are feasible for controlling the helical sense and strength of the CLC. The binaphthyl derivative is known to have restricted freedom of internal rotation along the carbon–carbon bond between the 1 and 1′ positions of the binaphthyl rings. The helicity of the binaphthyl derivative therefore depends on the dihedral angle (θ) between the two naphthyl rings: *cisoid* ($\theta < 90°$) or *transoid* ($\theta > 90°$) conformation affects the helical sense and twist intensity of doped CLCs [69]. Some examples from axially chiral azobenzene dopants have also been reported to exhibit photoinduced reversible handedness inversion. Li and coworkers reported on cyclic azobenzophane dopants bearing binaphthyl units (**20** in Table 4.1) with 425% of maximum HTP switching upon *cis*-isomerization, and a recovery of the original handedness upon visible light irradiation [62].

The authors also developed a type of azobenzene compound substituted with axially chiral binaphthyl units of both the same and opposite chiral configurations (**21**) shown in Figure 4.16A [63]. The handedness inversion in LC hosts (5CB, E7 and ZLI-1132) could be induced with CLC containing the dopants with opposite chiral configurations (*S,R,S*) upon UV light irradiation accompanying the chiral nematic and racemic nematic phases (Figure 4.16B), followed by a decent recovery of their original handedness and HTP (up to 90%) at photostationary state (PSS_{vis}) as described in Table 4.1, whereas a compound with the same chiral configurations (*R,R,R*) exhibited no handedness inversion. Therefore,

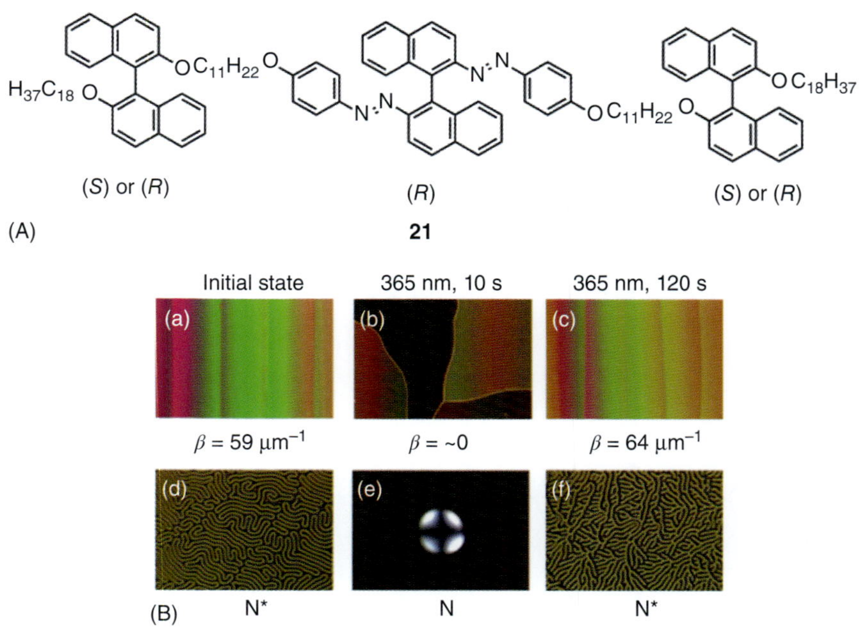

Figure 4.16 (A) Chemical structure of chiral azoarene dopants **21**. (B) Handedness inversion of 0.27 mol% (*S,R,S*)-**21** in 5CB observed in a wedge cell (a–c) and homeotropic cell (d–f) from right-handed CLC (a, d) to left-handed CLC (c, f) through a transient nematic phase (b, e), observed using crossed polarizing microscopy. Source: Li et al. 2013 [63]. Copyright 2013. Reproduced with permission from Wiley-VCH.

the introduction of chiral conflict at the molecular level plausibly influences the handedness inversion of CLC through an equilibrium shifting between opposite chiral configurations in a single chiral switch.

Moreover, Fu et al. have investigated a dopant system similar to that mentioned above, where additional chiral units were incorporated into a chiral azobenzene unit by hydrogen bond (**22** in Figure 4.17a) [64]. The binaphthyl azobenzene molecule was used as a proton acceptor, and binaphthyl acids with opposite chiral configuration were introduced as proton donors forming H-bonded chiral switches. As shown in Figure 4.17b, a phase transition from N* to N accompanying the helical inversion and the HTP increase or decrease in N* could be achieved upon light irradiation. The length of a terminal alkyl chain in proton donors of **22** affected the HTP and helix inversion characteristics of CLCs. Although the HTP values of **22** are quite small compared to those of **21**, the HTP switching ratio reaches over 800% (Table 4.1).

4.3.1.2 Diarylethenes

Diarylethenes are of particular interest as chiral dopants for photoresponsive CLCs because of their excellent fatigue resistance and thermal stability in both the open and closed forms [38, 57, 70, 71]. Binaphthyl moieties are known as powerful helicity inducers in achiral NLCs as described above. Thus, diarylethene dopants coupled with axially chiral binaphthyl moieties have been widely studied for photonic applications of CLCs taking advantage of the large HTP and its stable photoswitching properties [23, 24]. In terms of dopant molecular design, modification of the binaphthyl units by substituting mesogen units possessing similar conformation to NLC host and/or bridging naphthyl groups has been attempted and these approaches bring a significant difference in the miscibility with the nematic host or the rigidity of the chiral conformation, which leads to various HTPs and handedness switching behaviors [57, 70, 71].

Akagi et al. presented the first dynamic photoswitching of the helical inversion in CLCs based on a dithienylethene derivative bearing two axially chiral binaphthyl moieties (**23**) [65]. The photoisomerization of the dithienylethene moiety after irradiation with UV and visible light induces a reversible dihedral angle change of the binaphthyl rings. CLCs doped with the chiral dopant bearing the shortest alkyl chain in a binaphthyl moiety (**23**) exhibit reversible handedness switching via photoisomerization between the open and closed forms of the dithienylethene moiety (Figure 4.18).

Li and coworkers demonstrated bridged binaphthyl-substituted dithienylcyclopentenes shown in Figure 4.19A, which exhibit the helix inversion of CLCs with a bridge length of $n = 4$ (**24**) [66]. The bridged binaphthyl derivative normally has a higher HTP than that of the non-bridged moieties due to its narrow dihedral angles, and an increase in bridge length may lead to increase in molecular flexibility with larger phototunabilities. Indeed, those of shorter bridge lengths ($n = 2$ and 3) [57] exhibited smaller HTP switching without any inversion of handedness. According to their report, **24** doped in 5CB, E7, or ZLI-1132 underwent right- to left-handed inversion upon UV light irradiation as shown in Figure 4.19B. Moreover, it showed the highest initial HTP of 54 μm^{-1} in 5CB, while 517% of maximum HTP switching ratio was obtained in ZLI-1132.

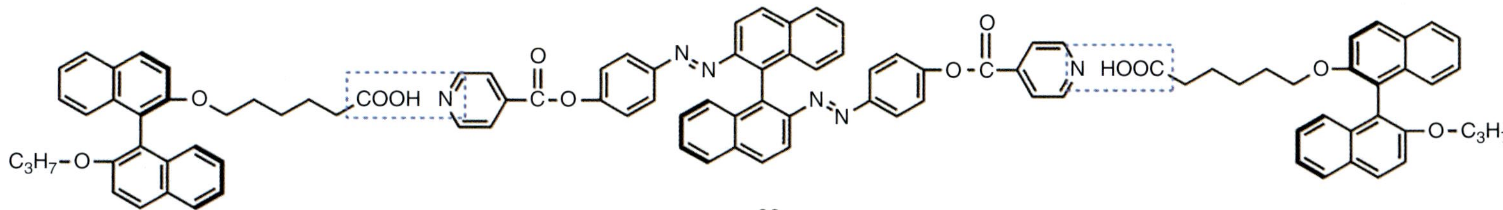

(a)

22

(*R*,*S*,*R*,)-**22**

(b)

Figure 4.17 (a) Structure of hydrogen-bonded chiral azobenzene dopant **22**. (b) POM images of the wedge cell of photoresponsive *N** phase including (*R*,*S*,*R*)-**22** (1.0 wt%) between the *trans* state (left image) and PSS_{UV} (right image) by irradiation with UV at 25 °C. Schematics along the corresponding reversible helical inversion between N*–N–N* phase transition sequence with opposite handedness. Source: Fu et al. 2015 [64]. Copyright 2015. Reproduced with permission from The Royal Society of Chemistry.

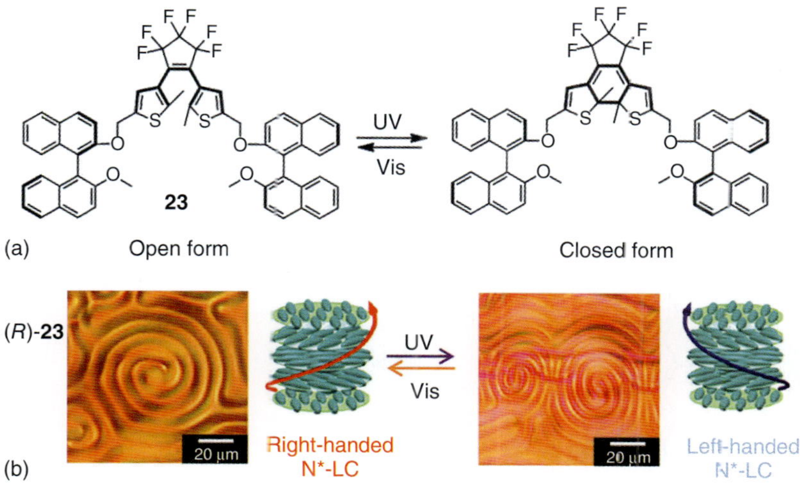

Figure 4.18 (a) Photoisomerization of dithienylethene derivative (**23**) with two axially chiral binaphthyl moieties. (b) Cholesteric texture transition of photoresponsive CLCs containing (*R*)-**23** (1.0 mol%) between the open form (left) and PSS (right) by irradiation with UV ($\lambda = 254$ nm) and vis ($\lambda > 400$ nm) light. Source: Hayasaka et al. 2012 [65]. Copyright 2012. Reproduced with permission from American Chemical Society.

Photoinduced large HTP switching and handedness inversion are attributable to the significant increase in the θ value of (*S*,*S*)-**24** from the open to the closed forms (Figure 4.19C). Especially, several optical applications were carried out based on this handedness invertible dithienylcyclopentene dopant (**24**) such as reflection color control by photoswitching of a cholesteric pitch length and NIR-light-induced handedness inversion. The authors utilized **24** and an upconversion nanoparticle in CLC as a handedness switching inducer and NIR transducer, respectively [72].

Furthermore, based on the unique helical sense inversion associated with dynamic reorganization of the entire superstructure, **24**-doped CLC was utilized for the three-dimensional control of the helical axis [73]. UV irradiation (wavelength of 310 nm) of the chiral switch in a planar cell transformed the helical superstructure from its original standing helices (SHs) through the unwound homogeneous state and the SH arrangement with opposite handedness to the lying helix (LH) arrangement (Figure 4.20A). The authors confirmed the resultant control of helical axis by the transmitted light diffraction through a bilayer CLC sample as shown in Figure 4.20B. The dimensionality transformation was directly visualized between two-dimensional, one-dimensional, and the diffraction-off states. In addition, the authors recently demonstrated the dynamically reconfigurable zigzag pattern deformation with (*S*,*S*)-**24** (3.7 wt% in E7) under the simultaneous influence of an applied electric field and light irradiation (Figure 4.21) [74]. The straight and zigzag-shaped patterns could be generated and terminated repeatedly on demand, and easily manipulated by alternating irradiation with ultraviolet and visible light [74]. The cohesive and facile phototuning of the helical axis can be promising for advanced 3D photonic devices.

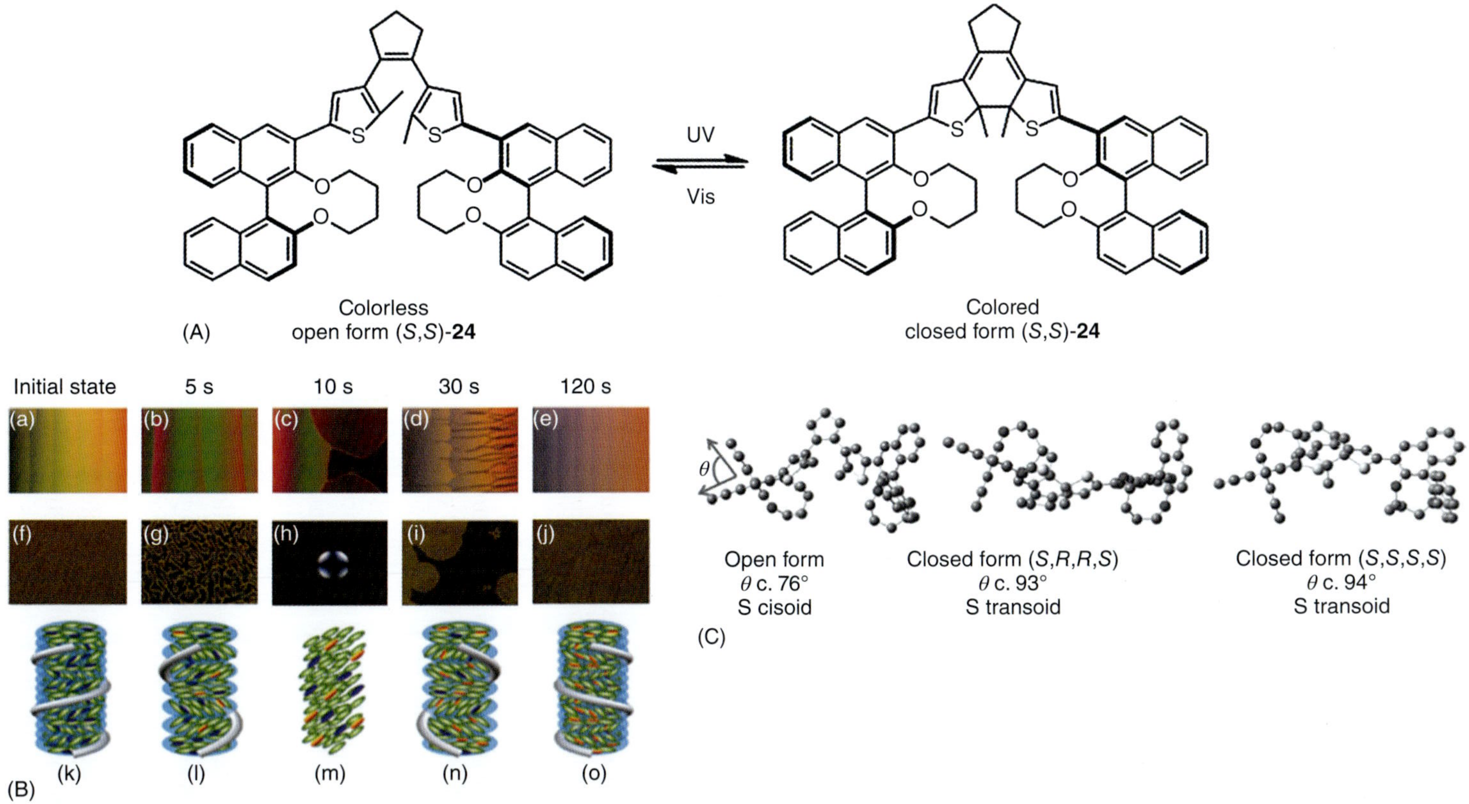

Figure 4.19 (A) Photoisomerization of bridged binaphthyl-substituted dithienylcyclopentenes **24**. (B) Helical pitch length change and handedness inversion of (*S*,*S*)-**24** in 5CB upon continuous UV irradiation. (C) Dihedral angles in its open and closed forms. Source: Li et al. 2013 [66]. Copyright 2013. Reproduced with from permission Wiley-VCH.

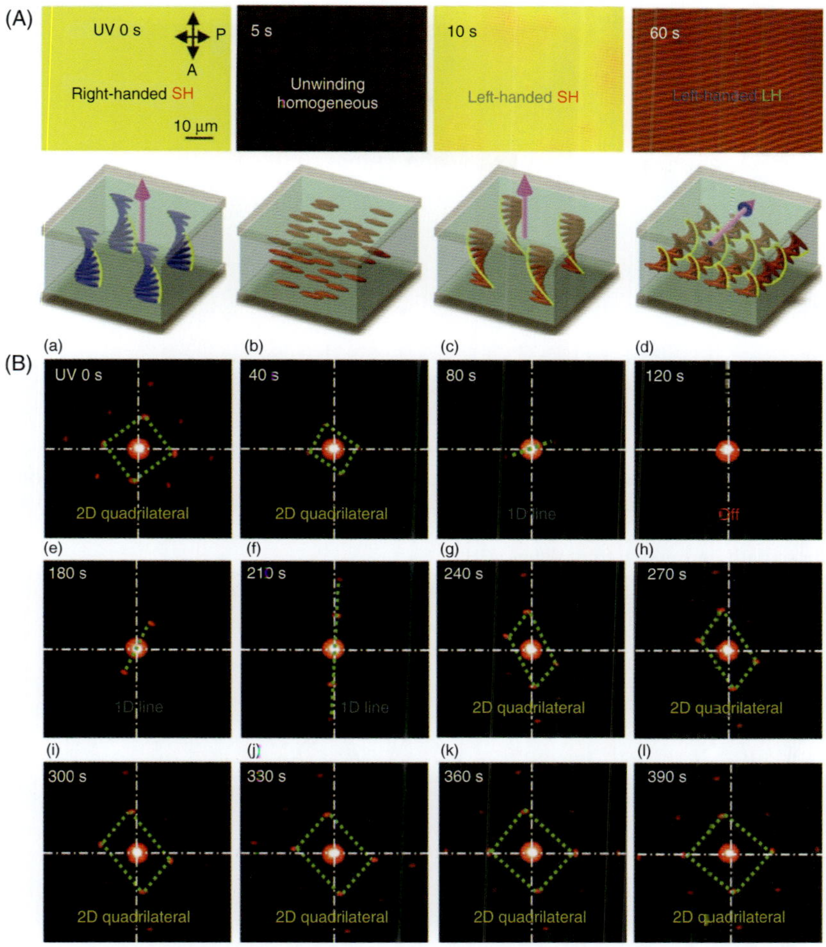

Figure 4.20 (A) UV irradiation to the chiral switch (**24**) in a planar cell transforms the helical superstructure from its original standing helix (SH) through the unwound homogeneous state and the SH arrangement with opposite handedness to the lying helix (LH) arrangement. (B) Light-induced diffraction dimensionality transformation of a bilayer CLC sample. Diffraction patterns were transformed from two-dimensional (2D; a, b), through one-dimensional (1D; c) and the diffraction off state (d), to 1D again (e, f), and finally to 2D (g–l). Source: Zheng et al. 2016 [73]. Copyright 2016. Reproduced with permission from Nature Publishing Group.

4.3.1.3 Overcrowded Alkenes

Feringa and coworkers have developed a series of sterically overcrowded alkenes with intrinsic helical conformations, which include a stereogenic center at the rotor part and a fluorene group at the stator part [16, 75–77]. The photoisomerization around the C=C double bond of this molecular motor is known to invert the molecular helicity.

Figure 4.22 represents a reversible isomerization process of a representative overcrowded alkene motor (**25**), which has broadened the application of

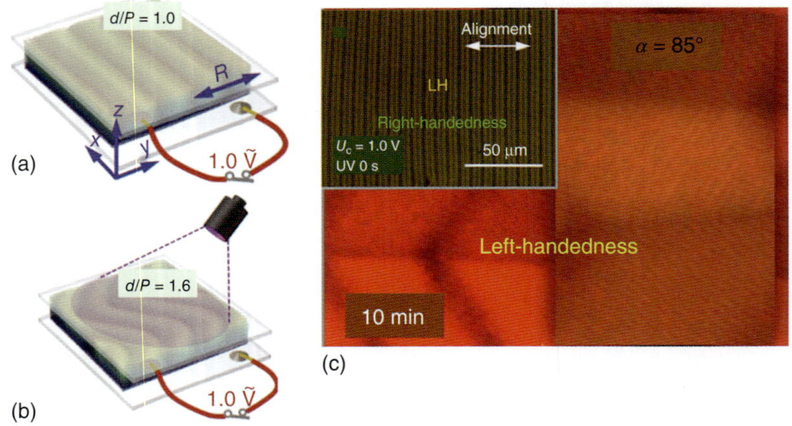

Figure 4.21 Generation of a straight (a, inset of c) and zigzag-shaped patterns (b, c) based on the CLC containing **24** through a combination of light irradiation and electric field application at a constant voltage $U_c = 1.0\,V$ (1 kHz). Source: Zheng et al. 2017 [74]. Copyright 2017. Reproduced with permission from Wiley-VCH.

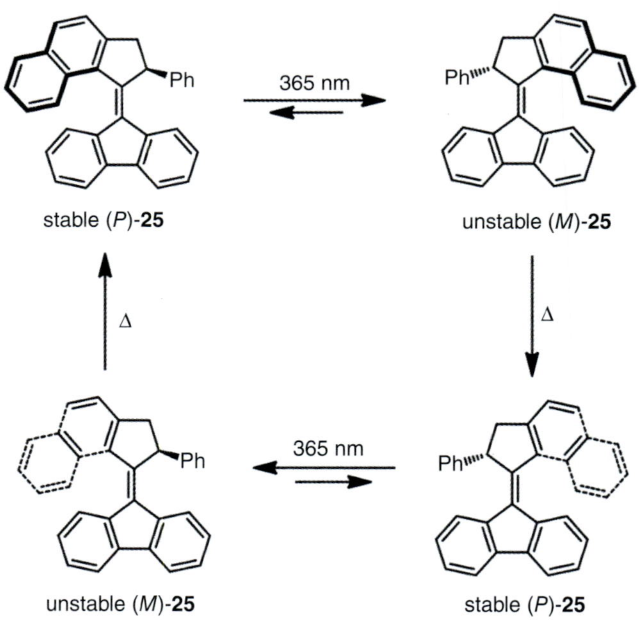

Figure 4.22 Unidirectional rotary cycle of fluorene-based motor **25**. Source: Eelkema et al. 2006 [78]. Copyright 2006. Reproduced with permission from American Chemical Society.

its unique photonic and mechanical functions [75, 78–80]. Irradiation with UV light results in its *P*-helicity switching to *M*-helicity, which accompanies the large HTP switching (Table 4.1). However, as the isomer with *M*-helicity is not thermally stable, non-photochemical transition from *M*- to *P*-helicity occurs. The

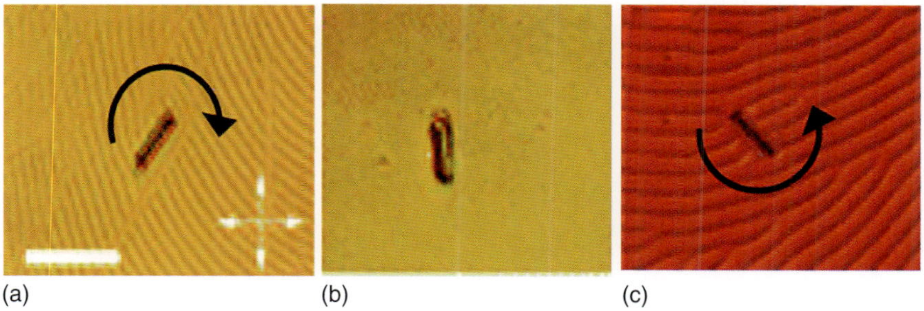

Figure 4.23 Optical micrographs of a glass rod rotating on an LC film doped with molecular motor **25**, upon irradiation with UV light ($\lambda = 365$ nm). (a) The rod and the cholesteric texture rotate in a clockwise manner. (b) No cholesteric texture is shown during helix inversion of the motor. (c) The rod and the cholesteric texture rotate in a counterclockwise manner after the helix inversion. Source: Eelkema et al. 2006 [78]. Copyright 2006. Reproduced with permission from American Chemical Society.

corresponding light-directing handedness inversion has been observed in a CLC fabricated from E7 doped with 1 wt% of **25** upon irradiation with UV light, with the reverse process occurring spontaneously at room temperature (Figure 4.23). After 135 seconds of irradiation with 365 nm light, it reached the helix inversion point and a non-chiral nematic phase (compensated nematic, Figure 4.23b) was formed without a cholesteric fingerprint texture. Further irradiation induced the rewinding of the cholesteric helix with an opposite handedness (Figure 4.23c). In particular, a micro-sized glass rod floated on the CLC film exhibited opposite rotational direction via non-rotational movement state according to the handedness change of CLC, reflecting that molecular level chirality change can be translated to the macroscopic level (Figure 4.23). White et al. [67] and Aßhoff et al. [81] confirmed the reversible and broad tunability of the helical pitch length and photoinvertible handedness of overcrowded-alkenes-based CLCs (Figure 4.24). Especially Aßhoff et al. showed that the thermal relaxation rate of the cholesteric helix can be fully determined through the helix inversion process of the overcrowded alkene motors.

Chen et al. reported on 10,11-disubstituted dibenzeosuberane (DBS)-based helicenes (**26**), which undergo molecular helicity inversion upon photoisomerization (Figure 4.25A) [68]. The induced CLC phase in E7 was found to exhibit light-driven handedness inversion through photoisomerization of the dopant by sequential irradiation with UV light of different wavelengths, 290, 310, and 330 nm. The selective photoswitching is driven by the conformation control of the top DBS template unit. Alternating irradiation at these three different wavelengths resulted in a diverse photomodulation of the pitch length, handedness reversal, and chirality turn-off mode of the cholesteric mesophase (Figure 4.25B). Moreover, the authors demonstrated a unique type of helicene (**27**), which is a derivative of **26** incorporating aromatic amide ("aramide") moieties in the top C2-symmetric dibenzosuberane template [82]. This C2-symmetric dibenzosuberane-based helicene bearing gallamide groups is converted into its pseudoenantiomeric helicene *P′* through the photoisomerization of the dibenzosuberane

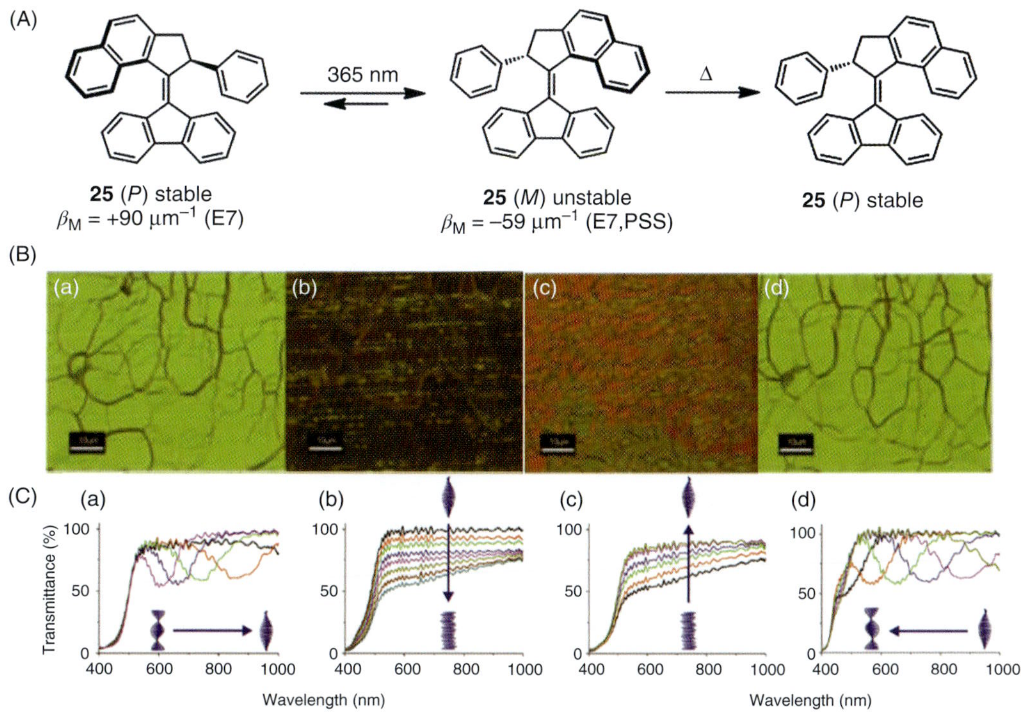

Figure 4.24 (A) Photoisomerization of **25**. (B) POM images of **25** (4.2 wt%)/1444 mixture during exposure to 365 nm light (*in situ*): (a) grandjean texture before exposure; (b) formation of nematic phase during helical inversion; (c) fingerprint texture after helical inversion; (d) grandjean texture restored in the dark. White scale bar in (B) corresponds to 10 μm. Source: Reproduced from Ref. [67] with permission. Copyright 2011 Wiley-VCH. (C) UV–vis spectral change during photoisomerization of CLC containing **25** upon continuous UV irradiation. Source: Aßhoff et al. 2013 [81]. Copyright 2013. Reproduced with permission from The Royal Society of Chemistry.

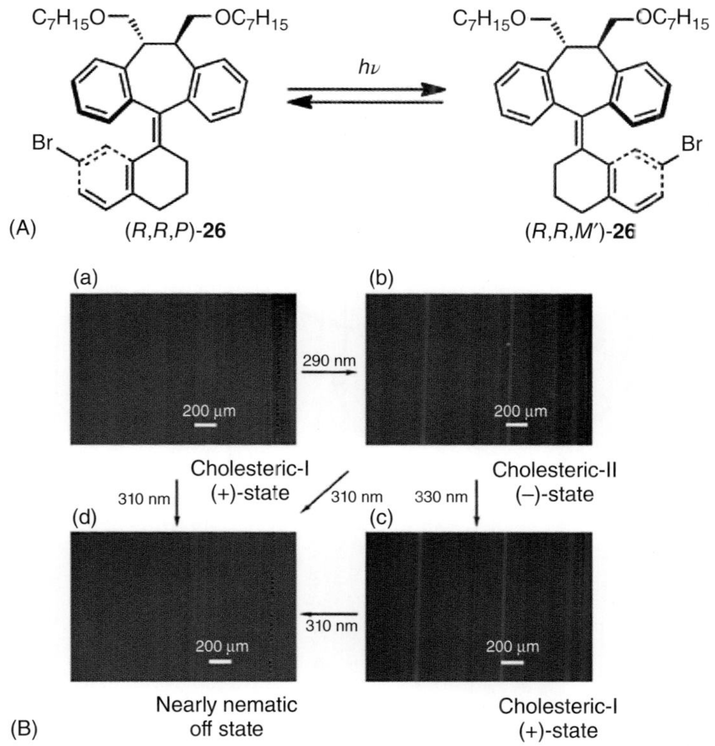

Figure 4.25 (A) Photoisomerization of 10,11-disubstituted dibenzeosuberane-based helicenes **26**. (B) POM images of (*P*)-**26** (1 wt%) doped E7 in a wedge cell upon UV irradiation: (a) the initial state; (b) after irradiation at 290 nm of (a) for 3 hours; (c) after irradiation of (b) at 330 nm for 2 hours; (d) after irradiation at 310 nm for 3 hours. Source: Chen et al. 2010 [68]. Copyright 2010. Reproduced with permission from Wiley-VCH.

residue upon UV irradiation. This complementary helicity inversion was reversible, and the *P/M* ratio could be regulated by irradiation of a selected wavelength in the solution state (Figure 4.26a). Interestingly, **27** could reversibly self-assemble to form a gel depending on the ratio of *P* and *M* based on the intermolecular amide hydrogen bonding and π–π stacking (Figure 4.26b). Either a left- or a right-handed helical superstructure was obtained as confirmed from transmission electron microscope (TEM) image and CD spectra (Figure 4.26c,d).

4.3.2 Chiral Polymers

In helical polymers, stereoselective interactions between photoresponsive chiral guest molecules and polymers can induce the formation of helical structures of polymer chains or trigger a conformational change in the helical structure. The photochemical control of the macromolecular helical handedness or the cholesteric pitch has been achieved using multiple photochromic molecular switches in the side chains [83, 84], or a single photoswitchable unit at the end of the polymer chain [85]. Upon photoisomerization of the pendant unit, the

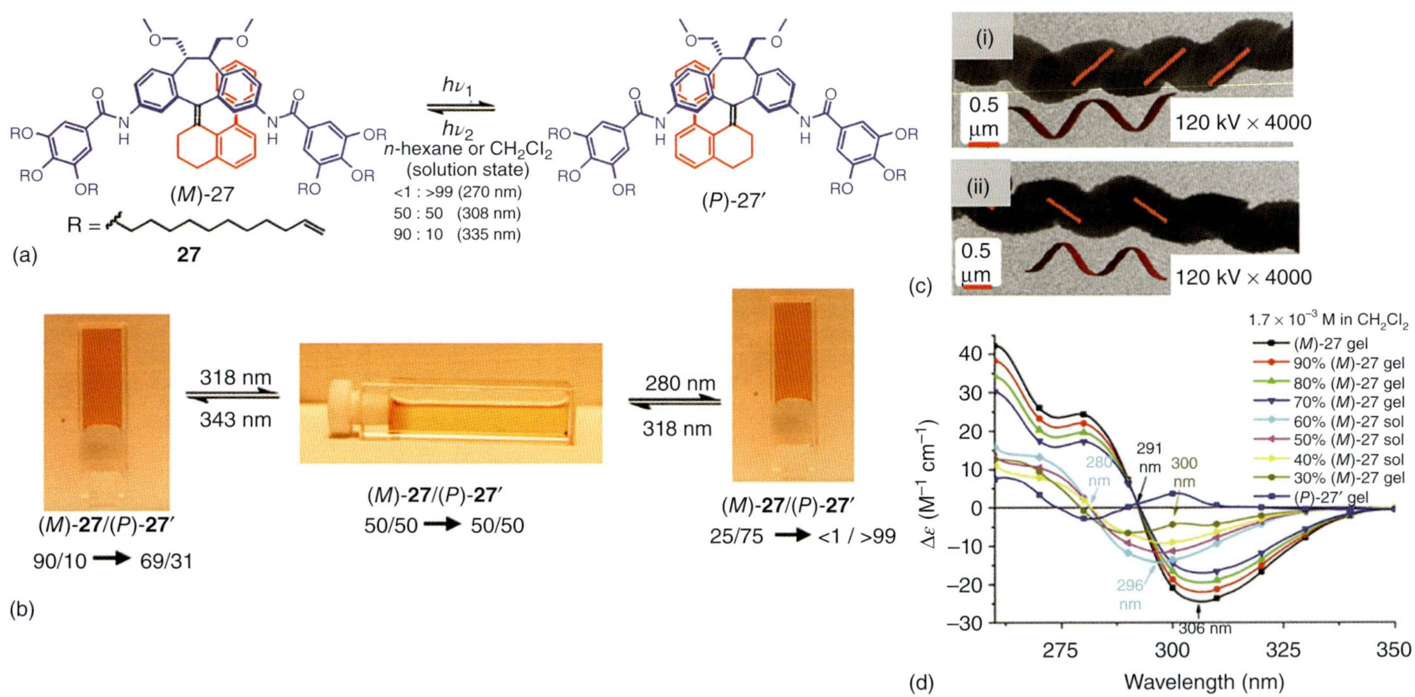

Figure 4.26 (a) Photoisomerization of **27**. (b) Reversible photoinduced gelation of **27** in CH_2Cl_2. (c) TEM images of superstructure of xerogels prepared from (*M*)-**27** (i) and (*P*)-**27** (ii). (d) CD spectra of the photoisomerization tracing on (*M*)-**27** gel. Source: Chen et al. 2013 [82]. Copyright 2013. Reproduced with permission from The American Chemical Society.

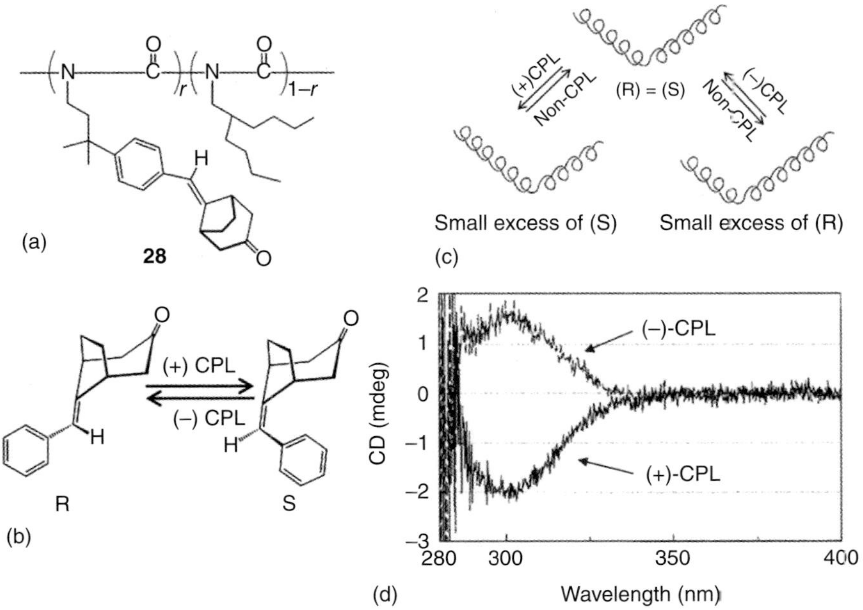

Figure 4.27 (a) Structure of a polyisocyanate with bicycloketone chromophore pendants and achiral alkyl groups (**28**). (b) Reversible photoresolution process of bicycloketone pedant group of **28**. (c) Schematic representation of transition between right- and left-handed backbone helices of **28** subjected to irradiation with (+) or (−) CPL or non-CPL. (d) CD spectra of **28** n–π^* absorption region upon CPL irradiation. Source: Li et al. 2000 [86] Copyright 2000. Reproduced with permission from The American Chemical Society.

handedness of the polymer backbone reverses, which in turn prompts handedness inversion of the whole supramolecular structure. The relatively low free energy difference between the right- and left-handed helical states allows the helical sense preference of the helical polymer to be switched into the opposite handedness dynamically. In 2000, Schuster et al. demonstrated a system in which chiral information of CPL is used to control the helical sense of a polymer [86] (Figure 4.27). A racemic mixture of a photoresolvable ketone [28] and an achiral alkyl group was appended to a polyisocyanate (**28**) having equal populations of dynamically interconverting *l*- and *r*-handed helical backbone conformations. Irradiation of the pendant ketone group with CPL results in its partial photoresolution (Figure 4.27b). The majority rule effect [87] transfers the chirality of the pendant group to the polymer backbone helix, resulting in amplification of the photoresolution as observed from CD signals that changed sign with a change in the sense of the CPL (Figure 4.27c,d). The small enantiomeric excess produced by the irradiation, even diluted with large proportions of achiral pendants, is capable of enforcing a disproportionate excess of one helical sense in the polymer [86]. Such helix-sense switchable polymers have potential applications for chirality switchable asymmetric catalysts, optical devices, and displays [3].

4.3.2.1 Azopolymers

In 1994, Zentel and coworker succeeded for the first time in inducing photochemically reversible shift of the equilibrium between *P* and *M* helices in polyisocyanates [88]. Later, the authors discovered that the helix-sense of an optically active polyisocyanate copolymer (**29** in Figure 4.28a) of a chiral isocyanate bearing an azobenzene side group containing two stereogenic centers with an achiral isocyanate can be switched by the photoisomerization of the azobenzene moiety from the *trans* to *cis* form (Figure 4.28b) [83, 89]. The helix inversion permits the reversible switching between *P* and *M* helices by controlling the proximity of the chiral center to the polymer backbone (Figure 4.28c).

Barberá et al. [90] reported on the photomodulation of the chiroptical properties of the intrinsically chiral polymer (**30**) in a thin film state. In this system, the polymer, in the smectic phase (Figure 4.29a,b), is provided chirality originating at the molecular level by the asymmetric functionality of the *l*-lactic acid unit. It is of highly homogeneous conformation with a prevailing chirality that appears related to the presence of H-aggregates having conformational dissymmetry of one prevailing screw sense. Upon irradiation with *l*-CPL, the CD spectra of the films show enhancement of ellipticity and a net inversion sign (Figure 4.29c). Additionally, the mirror image of the CD spectra could be restored by pumping with *r*-CPL (Figure 4.29d).

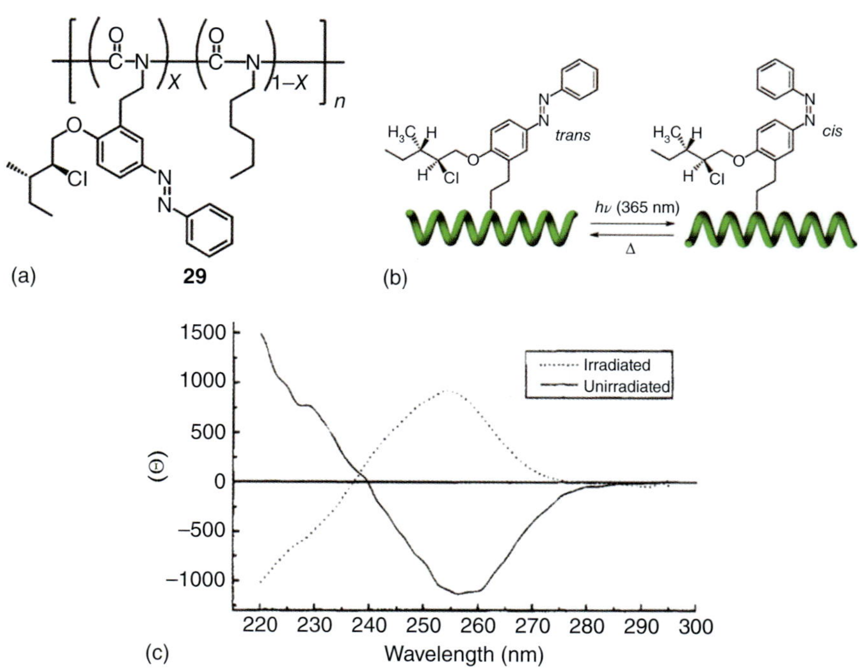

Figure 4.28 Chemical structure (a) and photoisomerization-induced helix inversion of polyisocyanates **29** (b). Source: Yashima et al. 2009 [89]. Copyright 2009. Reproduced with permission from The American Chemical Society. (c) CD spectra of **29** in tetrahydrofuran (0.5 mg ml^{-1}) before (solid line) and after (dotted line) UV irradiation. Source: Maxein and Zentel 1995 [83]. Copyright 1995. Reproduced with permission from The American Chemical Society.

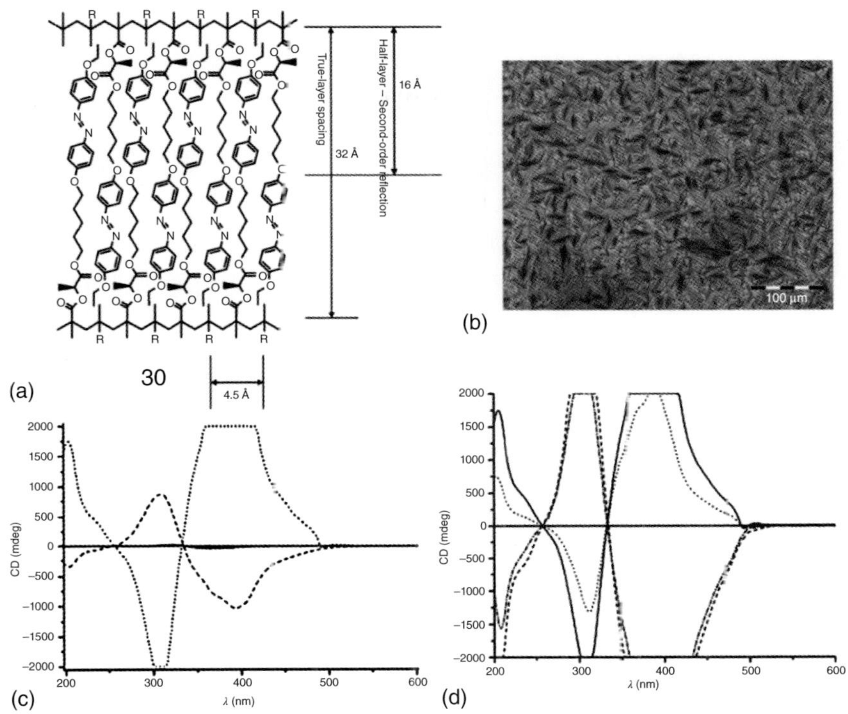

Figure 4.29 (a) Smectic layer spacings of a fully interdigitated side chain chromophoric configuration **30**. (b) Smectic phase from **30** observed with PCM. (c) CD spectra of a thin film in the virgin state (solid line), in smectic phase (dashed line) and after irradiation with *r*-CPL at 488 nm for 30 minutes (dotted line), and (d) irradiated with *r*-CPL (solid line) or *l*-CPL (dashed line) for 30 minutes after isotropization and annealing at 90 °C for 15 minutes. Source: Barberá et al. 2008 [90]. Copyright 2008. Reproduced with permission from Wiley-VCH.

Iamsaard et al. demonstrated a macroscopic helical motion control with a liquid crystalline chiral azopolymer [50]. It was prepared by photopolymerization of an azobenzene-containing monomer (**31**), an achiral liquid crystalline monomer, and chiral dopants (Figure 4.30a). By cutting the resultant polymer film in a specific direction, a polymer spring was formed. Interestingly, molecular motions induced by reversible *cis–trans* photoisomerization resulted in the macroscopic helix modulation, *i.e.* winding, unwinding, and helix inversion, which can be directly observed, and led to a light-driven mechanical work (Figure 4.30b,c).

4.3.2.2 Overcrowded Alkene-Based Polymers

In 2008, Pijper et al. [91] reported a light-directed reversible handedness inversion process in the polyisocyanate system functionalized with a molecular motor **32** (Figure 4.31). The authors covalently linked an overcrowded-alkene-based molecular switch to the terminus of a polyisocyanate polymer (Figure 4.31A). This terminal functionalization with molecular motor enabled light-driven reversible control over the handedness of the polyisocyanate backbone. The

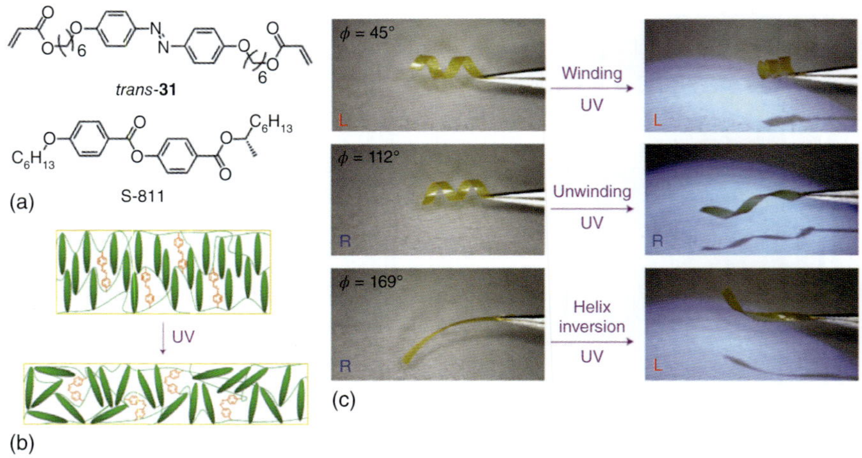

Figure 4.30 (a) Chemical structures of a photoresponsive azobenzene monomer **31** and a chiral dopant S-811. (b) Photoactuation modes of the polymer springs doped with S-811: Spiral ribbons irradiated for two minutes with UV light (365 nm) display isochoric winding, unwinding, and helix inversion as dictated by their initial shape and geometry. (c) Under irradiation with ultraviolet light, the ribbons contract along the director and expand in the perpendicular directions, as is consistent with an ultraviolet-induced increase of disorder. Source: Iamsaard et al. 2014 [50]. Copyright 2014. Reproduced with permission from Nature Publishing Group.

reversible handedness inversion by photoirradiation was obtained from photoresponsive lyotropic CLC as seen in Figure 4.31A. Cholesteric texture and pitch clearly made transitions upon UV or visible light irradiation, which resulted in the ratio change between (2′*S*)-(*M*)-**32** and (2′*S*)-(*P*)-**32** form, as observed from POM images in Figure 4.31B.

Meanwhile, Zhao et al. [92] reported on molecular-motor-based double helicates (**33** in Figure 4.32a) in which, for the first time, the multistage assembly process and helicity could be driven by light in a fully reversible manner. The responsive supramolecular systems with a stepwise change in self-assembly of metal complexes were induced by a light-driven rotary motor scaffold as described in Figure 4.32b. The sequence of events in this dynamic self-assembled system (from oligomers to monomers and a direction of helix inversion of $P' \rightarrow M'$ or $M' \rightarrow P'$) is governed by the unidirectional rotary motion of the motor unit. Initially, coordination oligomers of bipyridine–Cu(I) complex can be formed by treating **33** with Cu(I). Then, these oligomers can split into monomers with P'-helicity due to the unidirectional rotation of the molecular motor from the stable (*P,P*)-*trans* state to the less stable (*M,M*)-*cis* state upon light irradiation. The mechanism of disassembly of the oligomers or unfolding of the metal-helicates following irradiation could plausibly be a partial unwrapping of the complex, assisted by acetonitrile as a small monodentate ligand for Cu(I), that precedes the photoisomerization steps. The inversion of the chirality of the Cu-helicate from P' to M' is governed by a subsequent thermal helical inversion (from *M,M* to *P,P*) of the molecular motor core structure. Photoisomerization of stable (*P,P*)-*cis* state to (*M,M*)-*cis* state through the rotary cycle via the

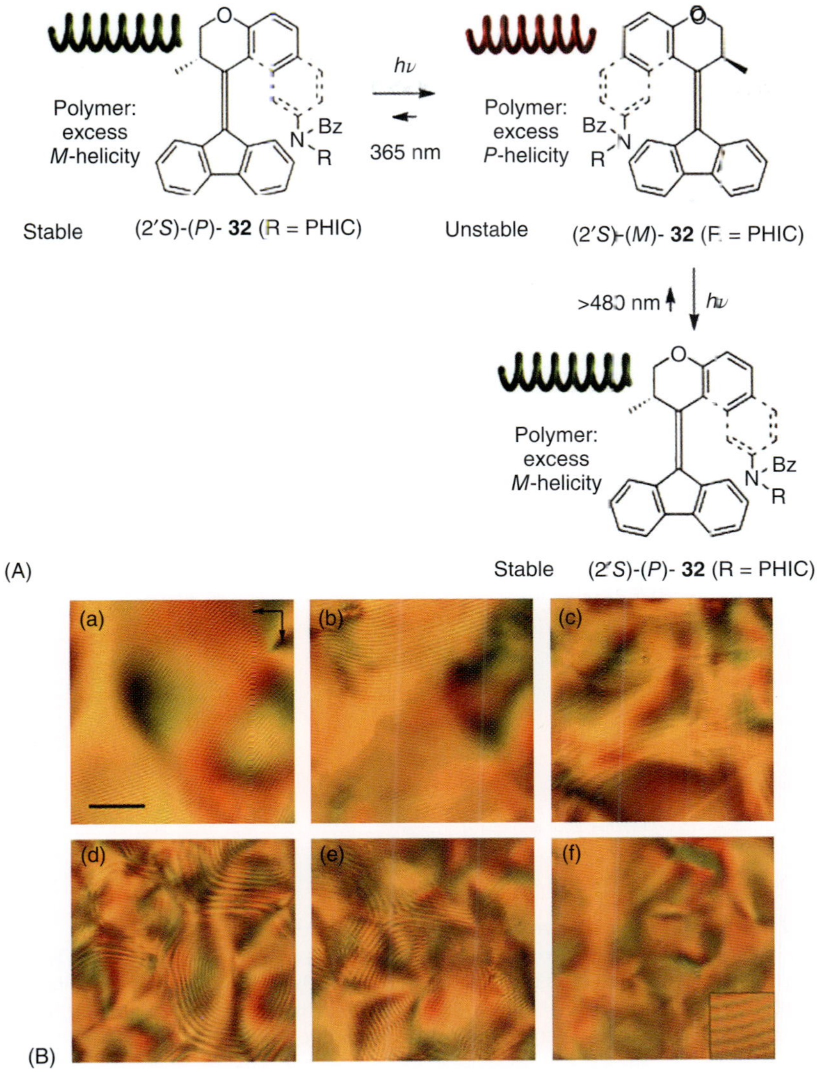

Figure 4.31 (A) Rotary motion and chiral handedness inversion of the overcrowded-alkene-based polymer **32**. (B) Optical micrographs of a thin film (thickness, 200 μm) of **32** in toluene (30 wt%) (a) before irradiation (p = 6.0 μm) and (b) after 15 minutes (p = 6.2 μm), (c) 45 minutes, (d) 90 minutes, and (e) 150 minutes of UV irradiation (λ = 365 nm), in which a PSS mixture is obtained consisting of a large excess of (2′*S*)-(*M*)-**32** over (2′*S*)-(*P*)-**32**, and (f) after leaving the irradiated sample in the dark overnight (p = 4.5 μm). Source: Pijper et al. 2008 [91]. Copyright 2008. Reproduced with permission from The American Chemical Society.

trans-oligomeric state inverts the chirality of the Cu-helicate back from M' to P'. In UV–vis absorption spectra (Figure 4.32c), the change in the visible light region is ascribed to a metal-to-ligand charge-transfer (MLCT) band, and the corresponding CD signals (Figure 4.32d) help determine the absolute configuration inversion of the dinuclear helicates (positive Cotton effect for P' helicity and negative for M' helicity).

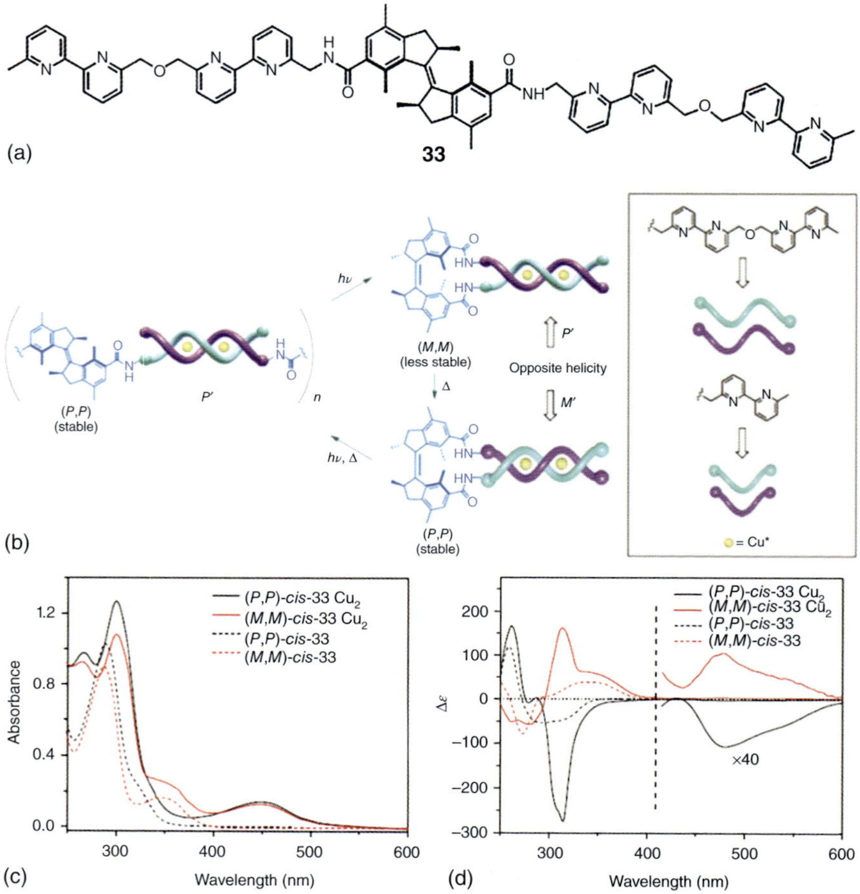

Figure 4.32 (a) Structure of overcrowded alkene based double-stranded helicates **33**. (b) Helicity inversion process of **33** coordinated with Cu(I). UV–vis absorption spectra (c) and CD spectra (d) in CH_3CN of (*P*,*P*)-*cis*-**33**-Cu_2 (black solid line), (*M*,*M*)-*cis*-**33**-Cu_2 (red solid line), (*P*,*P*)-*cis*-**33** (black dashed line), and (*M*,*M*)-*cis*-**33** (red dashed line). Source: Zhao et al. 2017 [92]. Copyright 2017. Reproduced with permission from Nature Publishing Group.

4.4 Summary and Outlook

The systems reviewed here highlight that light stimuli can be used to induce chirality from achiral systems and/or the inversion of the resultant chiral soft materials such as CLCs, self-assembled helical supramolecules, or helical polymers. A photochromic molecular switch (dopant) such as azobenzene, overcrowded alkene, and diarylethene is introduced as a photo-trigger/mediator of chiral information from the light to the host medium. It is demonstrated that either a right- or a left-handed CPL can be utilized to induce a preferred-handed helical conformation in a dynamically racemic helical superstructure in the solid state, which appears to be one of the promising approaches for the handedness-selective synthesis of chiral soft materials with an optical and mechanical function. In addition, the interconversion between enantiomeric chiral organizations in a reversible manner using orthogonal CPL or CPUL, or even turning off the photoinduced chiroptical properties by thermal treatment or using unpolarized light, is possible.

Moreover, photochemical helical inversion is achieved with chiral liquid crystals and polymers based on intrinsically chiral photoresponsive dopants with a certain handedness, and it also photochemically switches to the other handedness by UV or visible light irradiation in a reversible manner. Molecules of opposite handedness simultaneously influence the orientation of the host molecules, and a subtle unbalance between the two handednesses determines the helicity of the system. An initial HTP and its switching upon photoisomerization of a chiral dopant in CLC significantly affect the modulating degree of the helical pitch including the length, handedness, and director orientation.

One of the innovative aspects of the presented works is that photoinduced chiral transition is realized, and also the chiral information is amplified and expressed at various scales, which offers a significant advantage of photo-mediated chiral materials and their applications for advanced photonics such as CPL-utilized display, reflectors, optical storage, and chiral sensors, and for attaining macroscopic work.

Nevertheless, it should be remarked that precisely predicting which system would provide certain targeted chiroptical or mechanical properties still remains a challenge. The interplay of ongoing efforts to reveal the underlying principle linking molecular events and the macroscopic phenomena and to develop novel chiral functions from photochromic switches would pave the way to providing a general method for the development of chiral photoswitches and chiral supramolecules with tailor-made functions.

References

1 Mason, S.F. (1984). Origins of biomolecular handedness. *Nature* 311: 19.

2 Liu, M., Zhang, L., and Wang, T. (2015). Supramolecular chirality in self-assembled systems. *Chem. Rev.* 115: 7304.

3 Yashima, E., Ousaka, N., Taura, D. et al. (2016). Supramolecular helical systems: helical assemblies of small molecules, foldamers, and polymers with chiral amplification and their functions. *Chem. Rev.* 116: 13752.

4 Mateos-Timoneda, M.A., Crego-Calama, M., and Reinhoudt, D.N. (2004). Supramolecular chirality of self-assembled systems in solution. *Chem. Soc. Rev.* 33: 363.

5 Thomas, R. and Tamaoki, N. (2011). Chirality transfer from chiral solvents and its memory in an azobenzene derivative exhibiting photo-switchable racemization. *Org. Biomol. Chem.* 9: 5389.

6 Yin, L., Liu, M., Zhao, Y. et al. (2018). Supramolecular chirality induced by chiral solvation in achiral cyclic azo-containing polymers: topological effects on chiral aggregation. *Polym. Chem.* 9: 769.

7 Feringa, B.L., van Delden, R.A., and Koumura, N. (2000). Chiroptical molecular switches. *Chem. Rev.* 100: 1789.

8 Hembury, G.A., Borovkov, V.V., and Inoue, Y. (2008). Chirality-sensing supramolecular systems. *Chem. Rev.* 108: 1.

9 Tejedor, R.M., Oriol, L., Serrano, J.L., and Sierra, T. (2008). Chiral photochemical induction in liquid crystals. *J. Mater. Chem.* 18: 2899.

10 Breitenlechner, S. and Bach, T. (2008). A polymer-bound chiral template for enantioselective photochemical reactions. *Angew. Chem. Int. Ed.* 47: 7957.

11 Mes, T., van der Weegen, R., Palmans, A.R.A., and Meijer, E.W. (2011). Single-chain polymeric nanoparticles by stepwise folding. *Angew. Chem. Int. Ed.* 50: 5085.

12 Wu, S.-T., Cai, Z.-W., Ye, Q.-Y. et al. (2014). Enantioselective synthesis of a chiral coordination polymer with circularly polarized visible laser. *Angew. Chem. Int. Ed.* 53: 12860.

13 Yang, G., Hu, J., Xia, H. et al. (2017). Near-infrared circularly polarized light triggered enantioselective photopolymerization by using upconversion nanophosphors. *Chem. Eur. J.* 23: 8032.

14 Ruslim, C. and Ichimura, K. (2000). Conformational effect on macroscopic chirality modification of cholesteric mesophases by photochromic azobenzene dopants. *J. Phys. Chem. B* 104: 6529.

15 Ruslim, C. and Ichimura, K. (2001). Conformation-assisted amplification of chirality transfer of chiral Z-azobenzenes. *Adv. Mater.* 13: 37.

16 Eelkema, R. and Feringa, B.L. (2006). Amplification of chirality in liquid crystals. *Org. Biomol. Chem.* 4: 3729.

17 Jin, L., Li, Y., Ma, J., and Li, Q. (2010). Synthesis of novel thermally reversible photochromic axially chiral spirooxazines. *Org. Lett.* 12: 3552.

18 Yokoyama, Y. and Sagisaka, T. (1997). Reversible control of pitch of induced cholesteric liquid crystal by optically active photochromic fulgide derivatives. *Chem. Lett.* 26: 687.

19 Sagisaka, T. and Yokoyama, Y. (2000). Reversible control of cholesteric liquid crystals by photochromism of chiral fulgide derivatives. *Bull. Chem. Soc. Jpn.* 73: 191.

20 Kim, Y., Frigoli, M., Vanthuyne, N., and Tamaoki, N. (2017). Helical naphthopyran dopant for photoresponsive cholesteric liquid crystal. *Chem. Commun.* 53: 200.

21 Feringa, B.L. (2007). The art of building small: from molecular switches to molecular motors. *J. Org. Chem.* 72: 6635.

22 Irie, M., Fukaminato, T., Matsuda, K., and Kobatake, S. (2014). Photochromism of diarylethene molecules and crystals: memories, switches, and actuators. *Chem. Rev.* 114: 12174.

23 Wang, Y. and Li, Q. (2012). Light-driven chiral molecular switches or motors in liquid crystals. *Adv. Mater.* 24: 1926.

24 Bisoyi, H.K. and Li, Q. (2016). Light-driven liquid crystalline materials: from photo-induced phase transitions and property modulations to applications. *Chem. Rev.* 116: 15089.

25 Yan, X., Wang, F., Zheng, B., and Huang, F. (2012). Stimuli-responsive supramolecular polymeric materials. *Chem. Soc. Rev.* 41: 6042.

26 Qing, G. and Sun, T. (2012). The transformation of chiral signals into macroscopic properties of materials using chirality-responsive polymers. *NPG Asia Mater.* 4: e4.

27 Inoue, Y. (1992). Asymmetric photochemical reactions in solution. *Chem. Rev.* 92: 741.

28 Suarez, M. and Schuster, G.B. (1995). Photoresolution of an axially chiral bicyclo[3.3.0]octan-3-one: phototriggers for a liquid-crystal-based optical switch. *J. Am. Chem. Soc.* 117: 6732.

29 Hashim, P.K., Thomas, R., and Tamaoki, N. (2011). Induction of molecular chirality by circularly polarized light in cyclic azobenzene with a photoswitchable benzene rotor. *Chem. Eur. J.* 17: 7304.

30 Feringa, B.L. (ed.) (2001). Molecular Switches. Weinheim: Wiley-VCH.

31 Huck, N.P.M., Jager, W.F., de Lange, B., and Feringa, B.L. (1996). Dynamic control and amplification of molecular chirality by circular polarized light. *Science* 273: 1686.

32 Tamaoki, N. and Wada, M. (2006). Dynamic control of racemization rate through E-Z photoisomerization of azobenzene and subsequent partial photoresolution under circular polarized light. *J. Am. Chem. Soc.* 128: 6284.

33 Rijeesh, K., Hashim, P.K., Noro, S., and Tamaoki, N. (2015). Dynamic induction of enantiomeric excess from a prochiral azobenzene dimer under circularly polarized light. *Chem. Sci.* 6: 973.

34 Choi, S.-W., Izumi, T., Hoshino, Y. et al. (2006). Circular-polarization-induced enantiomeric excess in liquid crystals of an achiral, bent-shaped mesogen. *Angew. Chem. Int. Ed.* 45: 1382.

35 Vera, F., Tejedor, R.M., Romero, P. et al. (2007). Light-driven supramolecular chirality in propeller-like hydrogen-bonded complexes that show columnar mesomorphism. *Angew. Chem. Int. Ed.* 46: 1873.

36 Vera, F., Barber, J., Romero, P. et al. (2010). Orthogonal action of noncovalent interactions for photoresponsive chiral columnar assemblies. *Angew. Chem. Int. Ed.* 49: 4910.

37 Alam, M.Z., Yoshioka, T., Ogata, T. et al. (2007). Influence of helical twisting power on the photoswitching behavior of chiral azobenzene compounds: applications to high-performance switching devices. *Chem. Eur. J.* 13: 2641.

38 Li, Y., Wang, M., Urbas, A., and Li, Q. (2013). A photoswitchable and thermally stable axially chiral dithienylperfluorocyclopentene dopant with high helical twisting power. *J. Mater. Chem. C* 1: 3917.

39 Nikolova, L., Todorov, T., Ivanov, M. et al. (1997). Photoinduced circular anisotropy in side-chain azobenzene polyesters. *Opt. Mater.* 8: 255.

40 Weigert, F. (1919). Dichroism induced in a fine-grain silverchloride emulsion by a beam of linearly polarized light. *Verh. Dtsch. Phys. Ges.* 21: 479.

41 Tamaoki, N. and Kamei, T. (2010). Reversible photo-regulation of the properties of liquid crystals doped with photochromic compounds. *J. Photochem. Photobiol. C: Photochem. Rev.* 11: 47.

42 Tejedor, R.M., Oriol, L., Serrano, J.L. et al. (2007). Photoinduced chiral nematic organization in an achiral glassy nematic azopolymer. *Adv. Funct. Mater.* 17: 3486.

43 Royes, J., Provenzano, C., Pagliusi, P. et al. (2014). A bifunctional amorphous polymer exhibiting equal linear and circular photoinduced birefringences. *Macromol. Rapid Commun.* 35: 1890.

44 Choi, S.-W., Ha, N.Y., Shiromo, K. et al. (2006). Photoinduced circular anisotropy in a photochromic W-shaped-molecule-doped polymeric liquid crystal film. *Phys. Rev. E* 73: 021702.
45 Wang, L., Yin, L., Zhang, W. et al. (2017). Circularly polarized light with sense and wavelengths to regulate azobenzene supramolecular chirality in optofluidic medium. *J. Am. Chem. Soc.* 139: 13218.
46 Zou, G., Jiang, H., Kohn, H. et al. (2009). Control and modulation of chirality for azobenzene-substituted polydiacetylene LB films with circularly polarized light. *Chem. Commun.* 5627.
47 Kim, J., Lee, J., Kim, W.Y. et al. (2015). Induction and control of supramolecular chirality by light in self-assembled helical nanostructures. *Nat. Commun.* 6: 6959.
48 Royes, J., Polo, V., Uriel, S. et al. (2017). Chiral supramolecular organization from sheet-like achiral gel: study of chiral photoinduction. *Phys. Chem. Chem. Phys.* 19: 13622.
49 Lv, Z., Chen, Z., Shao, K. et al. (2016). Stimuli-directed helical chirality inversion and bio-applications. *Polymers* 8: 310.
50 Iamsaard, S., Aßhoff, S.J., Matt, B. et al. (2014). Conversion of light into macroscopic helical motion. *Nat. Chem.* 6: 229.
51 Russew, M.-M. and Hecht, S. (2010). Photoswitches: from molecules to materials. *Adv. Mater.* 22: 3348.
52 Kitzerow, H. and Bahr, C. (eds.) (2001). Chirality in Liquid Crystals. New York: Springer.
53 Katsonis, N., Lacaze, E., and Ferrarini, A. (2012). Controlling chirality with helix inversion in cholesteric liquid crystals. *J. Mater. Chem.* 22: 7088.
54 Kim, Y. and Tamaoki, N. (2014). A photoresponsive planar chiral azobenzene dopant with high helical twisting power. *J. Mater. Chem. C* 2: 9258.
55 Kausar, H.N., Kuwahara, Y., Ogata, T., and Kurihara, S. (2011). Photocontrolled manipulation of a microscale object: a rotational or translational mechanism. *Chem. Eur. J.* 17: 508.
56 Qin, L., Gu, W., Wei, J., and Yu, Y. (2018). Piecewise phototuning of self-organized helical superstructures. *Adv. Mater.* 30: 1704941.
57 Li, Y., Urbas, A., and Li, Q. (2012). Reversible light-directed red, green and blue reflections with thermal stability enabled by a self-organized helical superstructure. *J. Am. Chem. Soc.* 134: 9573.
58 Bisoyi, H.K. and Li, Q. (2016). Light-directed dynamic chirality inversion in functional self-organized helical superstructures. *Angew. Chem. Int. Ed.* 55: 2994.
59 Mathews, M. and Tamaoki, N. (2008). Planar chiral azobenzenophanes as chiroptic switches for photon mode reversible reflection color control in induced chiral nematic liquid crystals. *J. Am. Chem. Soc.* 130: 11409.
60 Thomas, R., Yoshida, Y., Akasaka, T., and Tamaoki, N. (2012). Influence of a change in helical twisting power of photoresponsive chiral dopants on rotational manipulation of micro-objects on the surface of chiral nematic liquid crystalline films. *Chem. Eur. J.* 18: 12337.
61 Mathews, M. and Tamaoki, N. (2009). Reversibly tunable helicity induction and inversion in liquid crystal self-assembly by a planar chiroptic trigger molecule. *Chem. Commun.* 3609.

62 Mathews, M., Zola, R.S., Hurley, S. et al. (2010). Light-driven reversible handedness inversion in self-organized helical superstructures. *J. Am. Chem. Soc.* 132: 18361.
63 Li, Y., Wang, M., White, T.J. et al. (2013). Azoarenes with opposite chiral configurations: light-driven reversible handedness inversion in self-organized helical superstructures. *Angew. Chem., Int. Ed.* 52: 8925.
64 Fu, D., Li, J., Wei, J., and Guo, J. (2015). Effects of terminal chain length in hydrogen-bonded chiral switches on phototunable behavior of chiral nematic liquid crystals: helicity inversion and phase transition. *Soft Matter* 11: 3034.
65 Hayasaka, H., Miyashita, T., Nakayama, M. et al. (2012). Dynamic photoswitching of helical inversion in liquid crystals containing photoresponsive axially chiral dopants. *J. Am. Chem. Soc.* 134: 3758.
66 Li, Y., Xue, C., Wang, M. et al. (2013). Photodynamic chiral molecular switches with thermal stability: from reflection wavelength tuning to handedness inversion of self-organized helical super structures. *Angew. Chem., Int. Ed.* 52: 13703.
67 White, T.J., Cazella, S.A., Freer, A.S. et al. (2011). Widely tunable, photoinvertible cholesteric liquid crystals. *Adv. Mater.* 23: 1389.
68 Chen, W.-C., Lin, P.-C., Chen, C.-H., and Chen, C.-T. (2010). Modulation of photoswitching profiles by 10,11-dialkoxymethyl substituents in C2-symmetric dibenzosuberane-based helicenes. *Chem. Eur. J.* 16: 12822.
69 Goh, M. and Akagi, K. (2008). Powerful helicity inducers: axially chiral binaphthyl derivatives. *Liq. Cryst.* 35: 953.
70 Li, Y. and Li, Q. (2012). Photochemically reversible and thermally stable axially chiral diarylethene switches. *Org. Lett.* 14: 4362.
71 Li, Y., Wang, M., Wang, H. et al. (2014). Rationally designed axially chiral diarylethene switches with high helical power. *Chem. Eur. J.* 20: 16286.
72 Wang, L., Dong, H., Li, Y. et al. (2015). Luminescence-driven reversible handedness inversion of self-organized helical superstructures enabled by a novel near-infrared light nanotransducer. *Adv. Mater.* 27: 2065.
73 Zheng, Z., Li, Y., Bisoyi, H.K. et al. (2016). Three-dimensional control of the helical axis of a chiral nematic liquid crystal by light. *Nature* 531: 352.
74 Zheng, Z., Zola, R.S., Bisoyi, H.K. et al. (2017). Controllable dynamic zigzag pattern formation in a soft helical superstructure. *Adv. Mater.* 29: 1701903.
75 Eelkema, R., Pollard, M.M., Vicario, J. et al. (2006). Molecular machines: nanomotor rotates microscale objects. *Nature* 440: 163.
76 Pijper, T.C., Pijper, D., Pollard, M.M. et al. (2010). An enantioselective synthetic route toward second-generation light-driven rotary molecular motors. *J. Org. Chem.* 75: 825.
77 Kistemaker, J.C.M., Stacko, P., Roke, D. et al. (2017). Third-generation light-driven symmetric molecular motors. *J. Am. Chem. Soc.* 139: 9650.
78 Eelkema, R., Pollard, M.M., Katsonis, N. et al. (2006). Rotational reorganization of doped cholesteric liquid crystalline films. *J. Am. Chem. Soc.* 128: 14397.
79 Bosco, M.G.M.J., Eelkema, R., Katsonis, N. et al. (2008). Photoinduced reorganization of motor-doped chiral liquid crystals: bridging molecular isomerization and texture rotation. *J. Am. Chem. Soc.* 130: 14615.

80 Chen, J., Leung, F.K.-C., Stuart, M.C.A. et al. (2018). Artificial muscle-like function from hierarchical supramolecular assembly of photoresponsive molecular motors. *Nat. Chem.* 10: 132.

81 Aßhoff, S.J., Iamsaard, S., Bosco, A. et al. (2013). Time-programmed helix inversion in phototunable liquid crystals. *Chem. Commun.* 49: 4256.

82 Chen, C.-T., Chen, C.-H., and Ong, T.-G. (2013). Complementary helicity interchange of optically switchable supramolecular-enantiomeric helicenes with (−)-gel-sol-(+)-gel transition ternary logic. *J. Am. Chem. Soc.* 135: 5294.

83 Maxein, G. and Zentel, R. (1995). Photochemical inversion of the helical twist sense in chiral polyisocyanates. *Macromolecules* 28: 8438.

84 Mayer, S. and Zentel, R. (2001). Chiral polyisocyanates, a special class of helical polymers. *Prog. Polym. Sci.* 26: 1973.

85 Pijper, D. and Feringa, B.L. (2008). Control of dynamic helicity at the macro- and supramolecular level. *Soft Matter* 4: 1349.

86 Li, J., Schuster, G.B., Cheon, K.-S. et al. (2000). Switching a helical polymer between mirror images using circularly polarized light. *J. Am. Chem. Soc.* 122: 2603.

87 Green, M.M., Garetz, B.A., Munoz, B., and Chang, H.P. (1995). Majority rules in the copolymerization of mirror image isomers. *J. Am. Chem. Soc.* 117: 4181.

88 Muller, M. and Zentel, R. (1994). Photochemical inversion of the helical twist sense in chiral polyisocyanates. *Macromolecules* 27: 4404.

89 Yashima, E., Maeda, K., Iida, H. et al. (2009). Helical polymers: synthesis, structures, and functions. *Chem. Rev.* 109: 6102.

90 Barberá, J., Giorgini, L., Paris, F. et al. (2008). Supramolecular chirality and reversible chiroptical switching in new chiral liquid-crystal azopolymers. *Chem. Eur. J.* 14: 11209.

91 Pijper, D., Jongejan, M.G.M., Meetsma, A., and Feringa, B.L. (2008). Light-controlled supramolecular helicity of a liquid crystalline phase using a helical polymer functionalized with a single chiroptical molecular switch. *J. Am. Chem. Soc.* 130: 4541.

92 Zhao, D., van Leeuwen, T., Cheng, J., and Feringa, B.L. (2017). Dynamic control of chirality and self-assembly of double-stranded helicates with light. *Nat. Chem.* 9: 250.

5

Soft Photoactuators in Microfluidics

Lu-Jian Chen[1], and Quan Li[2]

[1] *Xiamen University, Department of Electronic Engineering, College of Electronic Science and Technology, 422 Siming South Road, Xiamen, Fujian 361005, China*
[2] *Kent State University, Liquid Crystal Institute and Chemical Physics Interdisciplinary Program, 1425 Lefton Esplanade, Kent, OH 44242, USA*

5.1 Introduction

Microfluidics, also known as lab-on-a-chip (LOC) and micro-total-analysis-system (μTAS), cover a set of multidisciplinary technologies dealing with the manipulation of small amounts of liquids to perform reactions, analyses, or fundamental investigations in biology, physics, and at the interdisciplinary frontiers. In contrast to the conventional bulky bench-top instruments and the associated manual methods to handle large amounts of biological and chemical reagents that are pretty time consuming and not so environmentally friendly, a fine control over the motion of continuous fluids or nano/picoliter-sized discrete droplets in microscale channels is beneficial in miniaturized systems. Therefore, there is a growing interest in downscaling the corresponding processes within LOCs and still retaining cost-effectiveness and ease of fabrication. The microfluidic systems possess intrinsic features including minimized consumption of reagents, portability, increased automation, reduced time, and cost efficiency, which make them particularly attractive from a wide range of laboratory and industrial perspectives.

The successful operations of microfluidic devices typically rely on electrokinetic, mechanical, hydraulic, or pneumatic forces and are usually driven by external transducers, such as pumps, valves, and electrodes. Recent developments of novel stimuli-responsive soft materials (e.g., polymers, gels, and liquid crystals) in response to small variations of environmental conditions [1–4] provide an alternative choice and contribute to the rapid evolution of next-generation LOCs by integrating different functionalities onto the same chip, which comprises not only tiny channels but also multiple microactuators with mechanical properties (e.g., micropumps, micromixers, and microvalves) or optical properties (e.g., microlenses). In the sophisticated microsystems toward automotive control of liquid motions, micropumps and micromixers are essential for fluidic handling,

Photoactive Functional Soft Materials: Preparation, Properties, and Applications, First Edition.
Edited by Quan Li.

and microvalves are crucial for fluidic control. Less energy is required for the operation, and the dead volume is minimized through the elimination of interconnecting tubes between the functional components.

It is believed that the integration of microfluidic actuators using stimuli-responsive soft materials offers new possibilities to facilitate simplified designs of efficient chip configurations as a means of noncontact microfluidic flow control consequently, reducing the dimensions and complexity of the microfluidic platforms. In general, most of the functional stimuli-responsive soft materials respond to a set of physical, chemical, and biochemical stimuli ranging from temperature, pH value, light, electric and magnetic fields, and chemical analytes to biological components, and hence change their shape with respect to the configuration or dimension to exert mechanical force on surrounding fluids. Among these external stimuli, light is particularly appealing for LOC applications as it can provide contactless spatial and temporal control, thus allowing the microfluidic functions to be integrated in geometrically confined environments [5–8]. It is noteworthy that low-energy light stimulus could be biocompatible to elicit responses in biological systems.

The term "optofluidics" is engendered by anything that involves both light and fluidics. In this context, it is a new paradigm and relatively straightforward to consider two specific processes: (i) light can manipulate fluids and (ii) fluids can manipulate light. As depicted in Figure 5.1, most of the direct strategies to convert the light energy into liquid motions by using optical forces (through radiation pressure and optical tweezers) and light-induced capillary forces (from a light-induced wettability gradient and Marangoni effects) have been well categorized by Baigl [9]. Nevertheless, the former approach requires special optical setups, and the latter

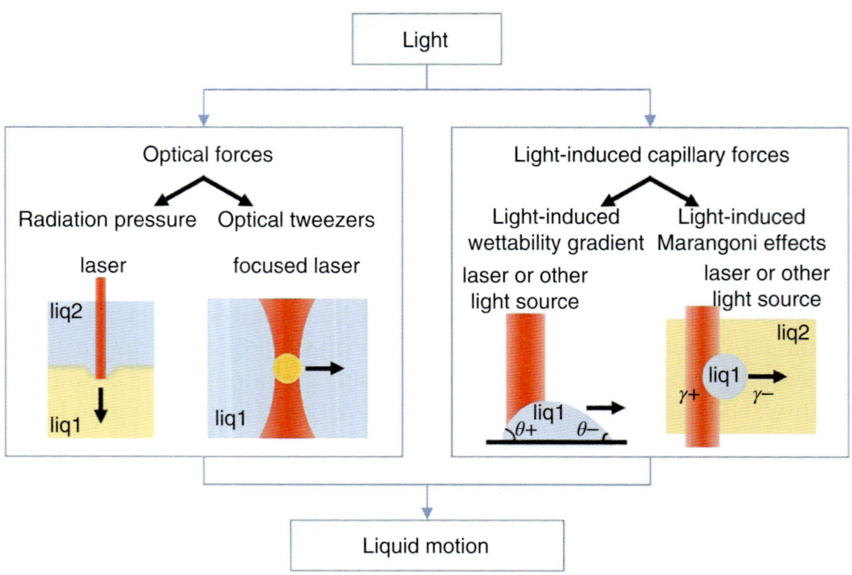

Figure 5.1 Main strategies to directly induce liquid motion driven by light stimulus.

one is not practical because of the deficiencies such as short-distance motion, local heating, and possible contamination problems.

Actually, a direct interaction between light and fluids is not necessarily required, thereby endowing the optofluidic devices with design flexibilities. To date, indirect interactions between light and fluids in microfluidics have been mostly mediated by means of photoactive functional soft materials that act as microactuators. Advanced LOC systems are integrated with these soft photoactive components for pumping, mixing, and valving, as well as the fluid channels typically in microscale feature dimensions. Compared with the abovementioned direct light–fluids interactions, we would like to highlight that further opportunities can be made available with the attempts that include and are not limited to the motion control of fluids by soft photoactuators.

The purpose of this chapter is to provide an overview on the advances and trends of photoactive functional soft materials aiming at the design and fabrication of microactuators, also known as soft photoactuators. In doing this, we occasionally cover potential microsystems that might find microfluidic applications in the future although it is a developing area. From the point of view of material types and working principles, the chapter is organized into four major parts, corresponding to the main application scopes of soft photoactuators in microfluidics. The discussion on the photoactive soft materials as generic microactuators in enclosed microfluidic channels is followed by the exploration of photoactive soft materials as optical microcomponents. Finally, we offer general conclusions and the perspective for future work at the end of this chapter.

5.2 Photoactive Soft Materials as Generic Microactuators

As shown in Figure 5.2, the basic principles of microfluidic actuation for the most popular classes of stimulus-responsive polymers have been classified by Hilber with respect to the two main applications in the generic microfluidic systems with enclosed channels: pumping/mixing (a–d) and valving (e–f) [10]. Micropumps/micromixers and microvalves are essential microactuators to actively handle and control the fluid flow, respectively. The basic physical changes under stimulation are illustrated as follows: (i) change in mechanical properties, (ii) nonuniform strain generation, and (iii) change in volume. We review herein how to employ light to regulate liquid flow by using photoactive soft materials as generic microactuators, and discuss the general principles to improve the function and efficiency of LOC devices by soft photoactuators, thus offer intriguing possibilities in the facilitation of overall design of microfluidic systems. Microvalves are discussed first, followed by micropumps and micromixers, since many micropumps could be built from a string of microvalves operated in sequence.

5.2.1 Light-Driven Microvalves

In the past decade, many breakthroughs have been made in the development of integrated microactuators with novel stimulus-active polymer hydrogels (PHs),

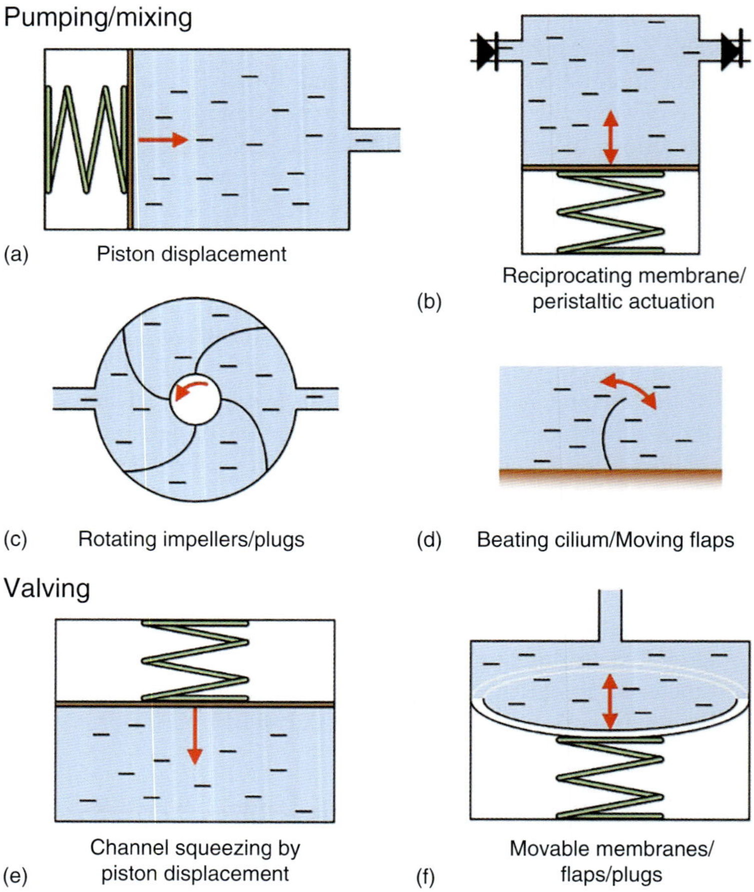

Figure 5.2 Generic actuation concepts based on stimulus-responsive polymers.

which are three-dimensional (3D) polymer networks with high water proportion. Hydrogel actuators do not require power, electronics, or control algorithms. With the variation of water proportion in the polymer network, which leads to almost instantaneous shrinkage or swelling of the polymer, PHs can undergo considerable volume changes that reach a factor of 10 or even more. Obviously, hydrogels can be operated only in aqueous media, which is ideal for water-based microfluidics. Existing studies of stimulus-active bulk PHs suggest that these materials should be suitable for microfluidic applications. In contrast to the off-chip mechanical valves, dimensions and complexity of microfluidic systems are greatly reduced by the use of hydrogels as on-chip microvalves to enable the regulation of fluid flow in enclosed channels. The *in situ* photoinitiated polymerization process allows the fabrication of multiple components either sequentially or simultaneously.

Automatic microfluidic systems are built based on the phase transition behavior of various stimulus-active hydrogels in response to a set of external stimuli from the surrounding medium, e.g., physical values such as temperature, light, electric, and magnetic fields, and chemical parameters such as pH value, solvent composition, and the concentration of specific ions. Recently, pH-responsive hydrogel materials have been used as valves in microfluidic devices. However, many applications are not compatible with a wide range of pH values. Particularly, the utilization of light stimulus as a remote control mechanism is attractive and has been demonstrated to achieve independent control of microvalves based on various photoactive hydrogel systems.

5.2.1.1 Hydrogel Microvalves Actuated by Photothermal Effect

Poly(*N*-isopropylacrylamide) (pNIPAAm) hydrogels are one of the most prominent PHs that can show a temperature-induced phase transition, relying on swelling and deswelling occurring at a certain, well-defined temperature called the lower critical solution temperature (LCST) of 32 °C. The first demonstration of thermally actuated valves was realized in the mid-1990s [11]. The potential use of *p*NIPAAm hydrogels as thermoreversible microactuators in LOC devices has been well explored as microvalves [12]. Above the LCST, *p*NIPAAm chains collapse (valve is open) while below the LCST they undergo a rapid and reversible entropy-driven phase transition to extended hydrated chains that block the flow (valve is closed). It was also found that fluid flowing in Y-shaped microchannels with one inlet and two outlets could be controlled directly by infrared (IR)-laser-heating-induced local sol–gel transformation of the fluid containing the block copolymer of poly(*N*-isopropylacrylamide-*co*-*n*-butyl methacrylate) and poly(ethylene glycol) [13].

The straightforward approach of incorporating light-absorbing photothermal nanoparticles or molecules [14–16] into a thermoresponsive hydrogel can also lead to structures with even faster swelling or shrinking rates by photoinduced heating [17, 18]. To achieve reversible changes in shape and size in response to different light wavelengths, Sershen et al. proposed two hydrogel systems with composites of thermoresponsive *p*NIPAAm and gold nanoparticles in the form of colloid and nanoshells, respectively [19]. The gold–colloid nanocomposite hydrogel collapses in response to green light, while the one with gold–nanoshell collapses in response to near-infrared (NIR) light. The laser light irradiation in the range of 1.6–2.7 W cm^{-2} was converted into heat by nanoparticles. The heat-induced shrinkage of the *p*NIPAAm hydrogels resulted in the opening of microvalves. As shown in Figure 5.3, they demonstrated independent control of two nanocomposite hydrogel valves at a T-junction within a microfluidic device by light irradiation with suitable wavelengths.

The photothermal deformation mechanisms involved in a thermoresponsive hydrogel with light-absorbing nanoparticles were investigated theoretically when considering the equilibrium thermodynamics of a swelling gel through a variational approach; in addition, a case study in which the composite was employed as a microvalve was carried out to validate the numerical approach [20]. Lee et al. found that the light-induced volume change of a hydrogel composite was maximized when the temperature of the medium was just below the LCST [21]. By copolymerization with hydrophobic *N*,*N*-diethylacrylamide (NDEAm), the LCST of *p*NIPAAm could be lowered to maximize light-induced volume change at room temperature. Knowing this relationship between light-

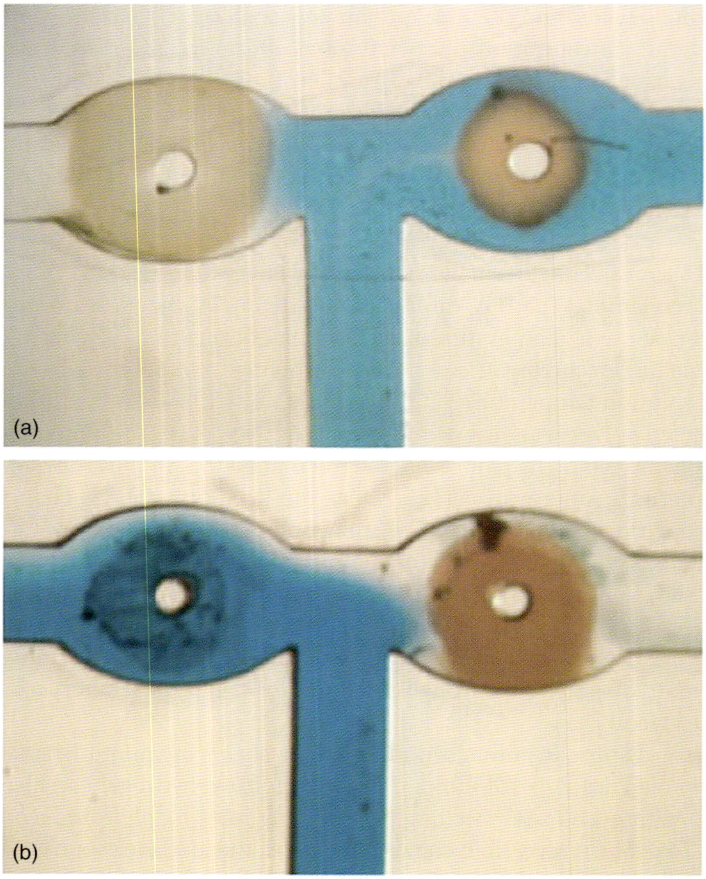

Figure 5.3 Two valves formed at a T-junction in a microfluidic device with a gold–colloid nanocomposite hydrogel and a gold–nanoshell nanocomposite hydrogel. Source: Sershen et al. 2005 [19]. Copyright 2005. Reprinted with permission from John Wiley and Sons.

induced volume changes and thermal behaviors, the light irradiation time could be minimized to achieve a desired volume decrease in the hydrogel.

Jadhav et al. demonstrated that a *p*NIPAAm hydrogel with polypyrrole (PPy) nanoparticles was used to achieve a photoresponsive microvalve for remote actuation with NIR laser irradiation for flow control in microfluidic devices [22]. As illustrated in Figure 5.4, the microvalve is based on in-channel trapping of microgel particles, which undergoes volume change and enables on/off switching of localized fluid flow upon NIR light irradiation. The response rate and the "open" duration of the microvalve can be simply controlled by adjusting the laser power and exposure time. The trapped nanocomposite microgel can be triggered to shrink sufficiently to open a channel within as low as approximately one to two seconds, while it swells to reseal the channel within approximately six to eight seconds. This feature allows speedy flow switching applications. They also employed soft actuators to control fluidic interface between laminar flow streams within a Y-junction device.

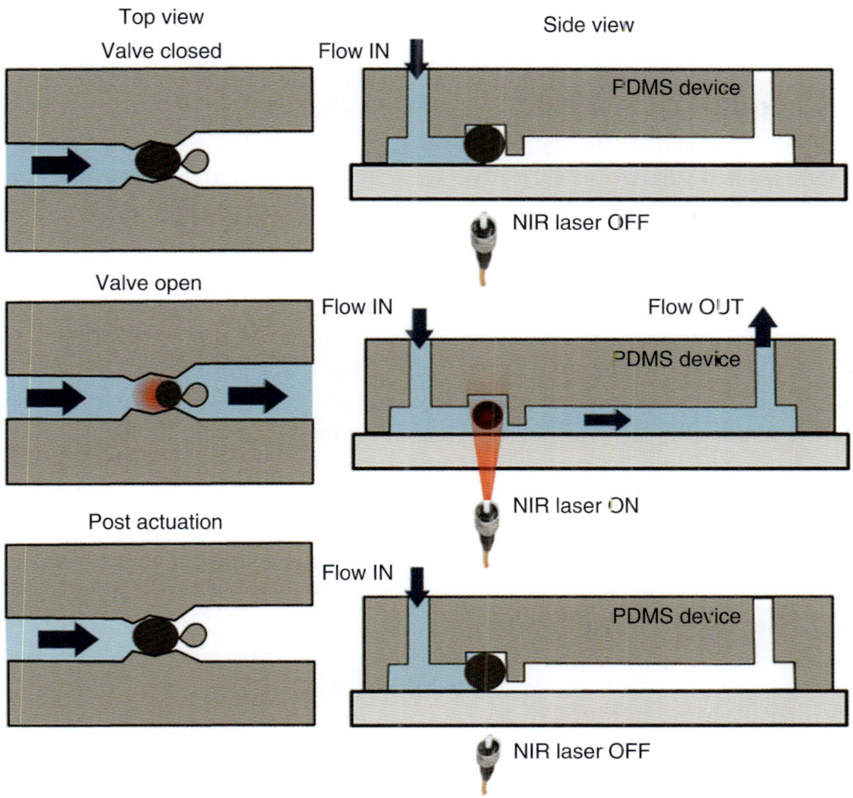

Figure 5.4 Schematic showing the structure and operation of the photoresponsive hydrogel microvalve with a 980 nm NIR laser, allowing flow switching in microfluidic devices. Source: Jadhav et al. 2015 [22]. Copyright 2015. Reprinted with permission from AIP Publishing LLC.

Two-dimensional (2D) materials with exceptional surface-area-to-mass ratio are promising NIR photothermal candidates that can be loaded with high cargo concentrations [16, 23]. Lo et al. demonstrated a *p*NIPAAm hydrogel incorporated with glycidyl methacrylate (GMA)-functionalized graphene oxide (GO) nanocomposite (GO–GMA) that underwent significant volume change in response to IR light irradiation [24]. The GO–GMA hydrogel carries a larger swelling ratio as well. In addition, the transition temperature is lowered by 10 °C due to the incorporation of GO. Zhu et al. reported a facile method to incorporate GO in *p*NIPAAm hydrogels by *in situ* γ-irradiation-assisted polymerization of an aqueous solution of *N*-isopropylacrylamide and GO [25]. Chemically exfoliated MoS_2 nanosheets were also successfully incorporated into *p*NIPAAm hydrogels by *in situ* polymerization [26]. Microvalves with excellent photothermal properties implementing such nanocomposite hydrogels with these 2D materials were fabricated to control the fluidic flow within the microchannel through remote NIR light actuation. The NIR-light-induced temperature change can be tuned effectively by variation of the GO doping concentration and the irradiation time.

5.2.1.2 Hydrogel Microvalves Actuated by Photoisomerization

An alternative popular approach was implemented by copolymerization to create photoactive PHs with covalent attachments between photochromic units and the polymer backbone. Only a few specific photochromic compounds that exhibit light-induced isomerization have been investigated, such as azobenzenes and spiropyrans. Spiropyran derivatives are well-known photochromes that can be reversibly isomerized between metastable states under light illuminations [27, 28]. In an acidic aqueous solution, the phenolate anion may be protonated in the dark, leading to a positively charged open-ring merocyanine (McH) form, whereas visible-light irradiation results in an effective photoisomerization of the open ring McH into closed-ring spirocyclic (Sp) form. When the irradiation is stopped, the Sp form spontaneously converts back to the thermodynamically stable McH form. As shown in Figure 5.5, since the Sp is a highly hydrophobic molecule whereas McH is an ion and thus a more hydrophilic molecule, *p*NIPAAm hydrogels grafted with spirobenzopyran chromophore (*p*SPNIPAAm) undergo drastic light-induced volume variation due to the hydration and dehydration effects of the copolymer and solubility changes in aqueous solution. Consequently, *p*SPNIPAAm hydrogels can swell in the acidic aqueous solutions in the dark and rapidly shrink by visible light irradiation at room temperature. Notably, the volume change from Sp to McH is reversible in the dark.

Sugiura et al. employed *p*SPNIPAAm hydrogels to fabricate microvalves in microfluidic systems [29], and on-demand fluid flow systems constituting both arbitrarily rewritable microchannels and simultaneously controllable microvalves [30]. Blue light irradiation is applied locally to the discrete microvalves to induce the shrinkage of hydrogels, leading to the open state of each microvalve. The microvalve composed of *p*SPNIPAAm hydrogels can be opened by 18–30 seconds light irradiation. As shown in Figure 5.6, the unique noncontact programmed method facilitates the independent control of soft photoactive microvalves on one single chip.

Unfortunately, the abovementioned light-mediated opened microvalve based on *p*SPNIPAAm hydrogels requires more than one hour to swell and be closed again although it is inherently compatible with microfluidics. For practical purposes, the valve is only for "one time" operation. It is believed that chemical modification of the *p*NIPAAm backbone is expected to optimize its response time. Chen et al. found that the mechanical strength and the pressure tolerance of *p*NIPAAm-based microvalves could be tuned by choosing a suitable amount of monomer and cross-linker for the polymer monolith inside the microchannel [31]. IR-light-induced heating was provided by a quartz halogen illuminator. The microvalves can withstand leakage at pressures up to 93 bar and feature response times of opening and closing in the order of only a few seconds.

Another solution toward fast, reversible, and repeatable open/close operations is to modify the chemical nature of photochromic molecules. Schenning and coworkers demonstrated that molecular design of spiropyran derivatives is important to improve the isomerization speed and the reversible swelling/shrinking behaviors of associated hydrogels [32] (Figure 5.7). Acrylic acid acts as a proton source in solution/polymer to stabilize the open-ring hydrophilic McH form in the self-protonating *p*NIPAAm hydrogel. These improved gel valves

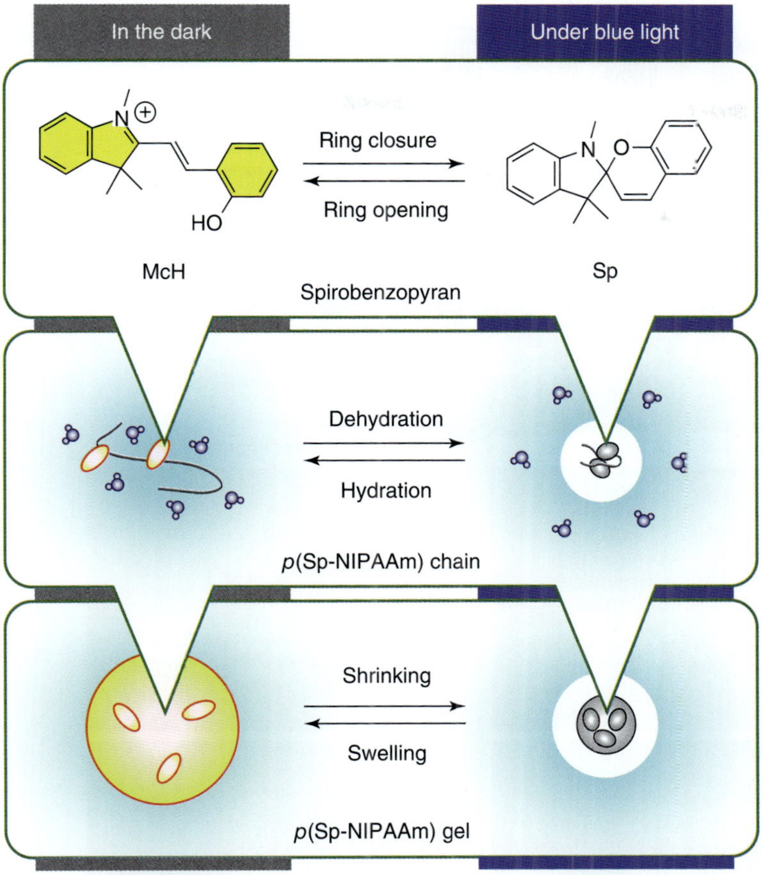

Figure 5.5 Relations between an isomerization of a spirobenzopyran into protonated merocyanine (top), the hydration behavior of a *p*SPNIPAAm chain (middle), and the volume change in a *p*SPNIPAAm gel in acidic aqueous solutions (bottom). Source: Satoh et al. 2011 [27]. Copyright 2011. Reprinted with permission from Royal Society of Chemistry.

integrated within microfluidic channels allow reversible and repeatable operation in minutes. Recently, Coleman et al. described a method for microfluidic flow control based on photoactive *p*SPNIPAAm valves with rapid and reversible actuation properties within practical timescales [33]. Application of pulsed light (450 nm) allows the valves to be held at an intermediate position for extended periods of time, resulting in a forced oscillation and control over the valve through photoisomerization of spiropyran between the McH and Sp forms. Varying the extent of light pulsing enables the instantaneous regulation of flow rate up to 0–27 $\mu l\,min^{-1}$.

Remarkably, a new concept in materials science is the incorporation of ionic liquids (ILs), organic salts that exhibit melting points below 100 °C, within polymer gels to generate ionogels with the inherent advantages of ILs. Compared with an equivalent IL-free hydrogel, the physical robustness of photoresponsive

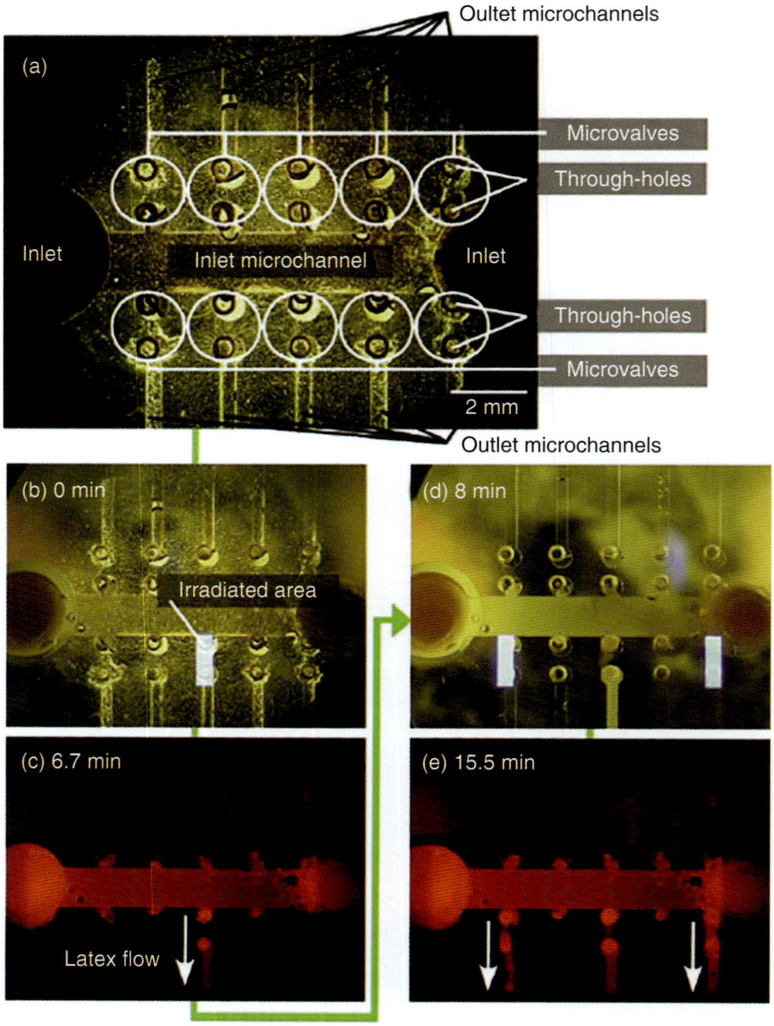

Figure 5.6 Independent and parallel flow control in the PDMS microchannel network equipped with 10 photoresponsive microvalves created from a *p*SPNIPAAm hydrogel sheet. White arrows indicate the flow direction of the latex bead suspension. (a) Microchannel network before irradiation. (b–e) Independent control of a single microvalve and parallel control of multiple microvalves by micropatterned light irradiation, as shown by the latex bead suspension flow after irradiation. Source: Sugiura et al. [30]. Copyright 2008. Reprinted with permission from Royal Society of Chemistry.

phosphonium-based ionogels was improved [34]. Upon irradiation with visible light, the ionogels decreased in volume due to McH to Sp switching and the associated loss of water from the polymer. These ionogels were then incorporated into a microfluidic device as photoactuated microvalves. A simple LED with a mild output of approximately 1 mW cm^{-2} can be used to actuate the microvalve. The swelling/shrinking properties can be controlled by alternating the anion of the phosphonium ILs entrapped within the gel matrix [35]. As shown in

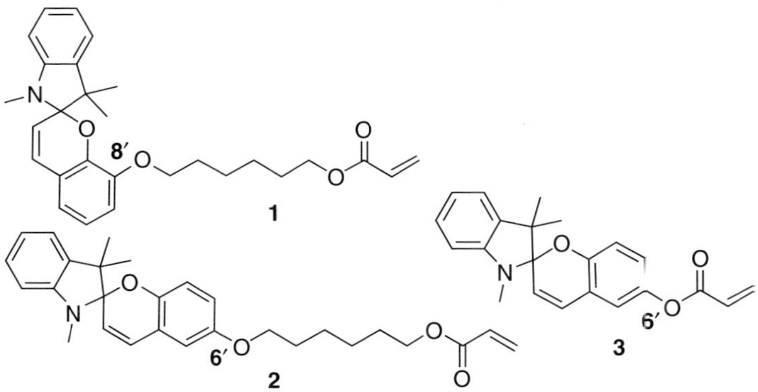

Figure 5.7 The molecular structures of spiropyran derivatives with an ester (3) or ether (2) at the 6′ position and an ether (1) at the 8′ position. Source: ter Schiphorst et al. 2015 [32]. Copyright 2015. Reprinted with permission from American Chemical Society.

Figure 5.8, the valve actuation time and liquid flow within each channel can be individually tuned, facilitating sequential actuation events under common illumination conditions. Moreover, the microvalve fabricated by cross-linked *p*NIPAAm ionogel can be reversibly actuated for up to eight times and resist pressures up to 1100 mbar [36].

5.2.2 Light-Driven Micropumps and Micromixers

Micropumps and micromixers are essential components for fluidic handling. In general, an off-chip micropump is connected to the microfluidic chip to introduce liquid samples and facilitate the fluid flow. The integration of micropumps offers potentials in miniaturized microfluidic systems. The involved mechanisms can be classified into two categories: (i) mechanical type, which includes electrostatic, piezoelectric, thermopneumatic, and electromagnetic effects, etc. or (ii) nonmechanical type, which includes magnetohydrodynamic, electrohydrodynamic, electro-osmotic, electrowetting, electrochemical, and evaporation effects, etc. Even though impressive progress on numerous designs and materials has been achieved to realize on-chip micropumps, the lack of remote on/off switching capabilities limits their applications in microfluidics.

Repeated light stimulus can be transformed into either mechanical or nonmechanical responses to pump fluids. It is found that light momentum can be transferred to trap and manipulate microscale-sized particles, which are used as mobile elements to actuate liquids. Pioneered by Terray et al., time-shared traps generated by optical tweezers were used to control the collective motion of colloidal particles inside microfluidic devices, working as a two-lobe gear pump or as a peristaltic pump [37]. The advantage of this approach is that no specific microparticle fabrication is required, so that it offers a quite large versatility. However, the particles must be continuously trapped to avoid the contamination problem when being washed away by the flow.

Chen et al. proposed another mechanical micropump by the combination of an elastic pump membrane and a photodeformable liquid crystalline polymer

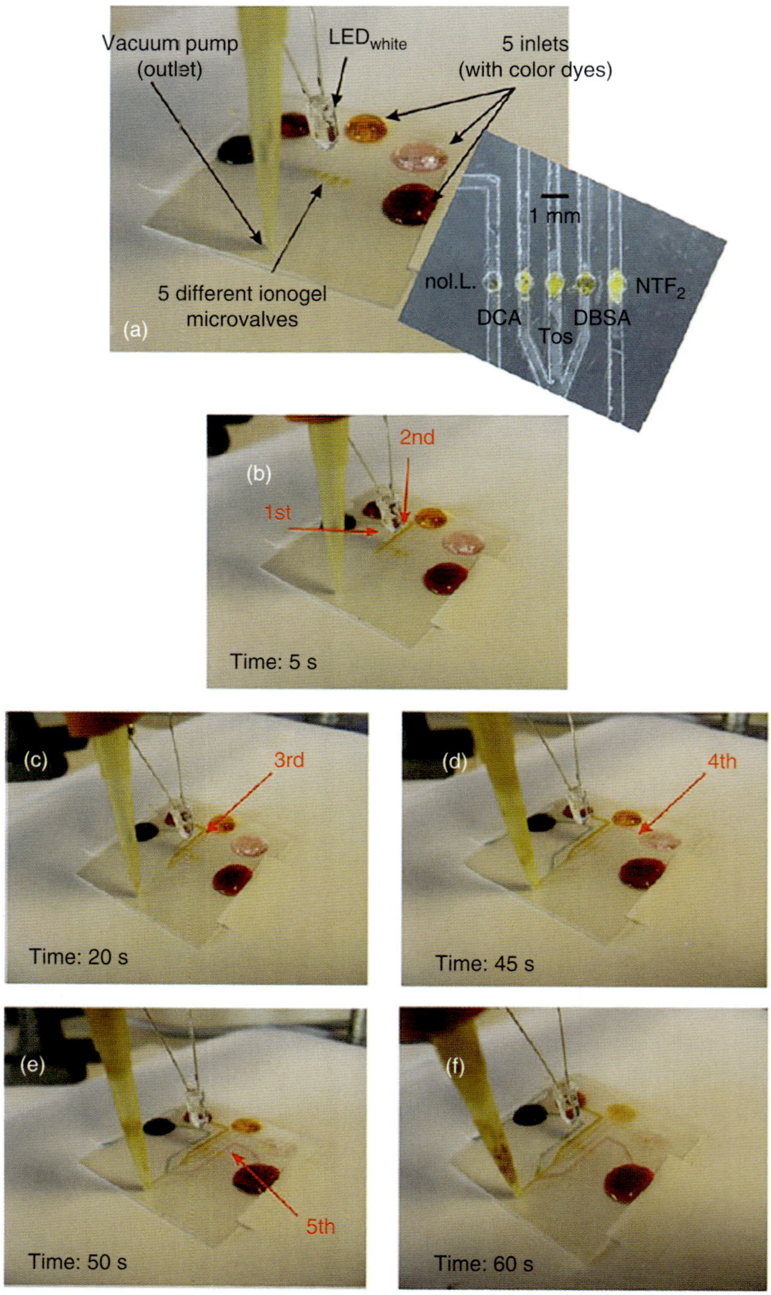

Figure 5.8 Performance of the ionogel microvalves: (a) microvalves closed. (b–e) Valves are actuated followed by "No I.L.," ionogels with $[dca]^-$, $[tos]^-$, $[dbsa]^-$, and $[NTf2]^-$. (f) All valves are open. The applied vacuum is unable to pull the dyes through the microchannels. White light is applied for the time specified in each picture. Numbers and arrows indicate when the channel is filled with the dye because of microvalve actuation. Source: Benito-Lopez et al. 2010 [34]. Copyright 2009. Reprinted with permission from Royal Society of Chemistry.

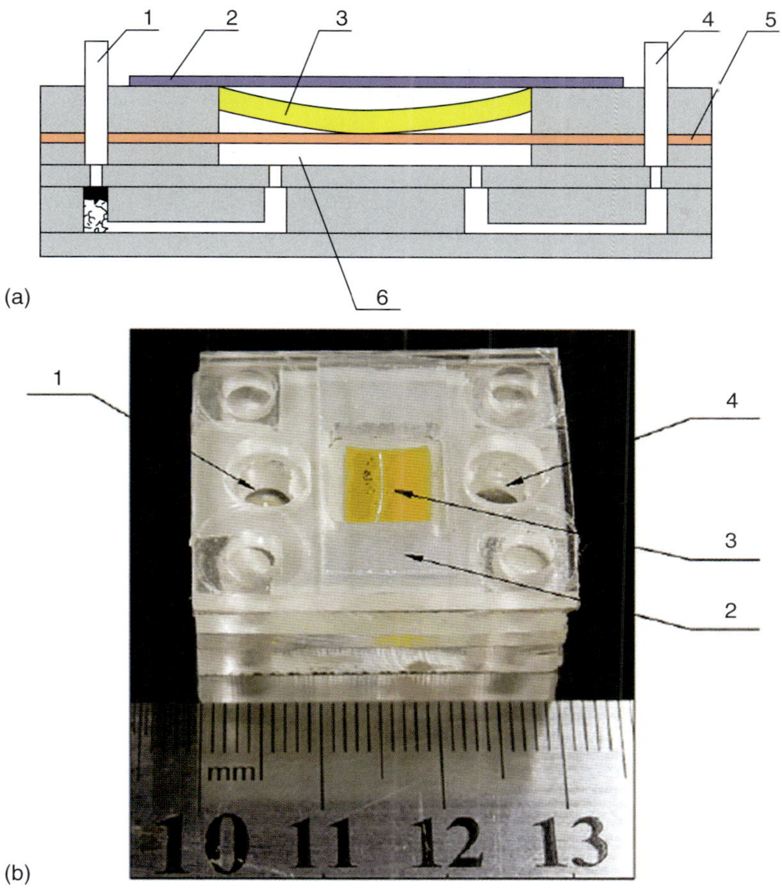

Figure 5.9 (a) The cross section of the assembled prototype micropump. (b) Picture of the experimental device (1. inlet, 2. press plate, 3. photodeformable material, 4. outlet, 5. pump membrane, 6. pump chamber). Source: Chen et al. 2010 [38]. Copyright 2010. Reprinted with permission from Springer Nature.

(LCP) film [38]. The assembled structure of the prototype device is shown in Figure 5.9. The bending behavior of the LCP film is utilized to work against a fixed lid and induces pressure strokes on the flexible membrane and thus the pump chamber.

Another example of the application of soft photoactuators in miniaturized device is the self-powered rechargeable micropumps developed by Patra et al. [39]. The micropump is based on the formation of a β-cyclodextrin/*trans*-azobenzene complex [40]. The complex can be activated by two different mechanisms, i.e. the *trans–cis* isomerization of azobenzene by ultraviolet (UV) light and the disassembly of association between cyclodextrin and azobenzene by introducing a competing guest molecule (adamantine, ADA). As shown in Figure 5.10, the light-induced *cis*-isomer of azobenzene has less binding affinity toward β-cyclodextrin and hence gets dissociated from the cyclodextrin cavity.

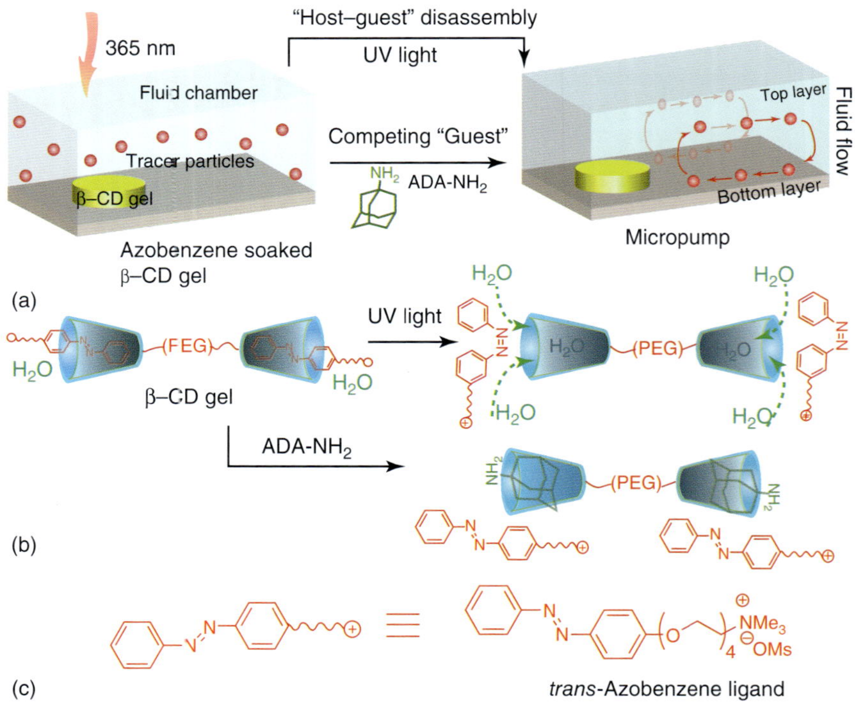

Figure 5.10 (a) Schematic representation of dual responsive micropump; β-CD-PEG micropump initiates fluid flow enabled by either photo or chemical stimuli. (b) Illustration of fluid motion originating from host–guest assembly/disassembly. (c) Chemical structure of *trans*-azobenzene guest molecule. Source: Sun et al. 2016 [40]. Copyright 2013. Reprinted with permission from American Chemical Society.

This dissociation leads to changes in the local fluid density, which is provided as a possible reason for the self-powered fluid flow in the gel. The gel could be reused when soaked in a solution of *trans*-azobenzene or left in the solution for some time. In this manner, *cis*-azobenzene turns into *trans*-azobenzene and forms the β-cyclodextrin/*trans*-azobenzene complex again. The micropumps might be of great help in fluid handling and circulation of culture media for the development of organs-on-a-chip devices.

Microfluidic paper-based analytical devices (μPADs) are a type of innovative, low-cost, equipment-free analytical platforms due to impressive properties comprising flexibility, lightness, and low thickness. Nevertheless, the effective methods to handle and control fluids on paper are still limited. Akyazi et al. provided a new concept for fluid flow manipulation in μPADs by introducing ionogels as passive micropumps [41]. Two types of *p*NIPAAm ionogels, with different chemical and physical properties, were drop-casted at the inlet of the μPADs by a PDMS-based ink-stamping process. The delaying effect of the ionogels depends on the characteristics (chemical and dimensions) of the ionogel, which determine the swelling degree of the gel. As a proof of concept, the fluid delay capability of the ionogel was tested on a Y-shaped μPAD. Furthermore, they integrated a gasket to ensure the fabrication of negative passive pumps with different volumes of ionogel in a single μPAD [42]. The flow behavior of the μPADs can be modulated, redirected, and even reversed in the presence of an ionogel negative passive pump.

Microfluidic channels possess several characteristics, such as small overall volumes, laminar flow, and a large surface-to-volume ratio. Fluid flow in these microchannels is normally nonturbulent due to low Reynolds numbers. The mixing of confined liquids is critical yet challenging in most chemical and biochemical processes. Passive mixers rely on diffusion or chaotic advection, while active mixers use the disturbance generated by an external field for the mixing process, and thus require external energy. Supplemental transducers such as pumps or electrodes were implemented in active mixers to enable dynamic and tunable mixing, thus increasing the complexity and fragility of the device while reducing its portability and flexibility.

Inspired by nature, a simple design of microactuators to facilitate pumping and mixing in microfluidic environments was adopted to mimic cilia, which are the greatest natural boon to living organisms; through them, organisms can manipulate, sense, and control their surrounding flow [43–45]. Artificial cilia driven by light have a promising future. van Oosten et al. used reactive liquid crystal monomer inks in combination with the self-assembling properties of an LCP to produce light-driven artificial cilia by an inkjet-printing process [46]. In order to remotely control the movement of the artificial cilia, they chose to use the photomechanical effect already developed using azobenzene-containing liquid crystalline elastomers (LCEs). Microactuators created with different subunits can be selectively addressed remotely, thus allowing well-controlled motions that have the potential to be more effective in mixing. Clearly, such a reversible photomechanical actuation requires a switching operation of dual wavelengths, which are not viable for many practical applications with a fast response. To improve the performance of LCE-based photoactuators, Yang et al. proposed a robust strategy for *in situ* grafting of graphene-enabled LCE nanocomposites with superior and tunable photodeformable properties upon NIR irradiation [47]. In this NIR-actuated system, graphene sheets play an important role as a photoactive constituent. The LCEs exhibit two distinct advantages, i.e. the self-organization of LCs and the entropy-driven elasticity that enables larger reversible deformations compared to conventional polymeric materials. The actuator performances, including deformation amplitude, actuation stress, and response rate, can be readily controlled by tailoring the quality, the loading content, and the degree of alignment of graphene in the LCE matrix.

Overall, there are still plenty of microscale-responsive polymer systems that have not yet been but can be utilized as soft photoactuators and find applications in microfluidics [48]. Among these potential systems, LCPs have been the major subject of ongoing scientific research [49, 50]. The underlying principle is that the mesogens act as an anisotropic solvent for the isotropic polymer chains. A delicate interplay of the effects of surface-induced ordering, elastic strain of LCs, and topological defects is involved in the study of LCP systems. Therefore, LCPs can undergo several forms of macroscopic shape deformations, including contraction (expansion), bending (unbending), twisting, and other complex forms, dictated by the change of molecular order from the aligned state under external stimuli [51]. Since stimuli such as heat and light are easily attainable, thermo-driven or photo-driven LCP actuators are a promising class of soft materials that exhibit a number of intriguing properties. We believe that the aforementioned approaches discussed in this section may extend the perspectives of the microfluidic systems for future applications when complex mixing processes and remote controllability are required, in particular for biological applications.

5.2.3 Light-Driven Emulsification and De-emulsification

High-throughput emulsification and de-emulsification are an important concern for droplet-based microfluidics. The interface between two immiscible liquids is increased and fragmented by mechanical actuation and hydrodynamic shear forces to generate microdroplets. For a given fluid composition and an appropriate channel junction geometry, droplet size and frequency are controllable by varying flow rates. Generally, surfactants are used to stabilize microdroplets, and also account for many valuable rheological proprieties of the emulsion system. Similarly to the previous fluid mixing operation, the surfactant-dependent strategies to merge droplets can also be passive and active.

With the development of optofluidics, there is a need to explore new approaches to achieve the aforementioned microfluidic actuations by using light as a non-contact stimulus. However, only a few light-triggerable processes have been reported to mix a multiphase flow in microchannels or to realize photoactive emulsification and de-emulsification.

Diguet et al. reported an intriguing method to induce reversible switches from a continuous two-phase laminar flow to a droplet-generating regime with a usual water-in-oil (W/O) flow-focusing geometry, by introducing a photosensitive surfactant in the aqueous phase [52]. The chemical structure of the cationic photosensitive surfactant azobenzene trimethylammonium bromide (AzoTAB) is shown in Figure 5.11. The isomerization of *trans*-AzoTAB to *cis* configuration upon UV illumination results in a more polar tail. As a result, the interfacial energy between liquids and the channel is modulated, leading to the fragmentation of the liquid into an emulsion with monodisperse droplets. The system can go back to the laminar regime as the light irradiation stops.

Later, Venancio-Marques et al. described the reversible control of microfluidic mixing based on the light-induced generation of W/O emulsion with water droplets, acting as reversible stirrers to mix two oil phase flows containing samples (Figure 5.12) [53]. Two continuous oil phases and a photosensitive water phase are injected in a flow-focusing device. The AzoTAB molecule dissolved in the water phase is responsible for the photoisomerization-induced change of wetting properties. Many cycles of reversible transitions between a nonmixing behavior and full homogenization of the two oil phases were demonstrated with the aid of light. Nurdin et al. reported a new dual flow-focusing microfluidic device configuration to generate stable dual jet and dual drop regimes [54]. After introducing the AzoTAB surfactant, the flow regimes were found to be dependent on illumination conditions. The range of flow rates was identified where the

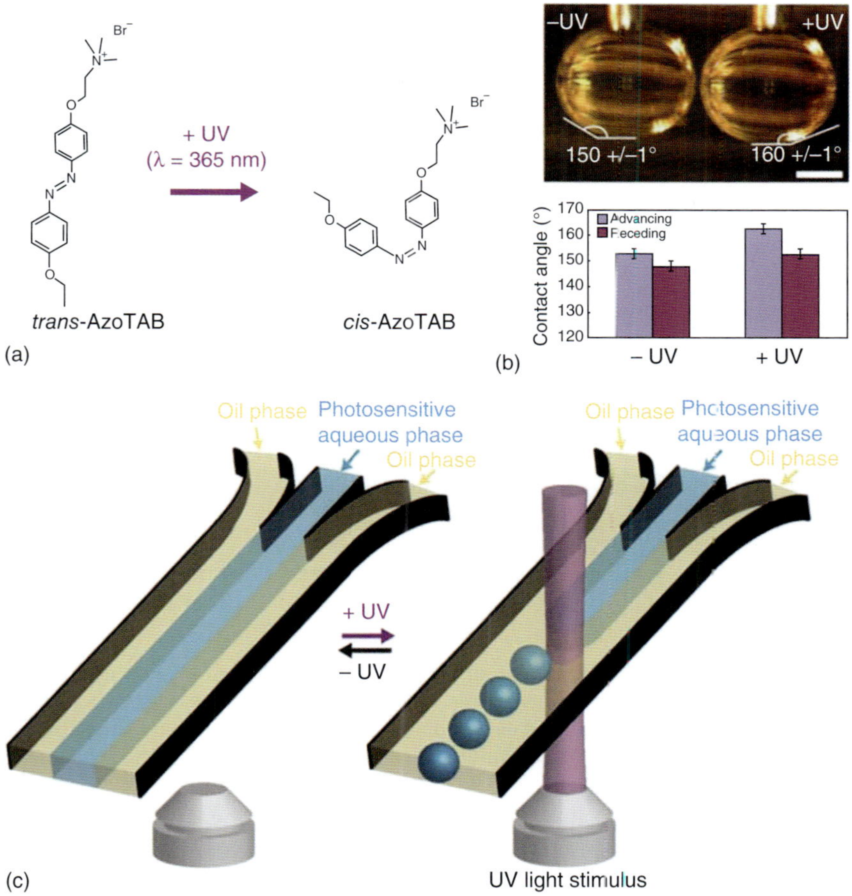

Figure 5.11 A photosensitive liquid in a two-phase microfluidic flow. (a) Molecular structure of AzoTAB. (b) Contact angle measurement of a 10 mM AzoTAB aqueous solution droplet immersed in an oleic acid bath on the substrate. (c) The UV-light-induced change of substrate wettability can be used to control liquid fragmentation. Source: Diguet et al. 2011 [52]. Copyright 2011. Reprinted with permission from Royal Society of Chemistry.

flow regime could be dynamically and reversibly switched upon UV illumination at 365 nm. The two phases were well separated and never mixed in the absence of illumination.

Recently, Takahashi et al. reported the photoinduced de-mulsification of emulsions with a photoresponsive gemini surfactant (C7-azo-C7) bearing an azobenzene skeleton as a spacer [55]. Light-induced isomerization of the azobenzene spacer allows reversible conversion between the *trans* isomer and the *cis* isomer. The modification of the interfacial tension leads to controlled coalescence. However, the de-mulsification experiment was carried out for the overall emulsion without spatial differentiation. The several minutes of irradiation time that is required is considered to be too slow for microfluidic applications, where a response in the sub-second range is more suitable.

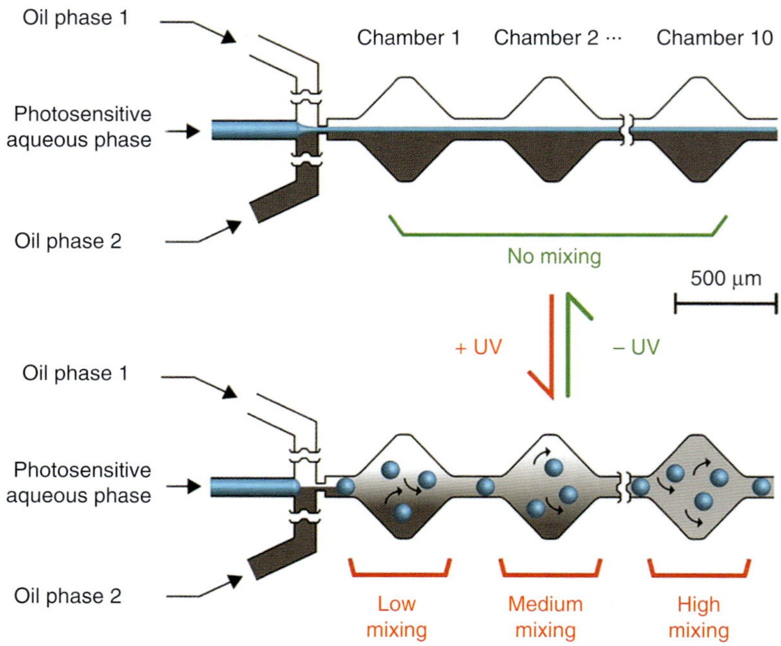

Figure 5.12 Concept and design of a reversible optofluidic mixer: Two oil phases and a photosensitive water phase are injected in a flow focusing device (drawn to scale) that contains a series of 10 regularly spaced expansion chambers. Source: Venancio-Marques et al. 2013 [53]. Copyright 2013. Reprinted with permission from American Chemical Society.

Dunkel et al. explored another light-controlled merging approach by the irreversible photolysis of photosensitive surfactants (Figure 5.13). The amino-quinoline (AQ)-derived photosensitive surfactants is composed of polyethylene-glycol/perfluorinated polyether (PEG/PFPE) diblock amphiphiles [56]. The light-induced partial or total depletion of surfactants result in increased surface tension that would be equilibrated by the uptake of extra surfactants either from the bulk or from a neighboring droplet in oil-in-water (O/W) emulsions. The destabilization may lead to coalescence of droplets if the surfactant concentration goes below the threshold of critical value. Photolysis of microdroplets by using 355 nm picosecond-pulsed laser light results in rapid, controlled fusion of vicinal droplets. Since the photolysis can be controlled precisely by adjusting the power, exposition time, and spot size of the laser beam, the method offers a new prospect of microdroplet merging with high stereospatial integrity in microfluidics.

5.2.4 New Conceptual Light-Driven Fluid Motion in Microchannels

Manipulating small amounts of liquids to perform reactions, analysis or fundamental investigations in biology, physics, or chemistry are important in both scientific research and practical applications. As has been depicted in Figure 5.1, a light stimulus can be directly converted into fluid motion by optical forces or light-induced capillary forces. Although these methods are effective to provide

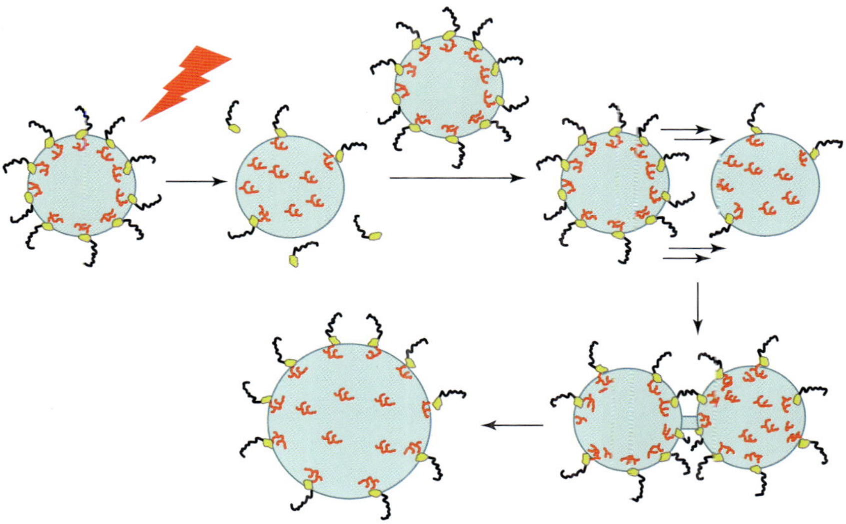

Figure 5.13 The mechanism behind the photolysis-driven droplet merging. Source: Dunkel et al. 2016 [56]. Copyright 2016. Reprinted with permission from Royal Society of Chemistry.

contactless, spatial, and temporal control, several drawbacks still exist. For instance, the wettability gradients induced by light might be too small to conquer the contact line pinning in these cases. The Marangoni effect requires either local heating or addition of photosensitive surfactants into liquids, which limits their potential biomedical application because of contamination problems. In practice, simple and straightforward techniques are required to realize high-speed fluid transportation over long distances.

Inspired by the lamellar structure of artery walls, Lv et al. proposed a conceptually novel way to control fluid motion by capillary force arising from photoinduced asymmetric deformation, instead of the aforementioned wettability gradients and Marangoni effect [57]. Robust tubular microactuators (TMAs) were prepared from a newly designed linear LCP (LLCP) that has a long flexible backbone containing double bonds and azobenzene moieties in side chains acting as both mesogens and photoresponsive groups. As shown in Figure 5.14, the asymmetric geometry change of the TMA from cylindrical to conical is induced upon attenuated 470 nm light irradiation with intensity gradient along the long axis of the TMA. The photoinduced reorientation of LLCP results in a decrease in the thickness of the TMA wall and an elongation of the perimeter of the TMA, which together cause an increase in the cross-sectional area of the TMA. By this means, the motion of a fully wetting liquid slug confined in the TMA microchannel can be driven by light. The TMAs show unique abilities to propel not only common liquids spanning a broad range of polarity, but also complex fluids widely used in biomedical and chemical engineering. Furthermore, efficient liquid mixing and the capture and movement of microspheres on the microscale in the photodeformable TMA can be triggered by external light, offering a versatile toolbox in microfluidic fields of microreactors, LOCs, and micro-optomechanical systems.

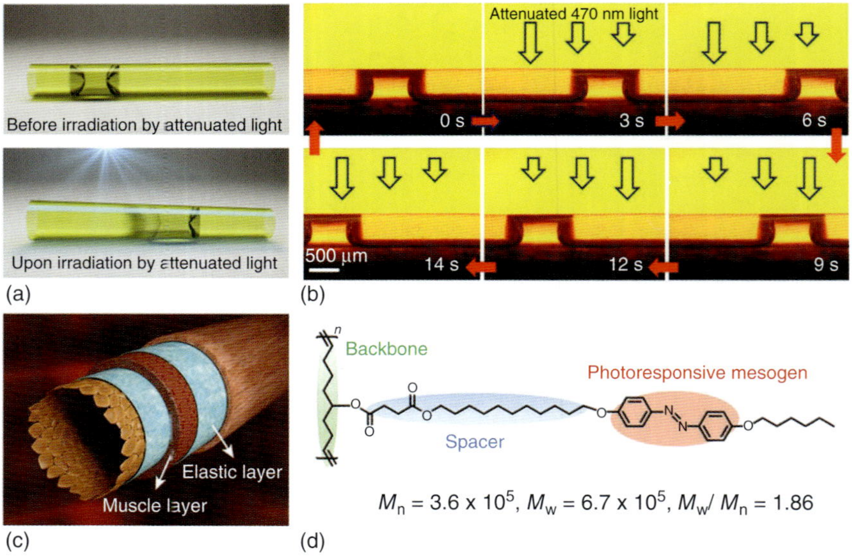

Figure 5.14 Design of tubular microactuators. (a) Schematic motion of a liquid slug driven by photodeformation of a TMA. (b) Lateral photographs of the light-induced motion of a silicone oil slug confined in a TMA. (c) Schematic illustration of the structure of artery walls. (d) Molecular structure of a novel LLCP. Source: Lv et al. 2016 [57]. Copyright 2016. Reprinted with permission from Springer Nature.

5.3 Soft Photoactuators as Optical Microcomponents

The last range of soft photoactuators relies on the use of light stimuli to control, tune, and reconfigure optical microcomponents in microfluidics. As we know, optical microcomponents, typically fabricated by micromolding or photolithographic processes, play a crucial role in imaging and display technology, biosensing, beam shaping, optical switching, wavefront analysis, and device miniaturization. Toward both industrial and academic pursuits with regard to enhanced light manipulation capabilities of optical systems, tunable microlenses are employed as responsive in-line, phase-modulating, intensity-preserving microcomponents. However, most of these microlens designs can only operate with fixed focal lengths, relatively large diameters, and/or slow focal length switching speeds. Currently, considerable research interests have been focused on the development of novel material systems with soft microstructures. Benefiting from the use of photoactive soft materials, the microlens array has the potential to allow for dynamically tunable focal lengths without mechanical parts. Notably, the focal length of an individual microlens can be addressed on demand and independently of other lenses. And the fabrication technology is facile, inexpensive, and scalable. In contrast to traditional optical systems, a higher degree of integration of optical microcomponents with microfluidics can be anticipated, representing an enabling technology for a new class of tunable optical microcomponents for adaptive optics, imaging devices, or sensors.

5.3.1 Tunable Microlenses Actuated by Photoactive Hydrogels

Hydrogel-based lensing materials can be dynamically reshaped through various external stimuli after the optical elements are formed. Kim et al. presented a simple wet-chemical method to fabricate microlens arrays, which comprised of fast-responding, reversible, photoactive hydrogel elements [58]. Glass substrates were functionalized by 3-aminopropyltrimethoxysilane (APTMS), Au nanoparticles (16 ± 1.6 nm), and the cationic polyelectrolyte poly(allylamine hydrochloride) (PAH) subsequently. Then, anionic *p*NIPAAm-*co*-AAc microgels were attached to the APTMS-Au-PAH-modified substrate through electrostatic interactions. In this manner, colloidal Au nanoparticles allowed for local heating of the sample through excitation of surface plasmon modes. Figure 5.15 shows a schematic depiction and a scanning electron microscope (SEM) image of this construction, in which the laser-power-dependent lensing ability and the apparent response time relate to the photothermal properties of Au nanoparticles and the thermoresponsivity of microgel lens elements. The unique micro-optical characteristics also include the ability to spatially confine the tunable region to the single-microlens scale with the aid of pulsed laser.

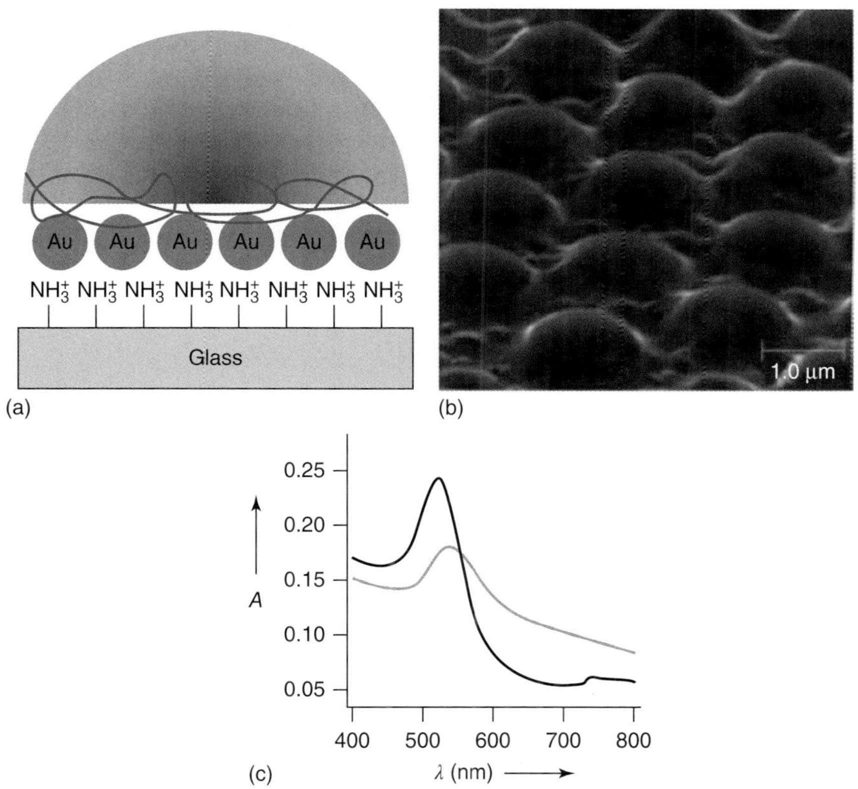

Figure 5.15 (a) Depiction of the multicomponent film used to create photoswitchable microlens arrays. (b) SEM image of the microlens array at grazing angle. (c) UV/vis spectra of Au nanoparticles in solution (black) and bound to substrate (gray). Source: Kim et al. 2005 [58]. Copyright 2005. Reprinted with permission from John Wiley and Sons.

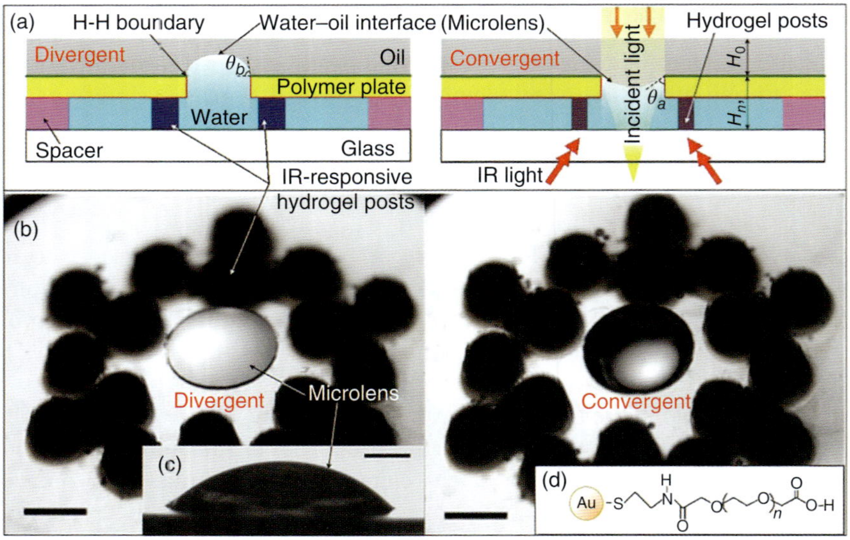

Figure 5.16 Schematic and optical images of an IR-light-actuated tunable microlens. (a) Schematics of the microlens when it is divergent and convergent, respectively. (b) Optical images of a microlens with 18 hydrogel microposts. (c) Side profile of water meniscus of the microlens in the divergent status at the starting point taken with a goniometer. (d) Chemical structure of the gold nanoparticles coated with thiolated PEG ligands. The scale bar is 1 mm. Source: Zeng and Jiang 2008 [59]. Copyright 2008. Reprinted with permission from AIP Publishing LLC.

In another attractive demonstration as shown in Figure 5.16, Zeng and coworker fabricated a liquid tunable microlens actuated by multiple micropost structures of Au-*p*NIPAAm nanocomposite hydrogels photopatterned around the lens aperture [59]. Benefitting from the heat converted from absorbed light by entrapped Au nanoparticles, the swelling state (expansion and contraction) of the hydrogel network was controlled. The curvature of a liquid–liquid interface forming the microlens at the aperture and its focal length were regulated in turn. The focal length of the microlens can be tuned from −17.4 to $+8.0 \pm 0.4$ mm in seconds under IR irradiation. This approach paves a potential way to replace current technology that uses mechanical or electrical signals and intrinsically integrates such light-controlled tunable microlenses into optofluidic systems.

5.3.2 Microlens Arrays Actuated by Photoactive Emulsions

As mentioned previously in Section 5.2.2, microfluidic emulsification is one of the most powerful and important techniques for mixing and dispersing immiscible fluid compositions within a continuous liquid phase. Complex emulsions, including multiple emulsions and Janus droplets (with faces of differing chemistries), are of growing interest, especially in dynamic micro-optics.

Taking advantage of the temperature-sensitive miscibility of hydrocarbon, silicone, and fluorocarbon liquids, Zarzar et al. described a one-step approach to fabricate complex emulsions with highly controllable and reconfigurable

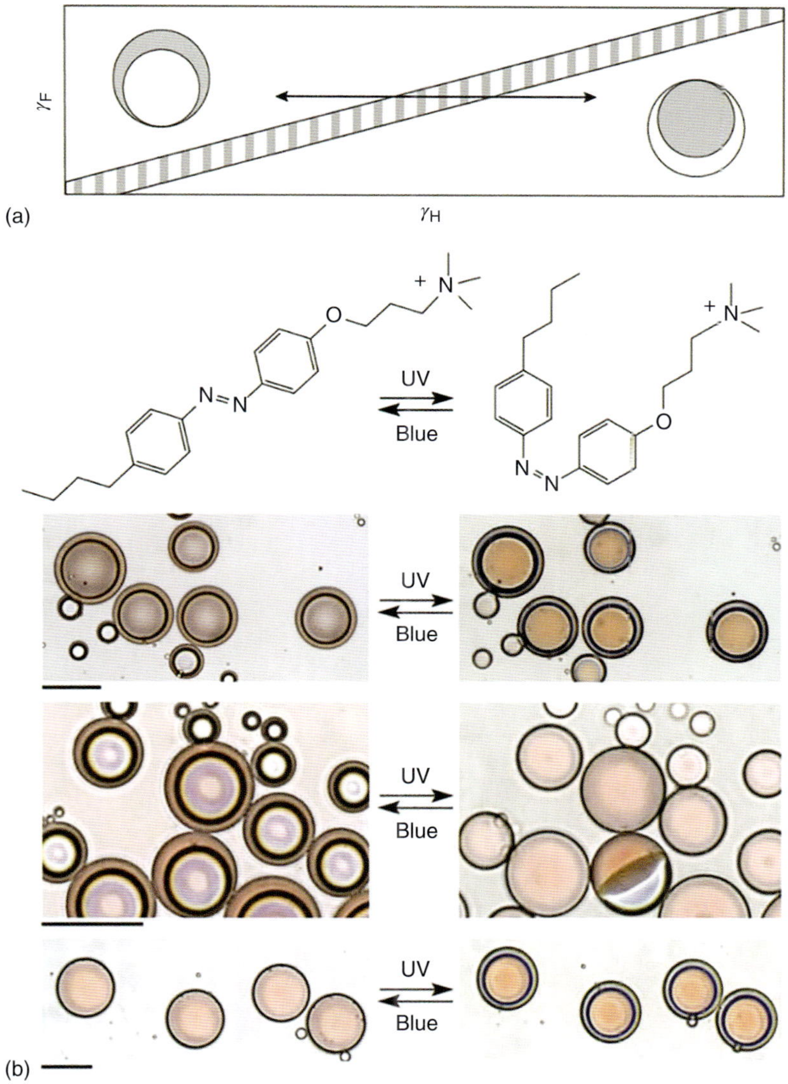

Figure 5.17 Emulsions reconfigure in response to light. (a) Sketch of the drop morphology and a phase-stability diagram. Gray represents hexane and white represents perfluorohexane. (b) Chemical structure of the light-responsive surfactant that reversibly isomerizes under UV and blue light. Aligned beneath are optical micrographs of hexane-perfluorohexane emulsions that are tuned to undergo specific morphological transitions in response to light. Hexane is dyed red, and the aqueous phase consists of Zonyl and the light-responsive surfactant. Droplets undergo complete inversion (top). F/H/W double-emulsion drops transition to Janus droplets (middle). Janus droplets transition to an H/F/W double emulsion (bottom). The scale bar is 100 mm. Source: Zarzar et al. 2015 [60]. Copyright 2015. Reprinted with permission from Springer Nature.

morphologies after emulsification [60]. To endow these emulsions with photoactivities, a photoresponsive surfactant consisting of an azobenzene moiety, which reversibly undergoes photoinduced isomerization between a more effective *trans* configuration and a less effective *cis* configuration, was synthesized and used in combination with a nonionic fluorosurfactant Zonyl FS-300 with suitable concentrations. As shown in Figure 5.17, the rapid and reversible morphologic switch of hexane-perfluorohexane droplets, between double emulsion (encapsulated) and Janus states (two-sided) or to invert states entirely, results from changes in the balance between interfacial tensions. Thus, the dynamic reconfiguration of droplet morphology was observed in response to UV (365 nm) and blue (470 ± 20 nm) light by varying the duration or intensity of light exposure. This yields a general strategy to create a wide range of photoresponsive elements for abundant applications of emulsions.

Very recently, Nagelberg et al. produced stable bi-phase emulsions with a highly uniform size distribution as tunable microlens arrays by glass-capillary-based microfluidics [61]. Such complex droplets are composed of optically distinct, immiscible hydrocarbons, heptane (n_{HP} = 1.387) or hexane (n_{HX} = 1.375), and fluorocarbons FC-770 (n_{FC} = 1.27) in the aqueous phase (n_W = 1.33). The

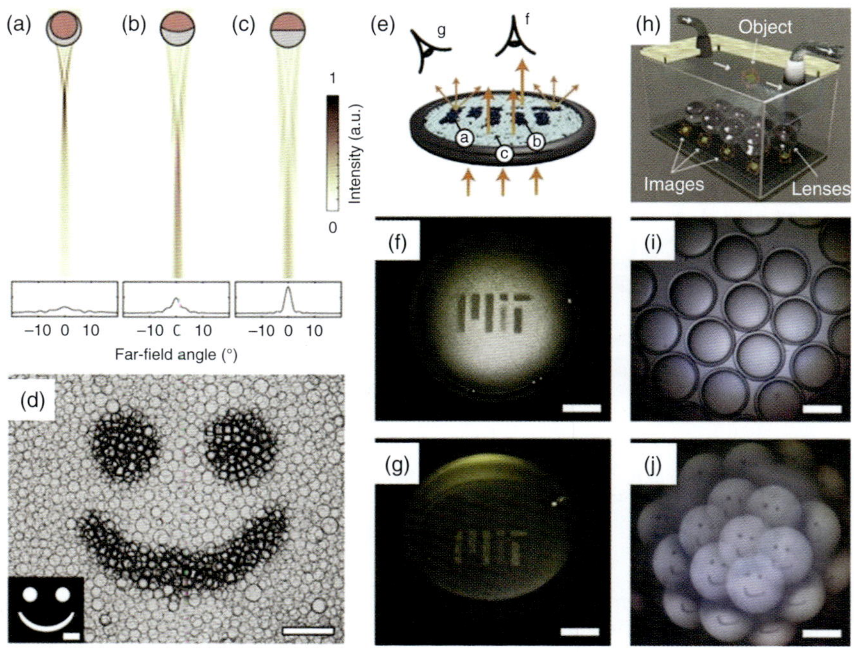

Figure 5.18 Toward potential applications. (a–c) 2D finite different time domain simulations of droplets of 5 mm radius for incident light of 500 nm wavelength. (d) Localized exposure of light-sensitive surfactants to UV light leads to a local variation in droplet morphology and scattering behavior to create an image. The dark zones in the image represent particles that have switched to the double emulsion state, (e) Schematic geometries for observing the photo-patterned droplet films, corresponding to the perceived images shown in (f, g). The labels a, b, and c correspond to the droplet geometries shown in panels (a–c). (f, g) Photo-patterned droplets viewed from above (f) and at an angle (g). The scale bars are 5 mm. (h) Concept sketch for tomographic imaging of microscale objects in a microfluidic system using the fluid compound lenses. (i, j) Monolayer array of fluid compound lenses and the images projected by the monolayer lenses. The scale bars are 100 mm. Source: Nagelberg et al. 2017 [61]. Copyright 2017. Reprinted with permission from Springer Nature.

photoactive emulsion can be dynamically reconfigured through light stimuli in the presence of an azo-surfactant, thus showing different optical properties depending on the configuration. Double emulsions with higher refractive index (RI) core phase strongly focus light, while those with higher RI shell phase show strong light scattering. However, Janus droplets do not significantly disturb the light wavefront when the light propagates along the surface normal of the internal interface.

The RI contrasts between hydrocarbons, fluorocarbons, and water, as well as the curvature of each interface contribute to the focusing power of refractive optical elements. By adjusting the droplet's interfacial tensions with the aqueous phase, the focal length of emulsion-based microlens can be varied from ±3.5× the drop diameter to infinity. Therefore, a light-controlled switch of the droplet morphology consequently results in a predictable variation of the droplets acting as converging lenses projecting real inverted images, or as diverging lenses forming virtual upright images. It is also confirmed that microscale optical tunability of the droplets could be translated to observable differences in macroscopic appearance, which can be used for the creation of patterns and images. Additionally, two potential applications of the microlenses' functionality were demonstrated to be integral in microscale imaging devices and light field display technology. As shown in Figure 5.18, this feature allows for the creation of microscopic and macroscopic patterns with tunable contrast, which could form the basis for light field displays capable of creating 3D images and projecting variable information content into different directions.

5.4 Summary and Outlook

Overviews of the photoactive soft materials, involving hydrogels, LCPs, and emulsions, have shown their unique microfluidic applications as photoactuators. Owing to the considerable scope of optofluidics, the related discussion in this chapter only focused on a few selected but significant topics. Basically, the structure and morphology changes of photoactuators are induced by light, thereby offering the on-demand control and handling of fluids, as well as the manipulation of light interactions. The mechanisms involved are also discussed in details. Until now, the practical implementations of light-responsive soft materials with microfluidic functions are not satisfying and should be extended to broader aspects.

For instance, microscale channels have a high surface-to-volume ratio, leading to different characteristics from bulk volumes. The wettability change of microchannels based on photoactive soft materials can be used to modify the surface interactions and influence fluid movements. The *trans–cis* isomerization of azobenzene leads to large changes in wettability when the structures of a surface topography are in the micrometer size regime [62]. As stated before, spiropyran can also undergo a hydrophilicity change upon isomerization. Polymer surfaces functionalized with these organic molecules can be more hydrophilic or hydrophobic upon light exposure [63]. However, to the best of our knowledge, the integration of these polymer coatings with tunable wettability into microfluidic channels was rarely reported to be based on photoswitchable surface topographies. In addition, *p*NIPAAm hydrogels are widely used as prototype actuators, which can be only responsive to light irradiations in the presence of aqueous solutions with relatively low ionic strength, significantly restricting the types of fluids that can be manipulated. Furthermore, the approaches to tune the flow control with intermediate states, rather than a simple stop/start function with these gel-based microvalves, are appealing and need to be further explored.

Fortunately, recent advances and trends show the possibilities to fully exploit the benefits of light stimulus by taking advantage of molecular design and tailoring, which enlighten the development of new soft material systems. From this point of view, it is reasonably expected that the interplay among four main areas within microfluidics, i.e. material properties, functions, integration, and applications, are worthy of being reinvestigated to realize enhanced performance as desired. Our specific emphasis herein is that the discovery of new conceptual soft photoactuators would pave a promising way and bridge the gap between remote-controllable LOC devices and many fields such as biomedical and pharmaceutical research, chemical reaction manifolds, and for chemical/bioanalysis, which is undoubtedly the primary objective of future optofluidics involving photoactive functional dynamic materials and systems.

Acknowledgments

Lu-Jian Chen thanks the support from the National Natural Science Foundation of China (NSFC) (No. 61675172), the Natural Science Foundation of Fujian Province, China (No. 2017J01124), and Shenzhen Science and Technology Project (No. JCYJ20170306142028457).

References

1 Li, Q. (ed.) (2013). *Intelligent Stimuli-responsive Materials: From Well-defined Nanostructures to Applications*. Hoboken, NJ: Wiley.
2 Wang, L. and Li, Q. (2018). Photochromism into nanosystems: towards lighting up the future nanoworld. *Chem. Soc. Rev.* 47: 1044.
3 Dong, L. and Jiang, H. (2007). Autonomous microfluidics with stimuli-responsive hydrogels. *Soft Matter* 3: 1223.
4 Wang, L. and Li Q. (2016). Stimuli-directing self-organized 3D liquid crystalline nanostructures: from materials design to photonic applications. *Adv. Funct. Mater.* 26: 10.
5 ter Schiphorst, J , Saez, J., Diamond, D. et al. (2018). Light-responsive polymers for microfluidic applications. *Lab Chip* **18**: 699.
6 Chen, L., Li, Y., Fan, J. et al. (2014). Photoresponsive monodisperse cholesteric liquid crystalline micro-shells for tunable omnidirectional lasing enabled by a visible light-driven chiral molecular switch. *Adv. Optical Mater.* 2: 845.
7 Fan, J., Li, Y., Bisoyi, H.K. et al. (2015). Light-directing omnidirectional circularly polarized reflection from liquid crystal droplets. *Angew. Chem. Int. Ed.* 54: 2160.

8 Wang, L., Chen, D., Gutierrez-Cuevas, K. et al. (2017). Optically reconfigurable chiral microspheres of self-organized helical superstructures with handedness inversion. *Mater. Horiz.* 4: 1190.
9 Baigl, D. (2012). Photo-actuation of liquids for light-driven microfluidics: state of the art and perspectives. *Lab Chip* 12: 3637.
10 Hilber, W. (2016). Stimulus-active polymer actuators for next-generation microfluidic devices. *Appl. Phys. A* 122: 751.
11 Peters, E.C., Svec, F., and Frechet, J.M.J. (1997). Thermally responsive rigid polymer monoliths. *Adv. Mater.* 9: 630.
12 Harmon, M.E., Tang, M., and Frank, C.W. (2003). A microfluidic actuator based on thermoresponsive hydrogels. *Polymer* 44: 4547.
13 Shirasaki, Y., Tanaka, J., Makazu, H. et al. (2006). On-chip cell sorting system using laser-induced heating of a thermoreversible gelation polymer to control flow. *Anal. Chem.* 78: 695.
14 Wang, L., Gutierrez-Cuevas, K.G., Bisoyi, H.K. et al. (2015). NIR light-directing self-organized 3D photonic superstructures loaded with mesogen-functionalized anisotropic plasmonic nanorods. *Chem. Commun.* 51: 15039.
15 Gutierrez-Cuevas, K.G., Wang, L., Xue, C. et al. (2015) Near infrared light-driven liquid crystal phase transition enabled by hydrophobic mesogen grafted plasmonic gold nanorods. *Chem. Commun.* 51: 9845.
16 Wang, L., Bisoyi, H.K., Zheng, Z. et al. (2017). Self-organized chiral liquid crystalline superstructures for adaptive windows enabled by mesogen-functionalized graphene. *Mater. Today* 20: 230.
17 Jiang, H.Y., Kelch, S., and Lendlein, A. (2006). Polymers move in response to light. *Adv. Mater.* 18: 1471.
18 Yoon, J., Bian, P., Kim, J. et al. (2012). Local switching of chemical patterns through light-triggered unfolding of creased hydrogel surfaces. *Angew. Chem. Int. Ed.* 51: 7146.
19 Sershen, S.R., Mensing, G.A., Ng, M. et al. (2005). Independent optical control of microfluidic valves formed from optomechanically responsive nanocomposite hydrogels. *Adv. Mater.* 17: 1366.
20 Toh, W., Ng, T.Y., Hu, J., and Liu, Z. (2014). Mechanics of inhomogeneous large deformation of photo-thermal sensitive hydrogels. *Int. J. Solids Struct.* 51: 4440.
21 Lee, E., Lee, H., Yoo, S.I., and Yoon, J. (2014). Photothermally triggered fast responding hydrogels incorporating a hydrophobic moiety for light-controlled microvalves. *ACS Appl. Mater. Interfaces* 6: 16949.
22 Jadhav, A.D., Yan, B., Luo, R.C. et al. (2015). Photoresponsive microvalve for remote actuation and flow control in microfluidic devices. *Biomicrofluidics* 9: 034114.
23 Yuan, Z., Xiao, X., Li, J. et al. (2018). Self-assembled graphene-based architectures and their applications. *Adv. Sci.* 5:1700626.
24 Lo, C.-W., Zhu, D., and Jiang, H. (2011). An infrared-light responsive graphene-oxide incorporated poly(*N*-isopropylacrylamide) hydrogel nanocomposite. *Soft Matter* 7: 5604.
25 Zhu, C.-H., Lu, Y., Peng, J. et al. (2012). Photothermally sensitive poly(*N*-isopropylacrylamide)/graphene oxide nanocomposite hydrogels as remote light-controlled liquid microvalves. *Adv. Funct. Mater.* 22: 4017.

26 Zhang, J.Y., Du, P., Xu, D.Y. et al. (2016). Near-infrared responsive MoS_2/poly(*N*-isopropylacrylamide) hydrogels for remote light-controlled microvalves. *Ind. Eng. Chem. Res.* 55: 4526.
27 Satoh, T., Sumaru, K., Takagi, T., and Kanamori, T. (2011). Fast-reversible light-driven hydrogels consisting of spirobenzopyran-functionalized poly(*N*-isopropylacrylamide). *Soft Matter* 7: 8030.
28 Jin, L., Li, Y., Ma, J., and Li, Q. (2010). Synthesis of novel thermally reversible photochromic axially chiral spirooxazines. *Org. Lett.* 12: 3552.
29 Sugiura, S., Sumaru, K., Ohi, K. et al. (2007). Photoresponsive polymer gel microvalves controlled by local light irradiation. *Sens. Actuators, A* 140: 176.
30 Sugiura, S., Szilagyi, A., Sumaru, K. et al. (2009). On-demand microfluidic control by micropatterned light irradiation of a photoresponsive hydrogel sheet. *Lab Chip* 9: 196.
31 Chen, G., Svec, F., and Knapp, D.R. (2008). Light-actuated high pressure-resisting microvalve for on-chip flow control based on thermo-responsive nanostructured polymer. *Lab Chip* 8: 1198.
32 ter Schiphorst, J., Coleman, S., Stumpel, J.E. et al. (2015). Molecular design of light-responsive hydrogels, for in situ generation of fast and reversible valves for microfluidic applications. *Chem. Mater.* 27: 5925.
33 Coleman, S., ter Schiphorst, J., Ben Azouz, A. et al. (2017). Tuning microfluidic flow by pulsed light oscillating spiropyran-based polymer hydrogel valves. *Sens. Actuators, B* 245: 81.
34 Benito-Lopez, F., Byrne, R., Raduta, A.M. et al. (2010). Ionogel-based light-actuated valves for controlling liquid flow in micro-fluidic manifolds. *Lab Chip* 10: 195.
35 Czugala, M., O'Connell, C., Blin, C. et al. (2014). Swelling and shrinking behaviour of photoresponsive phosphonium-based ionogel microstructures. *Sens. Actuators, B* 194: 105.
36 Benito-Lopez, F., Antonana-Diez, M., Curto, V.F. et al. (2014). Modular microfluidic valve structures based on reversible thermoresponsive ionogel actuators. *Lab Chip* 14: 3530.
37 Terray, A., Oakey, J., and Marr, D.W.M. (2002). Microfluidic control using colloidal devices. *Science* 296: 1841.
38 Chen, M., Xing, X., Liu, Z. et al. (2010). Photodeformable polymer material: towards light-driven micropump applications. *Appl. Phys. A* 100: 39.
39 Patra, D., Zhang, H., Sengupta, S., and Sen, A. (2013). Dual stimuli-responsive, rechargeable micropumps via "host-guest" interactions. *ACS Nano* 7: 7674.
40 Sun, R., Bisoyi, H., Xie, M., and Li, Q. (2016). Photo and redox dual-stimuli-directed reversible disassembly and reassembly of linear supramolecular polymer formed by orthogonal host-guest molecular recognition. *Dyes Pigm.* 132: 336.
41 Akyazi, T., Saez, J., Elizalde, J., and Benito-Lopez, F. (2016). Fluidic flow delay by ionogel passive pumps in microfluidic paper-based analytical devices. *Sens. Actuators, B* 233: 402.
42 Akyazi, T., Gil-Gonzalez, N., Basabe-Desmonts, L. et al. (2017). Manipulation of fluid flow direction in microfluidic paper-based analytical devices with an ionogel negative passive pump. *Sens. Actuators, B* 247: 114.

43 Khatavkar, V.V., Anderson, P.D., den Toonder, J.M.J., and Meijer, H.E.H. (2007). Active micromixer based on artificial cilia. *Phys. Fluids* 19: 083605.
44 Toonder, J., Bos, F., Broer, D. et al. (2008). Artificial cilia for active micro-fluidic mixing. *Lab Chip* 8: 533.
45 Wu, Y.-A., Panigrahi, B., Lu, Y.-H., and Chen, C.-Y. (2017). An integrated artificial cilia based microfluidic device for micropumping and micromixing applications. *Micromachines* 8: 260.
46 van Oosten, C.L., Bastiaansen, C.W., and Broer, D.J. (2009). Printed artificial cilia from liquid-crystal network actuators modularly driven by light. *Nat. Mater.* 8: 677.
47 Yang, Y.K., Zhan, W.J., Peng, R.G. et al. (2015). Graphene-enabled superior and tunable photomechanical actuation in liquid crystalline elastomer nanocomposites. *Adv. Mater.* 27: 6376.
48 Yang, H., Ye, G., Wang, X., and Keller, P. (2011). Micron-sized liquid crystalline elastomer actuators. *Soft Matter* 7: 815.
49 Yu, L., Yu, H., and Li, Q. (2014). Nanotechnology and nanomaterials in photodeformable liquid crystalline polymers, Chapter 10. In: *Nanoscience with Liquid Crystals: From Self-Organized Nanostructures to Applications* (ed. Q. Li). Heidelberg: Springer.
50 Bisoyi, H.K. and Li, Q. (2016). Light-driven liquid crystalline materials: from photo-induced phase transitions and property modulations to applications. *Chem. Rev.* 116: 15089.
51 Kularatne, R.S., Kim, H., Boothby, J.M., and Ware, T.H. (2017). Liquid crystal elastomer actuators: synthesis, alignment, and applications. *J. Polym. Sci. Polym. Phys.* 55: 395.
52 Diguet, A., Li, H., Queyriaux, N. et al. (2011). Photoreversible fragmentation of a liquid interface for micro-droplet generation by light actuation. *Lab Chip* 11: 2666.
53 Venancio-Marques, A., Barbaud, F., and Baigl, D. (2013). Microfluidic mixing triggered by an external LED illumination. *J. Am. Chem. Soc.* 135: 3218.
54 Nurdin, L., Venancio-Marques, A., Rudiuk, S. et al. (2016). High-throughput photocontrol of water drop generation, fusion, and mixing in a dual flow-focusing microfluidic device. *C.R. Chim.* 19: 199.
55 Takahashi, Y., Fukuyasu, K., Horiuchi, T. et al. (2014). Photoinduced demulsification of emulsions using a photoresponsive gemini surfactant. *Langmuir* 30: 41.
56 Dunkel, P., Hayat, Z., Barosi, A. et al. (2016). Photolysis-driven merging of microdroplets in microfluidic chambers. *Lab Chip* 16: 1484.
57 Lv, J.A., Liu, Y., Wei, J. et al. (2016). Photocontrol of fluid slugs in liquid crystal polymer microactuators. *Nature* 537: 179.
58 Kim, J., Serpe, M.J., and Lyon, L.A. (2005). Photoswitchable microlens arrays. *Angew. Chem. Int. Ed.* 44: 1333.
59 Zeng, X. and Jiang, H. (2008). Tunable liquid microlens actuated by infrared light-responsive hydrogel. *Appl. Phys. Lett.* 93: 151101.
60 Zarzar, L.D., Sresht, V., Sletten, E.M. et al. (2015). Dynamically reconfigurable complex emulsions via tunable interfacial tensions. *Nature* 518: 520.

61 Nagelberg, S., Zarzar, L.D., Nicolas, N. et al. (2017). Reconfigurable and responsive droplet-based compound micro-lenses. *Nat. Commun.* 8: 14673.
62 Groten, J., Bunte, C., and Ruhe, J. (2012). Light-induced switching of surfaces at wetting transitions through photoisomerization of polymer monolayers. *Langmuir* 28: 15038.
63 Xin, B. and Hao, J. (2010). Reversibly switchable wettability. *Chem. Soc. Rev.* 39: 769.

6

Liquid Crystal Polymer Networks and Elastomers for Light-Fueled Robotics

Hao Zeng, Markus Lahikainen, Owies M. Wani, Alex Berdin, and Arri Priimagi

Tampere University of Technology, Laboratory of Chemistry and Bioengineering, Korkeakoulunkatu 8, FI-33720, Tampere, Finland

6.1 Photoactuation: A New Paradigm for Soft Micro-robotics

"Robot" is a term first introduced by the Czech writer Karel Capek in a science fiction play "R.U.R." in 1920 [1]. Another science fiction writer (and scientist), Isaac Asimov (1920–1992), stated that "today's science fiction is tomorrow's science fact." After a century of evolution, this statement has become a reality when it comes to robots, which continue disrupting our ways of living as well as triggering our imagination on what tomorrow's technologies will look like. For several decades, scientists and engineers have focused their efforts on perfecting the mechanics and electronics of conventional robotic devices, to build evermore sophisticated, precise, and reliable functions with faster actuation speed. However, new challenges have risen recently, as the robots of tomorrow should, instead of serving merely as a labor alterative or automation tool, act autonomously and adapt their function to overcome unexpected obstacles without human intervention, and all this in a human-friendly and safe manner. Reaching such beautiful visions may be beyond the capabilities of conventional hard-bodied machines.

In contrast to the hard-bodied robots with rigid components, soft robots constitute soft actuators and flexible joints and links that can offer additional degrees of freedom and further sophistication in robotic movements [2]. Soft robots can adapt to unpredictable obstacles, and they are capable of continuous actuation and shape change. A comparison between a rigid- and a soft-gripping device is shown in Figure 6.1. The rigid gripper (Figure 6.1a) requires pre-programming in order to precisely reach the target and to control the force exerted on it. The soft gripper, composed of pneumatic actuators made of soft silicone-based polymer and controlled by air pressure [3], can bend and adapt to the curvature of the object. Without any precise programming, the device can autonomously adapt

Photoactive Functional Soft Materials: Preparation, Properties, and Applications, First Edition.
Edited by Quan Li.

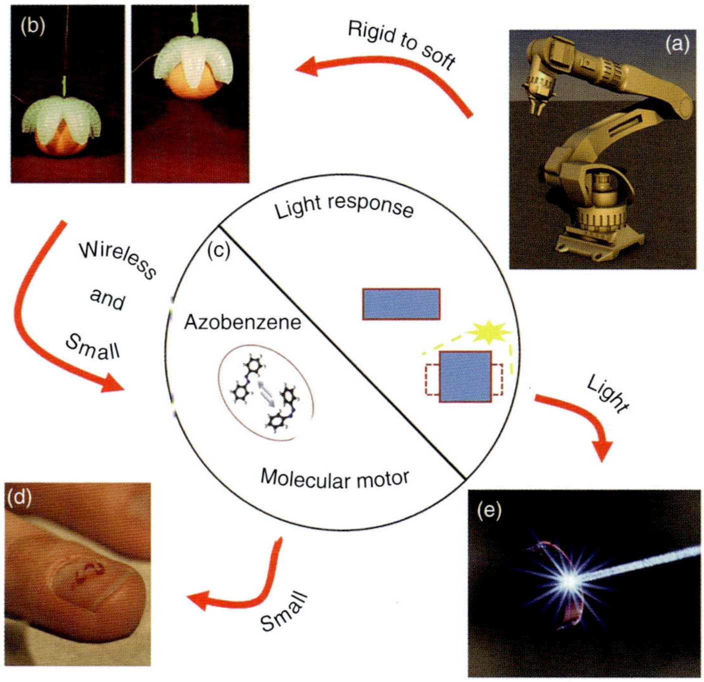

Figure 6.1 Paradigm shifts for robotics: from rigid to soft; from wired to wireless; from large to small. (a) Conventional robotic gripper based on rigid actuators relies on precise position programming to yield the desired operation. (b) A soft robot based on pneumatic actuation can adapt itself to object geometry, enabling manipulation of fragile objects. Source: Ilievski et al. 2011 [3]. Copyright 2011. Reprinted with permission from Wiley-VCH. (c) Schematic drawing of a concept behind wireless, light-fueled small-scale soft robot: light response in smart polymer enables actuation under an external energy source, while molecular motors can be used to induce photo-deformation and construct actuator with a miniature size. (d) A light-driven inching robot on a human nail. Source: Zeng et al. 2017 [4]. Copyright 2018. Reprinted with permission from Wiley-VCH. (e) An autonomous gripping device powered by light energy. Source: Wani et al. 2017 [5]. Copyright 2017. Reproduced with permission from Nature Publishing Group.

to unexpected geometries, and perform its task gently enough to handle also fragile objects (such as picking up an egg, as demonstrated in Figure 6.1b). However, hurdles remain in the pathway of soft robots when approaching practical applications. For instance, invasive surgery and task execution at small scales require the soft robot to be miniaturized and controlled by a wireless energy source. Most of the present-day soft robots comprise pneumatic actuators that rely on air/liquid tube powering, a concept which is extremely difficult to realize at the microscale.

As an alternative to pneumatic actuation, smart stimuli-responsive materials [6] stand out as particularly promising. In such materials, the powering source required for the actuation can be decoupled from the robot body, and the device can be fueled remotely. Various stimuli-responsive materials have been implemented into micro-robotics, using power supplies based on magnetic fields [7], light fields [8], and chemical reactions [9]. Among various energy sources, light stands out as particularly promising, because it is clean and tunable (in wavelength, intensity, polarization), enabling sophisticated control with high spatial and temporal resolution. Light-responsive smart materials can perform shape changes upon illumination (Figure 6.1c), thus paving way for miniaturized, remotely controlled soft-robotic devices. The photoactuation process in these materials is typically triggered by molecular motors that, upon proper materials design, translate the molecular-level movements into controlled actuation in the macroscopic scale [10]. This provokes a paradigm shift in soft robotics: the use of light-responsiveness of the material itself to build up robotic actuators and sensors, resulting in a remotely controlled actuation strategy and miniaturization of the entire system.

Liquid crystal (LC) polymer networks and elastomers (herein, we refer to them as liquid crystal networks (LCNs), without making a distinction between elastomeric and glassy polymer networks), are synthetic cross-linked polymer systems constituting liquid crystalline building blocks, or mesogens [11]. The inherent tendency of liquid crystalline materials to self-assemble opens up the possibility to pre-design the molecular orientation within the polymer network [12]. Importantly, this orientational order can be reversibly controlled (or even destroyed, to yield order–disorder transition) under external light stimuli, resulting in a shape change of the LCNs [13]. The most popular light-driven molecular motors in LCNs are azobenzene derivatives that can undergo reversible *trans*-to-*cis* isomerization upon irradiation with a proper wavelength, dictated by the chemical structure (Figure 6.1c) [14]. This molecular-level shape change can be amplified in LCNs to trigger large and reversible macroscopic deformation. To quote Zentel, one of the pioneers of LCN research, "the integration of photoactive mesogenic units (i.e. azobenzenes) into LCNs has given rise to remarkable results and is currently judged the most promising candidate for commercial applications" [15].

Light-driven, LCN-based soft robotics has been a nascent research field in the past few years. From light-controlled microscopic walkers to autonomous walking devices [16, 17], from swimmers in micrometer to centimeter length scales [18, 19] and from artificially manipulated grippers to intelligent autonomous gripping devices [5, 20], azobenzene-driven light-responsive systems serve as an important path to access small-scale soft robotics (see the examples in Figure 6.1d,e), yielding imaginative results that are not achievable with other materials. In this chapter, we provide an up-to-date overview of light-fueled LCN robotics. We start by introducing the photoactuation mechanisms, making a comparison between photothermal and photochemical actuation strategies, which often times yield distinct actuation behavior and shape changes, introduced subsequently. We then move from photoactuation to light robotics and discuss the basic physics that a small-scale LCN robot encounters, as the forces at the microscale are very different from our common everyday experience. Finally, we will draw the historical outline for important light-robotic achievements, and outline the future prospects of this exciting field of research.

6.2 Photoactuation in LCNs

To obtain light-fueled actuation, one must incorporate light-sensitive elements, i.e. photoswitches (azobenzenes) or nanoscopic heat generators (dyes or nanoparticles), into the material [21]. Both serve to absorb the light and introduce disorder into the initially ordered polymer network. In an ordered LCN, the mesogens are aligned in directions specified during the fabrication process, and the polymer chains connecting the mesogens are elongated. Upon light-induced order–disorder transition, the system loses this molecular order and becomes isotropic, and the polymer system adapts entropy-driven random-coil conformation. The presence of cross-links, in turn, ensures that the initial ordered state can be retained, and the process is reversible. During the phase transition, the volume of the structure is largely preserved. Therefore, the structure tends to contract along the molecular director (the direction in which the mesogens are aligned) and expand in the perpendicular direction(s), resulting in sometimes very pronounced macroscopic shape changes (especially in main-chain LCNs) [12, 22]. This process is schematically illustrated in Figure 6.2a.

From a chemical point of view, the methodologies to synthesize LCNs are often times divided into two classes, one utilizing polymeric and monomeric precursors (denoted as "Two-step reaction" in Figure 6.2b) and the other utilizing monomeric precursors only ("One-step reaction" in Figure 6.2b). These are extensively described in various review articles [12, 23], and we refer the reader to these sources for a more elaborate discussion. A prototype example of the two-step reaction is given by the pioneering work of Finkelmann and coworkers, who, in the beginning of the 1980s, demonstrated a two-step method for obtaining highly ordered, elastomeric LCNs based on siloxane chemistry [24]. The process is illustrated in Figure 6.2b, together with examples of the materials employed (Figure 6.2c). In the first step, the vinyl groups, reacting very fast, build up a weakly cross-linked polysiloxane gel connected by non-mesogenic cross-linkers. A mechanical load is then applied, in order to align the mesogens and obtain uniform director orientation, after which a second cross-linking step is applied to ensure the formation of the (nematic or smectic) polymer network [25]. The method yields monodomain alignment with high-order parameter over a large area, which further translates into efficient macroscopic actuation [26, 27]. The first example of a photomechanical actuator, which was demonstrated in 2001, was fabricated using this methodology [27]. The main drawback is limited control over the molecular alignment within the polymer network. Through proper materials engineering, the two-step method is amenable to complex thermomechanical actuation modes [28], but in vast majority of cases, simple uniaxial contraction–expansion is demonstrated.

The one-step reaction, developed by Broer et al. [29, 30] is based on photopolymerization of an LC monomer mixture, within the temperature range where it exhibits LC phase. The process often times leans on radical polymerization of acrylated monomers (Figure 6.2c), yet recently also other reactions such as thiol-ene polymerization, have been used [31, 32]. This method has one great advantage compared to the previously described one: as the polymerization is conducted in the low-viscosity monomeric state, techniques developed for LC

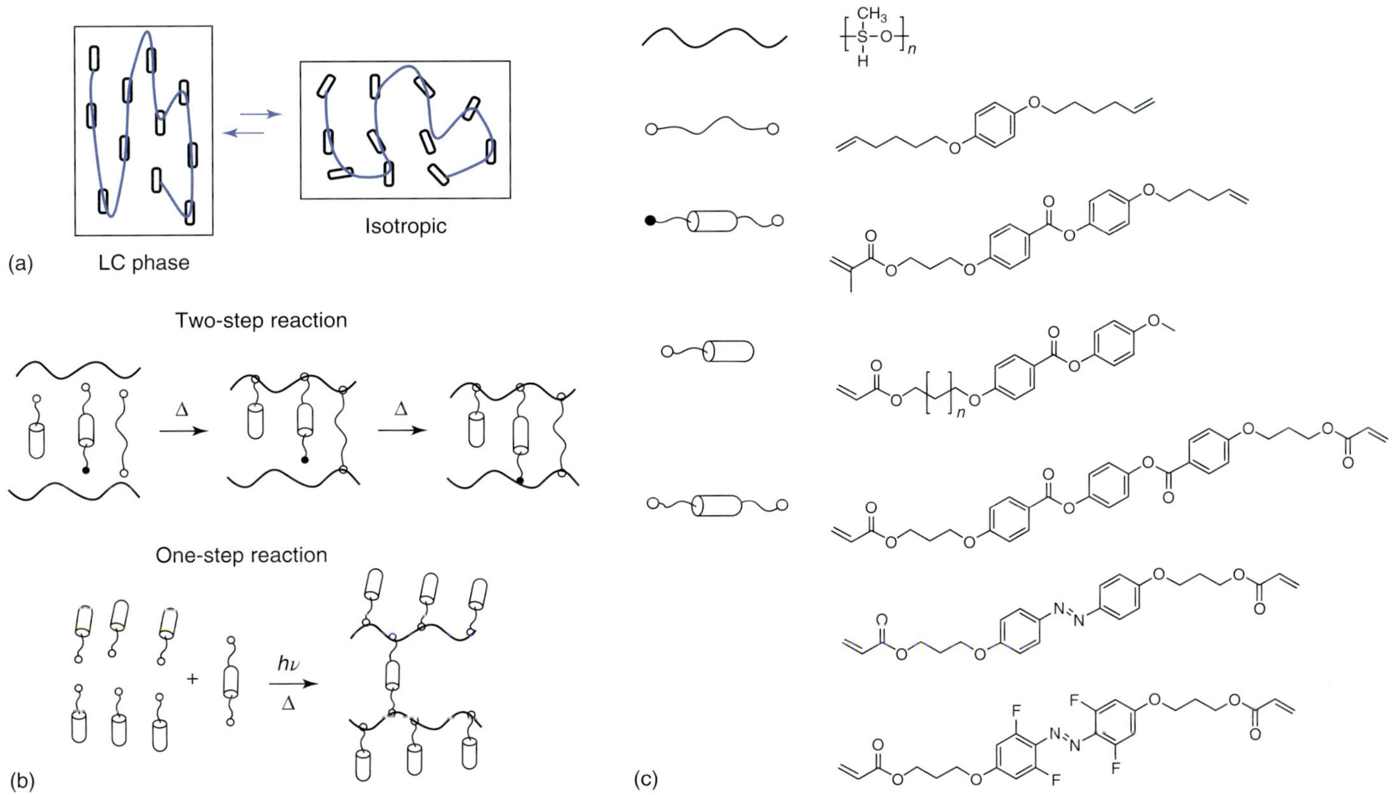

Figure 6.2 (a) Schematic illustration of the order–disorder phase transition in LCNs. (b) The two-step and one-step polymerization methods to fabricate LCNs, together with (c) some example molecules commonly used in LCN chemistry. From the top: polyhydrosiloxane chain; an isotropic cross-linker; an end-on mesogen with two polymerizable groups; an end-on mesogen with one polymerizable group; an LC cross-linker; a photoswitchable LC cross-linker; an *ortho*-fluorinated photoswitchable cross-linker. Source: Ohm et al. 2010 [11]. Copyright 2010. Reproduced with permission from Wiley-VCH.

displays (such as surface rubbing or photoalignment) can be adapted to control the molecular alignment prior to polymerization. As a result, the director orientation can be spatially patterned with high resolution, and separately controlled at the two LCN surfaces to yield complex, pre-programmed molecular alignment distributions within the polymer network [33]. This unique feature allows one to go beyond simple contraction/expansion or one-dimensional bending deformations, and depending on the director distribution, LCNs with exactly the same composition may exhibit diverse actuation modes. The relation between the molecular alignment and the light-induced shape change is further discussed in Section 6.3, after introducing photochemical and photothermal actuators and comparing their main differences.

6.2.1 Photochemical Actuation

Photochemical actuation of LCNs is triggered by photoswitchable molecules, which can undergo a reversible shape change upon photon absorption. The most popular photoswitches in the context of LCNs are photoisomerizable azobenzene derivatives [11, 34]. They are aromatic compounds, in which two phenyl rings are bridged by an azo (—N=N—) group. Their attractiveness is based on several features. First, a huge range of azobenzene derivatives with different activation wavelengths and photochemical properties can be relatively easily synthesized. Secondly, the rod-like *trans*-azobenzenes (Figure 6.3a) are well miscible with LCs, often times exhibiting LC phases themselves. Therefore, they can be mixed with several LC systems without problems of macroscopic phase separation or crystallization. Finally, *cis*-azobenzenes (Figure 6.3a), which are more globular in shape, are incompatible with LC phases, and tend to disrupt or destabilize them [37]. When incorporated into LCNs, the molecular-level *trans–cis* isomerization process (which gives rise to huge change in geometry, the length of the molecule changing from about 9.0 to 5.5 Å upon *trans–cis* isomerization) very efficiently modulates the molecular alignment within the polymer network, and triggers the photomechanical actuation through isothermal, photochemically induced order-to-disorder, or LC-to-isotropic, transition [34, 37].

The lifetime of the metastable *cis*-azobenzene determines the stability of the deformed state of the LCN photoactuator. Depending on the chemical substitution, the *cis*-lifetime can be controlled over a huge range, from sub-milliseconds in tautomerizable push–pull azobenzenes [38] to months or even years in heterocyclic or *ortho*-substituted azobenzenes [39, 40]. Using the latter, bi-stable photoactuators that preserve the photodeformed shape at least for several days have been reported [41]. Importantly, however, when exposed to a wavelength that triggers the reverse *cis–trans* isomerization, the initial shape can be retained at will. Chemical substitution also plays an important role in tuning the activation wavelength of the azobenzenes, and there's a thrust to replace conventional UV-absorbing azobenzenes with ones absorbing in the visible wavelength range, an important goal especially in applications in sunlight-energy harvesting and photobiology [42, 43].

Azobenzene-based photochemical actuators often times exhibit photoinduced bending as opposed to in-plane photocontraction. This is due to the strong

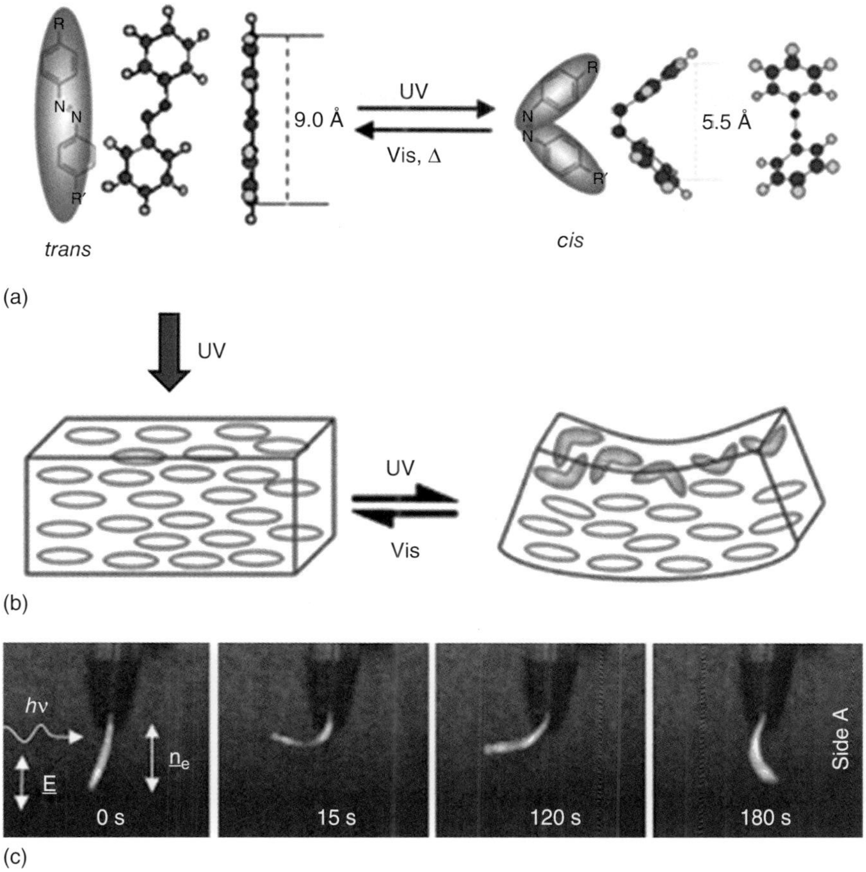

Figure 6.3 (a) Photoisomerization of azobenzene. Source: Yu and Ikeda 2011 [34]. Copyright 2011. Reprinted with permission from Wiley-VCH. (b) Illustration of photochemically induced bending in a planar-aligned LCN film. Source: Yu 2014 [35]. Copyright 2014. Reprinted with permission from Royal Society of Chemistry. (c) Photoinduced bending and subsequent unbending of a planar-aligned LCN actuator under stationary illumination Source: van Oosten et al. 2008 [36]. Copyright 2008. Reprinted with permission from American Chemical Society.

absorption of the azobenzene moieties, which constrains the light penetration into the LCN and confines the isomerization-induced loss in alignment order only to the surface of the films [35]. Therefore, the *cis*-azobenzene concentration depends on the distance from the sample surface, which gives rise to nonuniform stress distribution across the thickness and bending toward the illumination source (Figure 6.3b). Interestingly, the light-induced photochemical strain may display strongly nonlinear time response. This is because the *trans–cis* isomerization gives rise to significant spectral changes and reduction of the absorption coefficient at the pump (UV) wavelength. Thereby the LCN gradually "photobleaches," which affects the equilibrium shape of the photodeformed actuator. The nonlinear dynamics has been theoretically studied by Corbett and Warner [44, 45], and it has been also experimentally observed in homogeneously aligned, strip-like photochemical actuators (Figure 6.3c) [36, 46].

6.2.2 Photothermal Actuation

Another route toward photoactuation in LCNs is to employ photothermal heating to induce phase transition of the material. The mechanism behind the photothermal actuation is straightforward: the LCN is loaded with organic dyes or nanoparticles that, upon photon absorption, convert the absorbed light energy into heat through non-radiative thermal relaxation processes. This heat is transferred into the polymer environment, triggering the light-induced order–disorder transition and the subsequent shape change of the LCN [21]. An example of photothermal heating in a strip-like actuator is given in Figure 6.4a, where irradiation with green light gives rise to temperature increase from 23 to 43 °C and resultant deformation of the LCN actuator.

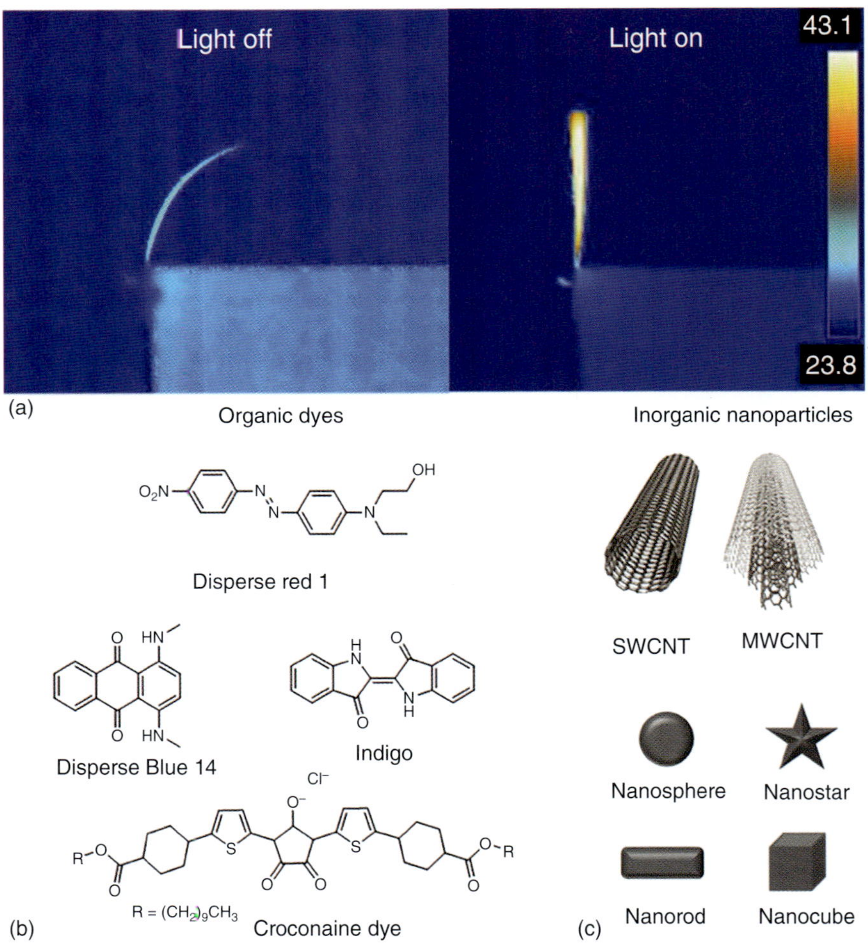

Figure 6.4 (a) Infrared images of a photothermal LCN strip actuator in dark conditions and under illumination. (b) Examples of light-absorbing organic moieties used in devising photothermal actuators. (c) Also carbon nanotubes (top) and plasmonic nanoparticles (bottom) can be used as efficient nanoscopic heat generators. Source: Choudhary and Gupta [47]. Copyright 2011. Reprinted with permission from InTech.

As the role of the photosensitive moieties incorporated into the LCN is merely to transfer the light energy into heat, a plethora of organic dyes and inorganic nanoparticles can be used for this purpose, some of which are shown in Figure 6.4. Azobenzenes with sufficiently short lifetime of the *cis*-isomer or simultaneous activation of both *trans–cis* and *cis–trans* isomerization reactions are efficient photothermal heat generators, a feature that has been recently utilized in various elegant photoactuator designs [16, 17, 48]. Broer and coworkers demonstrated that several non-photoisomerizable organic dyes and photostabilizers doped into the LCN serve the same purpose, and efficiently trigger the photothermal actuation [49]. The main requirement is that the dopants should be soluble in the LC mixture, which in most cases is not a problem because only a minor amount of the absorbers is enough to trigger significant photothermal heat generation. By proper dye selection, the LCN can be activated over the whole visible–near infrared (NIR) spectrum, rendering the photothermal effect suitable for human-safe operation and biological applications where UV light should be avoided and the NIR wavelengths are particularly attractive [50]. By doping the LCN with several dyes, one can cover the whole sunlight spectrum, providing a pathway toward sunlight-driven actuators and solar energy harvesting.

The second focal area in devising photothermally driven light-fueled actuators is doping inorganic carbon-based or metallic nanoparticles into the LCN [51]. The most prominent examples in this context are carbon nanotubes, which can be aligned along the direction of the LC mesogens [35], and actuated by both light and electric fields [52–56]. Another approach to utilize carbon nanomaterials is to incorporate graphene oxide flakes into a polymer environment to trigger the photothermal effect [57–59]. Gold nanoparticles are also an excellent alternative, due to the wealth of sizes and shapes (spheres, rods, stars, etc.) that can be synthesized, which in turn determine the plasmon resonance (and thereby the photothermal activation) wavelengths. Gold nanorods have been shown in some cases to self-align along the LC director, giving rise to reconfigurable photomechanical movements of LCN films [60] or micropillar structures [61]. Despite the fact that only a minor amount of nanoinclusions is needed to bring about the light-triggered shape change (in the order of 0.1 wt%) [62, 63], the main challenge in utilizing inorganic nanoparticles in LCNs is the poor miscibility, leading to deterioration in mechanical properties [51], decrease in order parameter, and increased light scattering. Even if these problems can be addressed by proper surface functionalization of the nanoparticles to increase the compatibility between the nanoparticles and the LC polymer network, at present we foresee organic dyes as more promising candidates as photothermal agents in light-fueled LCN robotics.

6.2.3 Comparison Between the Photochemical and Photothermal Effects

As described above, the photochemical and photothermal effects can both be utilized to realize LCN photoactuators, being potentially useful for future microrobotic applications. Both mechanisms have their own strengths, and are likely to

Table 6.1 Comparison between the photochemical and photothermal actuators.

		Photochemical	**Photothermal**
Light-absorbing moieties	Connection to the polymer backbone	Chemically bonded	Doped/bonded
	Absorbing wavelength	Typically UV/blue	UV–vis–NIR
	Concentration of the photoactive units	Relatively high	Low
Actuation	Form	Bi-stable (determined by *cis*-lifetime)	Reversible upon ceasing the irradiation
	Applied environment	Air, liquid	Air
	Speed	Seconds to minutes	Milliseconds to seconds

find distinct use cases when devising light-fueled robots. Herein we attempt to highlight their most important differences, and some of the characteristic features of photochemical versus photothermal actuators are given in Table 6.1.

From a chemical point of view, the main difference between photochemically and photothermally activated LCNs lies in the position of the active units within the polymer network. For efficient photochemical actuation, the azobenzenes should be covalently (or noncovalently) bonded, preferably as cross-links within the polymer network. Conversely, photothermal agents can be simply doped into the polymer network to yield efficient photoactuation. Covalent bonding between the photothermal units and the polymer network may allow increasing the concentration of the former, but to the best of our knowledge, it has not been shown that the bonding would be important from the viewpoint of efficiency of actuation. Secondly, typical azobenzenes are activated using ultraviolet or deep blue light, the reverse *cis–trans* isomerization taking place upon irradiation within the 450–550 nm range. Efforts have been taken toward shifting the activation wavelength to the green [64, 65] or even NIR wavelengths (using up-converting nanoparticles) [66]; yet, examples on this are still but few. In terms of efficient photoactuation using "human-friendly," low-intensity visible/NIR wavelengths, photothermal actuators seem superior compared to their photochemical counterparts. Thirdly, the concentration of the photoactive moieties may be very different in these two cases. Generally, it is thought that by increasing the amount of azobenzene within a photochemically actuated LCN matrix, one can generate larger absorption/strain gradients, thereby boosting the photoactuation. In fact, the first examples of photoinduced bending reported by the Ikeda group solely used azobenzene moieties (as both cross-links and side chains) to fabricate their actuators [67, 68]. Contrarily, photothermal actuators yield rapid actuation even if the dye concentration would only be in the range of a few mole percentage, which is enough to absorb majority of the incoming light [69].

Important distinctions between the photochemical and photothermal actuators can also be made from the physical perspective. The first is the actuation speed. For a photochemical actuator, although the molecular-level photoisomerization reaction occurs almost instantaneously upon irradiation, the macroscopic deformation of the polymer network takes seconds or even minutes [34]. The slow response can be considered as a drawback, but on the other hand, long lifetime of *cis*-isomer enables bi-stable performance of the actuator (i.e. the deformed geometry being preserved when the excitation is ceased) [41]. Photothermal actuators deform rapidly, within milliseconds to seconds, depending on the heat capacity. However, the original shape is also retained rapidly once the irradiation is ceased and the actuator cools down. A clear advantage of photochemical actuators is the ability to function in several environments, including aqueous media [70, 71]. This is not the case for photothermal actuators, since the process relies on heat conduction, and the thermal conductivity of water is 20 times higher than for air, thus efficiently reducing the temperature of the actuator in aqueous environment [72]. This poses great challenges for photothermal actuation in liquids, at least using light intensity levels that would be practically useful. The third difference we would like to point out relates to the fact that the photochemical effect relies on absorption gradients, which (i) makes it challenging to yield in-plane photocontraction (as opposed to out-of-plane bending) and (ii) may be accompanied with nonlinear bending dynamics due to "photobleaching," as described earlier [36]. Photothermal actuation in typical LCNs (thickness limited to tens of microns), where temperature is rapidly equilibrated, is dictated by molecular alignment distribution across the sample: splay alignment gives rise to out-of-plane bending (toward the planar-oriented film surface) and planar alignment to photocontraction, whereas in photochemically driven systems both cases yield bending deformation.

To summarize this part, by perfecting the material composition (e.g. cross-link density to determine the stiffness of the material; photoactive units to choose between photothermal or photochemical actuation, or utilize them synergistically), the molecular alignment within the LCN of desired thickness and the properties of the light field triggering the actuation process, efficient photoactuation can be attained. What renders LCN robotics unique is the ability to obtain multiple deformation modes using the same material composition and the same stimuli, simply by controlling the molecular alignment within the LCN. Such "programmability" yields a huge diversity in the light-induced shape changes, as will be demonstrated in Section 6.3.

6.3 Diversity of Shape Changes in LCNs

6.3.1 Uniaxial Contraction/Expansion

Uniaxially aligned LCNs can undergo reversible contraction–expansion along the molecular director axis n, thus presenting muscle-like motion under external stimulus [26, 73]. The LC molecular alignment order is described by the order parameter $S = ½(\cos^2\theta - 1)$ [74], where θ is the average angular deviation of the LC molecules from n. Under light irradiation, the LCN becomes more disordered, resulting in an increase in θ and decrease in S. Owing to the anisotropic nature of LCNs and the presence of cross-links, increased θ manifests itself as

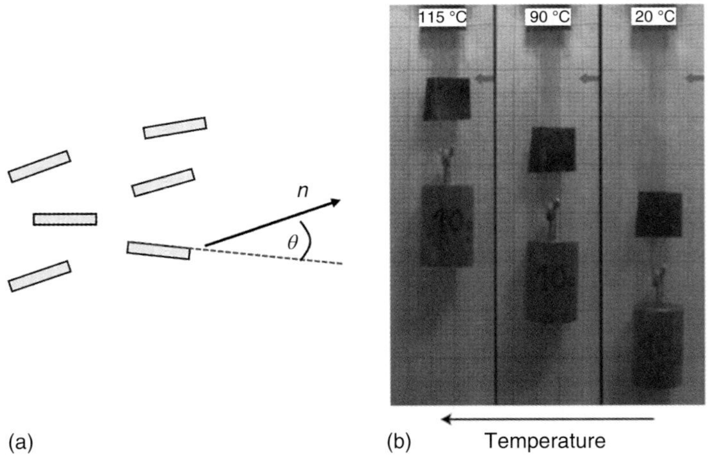

Figure 6.5 (a) Schematic representation of molecular orientation in nematic liquid crystals and the director axis, *n*. (b) Thermally induced deformation in uniaxially aligned LCNs, leading to contraction along the director and expansion perpendicular to it. Source: Ohm et al. 2010 [11]. Copyright 2010. Reprinted with permission from Wiley-VCH.

macroscopic contraction along the director axis, and expansion in the perpendicular directions (Figure 6.5a). This macroscopic deformation, arising from the change in molecular orientation, was firstly theoretically predicted by de Gennes [75], and experimentally demonstrated by Finkelmann et al. [24]. The degree of stimuli-induced contraction depends on the type of monomer as well as on the properties of the polymer chains forming the network. Different polymeric architectures can be adopted, e.g. main-chain, end-on, and side-on polymers, among which the main-chain LCNs display the highest strains and active forces (Figure 6.5b) [11, 12].

6.3.2 Bending and Coiling

Bending deformation is by far the most studied photoactuation mode in LCNs. What makes LCNs unique compared to other light-responsive materials is the ability toward bending motion in a monolithic material, i.e. without using a bilayer-construction strategy [76, 77], alike in bi-metallic cantilevers. This can be obtained by programming the molecular alignment, or by implementing light-absorption gradients that result in unhomogeneous stresses across the actuator thickness [74]. The typical molecular-alignment configurations used (in addition to the homogeneous planar and homeotropic alignments) are twisted alignment (in-plane rotation of the director by 90° from one surface to the other) and splay alignment (gradually tilting the director from horizontal to vertical), as illustrated in Figure 6.6a, respectively. Both twisted and splay alignment can trigger inhomogeneous inner stress across the film thickness when exposed to homogeneous stimuli, leading to contraction along *n* and expansion in the other directions, and bending toward the planar-aligned surface (in strip geometry with twisted alignment, the bending is toward the direction where the director is parallel to the strip).

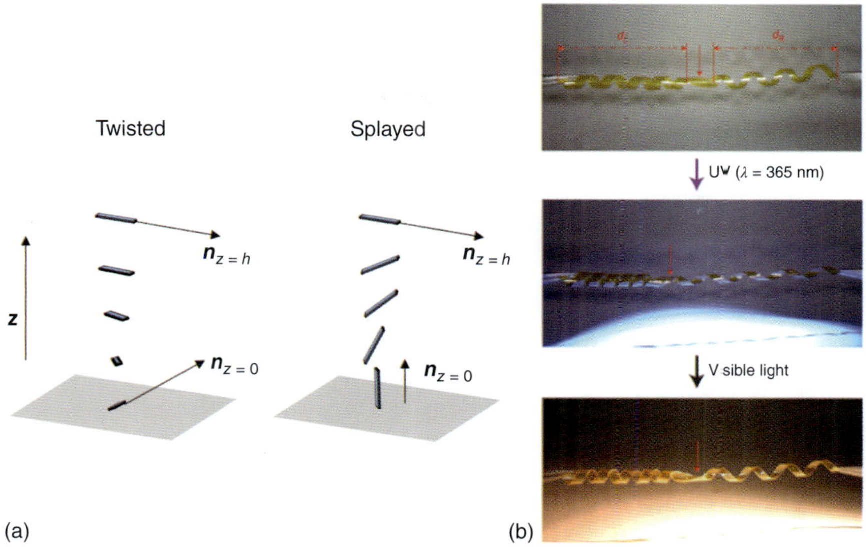

Figure 6.6 (a) Schematic representation of twisted and splayed molecular alignment. (b) A spiral ribbon with both right- and left-handed regions, where upon UV illumination one region contracts and the other expands. Source: Iamsaard et al. 2014 [78]. Copyright 2014. Reproduced with permission from Nature Publishing Group.

In order to observe pure bending deformation in splay-aligned and twist-aligned LCNs, the angle between the director and cutting axis (the long axis of the strip), denoted here as ϕ, should be close to zero. If ϕ is nonzero, the splay and twisted alignments can give rise to coils and helicoids, depending on ϕ. It is reported that a twist-aligned strip with high aspect ratio having $\phi = \pm 45°$ exhibits helicoidal deformation, whereas a wider one tends to coil [79]. The sign of ϕ dictates the handedness of the structure, and deviation from ±45° results in different pitch of the coiling. Making use of this chirality, Katsonis and coworkers demonstrated the fabrication of right-handed and left-handed spiral ribbons that, upon light illumination, could undergo complex light-induced motions and perform work (Figure 6.6b) [78, 80, 81].

6.3.3 From Flat Sheets to Cones

The examples above represent one-dimensional strip-like actuators, whose alignment difference across the thickness yields bending or coiling deformations. A two-dimensional LCN film with uniform director across the film thickness but spatially varying in the plane of the film yields very different actuation. Modes and Warner have theoretically predicted that if an azimuthally or radially aligned defect is adopted into an LCN, a flat film deforms out of plane, forming a cone or a saddle [82]. The existence of such a buckling deformation due to defect-like molecular alignment was confirmed experimentally by Broer and coworkers and is presented in Figure 6.7 [83]. These pioneering studies have led to many follow-ups of inscribing defect structures onto LCN films by using different fabrication

Figure 6.7 Schematic illustration of director orientation of radial and azimuthal defects, and deformation of respective LCN films upon thermal actuation (film diameter approximately 10 mm). Source: De Haan et al. 2012 [83]. Copyright 2012. Reproduced with permission from Wiley-VCH.

techniques. For example, laser scanning [32], microchannels [84], and laser projector patterning [85], have been used to inscribe similar LC defects into different LCNs.

6.3.4 Shape Changes via Complex Alignment Patterning

An elegant way of controlling the light-induced shape changes in LCN films is to locally pattern the director alignment across the film area. To achieve this, photoalignment is a powerful technique [86]. Photoalignment involves controlling the orientation of LC molecules by exposing a "command layer" with polarized light. The command layer contains light-sensitive units such as azobenzenes that tend to orient perpendicular to the polarization of light, or photopolymers that give rise to LC alignment parallel to polarization [87]. Molecules in the command layer guide the orientation of the LC monomers. Thus, by spatially controlling the light intensity or polarization, an arbitrary alignment profile can be patterned onto an LCN film [33]. In the following text we present some examples of such photoalignment patterning and the light-induced shape changes obtained thereof.

Schenning and coworkers reported an accordion-shaped actuator by patterning directors into four distinct orientation areas [88]. Each area consists of a 90° twist across the film thickness, however, exhibiting opposite twist directions in adjacent areas as shown in Figure 6.8a. Bending occurs under light illumination, while the alignment patterning ensures that the strip bends toward alternate directions in the adjacent regions, resembling an accordion.

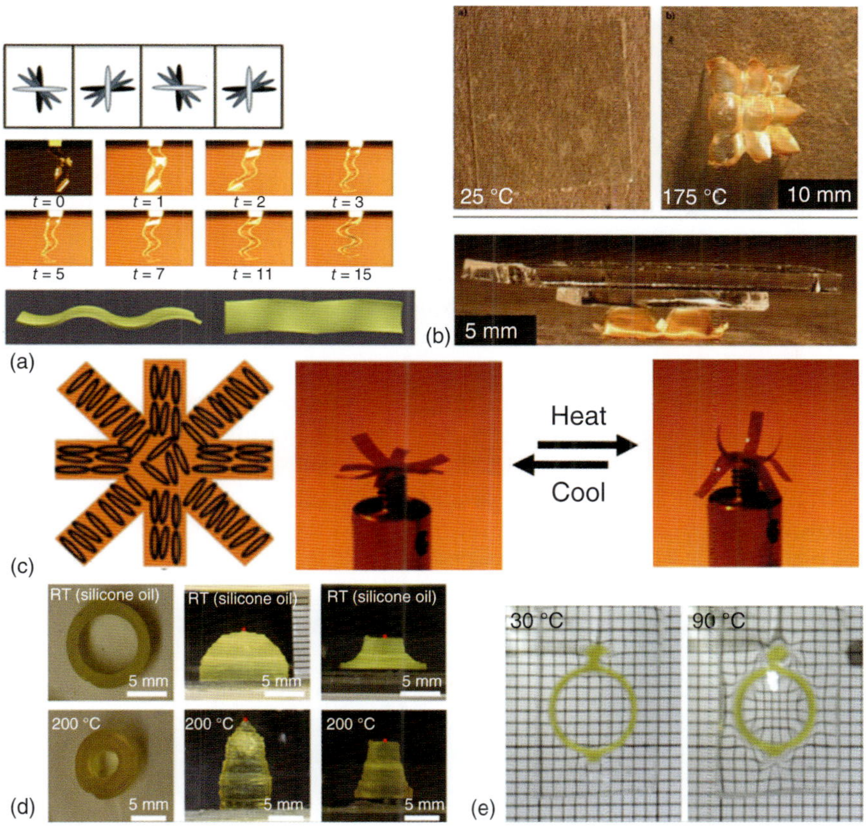

Figure 6.8 (a) Accordion-like deformation. From top to bottom: schematics of the molecular orientation, representing an alternating twist pattern; a series of pictures showing the deformation upon photothermal heating (*t* = exposure time in seconds); the deformation based on finite-element simulation. Source: De Haan et al. 2014 [88]. Copyright 2014. Reprinted with permission from Wiley-VCH. (b) An LCN film with a 3 × 3 array of radial defects results in deformation from a flat sheet into an array of pyramids. Source: Kularatne et al. 2017 [23]. Copyright 2017. Reprinted with permission from Wiley-VCH. (c) Schematics of molecular orientation (left) for an eight-legged octopod robot with photoinduced deformations (right). Source: Wani et al. 2018 [85]. Copyright 2018. Reprinted with permission from Wiley-VCH. (d) 3D-printed LCN structures with controlled molecular orientation, as dictated by the direction of LC ink flow. The structures can undergo reversible 3D shape changes upon thermal stimulus. Source: Ambulo et al. 2017 [89]. Copyright 2017. Reprinted with permission from American Chemical Society. (e) A plastic grid observed through 3D printed LCN–PDMS lens element at different temperatures. Source: López-Valdeolivas et al. 2017 [90]. Copyright 2017. Reprinted with permission from Wiley-VCH.

Broer and coworkers patterned an array of defects onto an LCN film, which they carefully cut along some defect lines, thus obtaining a flat film with a series of apertures. Those apertures could be opened and closed in a reversible fashion under light irradiation [91]. White and coworkers used a scanning laser beam with controllable polarization to pattern a flat LCN film with a 3 × 3 array of defect-type alignment patterns, which deformed into pyramids upon heating (Figure 6.8b) [92]. This concept has been further optimized by the same group in several follow-ups [93, 94]. Our group has used a commercial miniature laser projector to locally pattern the LCN alignment to obtain an octopus-like actuator, able to bend its arms in opposite directions, as dictated by the alignment patterning (Figure 6.8c) [85].

Despite all the significant advancements mentioned above, a major limitation associated with film-like actuators is the restriction to two dimensions. Recently, this limitation has been overcome by introducing 3D printing in the context of LCNs [89, 90, 95]. Ware and coworkers demonstrated a 3D-printed LCN where the LC alignment direction was controlled by the material flow through the nozzle. The structure could be efficiently actuated and reshaped through thermal triggering (Figure 6.8d) [89]. 3D printing also provides access to tunable actuators with photonic/robotic functions, as shown by Sánchez-Somolinos and coworkers who fabricated an adaptive lens by integrating a 3D-printed LCN ring with polydimethylsiloxane (PDMS) (Figure 6.8e) [90]. These are but the first steps taken with 3D-printable LCN structures, and we expect several other breakthroughs to follow in the near future.

6.4 Physics and Dynamics of Small-Scale Robots

We have discussed diverse photoactuation modes in LCNs, due to pre-designed distribution of molecular alignment. But, the question is: how to proceed from actuation to robotics? In other words, how to build a light-fueled device that can not only move its legs, but also walk? To answer this question, the interaction between the actuator and its environment must be considered, which is the scope of this section.

In order to understand the dynamics of small-scale robots, it is important to examine the forces involved in the system. Although the physics remains the same for small-scale objects and the ones acting at a macroscopic scale, the ratio between different forces and their contribution to the robotic actuation, can differ a lot. The forces that dominate depend highly on the characteristic length L of the object, a feature known as the scaling effect [96]. For example, the gravity force G and inertia are related to the entire volume of the object that scales with L^3, while active forces F_a, such as the muscular strength of an animal and the light-induced elastic force of an LCN robot, depend on the area, scaling with L^2. The ratio between F_a and G thus scales with L^{-1}, rendering the overall force strongly size-dependent. For a human weighing 80 kg (~800 N), the maximum pushing force from the legs is about 2000 N. Thus the ratio between F_a and G is small, <3. For an LCN robot with dimensions of, say, $5 \times 1 \times 0.02\,mm^3$, the gravitational force is c. 1 μN (for a material with density of about $1.1\,g\,cm^{-3}$) and the light-driven elastic force can reach 2 mN (cross-section of $1 \times 0.02\,mm^2$, active stress pressure c. 100 kPa), corresponding to $F_a/G > 1000$. The different F_a/G ratios at the different size scales point out that the design concepts for achieving efficient movement are distinct at the microscopic and macroscopic scales. This is evident in natural systems: big mammals build up strong skeleton to support their weight on land, and use two or four limbs to be able to walk or run; small animals such as spiders or insets can easily jump over very long distances due to the minor role played by gravitation, yet they have evolved with four or six legs with specific gaits because of the enhanced adhesion at the microscale. Even the simplest type of motion – walking – requires very different strategies at different scales [97].

Walking is locomotion at an interface, where a robot encounters forces, most notably adhesion and friction. Adhesion can be defined as the force required to separate two objects that are close together (a pull-off force). At the microscale, this force arises from a combination of van der Waals, capillary, and other forces. Van der Waals forces originate from fluctuating dipoles, being present in all materials [98]. They are only significant over very short distances (a few nanometers), thus being strongly affected by the contact area and surface roughness. When liquid exists at the interface, the capillary force becomes important. This attractive force originates from the desire to minimize the liquid surface energy between the surfaces [99]. The estimation of friction f for a soft-bodied micro-robot becomes very tricky, since the classical equation $f = \mu N$ (μ: friction coefficient, N: the normal force) does not properly describe the forces between soft materials. With negligible gravitation and enhanced adhesion, the friction depends highly on the material rigidity [100]. Upon light illumination, an LCN is softened, increasing the efficient contact area leading to a higher adhesion. Because of this, a microscopic walker gets easily stuck on the substrate, and cyclic shape change can no longer generate walking motion. Several methods have been developed to decrease the friction experienced by a walking LCN robot. For example, one can use additional elements (legs) made of rigid materials and with a conical tip to minimize the contact area. One may also isolate the soft body from the substrate, using, e.g. spacer particles or silicon oil.

Swimming can be considered as locomotion in a homogeneous liquid medium. Also in this case, the interaction between the object (swimmer) and the surrounding environment plays a significant role. The ratio between the inertial forces and the viscous forces is given by the Reynolds number [101], $Re = \rho v d/\mu$, where ρ and μ are the density and viscosity of the liquid, respectively, and v and d the velocity and the characteristic size of the object. Since v and d both depend linearly on L, Re scales with L^2. Therefore, a gigantic whale in an ocean can reach an Re of 10^7, a typical fish usually has an Re around 1–10, and bacteria have an Re in the range of 10^{-5} [102]. A representation of the Reynolds numbers experienced by different systems is given in Figure 6.9. For small-scale LCN robots, using a micro-swimmer demonstrated by Fischer's group as an example (cylindrical structure, 1 mm in length and 200 μm in diameter, v: ~1 μm s^{-1}) [19], the Re is in the range of 10^{-4} in water, decreasing significantly, to 10^{-8} or so, in a viscous liquid such as glycerol.

Qualitatively, the swimming mechanics is governed by inertial forces for large Reynolds numbers, and when the Re is small, viscous forces dominate the motion. Clearly, LCN robots belong to the latter class. Quantitatively, for $Re << 1$, the flow of a fluid is described by the Stokes equation [101]:

$$\nabla P_f = \mu \nabla^2 v + f$$

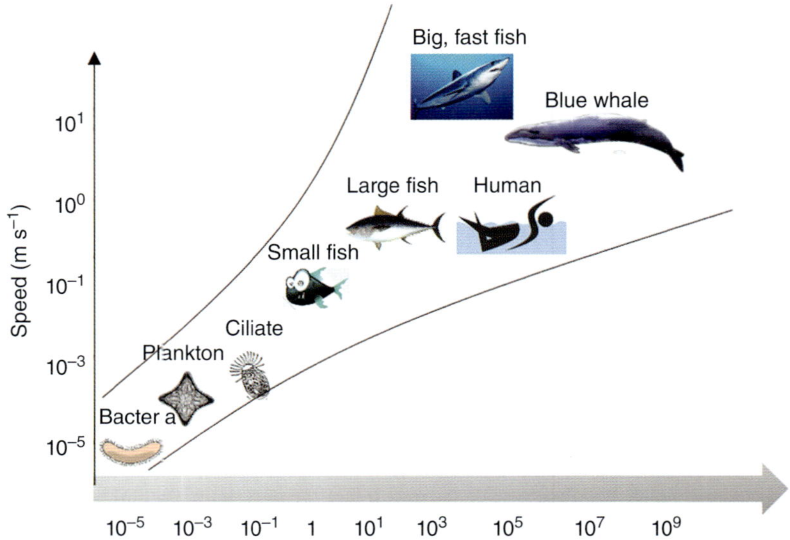

Figure 6.9 Reynolds number in different systems.

where, P_f is the fluid pressure and f is the external force. The equation does not contain any inertial terms and it is fully time-reversible. This implies that a perfect time-reciprocal motion cannot create a net motion. Thus, in order to construct an LCN-based micro-swimmer, a specific movement sequence must be chosen in actuation. Generally, in a nonreciprocal movement series, such as jellyfish type, travelling-wave-like, or spiral movement, is adopted [103].

Concerning the energy and control in LCN micro-robotics, a wireless approach by using a light field provides important advantages in delivering energy into the robotic system. The powering intensity typically ranges from few tens of milliwatt per square centimeter (a non-focused light-emitting diode (LED) source), to $>10^2\,W\,cm^{-2}$ when focused laser beams are used. The speed of photothermal heating of an LCN robot is governed by its heat capacity C, i.e. a measured physical quantity that estimates how fast the heat loss/gain is conducted in the system. Generally, C scales with the volume (L^3) of the object. Thus if the size decreases, the LCN robot exhibits a rapidly increasing speed in actuation. Examples are shown in two LCN robots constituting the same material: an inching walker with body length of 1 cm that deforms in few seconds, while a 100-μm-sized LCN robot undergoes complete shape change within only 30 ms [72].

In an LCN, the actuation force originates from the light-responsive elastic force, which scales with the cross-section (L^2), while the mass scales with L^3. As a consequence, the acceleration, $a = F/m$ (a: acceleration, F: active force, and m: mass) scales with L^{-1} and the robot's movement speed is boosted by decreasing its dimensions. This brings about some unusual characteristics of LCN micro-robots. Namely the light-induced force exerted by it can be remarkably huge: a vortexed LCN can lift up a load 150 times heavier than its own weight [92], and a 1.8-cm sized LCN "catapult" can launch a cotton ball with a speed of $30\,cm\,s^{-1}$ [104]. The elastic force, assisted by mechanical snapping motion, can exceed the weight of the actuator, leading to a jumping motion [89]. At the microscopic scale, the jumping can be very pronounced: being stuck on a rough substrate (e.g. paper), once the elastic energy is released due to some random fluctuations, the robot can jump over a distance of tens of its body lengths [16].

6.5 A Historical Overview of Light-Fueled Micro-robots

After the invention of LCNs, extensive research effort has been put to comprehend the diverse possibilities offered by this intriguing class of materials. Actuators with different chemical composition, responses to different stimuli, and with increased sensitivity, have been devised, to provide ever-more sophisticated control over their movements. At the same time, several fabrication techniques such as photoalignment [92], direct laser writing [105], and 3D printing [89] have been perfected to create diverse forms of actuators capable of fascinating shape changes for robotic applications. The pathway from an actuator to a robot is rather long as the material design parameters have to be re-thought, and the concepts taken from existing robots cannot be directly generalized to the context of LCN robots. Yet significant developments have been made over the past decade. Herein, we have collected the achievements we deem the most important in light-fueled LCN devices that perform specific robotic functions. Most of the highlighted results, comprising, e.g. walkers, swimmers, and grippers, are compiled in Figure 6.10, together with their publication years. This historical timeline reveals a clear trend: since 2008, which marked the beginning of light-fueled robotics as we define it, the field is drawing more and more attention, and has reached a booming phase in the past three years.

In 2008, Ikeda and co-workers reported a ring-like LCN motor that could perform rolling motion by irradiating the motor simultaneously with UV and visible light at different positions (Figure 6.10a) [106]. This motor served as a plastic belt to drive the motion of a pulley system, transferring light energy directly into perpetual mechanical motion. This pioneering work has acted as a source of inspiration for much of the light robotics research to follow. For example, in 2017 Broer and coworkers fabricated a light-driven "windmill device," converting light energy directly into rotating motion, obtaining rotation frequency of about 1 Hz (Figure 6.10p) [113]. An important feature of these achievements as compared to robots introduced later on is that the systems operate "autonomously," producing continuous movement under stationary illumination.

Shortly after introducing the plastic motor, the Ikeda group devised a bilayer actuator by laminating a polyethylene film with the same LCN material [107]. Using such a hybrid actuator, several deformation modes such as robotic-arm movement, were demonstrated. Combined with an asymmetric plastic substrate, the bilayer actuator experienced friction bias at the robot–ground interface, thereby being capable of inchworm-like movement (Figure 6.10b). To the best of our knowledge, this is the first example of an LCN-based walking robot. In 2013, Kohlmeyer and Chen reported a carbon-nanotube-doped LCN–polycarbonate bilayer actuator that could be driven with infrared illumination [108]. They devised a light-driven inchworm capable of crawling up a surface at 50° inclination, assisted by ratcheted wood surface and asymmetric geometry of the polycarbonate (Figure 6.10c). In 2015, the Wiersma group presented the first microscopic LCN robot (Figure 6.10d) [16]. The robot was fabricated using direct laser writing, and was composed of photodeformable LCN body and conical polymer legs made of photo-resin. The length of the robot was only 60 μm (less than the thickness of human hair), and it performed rotation, directional walking, and jumping, depending on the substrate conditions.

The above examples relied on temporal control of the light stimulus: the motion was driven by switching the light source on and off, which induced cyclic shape changes to the robot, while the pre-patterned asymmetry gave rise

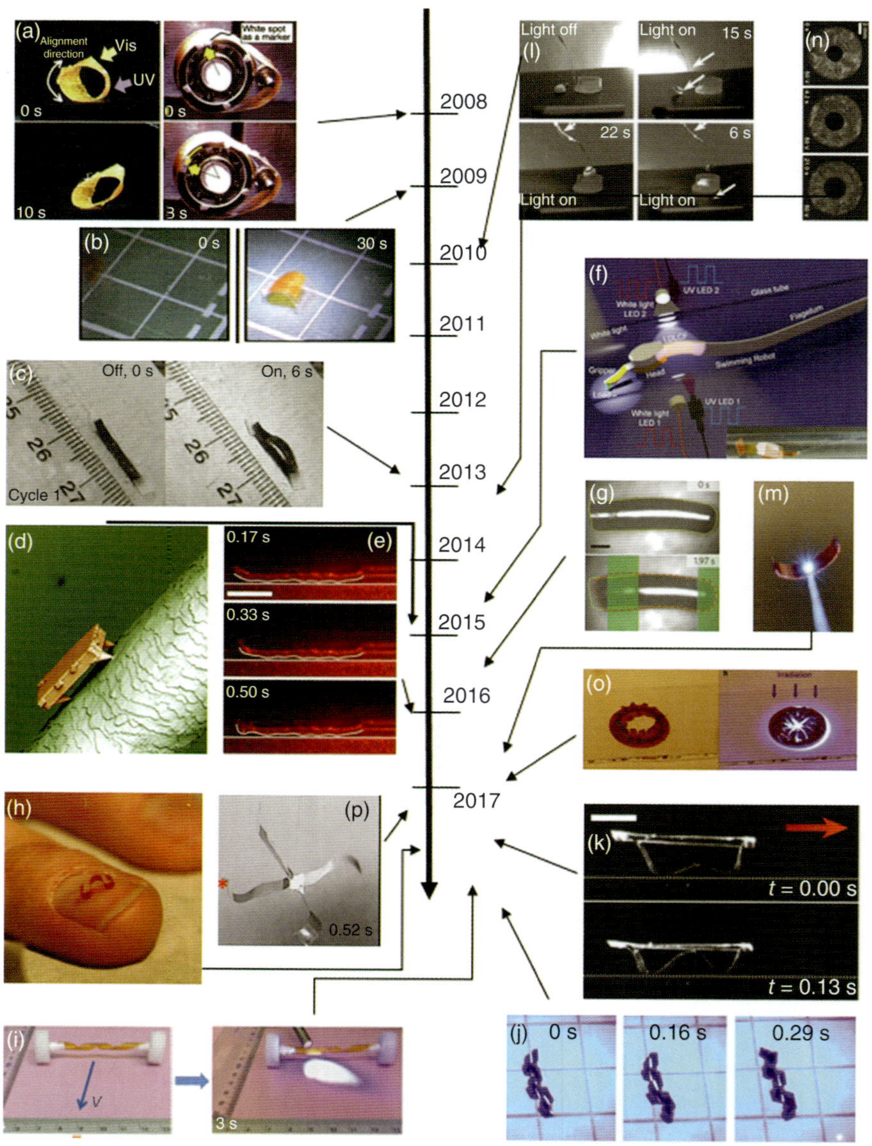

to friction bias required for locomotion. Another approach toward light-driven locomotion relies on spatial modulation of the light field. This was demonstrated in 2016 by Wasylczyk and coworkers, who devised a caterpillar-mimicking walker whose walking direction could be controlled by scanning a laser beam across the laser body (Figure 6.10e) [109]. In their device, the friction bias was not brought about by pre-patterned structural asymmetry, but by the momentum transfer from the scanned laser beam to the robot body, allowing locomotion to both forward and backward directions. The same light actuation strategies (temporal versus spatial modulation of the light fields) can also be applied to realize swimming robots. In 2015, a hybrid LCN swimmer was reported by Yu and coworkers [18]. They connected an LCN element to a flexible flagellum, and temporally switched light-actuation-triggered cyclic flipping motion, mimicking the motion of a fish and propelling itself in water (Figure 6.10f). In 2016, Fischer and coworkers reported a microscopic LCN swimmer driven by patterned light field obtained using a digital mirror array [19]. Similar to the laser scanning used for the caterpillar walker (Figure 6.10e), the modulated light field triggered travelling-wave deformation inside the LCN filament, enabling swimming motion in a high-viscosity liquid (Figure 6.10g).

The fascinating story of walking light-fueled robot continues, and the results published in 2016 and 2017 demonstrate, for example, a human-friendly millimeter-sized walker mimicking the inching motion of a caterpillar larva (Figure 6.10h) [4], and a spiral roller built in between two plastic wheels (Figure 6.10i) [110]. The most remarkable advancements of 2016 and 2017 relate to light-driven robots capable of autonomous locomotion. In 2016, White and coworkers reported an LCN spiral ribbon that can roll over large distances under spatially and temporally stationary illumination (Figure 6.10j) [111]. In such a device, the upper part of the polymer tends to exhibit a larger twisting

Figure 6.10 Historical development of light-fueled LCN robots. (a) Light-driven plastic motor. Source: Yamada et al. 2008 [106]. Copyright 2008. Reprinted with permission from Wiley-VCH. (b) The first light-driven inching walker. Source: Yamada et al. 2009 [107]. Copyright 2009. Reprinted with permission from Royal Society of Chemistry. (c) An IR-actuated inchworm. Source: Kohlmeyer and Chen 2013 [108]. Copyright 2013. Reprinted with permission from Wiley-VCH. (d) The first microscopic LCN walker. Source: Zeng et al. 2015 [16]. Copyright 2015. Reprinted with permission from Wiley-VCH. (e) Caterpillar-mimicking robot. Source: Rogóż et al. 2016 [109]. Copyright 2016. Reprinted with permission from Wiley-VCH. LCN swimmers at centimeter (f) [18] and micrometer (g) [19] scales. Source: Huang et al. 2015 [18] and Palagi et al. 2016 [19]. Copyright 2015/2016. Reprinted with permission from Nature Publishing Group. (h) Human-friendly inching walker. Source: Zeng et al. 2017 [4]. Copyright 2017. Reprinted with permission from Wiley-VCH. (i) Rolling robot. Source: Lu et al. 2017 [110]. Copyright 2017. Reprinted with permission from Wiley-VCH. Autonomous spiral roller (j) and self-propelling walker (k). Source: Gelebart et al. 2017 [17] and Wie et al. 2016 [111]. Copyright 2017/2016. Reprinted with permission from Nature Publishing Group. LCN gripper operated (l) under human control and (m) autonomously. Source: Wani et al. 2017 [5] and Cheng et al. 2010 [20]. Copyright 2017/2010. Reprinted with permission from RSC/Nature Publishing Group. The first LCN iris device (n) and its self-regulating counterpart (o). Source: Zeng et al. 2017 [69] and Schuhladen et al. 2014 [112]. Copyright 2017/2014. Reprinted with permission from Wiley-VCH. (p) A light-driven LCN windmill. Source: Vantomme et al. 2017 [113]. Copyright 2017. Reprinted with permission from Elsevier.

deformation compared to the bottom part which has been shadowed by the upper part. The asymmetric deformation brings about dynamic strains that continuously deform the spiral geometry and give rise to a net twisting moment. In 2017, Broer and coworkers constructed a photothermally driven LCN confined between the walls of a two-walled structure performing wave-like oscillation under static illumination. They harnessed such an autonomous wave-like oscillation to achieve light-fueled walking without any temporal or spatial modulation of the light field (Figure 6.10k) [17].

Considering the light robots of tomorrow, the LCN walkers are likely to evolve toward fully automated movements. This is already happening, as the examples ranging from spatial (scanning the laser beam, pointing the illumination source at a specific location on the robot body) and temporal light modulation (switching the light source on and off) to robots moving under stationary light fields demonstrate. Further, one may utilize yet another unique feature of LCNs, which is the coupling between the properties of the light field and the material shape changes. The light-induced deformation can be dynamic, in the sense that it provides feedback to the original light field through changes in the light absorption, which in turn further modifies the deformation. This special feedback loop may provide a paradigm shift for "traditional light robotics" with structured light fields: just let the robot sense the illumination conditions and act autonomously based on the feedback received from the environment. In this context two examples are highlighted, one based on gripping devices and one on iris-like devices. In 2010, Yu and coworkers realized a visible-light-controlled actuator based on an LCN–polyethylene bilayer structure [20]. They fabricated a gripping device consisting of "hand," "wrist," and "arm," by integrating photoactuators into three segments, and demonstrating several photodeformation modes. This way, for the first time they demonstrated transportation of a micro-object by a series of light illumination sequences exposing different parts of the robot (Figure 6.10l). About seven years later, another gripping device has been proposed by Priimagi and coworkers, where the gripping action is triggered by optical feedback, without the need for human intervention (Figure 6.10m) [5]. The LCN actuator is fabricated onto the tip of an optical fiber, which emits light and forms a "field of view" to the front of the device. The gripping action only takes place when an object in the field of view reflects or scatters enough light to the LCN actuator, thereby "recognizing" and selecting different targets based on the optical feedback. A similar concept was used by Wiersma and coworkers, who fabricated a microscopic LCN hand that autonomously closes on hot particles, with the ability to recognize different colors [48]. The second example, the iris-like actuation, dates back to 2014 when Zappe and coworkers presented an LCN-based iris device in which the aperture dimension could be tuned by electromagnetic heating (Figure 6.10n) [112]. To function like the human iris, the device was operated via a specific control circuit: the transmitted light intensity was measured with a photodetector, and the information was sent to the powering source of the actuator, which in turn tuned the aperture size. In 2017, Priimagi and coworkers demonstrated an autonomous LCN-based approach for an iris-mimicking device acting without a control circuit. The device is made by using radial molecular alignment and a structure with 12 petals that are open in dark conditions, and gradually close under light illumination, thereby controlling light transmission similarly to the human eye (Figure 6.10o) [69]. From the above examples, we can draw the conclusion that interplay between material properties and light fields results in autonomy, which may pave way toward small-scale robots that not only mimic the movements of natural systems, but also their intelligence.

6.6 Outlook

After decades of work toward comprehending and perfecting the chemical design and physical control of LCN actuators, the road is paved for a transition from photoactuation to light-fueled robotics. It can be perceived that the time is ripe for taking LCN actuators toward robotic applications, as exemplified by the significant advancements taken during 2016 and 2017 (see Figure 6.10). What makes LCN photoactuators particularly attractive is their intrinsically dynamic light response: light absorption induces material shape change, while at the same time the deformed shape modifies the material absorption, thereby self-adjusting its own deformation. Such optical–mechanical feedback provides the LCN actuators a very unique capacity, namely self-regulation, which can be used to realize sophisticated, dynamic robots such as light-driven oscillators and autonomous propellers and walkers. We can foresee such autonomous actuation, driven by dynamic light-response in the material, to adopt a significant role in the design of future LCN robots.

Within the next years, significant breakthroughs can be anticipated in both fabrication technologies and materials design. Regarding the former, we especially point out the recently demonstrated 3D printing of LCNs, which we expect to yield monolithic, complex robotic structures with delicate motion control [89, 90, 95]. From the materials point of view, reconfigurable materials based on dynamic covalent bonds [114–116], actuators responding to multiple colors of light [117], and LCNs whose stiffness can be patterned [28, 118], are deemed particularly promising. Such studies pave way for adaptive LCN actuators, whose performance can be reprogrammed at will, and that respond not only to light, but to multiple stimuli (or distinctly to different colors).

Finally, great efforts should be put into devising efficient robotic movement strategies. For movements at interfaces, controlling the interactions between the LCN device and the substrate (friction and adhesion) is the key, and the dream case would be if these interactions could be dynamically controlled to yield, e.g. locomotion steering. For movements in homogeneous medium such as swimming, nonreciprocal actuation strategies should be further developed. Although LCN robotics is inspired by nature, it should strongly interface with soft-robotic research conducted with more traditional materials. To put the opportunities offered by LCNs in next-generation soft robots into broader soft robotics context [119], it is important that the LCN researchers outreach the whole soft-robotics community, increasing the awareness on the unique features offered by these materials and finding the niche where LCNs offer a breakthrough for the future.

References

1 Roberts, A. (2006). *The History of Science Fiction*. New York: Palgrave macmillan.
2 Rus, D. and Tolley, M.T. (2015). Design, fabrication and control of soft robots. *Nature* 521: 467.
3 Ilievski, F., Mazzeo, A.D., Shepherd, R.F. et al. (2011). Soft robotics for chemists. *Angew. Chem. Int. Ed.* 50: 1890.
4 Zeng, H., Wani, O.M., Wasylczyk, P., and Priimagi, A. (2017). Light-driven, caterpillar-inspired miniature inching robot. *Macromol. Rapid Commun.* 39: 1700224.
5 Wani, O.M., Zeng, H., and Priimagi, A. (2017). A light-driven artificial flytrap. *Nat. Commun.* 8: 15546.
6 Stuart, M.A C., Huck, W.T.S., Genzer, J. et al. (2010). Emerging applications of stimuli-responsive polymer materials. *Nat. Mater.* 9: 101.
7 Tottori, S., Zhang, L., Qiu, F. et al. (2012). Magnetic helical micromachines: fabrication, controlled swimming, and cargo transport. *Adv. Mater.* 24: 811.
8 Mourran, A., Zhang, H., Vinokur, R., and Möller, M. (2017). Soft microrobots employing nonequilibrium actuation via plasmonic heating. *Adv. Mater.* 29: 1604825.
9 Maeda, S., Hara, Y., Sakai, T. et al. (2007). Self-walking gel. *Adv. Mater.* 19: 3480.
10 White, T.J. (ed.) (2017). *Photomechanical Materials, Composites, and Systems: Wireless Transduction of Light into Work*. Chichester, UK: Wiley.
11 Ohm, C., Brehmer, M., and Zentel, R. (2010). Liquid crystalline elastomers as actuators and sensors. *Adv. Mater.* 22: 3366.
12 White, T.J. and Broer, D.J. (2015). Programmable and adaptive mechanics with liquid crystal polymer networks and elastomers. *Nat. Mater.* 14: 1087.
13 Ikeda, T., Mamiya, J.I., and Yu, Y. (2007). Photomechanics of liquid-crystalline elastomers and other polymers. *Angew. Chem. Int. Ed.* 46: 506.
14 Bandara, H.M.D. and Burdette, S.C. (2012). Photoisomerization in different classes of azobenzene. *Chem. Soc. Rev.* 41: 1809.
15 Fleischmann E.K. and Zentel, R. (2013). Liquid-crystalline ordering as a concept in materials science: from semiconductors to stimuli-responsive devices. *Angew. Chem. Int. Ed.* 52: 8810.
16 Zeng, H., Wasylczyk, P., Permeggiani, C. et al. (2015). Light-fueled microscopic walkers. *Adv. Mater.* 27: 3883.
17 Gelebart, A.H., Mulder, D.J., Varga, M. et al. (2017). Making waves in a photoactive polymer film. *Nature* 546: 632.
18 Huang, C., Lv, J., Tian, X. et al. (2015). Miniaturized swimming soft robot with complex movement actuated and controlled by remote light signals. *Sci. Rep.* 5: 17414.
19 Palagi, S., Mark, A.G., Reigh, S.Y. et al. (2016). Structured light enables biomimetic swimming and versatile locomotion of photoresponsive soft microrobots. *Nat. Mater.* 15: 647.
20 Cheng, F., Yin, R., Zhang, Y. et al. (2010). Fully plastic microrobots which manipulate objects using only visible light. *Soft Matter* 6: 3447.

21 Hu, Y., Li, Z., Lan, T., and Chen, W. (2016). Photoactuators for direct optical-to-mechanical energy conversion: from nanocomponent assembly to macroscopic deformation. *Adv. Mater.* 28: 10548.

22 Ohm, C., Brehmer, M., and Zentel, R. (2012). Applications of liquid crystalline elastomers. In: *Liquid Crystals Elastomers: Materials and Applications.* Heidelberg: Springer-Verlag.

23 Kularatne, R.S., Kim, H., Boothby, J.M., and Ware, T.H. (2017). Liquid crystal elastomer actuators: synthesis, alignment, and applications. *J. Polym. Sci., Part B: Polym. Phys.* 55: 395.

24 Finkelmann, H., Kock, H.J., and Rehage, G. (1981). Investigations on liquid crystalline polysiloxanes. *Makromol. Chem. Rapid Commun.* 2: 317.

25 Küpfer, J. and Finkelmann, H. (1991). Nematic liquid single-crystal elastomers. *Makromol. Chem. Rapid Commun.* 726: 717.

26 Wermter, H. and Finkelmann, H. (2001). Liquid crystalline elastomers as artificial muscles. *E-Polymers* 1: 013.

27 Finkelmann, H., Nishikawa, E., Pereira, G.G., and Warner, M. (2001). A new opto-mechanical effect in solids. *Phys. Rev. Lett.* 87 (1).

28 Yang, R. and Zhao, Y. (2017). Non-uniform optical inscription of actuation domains in a liquid crystal polymer of uniaxial orientation: an approach to complex and programmable shape changes. *Angew. Chem. Int. Ed.* 56: 14202.

29 Broer, D.J., Finkelmann, H., and Kondo, K. (1988). In-situ photopolymerization of an oriented liquid-crystalline acrylate. *Makromol. Chem.* 189: 185.

30 Broer, D.J., Boven, J., and Mol, G.N. (1989). In-situ photopolymerization of an oriented liquid-crystalline acrylate, 3. Oriented polymer networks from a mesogenic diacrylate. *Makromol. Chem.* 190: 3201.

31 Yakacki, C.M., Saed, M., Nair, D.P. et al. (2015). Tailorable and programmable liquid-crystalline elastomers using a two-stage thiol-acrylate reaction. *RSC Adv.* 5: 18997.

32 Ware, T.H., Perry, Z.P., Middleton, C.M. et al. (2015). Programmable liquid crystal elastomers prepared by thiol-ene photopolymerization. *ACS Macro Lett.* 4: 942.

33 Kowalski, B.A., Guin, T.C., Auguste, A.D. et al. (2017). Pixelated polymers: directed self assembly of liquid crystalline polymer networks. *ACS Macro Lett.* 6: 436.

34 Yu, H. and Ikeda, T. (2011). Photocontrollable liquid-crystalline actuators. *Adv. Mater.* 23: 2149.

35 Yu, H. (2014). Recent advances in photoresponsive liquid-crystalline polymers containing azobenzene chromophores. *J. Mater. Chem. C* 2: 3047.

36 van Oosten, C.L., Corbett, D., Davies, D. et al. (2008). Bending dynamics and directionality reversal in liquid crystal network photoactuators. *Macromolecules* 41: 8592.

37 Ikeda, T. (2003). Photomodulation of liquid crystal orientations for photonic applications. *J. Mater. Chem.* 13: 2037.

38 García-Amorós, J. and Velasco, D. (2012). Recent advances towards azobenzene-based light-driven real-time information-transmitting materials. *Beilstein J. Org. Chem.* 8: 1003.

39 Knie, C., Utecht, M., Zhao, F. et al. (2014). Ortho-fluoroazobenzenes: visible light switches with very long-lived Z isomers. *Chem. A Eur. J.* 20: 16492.

40 Calbo, J., Weston, C.E., White, A.J.P. et al. (2017). Tuning azoheteroarene photoswitch performance through heteroaryl design. *J. Am. Chem. Soc.* 139: 1261.

41 Iamsaard, S., Anger, E., Aßhoff, S.J. et al. (2016). Fluorinated azobenzenes for shape-persistent liquid crystal polymer networks. *Angew. Chem. Int. Ed.* 55: 9908.

42 Bléger, D. and Hecht, S. (2015). Visible-light-activated molecular switches. *Angew. Chem. Int. Ed.* 54: 11338.

43 Dong, M., Babalhavaeji, A., Samanta, S. et al. (2015). Red-shifting azobenzene photoswitches for in vivo use. *Acc. Chem. Res.* 48: 2662.

44 Corbett, D. and Warner, M. (2007). Linear and nonlinear photoinduced deformations of cantilevers. *Phys. Rev. Lett.* 99: 174302.

45 Corbett, D. and Warner, M. (2009). Changing liquid crystal elastomer ordering with light – a route to opto-mechanically responsive materials. *Liq. Cryst.* 36: 1263.

46 Shimamura, A., Priimagi, A., Mamiya, J. et al. (2011). Simultaneous analysis of optical and mechanical properties of cross-linked azobenzene-containing liquid-crystalline polymer films. *ACS Appl. Mater. Interfaces* 3: 4190.

47 Choudhary, V. and Gupta, A. (2011). Polymer/carbon nanotube nanocomposites. In: *Carbon Nanotubes: Polymer Nanocomposites* (ed. S. Yellampalli). InTech.

48 Martella, D., Nocentini, S., Nuzhdin, D. et al. (2017). Photonic microhand with autonomous Action. *Adv. Mater.* 29: 1704047.

49 Gelebart, A.H., Vantomme, G., Meijer, E.J., and Broer, D.J. (2017). Mastering the photothermal effect in liquid crystal networks: a general approach for self-sustained mechanical oscillators. *Adv. Mater.* 29: 1606712.

50 Guo, L.-X., Liu, M.-H., Sayed, S.M. et al. (2016). Calamitic mesogenic near-infrared absorbing croconaine dye/liquid crystalline elastomer composite. *Chem. Sci.* 7: 4400.

51 Ji, Y., Marshall, J.E., and Terentjev, E.M. (2012). Nanoparticle-liquid crystalline elastomer composites. *Polymers (Basel)* 4: 316.

52 Ji, Y., Huang, Y.Y., Rungsawang, R., and Terentjev, E.M. (2010). Dispersion and alignment of carbon nanotubes in liquid crystalline polymers and elastomers. *Adv. Mater.* 22: 3436.

53 Marshall, J.E , Ji, Y., Torras, N. et al. (2012). Carbon-nanotube sensitized nematic elastomer composites for IR-visible photo-actuation. *Soft Matter* 8: 1570.

54 Ahir, S.V., Squires, A.M., Tajbakhsh, A.R., and Terentjev, E.M. (2006). Infrared actuation in aligned polymer-nanotube composites. *Phys. Rev. B Condens. Matter Mater. Phys.* 73: 085420.

55 Sun, X., Wang, W., Qiu, L. et al. (2012). Unusual reversible photomechanical actuation in polymer/nanotube composites. *Angew. Chem. Int. Ed.* 51: 8520.

56 Yang, L., Setyowati, K., Li, A. et al. (2008). Reversible infrared actuation of carbon nanotube-liquid crystalline elastomer nanocomposites. *Adv. Mater.* 20: 2271.

57 Yu, L., Cheng, Z., Dong, Z. et al. (2014). Photomechanical response of polymer-dispersed liquid crystals/graphene oxide nanocomposites. *J. Mater. Chem. C* 2: 8501.

58 Yu, L. and Yu, H. (2015). Light-powered tumbler movement of graphene oxide/polymer nanocomposites. *ACS Appl. Mater. Interfaces* 7: 3834.

59 Cheng, Z., Wang, T., Li, X. et al. (2015). NIR–vis–UV light-responsive actuator films of polymer-dispersed liquid crystal/graphene oxide nanocomposites. *ACS Appl. Mater. Interfaces* 7: 27494.

60 Hauser, A.W., Liu, D., Bryson, K.C. et al. (2016). Reconfiguring nanocomposite liquid crystal polymer films with visible light. *Macromolecules* 49: 1575.

61 Liu, X., Wei, R., Hoang, P.T. et al. (2015). Reversible and rapid laser actuation of liquid crystalline elastomer micropillars with inclusion of gold nanoparticles. *Adv. Funct. Mater.* 25: 3022.

62 Yang, H., Liu, J.J., Wang, Z.F. et al. (2015). Near-infrared-responsive gold nanorod/liquid crystalline elastomer composites prepared by sequential thiol-click chemistry. *Chem. Commun.* 51: 12126.

63 Zhang, H., Zhang, J., Tong, X. et al. (2013). Light polarization-controlled shape-memory polymer/gold nanorod composite. *Macromol. Rapid Commun.* 34: 1575.

64 Yin, R., Xu, W., Kondo, M. et al. (2009). Can sunlight drive the photoinduced bending of polymer films? *J. Mater. Chem.* 19: 3141.

65 Cheng, F., Zhang, Y., Yin, R., and Yu, Y. (2010). Visible light induced bending and unbending behavior of crosslinked liquid-crystalline polymer films containing azotolane moieties. *J. Mater. Chem.* 20: 4888.

66 Wu, W., Yao, L., Yang, T. et al. (2011). NIR-light-induced deformation of cross-linked liquid-crystal polymers using upconversion nanophosphors. *J. Am. Chem. Soc.* 133: 15810.

67 Yu, Y., Nakano, M., and Ikeda, T. (2003). Directed bending of a polymer film by light. *Nature* 425: 145.

68 Ikeda, T., Nakano, M., Yu, Y. et al. (2003). Anisotropic bending and unbending behavior of azobenzene liquid-crystalline gels by light exposure. *Adv. Mater.* 15: 201.

69 Zeng, H., Wani, O.M., Wasylczyk, P. et al. (2017). Self-regulating iris based on light-actuated liquid crystal elastomer. *Adv. Mater.* 29: 1701814.

70 Gelebart, A.H., Mc Bride, M., Schenning, A.P.H.J. et al. (2016). Photoresponsive fiber array: toward mimicking the collective motion of cilia for transport applications. *Adv. Funct. Mater.* 26: 5322.

71 van Oosten, C.L., Bastiaansen, C.W.M., and Broer, D.J. (2009). Printed artificial cilia from liquid-crystal network actuators modularly driven by light. *Nat. Mater.* 8: 677.

72 Zeng, H., Wasylczyk, P., Wiersma, D.S., and Priimagi, A. (2017). Light robots: bridging the gap between microrobotics and photomechanics in soft materials. *Adv. Mater.* 1703554.

73 Thomsen, D.L., Keller, P., Naciri, J. et al. (2001). Liquid crystal elastomers with mechanical properties of a muscle. *Macromolecules* 34: 5868.

74 De Haan, L.T., Schenning, A.P.H.J., and Broer, D.J. (2014). Programmed morphing of liquid crystal networks. *Polym.* 55: 5885.

75 de Gennes, P.-G. (1975). Réflexions sur un type de polymères nématiques. *C. R. Acad. Sci. Paris, Ser. B* 281: 101.

76 Zhang, X., Yu, Z., Wang, C. et al. (2014). Photoactuators and motors based on carbon nanotubes with selective chirality distributions. *Nat. Commun.* 5: 2983.

77 Hu, Y., Liu, J., Chang, L. et al. (2017). Electrically and sunlight-driven actuator with versatile biomimetic motions based on rolled carbon nanotube bilayer composite. *Adv. Funct. Mater.* 27: 1704388.

78 Iamsaard, S., Aßhoff, S.J., Matt, B. et al. (2014). Conversion of light into macroscopic helical motion. *Nat. Chem.* 6: 229.

79 Sawa, Y., Ye, F., Urayama, K. et al. (2011). Shape selection of twist-nematic-elastomer ribbons. *Proc. Natl. Acad. Sci.* 108: 6364.

80 Aßhoff, S.J., Lancia, F., Iamsaard, S. et al. (2017). High-power actuation from molecular photoswitches in enantiomerically paired soft springs. *Angew. Chem. Int. Ed.* 56: 3261.

81 Iamsaard, S., Villemin, E., Lancia, F. et al. (2016). Preparation of biomimetic photoresponsive polymer springs. *Nat. Protoc.* 11: 1788.

82 Modes, C.D. and Warner, M. (2011). Blueprinting nematic glass: Systematically constructing and combining active points of curvature for emergent morphology. *Phys. Rev. E* 84: 021711.

83 De Haan, L.T., Sánchez-Somolinos, C., Bastiaansen, C.M.W. et al. (2012). Engineering of complex order and the macroscopic deformation of liquid crystal polymer networks. *Angew. Chem. Int. Ed.* 51: 12469.

84 Xia, Y., Cedillo-Servin, G., Kamien, R.D., and Yang, S. (2016). Guided folding of nematic liquid crystal elastomer sheets into 3d via patterned 1d microchannels. *Adv. Mater.* 28: 9637.

85 Wani, O.M., Zeng, H., Wasylczyk, P., and Priimagi, A. (2018). Programming photoresponse in liquid crystal polymer actuators with laser projector. *Adv. Opt. Mater.* 6: 1700949.

86 Seki, T. (2014). New strategies and implications for the photoalignment of liquid crystalline polymers. *Polym. J.* 46: 751.

87 Yaroshchuk, O. and Reznikov, Y. (2012). Photoalignment of liquid crystals: basics and current trends. *J. Mater. Chem.* 22: 286.

88 De Haan, L.T., Gimenez-Pinto, V., Konya, A. et al. (2014). Accordion-like actuators of multiple 3D patterned liquid crystal polymer films. *Adv. Funct. Mater.* 24: 1251.

89 Ambulo, C.P., Burroughs, J.J., Boothby, J.M. et al. (2017). Four-dimensional printing of liquid crystal elastomers. *ACS Appl. Mater. Interfaces* 9: 37332.

90 López-Valdeolivas, M., Liu, D., Broer, D.J., and Sánchez-Somolinos, C. (2017). 4D printed actuators with soft-robotic functions. *Macromol. Rapid Commun.* 1700710.

91 Modes, C.D., Warner, M., Sanchez-Somolinos, C. et al. (2013). Angular deficits in flat space: remotely controllable apertures in nematic solid sheets. *Proc. R. Soc. A Math. Phys. Eng. Sci.* 469: 20120631.

92 Ware, T.H., McConney, M.E., Wie, J.J. et al. (2015). Voxelated liquid crystal elastomers. *Science* 347: 982.

93 Kowalski, B.A., Tondiglia, V.P., Guin, T., and White, T.J. (2017). Voxel resolution in the directed self-assembly of liquid crystal polymer networks and elastomers. *Soft Matter* 13: 4335.

94 Guin, T., Kowalski, B.A., Rao, R. et al. (2017). Electrical control of shape in voxelated liquid crystalline polymer nanocomposites. *ACS Appl. Mater. Interfaces* 10: 1187.

95 Kotikian, A., Truby, R.L., Boley, J.W. et al. (2018). 3D printing of liquid crystal elastomeric actuators with spatially programed nematic order. *Adv. Mater.* 30: 17064.

96 Shimoyama, I. (1995). Scaling in microrobots. In: *Proc. IEEE/RSJ Int. Conf. Intell. Robots Syst*, 208–211.

97 Hermanson, J.W. (2004). Principles of Animal Locomotion. *J. Mammal.* 85: 584–584.

98 Leckband, D. and Israelachvili, J. (2001). Intermolecular forces in biology. *Q. Rev. Biophys.* 34: 105.

99 Israelachvili, J. (2011). *Intermolecular and Surface Forces*. Waltham, MA: Academic Press.

100 Tas, N.R., Gui, C., and Elwenspoek, M. (2003). Static friction in elastic adhesion contacts in MEMS. *J. Adhes. Sci. Technol.* 17: 547.

101 Purcell, E. (1977). Life at low Reynolds number. *Am. J. Phys.* 45: 3.

102 Salta, M., Wharton, J.A., Stoodley, P. et al. (2010). Designing biomimetic antifouling surfaces. *Philos. Trans. R. Soc.* 368: 4729.

103 Abbott, J.J., Peyer, K.E., Dong, L.X., and Nelson, B.J. (2010). How should microrobots swim? *Springer Tracts in Advanced Robotics* 66: 157.

104 Lee, K.M., Koerner, H., Vaia, R.A. et al. (2011). Light-activated shape memory of glassy, azobenzene liquid crystalline polymer networks. *Soft Matter* 7: 4318.

105 Zeng, H., Martella, D., Wasylczyk, P. et al. (2014). High-resolution 3d direct laser writing for liquid-crystalline elastomer microstructures. *Adv. Mater.* 26: 2319.

106 Yamada, M., Kondo, M., Mamiya, J. et al. (2008). Photomobile polymer materials: towards light-driven plastic motors. *Angew. Chem. Int. Ed.* 47: 4986.

107 Yamada, M., Kondo, M., Miyasato, R. et al. (2009). Photomobile polymer materials – various three-dimensional movements. *J. Mater. Chem.* 19: 60.

108 Kohlmeyer, R.R. and Chen, J. (2013). Wavelength-selective, IR light-driven hinges based on liquid crystalline elastomer composites. *Angew. Chem. Int. Ed.* 52: 9234.

109 Rogóż, M., Zeng, H., Xuan, C. et al. (2016). Light-driven soft robot mimics caterpillar locomotion in natural scale. *Adv. Opt. Mater.* 4: 1689.

110 Lu, X., Guo, S., Tong, X. et al. (2017). Tunable photocontrolled motions using stored strain energy in malleable azobenzene liquid crystalline polymer actuators. *Adv. Mater.* 29: 1606467.

111 Wie, J.J., Shankar, M.R., and White, T.J. (2016). Photomotility of polymers. *Nat. Commun.* 7: 13260.

112 Schuhladen, S., Preller, F., Rix, R. et al. (2014). Iris-like tunable aperture employing liquid-crystal elastomers. *Adv. Mater.* 26: 7247.

113 Vantomme, G., Gelebart, A.H., Broer, D.J., and Meijer, E.W. (2017). A four-blade light-driven plastic mill based on hydrazone liquid-crystal networks. *Tetrahedron* 73: 4963.

114 Pei, Z., Yang, Y., Chen, Q. et al. (2014). Mouldable liquid-crystalline elastomer actuators with exchangeable covalent bonds. *Nat. Mater.* 13: 36.

115 Ube, T., Kawasaki, K., and Ikeda, T. (2016). Photomobile liquid-crystalline elastomers with rearrangeable networks. *Adv. Mater.* 28: 8212.

116 McBride, M.K., Hendrikx, M., Liu, D. et al. (2017). Photoinduced plasticity in cross-linked liquid crystalline networks. *Adv. Mater.* 29: 1606509.

117 Gelebart, A.H., Mulder, D.J., Vantomme, G. et al. (2017). A rewritable, reprogrammable, dual light-responsive polymer actuator. *Angew. Chem. Int. Ed.* 56: 13436.

118 Kim, H., Boothby, J.M., Ramachandran, S. et al. (2017). Tough, shape-changing materials: crystallized liquid crystal elastomers. *Macromolecules* 50: 4267.

119 Yang, G.-Z., Fischer, P., and Nelson, B. (2017). New materials for next-generation robots. *Sci. Robot.* 2: eaap9294.

7

Light-Driven Phase Transitions in Liquid Crystals and Their Applications

Ammathanadu S. Amrutha[1], *Ammathanadu S. Achalkumar*[1], *and Quan Li*[2]

1 Indian Institute of Technology Guwahati, Department of Chemistry, Near Doul Gobindc Road, Amingaon, North Guwahati, Guwahati, 781039, India
2 Kent State University, Liquid Crystal Institute and Chemical Physics Interdisciplinary Program, 1425 Lefton Esplanade, Kent, OH 44242, USA

7.1 Introduction to Liquid Crystals

The liquid crystal (LC) phase can be generally attained either by the action of heat on a mesogen or by dissolving an amphiphilic compound into an appropriate solvent [1–4]. It is also known as "mesophase," which refers to the amalgamation of the intermediate order and mobility of LC phases as compared to those of solids and liquids. The unique combination of facile mobility with the inter- and intramolecular order makes mesophases extremely sensitive to external stimuli such as thermal, mechanical, chemical, electrical, light, or magnetic fields. Chemically, these mesogens can be organic, organometallic, or inorganic in nature. The mesophases obtained by temperature variation are called *thermotropic*. Thermodynamically stable mesophases that appear on both heating and cooling are termed *enantiotropic*, while the thermotropic mesophases that appear only on cooling are referred to as *monotropic*. On the other hand, LC phases formed by dissolving amphiphilic compounds in an appropriate solvent (under given concentration and temperature conditions) are known as *lyotropic*. Besides, there are some molecules that exhibit LC phases under the influence of both heat and solvent; such systems are referred to as amphotropic LCs. Lyotropic LC phases are frequently encountered in everyday life, and most importantly, life itself is critically based on such ordered fluids [5–8]. Despite the significance of lyotropic LCs, thermotropic LCs have claimed relatively greater attention, firstly because they are simple to realize and handle, and secondly because they serve as an important component in the fabrication of low-power display devices. Notably, both kinds of LCs, in part, have facilitated our understanding of the supramolecular chemistry driven by secondary interactions [5, 6]. The classification of mesogens and mesophases has always been a complicated task; this is because over the past three decades a rich variety of mesophases have been discovered

Photoactive Functional Soft Materials: Preparation, Properties, and Applications, First Edition.
Edited by Quan Li.

through conventional, nonconventional, or new molecular architecture. Nonetheless, some basic principles are followed to categorize them. First and foremost, as mentioned above, they are fundamentally classified as being either thermotropic or lyotropic given the fact that the method of their realization and therefore, the resulting mesomorphism differs significantly. In particular, the mode of mesophase formation in thermotropics is by the organization of the individual molecules in contrast to that in lyotropics wherein the constituent (solute) molecules firstly aggregate and these fine structures then form different mesophases that generally depend on the temperature and concentration. Besides, there are different ways to classify these materials; for example, in terms of their structure (into amphiphilic and non-amphiphilic molecules); molecular shape (calamitic, discotic, banana mesogens, etc.); molecular size (as low- and high-molecular weight compounds); and with respect to the type of mesophase formed (nematic, cholesteric, smectic, columnar, cubic mesophases, etc.). The work presented in this chapter focuses on thermotropic LCs and hence our discussion is restricted only to this topic .

Most of the crystals on heating transform into the isotropic liquid phase by simultaneous loss of the long-range positional and orientational orders. If the molecules possess a certain amount of shape anisotropy, then the loss in one, two, or three dimensions of long-range translational order in the crystal may precede the complete breakdown of the long-range orientational order. Such compounds may show a series of mesomorphic transitions into fluid states that exhibit varying degrees of the structural symmetries. The temperature at which the crystal transforms into the mesophase is called the *melting point* while that from the mesophase to the isotropic state is called the *clearing point*.

Materials displaying thermotropic LC property are mostly organic or metal-containing organic compounds. Among the enormous number of organic compounds known, only a small fraction shows this LC behavior. Such types of compounds usually comprise hard (rigid) and soft (flexible) regions. The aromatic cores and some nonaromatic cores account for the rigidity while paraffinic chains are the soft regions of the molecule. However, these two distinct parts are combined in a specific way so as to attain a particular anisotropic shape of the molecule. The mobility in these systems is provided by large amplitude motions associated with molecules or molecular segments, namely the flexible tails/chains. The orientational order arises from the parallel alignment of anisometric molecules, while the positional order is predominantly the outcome of specific attractive forces and amphiphilic nature. Further, these interactions are also influenced by the presence of permanent dipole moments, their magnitude or the anisotropy of the molecular polarizability. Thus the anisotropic shape of the molecule plays a very important role in determining the formation and type of the LC phases.

Conventionally, rod-like and disk-shaped mesogens exhibit thermotropic mesomorphism; they are popularly known as calamitics and discotics respectively. For many years, it was believed that only long or rod-shaped molecules could be mesogenic (Figure 7.1a). However, in 1977 Chandrasekhar et al., demonstrated that even disc-like molecules (Figure 7.1b) such as benzene-1,2,3,4,5,6-hexayl hexaoctanoate display mesomorphism [9].

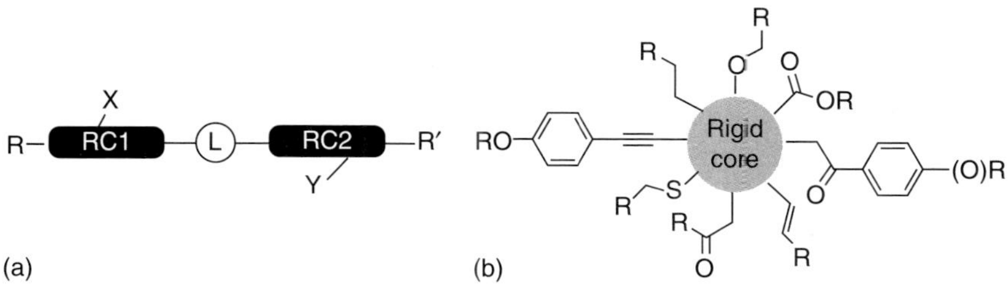

Figure 7.1 General templates for design of the rod-shaped (a) and disk-shaped (b) mesogens.

In the following section, a brief account on the general molecular structural features of rod-like mesogens is presented. In rod-like molecules, the axial dimension is larger than the radial one, while it is the other way round in discotics. As shown in a general template of a rod-like mesogen, **RC1** and **RC2** are the rigid cores that are often aromatic in nature but may also be alicyclic. These two cores are connected either directly through a covalent bond or via a linking unit **L** (e.g. —COO—, $—CH_2—CH_2$, —CH=N— and —N=N—). The terminal substituents **R** and **R′** are usually either alkyl or alkoxy chains. In many cases, one of the terminal units is a polar substituent such as CN, F, NCO, NCS, and NO_2. In some special cases, the lateral units X and Y (e.g. F, Cl, CN, and CH_3) are incorporated in the main molecular structure. Further, by employing rod-like mesogenic segments, both the main chain and side chain liquid crystal polymers (LCPs) have been designed and synthesized [2]. In the main chain LCPs, several rod-like anisotropic segments (the repeating units) join axially to form a long chain (the back bone). In side chain LCPs, rod-like mesogens are attached to the side of the polymer backbone. Inclusion of molecular chirality in rod-like molecules furnishes chiral LCs, which are known to exhibit various chiral structures [10].

Driven by curiosity and also in a quest to obtain novel mesophases, a great deal of attention has been paid to the design and synthesis of several different novel molecular architectures wherein the anisometric shape deviates from the conventional rod or disk forms [11–14]; they are collectively known as "nonconventional liquid crystals." A common feature of a majority of such materials is the structural contrast within a molecule i.e. these molecules are made up of chemically different molecular parts that are incompatible/immiscible with each other. Some of the important examples of nonconventional systems are oligomers, polycatenars, bent-core molecules, polyhydroxy amphiphiles, octahedral complexes, star-shaped molecules, rod-coil molecules, and dendrimers [2]. Based on the structural diversity or shape anisotropy and external stimuli such as temperature and pressure, a rich variety of LC phases have been observed. The structural differences between these phases are quite narrow and therefore the precise characterization of LC phases requires the use of different techniques such as polarizing optical microscopy (POM), differential scanning calorimetry (DSC), and X-ray diffraction (XRD) . Other experimental techniques such as electro-optic measurements and nuclear magnetic resonance (NMR) spectroscopy are also in use for the characterization of mesophases. In the following section, mesomorphism exhibited by conventional calamitic and discotic molecules will be discussed. Some of the nonconventional molecular structures mentioned in the previous section, which show a distinctive phase behavior, are discussed in recent review articles [5, 6, 15, 16].

7.2 Classification of Liquid Crystals

7.2.1 Calamitic Liquid Crystals: Phase Types and Structures

Originally, the mesophases exhibited by calamitic mesogens were classified into nematic, cholesteric, and smectic varieties based on the degree of positional and orientational orders (Figure 7.2). The discovery of new LC phases turns out to be usually associated with increased complexity. As discussed in previous sections, till date a number of liquid crystalline phases have been discovered with the help of several complementing characterization techniques [17].

7.2.1.1 Nematic (N) and Cholesteric (N*) Mesophase

The simplest LC phase is the nematic phase and it generally occurs just below the isotropic phase with a viscosity comparable to those of isotropic liquids. Depending on the substrate surface conditions, the N phase exhibits schlieren, marble-type, and pseudoisotropic textures. In this phase, the constituent molecules have no positional order but are on an average, oriented about a particular direction called the director, $\boldsymbol{n}$. The molecular organization in the nematic phase is illustrated in Figure 7.2. Although the preferred direction of molecules varies from point to point in the medium, a uniformly aligned sample is optically uniaxial (N_u) and birefringent. However, the biaxial modification of this phase N_b is also known [2]. Cholesteric mesophase is the chiral variant of the nematic mesophase and therefore it is also called the chiral nematic phase. It occurs in systems where the constituent molecules are chiral [10]. It can also be obtained by doping the nematic LC with optically active molecules. In these systems, there is a weak tendency for neighboring molecules to align at a slight angle to one another.

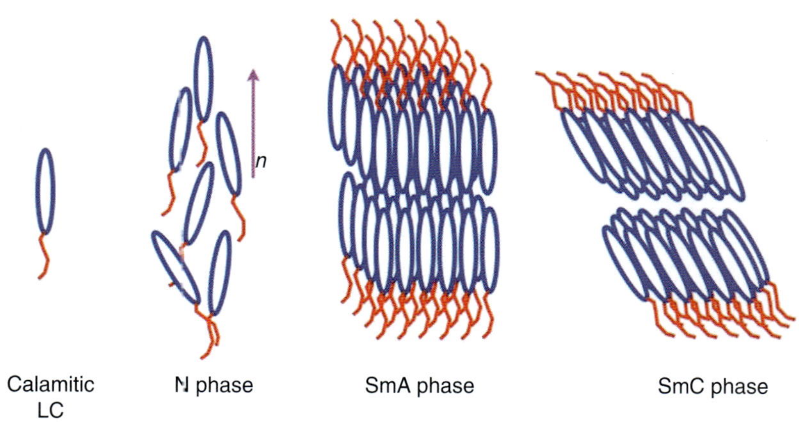

Figure 7.2 Schematic showing the organization of calamitic LCs into nematic and smectic phases.

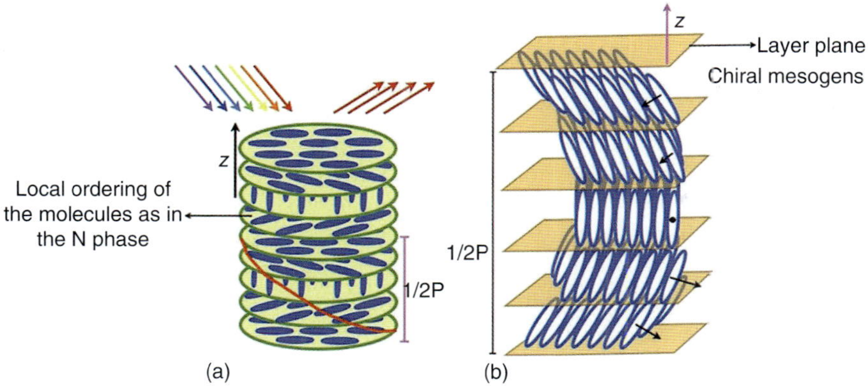

Figure 7.3 (a) Schematic showing the helical structure of the N* phase and selective reflection; (b) helical macroscopic structure of the SmC* phase in which the consecutive layers precess around the layer normal (*z*).

This leads the local director to form a helix (either right- or left-handed depending on the molecular conformation) in space with a well-defined pitch that is much longer than the size of a single molecule. The cholesteric state is illustrated in Figure 7.3a. This helical arrangement is responsible for the unique optical properties of the phase, such as selective reflection [18]. When plane-polarized light interacts with this chiral macroscopic structure, its plane of polarization gets rotated along the helix direction. When the pitch of the helix corresponds to a wavelength falling in the visible region of the spectrum (~400–800 nm) the chiral mesophase is colored. Importantly, the pitch of the helix is sensitive to a small change in temperature, and therefore, also to the reflected wavelength. Furthermore, these helical structures can be untwisted by the application of an electric field, which drives the reorientation of the molecular axis along the field direction. Depending on the surface conditions, the N* phase exhibits finger print and oily streak textures.

7.2.1.2 Smectic (Sm) Mesophase

In some phases, the average molecular center of gravity is arranged in equidistant planes, and hence in addition to orientational order, a positional order also exists, leading to a layered structure. Such phases are known as smectics (Sm) (Figure 7.2). Depending on the molecular arrangement within the layer and the extent of inter-layer correlations, smectic mesophases are further classified into different types [19]. Fluidity of the layer is an essential characteristic of all smectic phases. A more rigorous way to treat the layering arrangement is to consider a one-dimensional mass density wave [20]. There are many different smectic LC phases discovered thus far. They are smectic A (SmA), smectic B (SmB, also known as hexatic B) (non-tilted Sm phases), smectic C (SmC), smectic F (SmF), and smectic I (SmI) (tilted Sm phases). The chiral smectic mesophases are formed when the constituent molecules are chiral. There are chiral SmA, chiral smectic C (SmC*), chiral smectic I (SmI*), and chiral smectic F (SmF*) phases. The SmA and SmC phases and their chiral variants are the most commonly observed and extensively investigated fluid phases. In the following section, the structural details of some of the important smectic phases have been discussed briefly.

Smectic A (SmA) and Chiral SmA Phases In the smectic A phase, the director lies along the layer normal. The molecular packing within the layers is liquid-like and has no long-range positional correlation. Likewise, there is no correlation between the lateral positions of the molecules in successive layers. Similarly to biaxial nematic phase, SmA can also possess biaxial symmetry that has been termed the biaxial smectic A (SmA_b) phase wherein the molecules are along the layer normal, but have an additional director in the plane of the layers.Although the chiral version of the smectic A phase (observed with optically active molecules) is structurally the same as that of achiral SmA, owing to molecular chirality the symmetry is reduced from $D_{\infty h}$ to D_{∞}. As a consequence, in the SmA phase, when an electric field is applied orthogonal to the layer normal direction there will be a coupling of the electroclinic susceptibility to the field and the long molecular axes of the molecules will tilt with respect to the layer planes. For relatively low applied fields, the tilt angle varies linearly with the field. This linear electro-optic phenomenon is called the electroclinic effect [21]. Both SmA and chiral SmA phases exhibit characteristic focal-conic texture in slides treated for planar orientation and a dark field of view in slides treated for homeotropic orientation.

Smectic C (SmC) and Chiral Smectic C (SmC*) Phases The smectic C phase differs from the SmA phase, in that the director of each layer is inclined at an angle to the layer normal and this angle is identical for all layers (Figure 7.2) [22]. While the SmA phase is generally optically uniaxial, the SmC phase is optically biaxial. The tilt angle in this phase has been shown to vary with temperature and it often increases monotonically with decreasing temperature. X-ray studies indicate that the layer thickness in this phase is considerably less than the molecular length. The symmetry elements of this phase correspond to the point group C_{2h}. In the chiral smectic C (SmC*) phase, which is formed by optically active molecules, a macroscopic helical arrangement of the molecules exist. The helix occurs as a result of a gradual change in the molecular tilt direction ($\boldsymbol{n}$) from layer to layer, about an axis perpendicular to the layer planes (Figure 7.3b). Owing to the presence of chiral molecules, the SmC* phase has reduced C_2 symmetry, which creates inequivalence in the dipole moment along the C_2 axis normal to the tilt direction, resulting in spontaneous polarization (Ps) in each layer. However, the presence of helix averages out the Ps to zero [22]. Nevertheless, to obtain a macroscopic Ps (ferroelectric behavior) the helix can be unwound by surface interactions or by the application of an electric field [23].

7.2.1.3 Chiral Frustrated Phases

Chirality effectively reduces the molecular symmetry, which in turn can significantly modify the mesomorphic behavior. Chiral molecules prefer to pack in a certain manner, but such a structure cannot be adopted everywhere due to molecular constraints. Such a situation leads to the stabilization of frustrated phases [17, 18]. Examples include blue phases (BP) and twist grain boundary (TGB) phases, the structures of which are briefly described further.

Blue Phases (BP) These are distinct thermodynamic phases that appear over a narrow temperature range between chiral nematic and isotropic liquid phases in short pitch (<5000 Å) chiral nematic LCs [10] that are usually single component optically active rod-like mesogens or mixtures composed of achiral and optically active mesogens. Recently, they have been also reported to occur in discotics [24, 25] and LC dimers [26]. Three distinct blue phases have been identified, i.e. BPI, BPII, and BPIII that occur in the same order with increasing temperature. The naming of these phases as blue is due to their Bragg reflections in the blue wavelengths indicating a partially periodic structure with lattice parameters of several hundred nanometers. Bragg reflection studies indicate that BPI has a body- centered cubic lattice, BPII a simple cubic lattice, and BPIII is probably an amorphous one. The structure of the BPs involves a radial twist of the director called a double cylinder. However, this double twisted structure cannot extend perfectly into three-dimensional space. The structure of BPIII is proposed with a spaghetti model having randomly oriented wriggling double twist cylinders, while geometrical models of cubic blue phases (BPI and BPII) consist of cubic networks of double twist cylinders separated by disclination lines [17]. Combined interpretation of various experimental investigations on the structure of these phases has revealed that the BPII phase has a simple cubic unit cell of lattice defects, while the BPI phase is body-centered cubic, both with lattice parameters in the order of several hundred nanometers (the wavelength of visible light). The blue phases are thus optically isotropic, i.e. linear birefringence is absent. The optical texture of the cubic blue phases is quite characteristic, generally with a platelet texture. Amorphous BPIII is characterized by its foggy appearance.

Twist Grain Boundary (TGB) Phases These phases usually appear at the transition from the SmA or SmC* phase to the N* or isotropic phase [10, 27]. For short pitch chiral materials, there can be a competition between the need for the molecules to form a helical structure due to their chiral packing requirements and the need for the phase to form a layered structure. Molecules overcome this frustration by forming the TGB phase that consists of blocks of smectic layers, the successive blocks being rotated by a constant angle to give a helical structure with helix axis orthogonal to the layer normal direction. Since the smectic layers cannot be continuously twisted, they are separated from one another by screw dislocations to form grain boundaries which themselves are periodic, leading to the presence of layering as well as helical twist. Further, depending upon whether the smectic blocks are of smectic A, smectic C, or smectic C*, three different types of TGB phases i.e. TGBA, TGBC, and TGBC* are possible [28] and are observed experimentally [27, 29, 30]. TGBA phase exhibits a filament texture on a homeotropically aligned slide, while the TGBC* phase exhibits an undulated filament on the same. A planar texture is noticed for TGBA, while a square grid pattern is noticed for TGBC* on planar alignment condition.

7.2.2 Discotic Liquid Crystals: Phase Types and Structures

Discotic molecules primarily exhibit two types of mesophases, namely (i) nematic (N) phase and (ii) columnar (Col) phase.

7.2.2.1 Nematic (N) Phase

Discotic nematic mesophase is the least ordered mesophase formed by disk-like molecules where the mesophase possess only orientational order, while long-range positional order is absent (Figure 7.4a). In contrast to their calamitic counterparts, the short axis of the disk-shaped mesogens orient along the director (n) of the nematic phase. Hence the disks align perpendicular to the director. Consequently, they show negative birefringence. The nematic phase of disk-shaped molecules is usually not miscible with the nematic phase of rod-shaped molecules. The symbol "N" is often used to denote the discotic nematic mesophase [1, 2]. Similarly discotic analogue of cholesteric phase, chiral discotic nematic (N*) mesophase also exists. The nematic columnar (N_{Col}) phase is characterized by a columnar stacking of the molecules over a short range. However, these columns do not form two-dimensional (2D) lattice structures. They display a positional short-range order and an orientational long-range order. Similarly, the lateral nematic (N_L) phase is built of aggregates formed by multiple discotic mesogens. These supramolecular aggregates then organize into a nematic phase [1, 2].

7.2.2.2 Columnar (Col) Mesophases

Stacking of disk-like molecules one above the other results in the formation of columnar (Col) mesophases and these are the most common phases observed in discotics (Figure 7.4b). These show orientational order and positional order in two dimensions. They show rich poly-mesomorphism and are generally distinguished by the order within the columns and the order between the columns. Both are indicated in a subscripted suffix. Based on the intra- and intercolumnar order they are classified as: (i) Columnar hexagonal (Col_h); (ii) Columnar square (Col_{sq}); (iii) Columnar rectangular (Col_r); (iv) Columnar oblique (Col_{ob}); (v) Columnar plastic (Col_p); (vi) Columnar helical (H); and (vii) Columnar lamellar (Col_L) phases. However, it is difficult to determine the type of Col phase exactly, because optical textures are ambiguous and X-ray diffraction

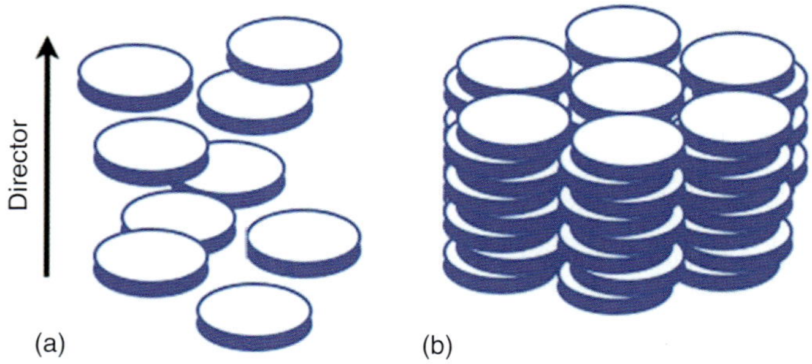

Figure 7.4 Schematics of discotic nematic (N) phase (a) and columnar (Col) phase (b).

frequently offers insufficient structural details. Chiral discotic molecules may organize in a helical fashion to form helical columnar structures.

7.3 Light-Driven Phase Transitions in Liquid Crystals

As discussed in Section 7.1, the molecular shape anisotropy is the major factor accountable for the formation of LC phases [1, 2]. Traditionally, controlling the shape anisotropy was carried out by altering the chemical structure of the molecules, which may stabilize/destabilize LC phases. While this has been a very popular method that has resulted in varieties of conventional and nonconventional LCs exhibiting rich mesophase behavior, the possibility to control the shape anisotropy by a nonchemical external stimulus is challenging and appealing. Photoisomerization is one such method, where the change in the molecular shape is brought by the photoisomerization-driven conformational change in response to light of a particular wavelength, known as "actinic light." Photochemical pathways have the advantage of specificity and selectivity over thermal and catalytic methods that result in giving mixtures of energetically unstable isomers. Photoisomerization can be defined as a photochemical process leading to isomerization of a certain chemical entity, either by bond rotation, structural rearrangement, or atom/functional group-transfer [31] in chemical compounds. Several classes of organic compounds like azobenzenes, chalcones, stilbenes, and alkenes are good examples of photoisomerizable molecular systems. These are also sometimes referred to as photochromic/photoactive/photoresponsive compounds. Several reviews on the topic of light-driven phase transitions in LCs exist in literature [32–37]. The following sections deal with a variety of photoresponsive systems that have been successfully exploited to bring about light-driven isothermal phase transitions of LCs by reversible structural changes in their photoresponsive constituent(s).

7.3.1 Azobenzenes

Among the various photoresponsive systems known, azobenzene and its derivatives form a major and attractive class. They have been significantly used to bring about photoinduced phase transitions. This is due to the ability of azobenzenes to exhibit fast, efficient, and fully reversible photoisomerization, and to remain stable over a large number of isomerization cycles [32]. Therefore, most of the results discussed in this chapter are based on azobenzenes and their derivatives. The photoisomerization of azobenzene occurs between two geometric isomers, in the *trans* and *cis* forms, which are also known as *E* and *Z* forms, respectively (Figure 7.5). These two forms differ from each other with respect to the orientation of aryl groups substituted on the central —N=N— moiety. The *E* form, where the aryl groups are parallel to each other leading to an elongated molecular shape, is energetically more stable, as compared with the metastable *Z* form with a bent shape. These forms have different absorption spectra and dipole moment. The *E* form exhibits absorption maxima centered at around 350 nm (π–π^* electronic transition) and another at around 440 nm (n–π^* electronic transition).

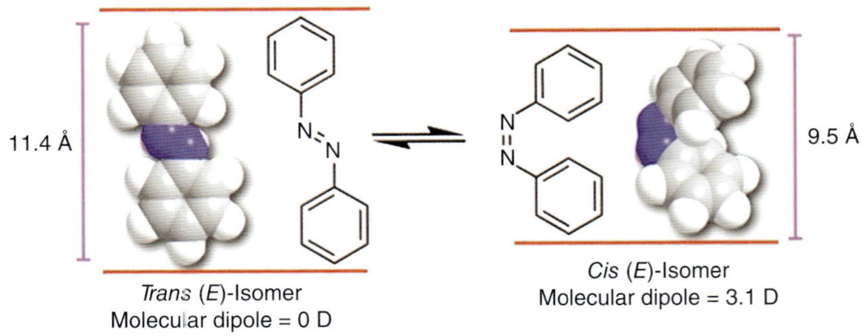

Figure 7.5 Schematic showing the photoinduced geometric isomerization of azobenzene about the azo bond, converting the *trans* (*E*) molecule into *cis* (*Z*) isomer by the actinic light irradiation and its reversal to the *E* form by thermal back relaxation or with the visible light irradiation. (Note that this isomerization results in exceptionally large changes in conformation [shape], size [length], and molecular dipole [polarity].)

The *E*-azobenzene possesses a zero dipole moment, while the *Z*-azobenzene is known to have a dipole moment of 3.1 D. These two forms are interconvertible by the use of actinic light. Usually, ultraviolet (UV) light ($\lambda \approx 350$ nm) irradiation results in the *E* to *Z* transformation, which can be followed by the change in the absorption spectra of the compound. Interestingly, the shining of visible light ($\lambda \approx 440$ nm) results in the spontaneous reversal to the original *E* form. However, heat can also drive the *Z* to *E* isomerization process, which is known as thermal back relaxation. Although light plays a major role in photoisomerization, the molecular constitution of the azobenzene derivative, temperature, and host environment can also have a significant role. Azobenzene and its derivatives can be used as dopants. In addition, the azobenzene core can be efficiently tailored to obtain photoresponsive low molar mass and polymeric LCs. The methodology, wherein azobenzene and its derivatives are employed as dopants or as pendant moieties in a copolymer, provides a valuable opportunity to strategize different guest–host combinations, and thereby control various properties such as order parameter, birefringence, color, viscosity, and polarity of the resulting photoresponsive system. The following sections will deal with the various general aspects of photoinduced effects on LCs with a special focus on phase transitions.

7.3.1.1 Photoinduced Nematic to Isotropic Phase Transition

Photoinduced phase transitions, where the nematic LCs are further disordered to the isotropic phase (N–I transition) is one of the thoroughly investigated areas involving azobenzene-derived systems. Azobenzene can be suitably functionalized to show mesomorphic behavior, it can be covalently linked to a mesogen, or it can be physically mixed as a dopant in an LC system. In all these cases, the *E* form will be more compatible with the LC phase due to its elongated rod-like structure. However, the *Z* form is not favorable for the stabilization of the liquid crystallinity due to the reduced shape anisotropy and bent shape. As shown in Figure 7.6, the molecular length of the energy-minimized structure of the *E* isomer of 4,4′-dipentylazobenzene obtained from CS Chemdraw 3D (version 12) program is 22 Å, while that of the *Z* isomer is 10.6 Å, which is less

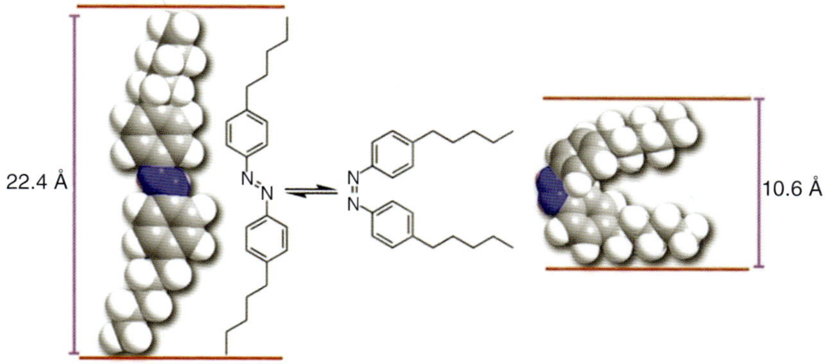

Figure 7.6 Molecular structures and the space filling models of *E*–*Z* isomers of 4,4′-dipentylazobenzene. Note the considerable change in the geometry, length, and the shape anisotropy of the molecules during the photoisomerization.

than half of the length of the *E* isomer. This combination of a drastic change in the shape and overall molecular length, in association with the high photochemical stability of azobenzenes, makes them highly suitable candidates to bring about photoinduced phase transitions. In fact, the *Z* form due to its bent shape is incompatible with the LC phase and thus acts like an "impurity." This destabilizes the LC phase and hence the *E* to *Z* photoisomerization leads to a lowering of the transition temperature. For example, consider a material containing the azobenzene derivative exhibiting an N–I phase transition. If the UV irradiation is carried out in the nematic phase, it results in the gradual lowering of the N–I transition temperature. This is due to the increase of *cis*-isomers, which destabilizes the N phase and finally leads to a collapse of the isotropic phase at a lower temperature. It must be emphasized that this transition is photochemical in nature and not driven by a change in temperature and hence it is known as an isothermal phase transition. Further, the photoinduced phase transitions in LC materials generally employ one-photon excitation of the azobenzene group. Nonetheless, recently, a photoinduced phase transition has been accomplished through two-photon excitations [32–34]. Such two-photon-driven photoisomerization holds promise in controlling the location and spatial extent of photoinduced structural changes and is therefore important for optoelectronic and photonic applications.

The high contrast in the birefringence of the LC phase and dark field view of the isotropic phase can be easily followed with the help of POM and the photoinduced isothermal N–I transition is easily visualized. For example, if the sample in the N phase placed between two crossed polarizers is irradiated with UV light at a certain region with the help of a mask, the region that is irradiated with the UV radiation undergoes photoinduced isothermal N–I transition and therefore appears dark indicating the isotropic phase. In comparison, the central masked portion, which is unirradiated, stays as such in the birefringent nematic state. LCs are a kind of elastic soft matter that are very sensitive; thus, if a small portion of LC molecules change their orientation in response to an external stimulus, the other LC molecules are induced to follow that change. Similarly,

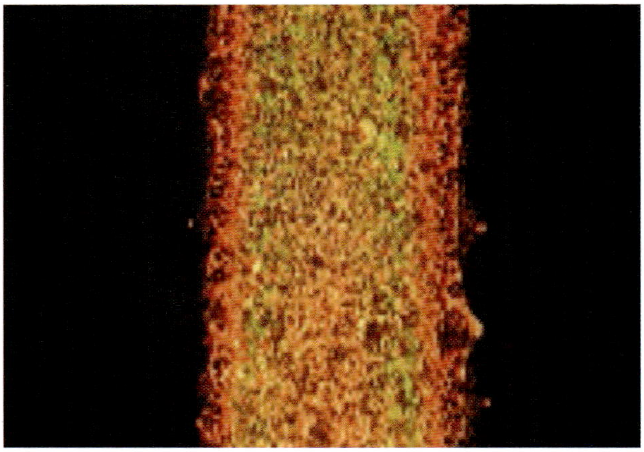

Figure 7.7 Photomicrograph of the sample in the nematic phase, exposed to UV radiation through a mask. After the UV illumination, the central region that is covered by the mask continues to have a bright birefringent appearance characteristic of a nematic phase, whereas the adjoining UV-exposed regions appear dark as the sample in those regions is in the isotropic phase caused by the photoinduced N–I transition. Source: Prasad et al. 2005 [36]. Reproduced with permission of Taylor and Francis.

during the photoisomerization, the induced bent form of the photoactive molecules initiates a local region of higher orientational disorder as compared to the regions of the host (rod-like non-photoactive) molecules. If the generated disorder is large enough, it leads to the formation of the isotropic phase domains at the irradiated sites, which appear as dark regions under POM (Figure 7.7). The zone of influence of the azobenzene molecules could become very large making the entire sample to undergo photoinduced N–I transition. It is interesting to note that if care is taken to block the light leakage, very sharp boundaries of the isotropic and nematic phases can be observed. Notably, if not illuminated with 450 nm light, it remains as such until the thermal back relaxation occurs [35, 36, 38]. Although the LC media is fluidic, the high nematic potential serves as a barrier for the diffusion of *Z* isomers from the isotropic region to the unirradiated region. The sample used in the study is a host nematic LC mixture and an azobenzene derivative as a dopant (**E7-EPH** mixture, Figure 7.8) [38].

Further, Prasad et al. showed that when they used LC dimers based on azobenzene derivatives, they observed that the photoinduced N–I transition was dependent critically on the length and parity (odd or even) of the flexible spacer. They observed a pronounced odd–even effect, with the even members of the series exhibiting higher clearing temperatures of the N phase, although the magnitude of the alternation decreases on increasing the spacer length. This type of alternation is often observed with increasing length of the terminal chain in case of monomeric LC materials, but the magnitude of the effect is very large for the LC dimers. The observed behavior has been rationalized in terms of the overall molecular shape in its all-*trans* conformation, i.e. the dimers with even parity for the spacer have a more elongated or rod-like structure, while those with odd parity possess a bent shape.

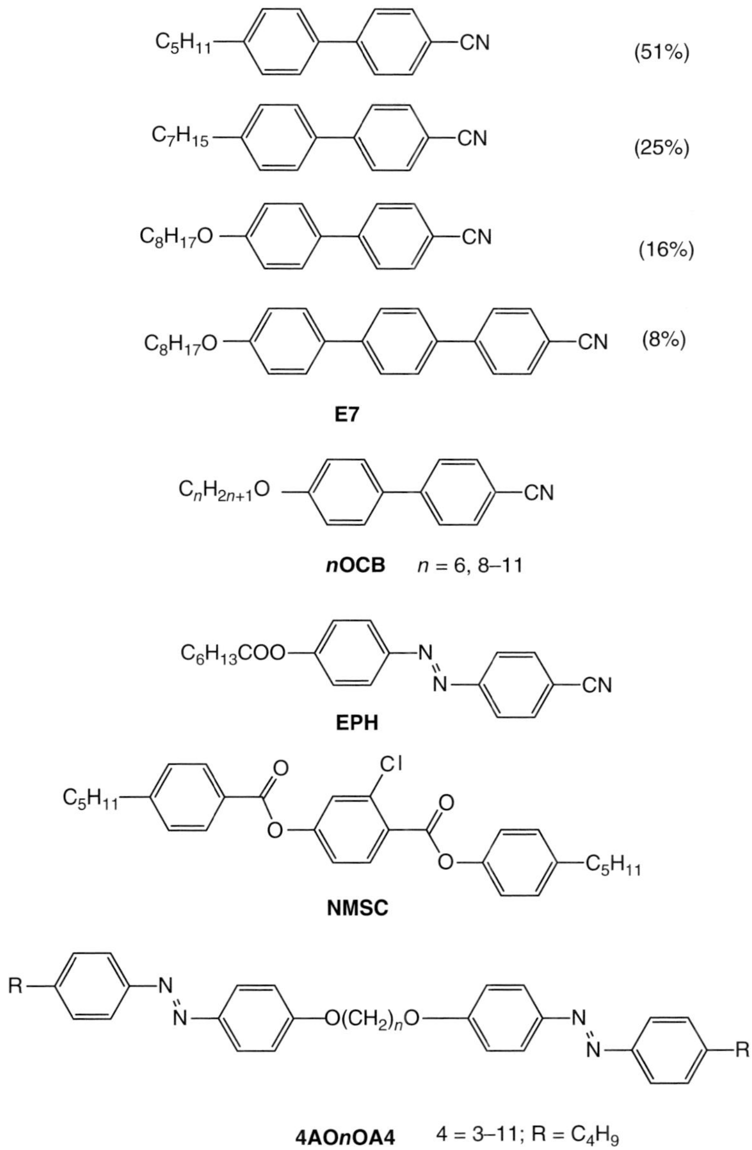

Figure 7.8 Molecular structures of the materials used for the above study. Source: Prasad et al. 2005 [36]. Reproduced with permission of Royal Society of Chemistry.

They have also investigated [38] the influence of the parity and length of the flexible spacer of the photoactive LC dimers in a guest–host system (**E7-4AOmOA4**) again for the photoinduced N–I transition. They reported a fascinating feature in this system: i.e. even when the guest azo dimers are present in a small concentration of ≈5 wt%, an odd–even spacer parity effect in a striking alternation of ΔT [$=T_{\mathrm{NI(without\ UV)}} - \Delta T_{\mathrm{NI(with\ UV)}}$] is exhibited or a change in the magnitude of the photoinduced shift in the nematic to isotropic transition is observed with respect to the spacer length of the **4AOmOA4** dimer. In other words, they observed that the spacer parity effect seen in the case of N–I transition temperature of azobenzene dimers was also extended when these dimers were doped in nematic host **E7**. It is interesting to note that such a large effect is seen despite the small concentration (≈5 wt%), which is due to the large conformational change in the photoinduced isomerization. They explained these results through the equilibrium configuration of the dimer dopant molecules in their *E* and *Z* conformations using the MM2 module in CS Chem3D software (Figure 7.9). The findings were: (i) as expected, the *E* conformers of the even parity spacers have an elongated shape, while the ones with odd spacers have a significantly bent shape; (ii) a remarkable change in the shape was noticed for the *Z* conformer, with the even member exhibiting a “Z” shape and the odd member exhibiting a “C” shape; and (iii) the projected length *l* shows an alternation with *m*, the extent

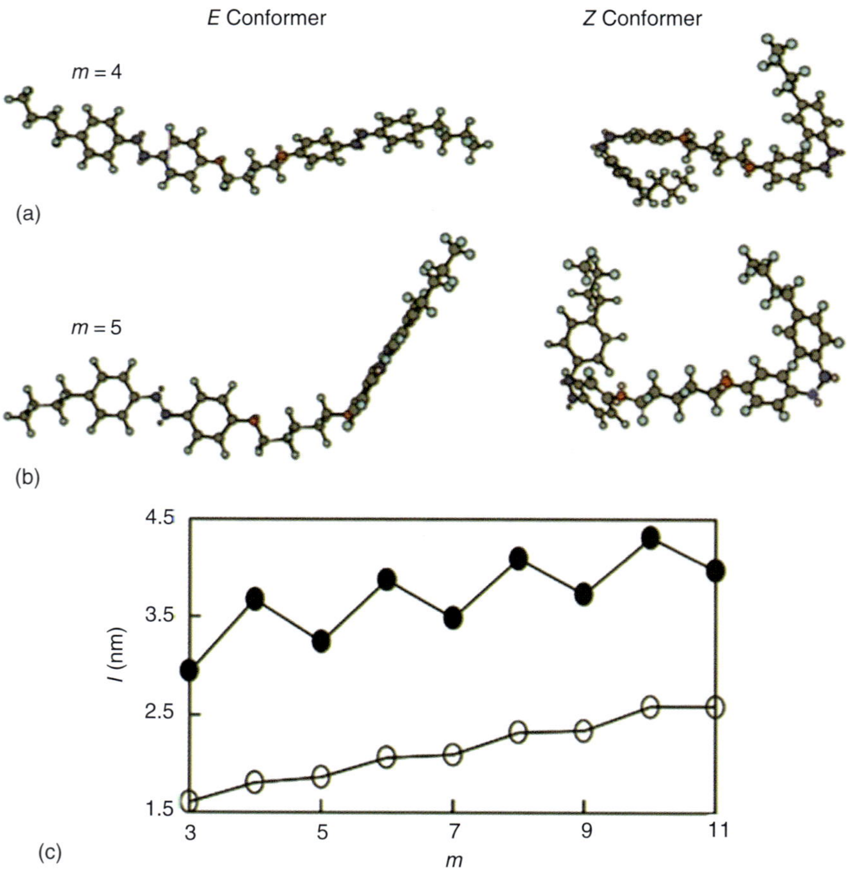

Figure 7.9 Minimum energy configurations in the *E* and *Z* conformer states for a representative (a) even (*m* = 4) spacer and (b) odd (*m* = 5) dimer molecule. (c) The projected length of the dimer molecule in the *E* (filled circle) and *Z* (empty circle) conformer states as a function of *n*, the spacer length. Source: Prasad et al. 2005 [36]. Reproduced with permission of Taylor and Francis.

of the alternation being much larger for the *E* conformers than for the *Z* conformers. In other words, upon isomerization, the change in the effective molecular length is much larger for the even (1.87 nm for $m = 4$) member than for the odd (1.39 nm for $m = 5$) member. This greater change in the dimension of the even-membered molecules is the cause for the larger value of ΔT observed for them. Thus, by designing molecules with a suitable shape in the all-*trans* conformation, it is possible to control the magnitude of the photoinduced shift in the transition temperature. Over the years the results obtained from various groups suggested that the photoinduced N–I transition depends on several aspects including the chemical nature of the photoactive molecule. This was even reported in one of the earlier reports on photoinduced N–I transition, where a small amount of azobenzene or its derivative, 4-*n*-butyl-4′-methoxyazobenzene (BMAB), was doped in a standard nematic LC, 4-cyano-4′-*n*-pentylbiphenyl (5CB) (Figures 7.10 and 7.11). These binary mixtures exhibited photo-driven N–I transition and the *trans–cis* isomerization was found to be nearly linear with the reaction time. A parameter termed as amplification, specified by the authors was found to be superior for **BMAB** than for the unsubstituted azobenzene [39, 40]. In the past five decades, numerous single component low molecular weight/polymeric and guest–host nematic LCs exhibiting light-driven N–I transition have been studied toward comprehending structure–property relationships [39–43].

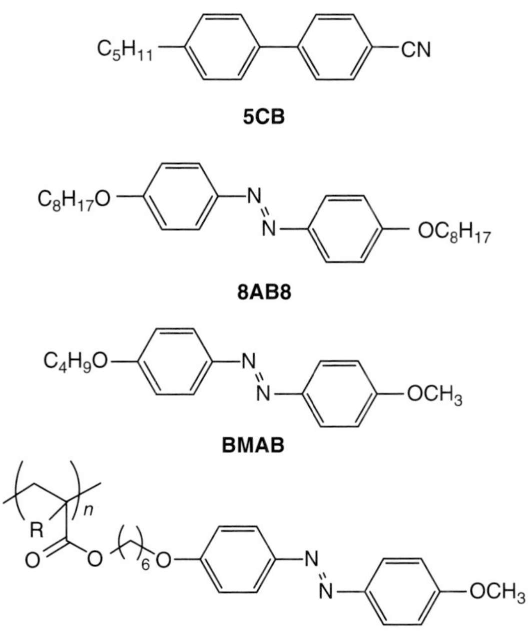

PA6AB2: R = H; X = OC_2H_5; **PM6AB2:** R = CH_3; X = OC_2H_5
PA6ABCN: R = H; X = CN; **PA6AB2:** R = CH_3; X = CN

Figure 7.10 Molecular structures of nematic host LC (**5CB**), photoactive LC monomeric azobenzenes (**BMAB** and **8AB8**), and photoactive LC side-chain polymers (**PA6AB2, PM6AB2, PA6ABCN**, and **PM6ABCN**) used in early important studies [38–40].

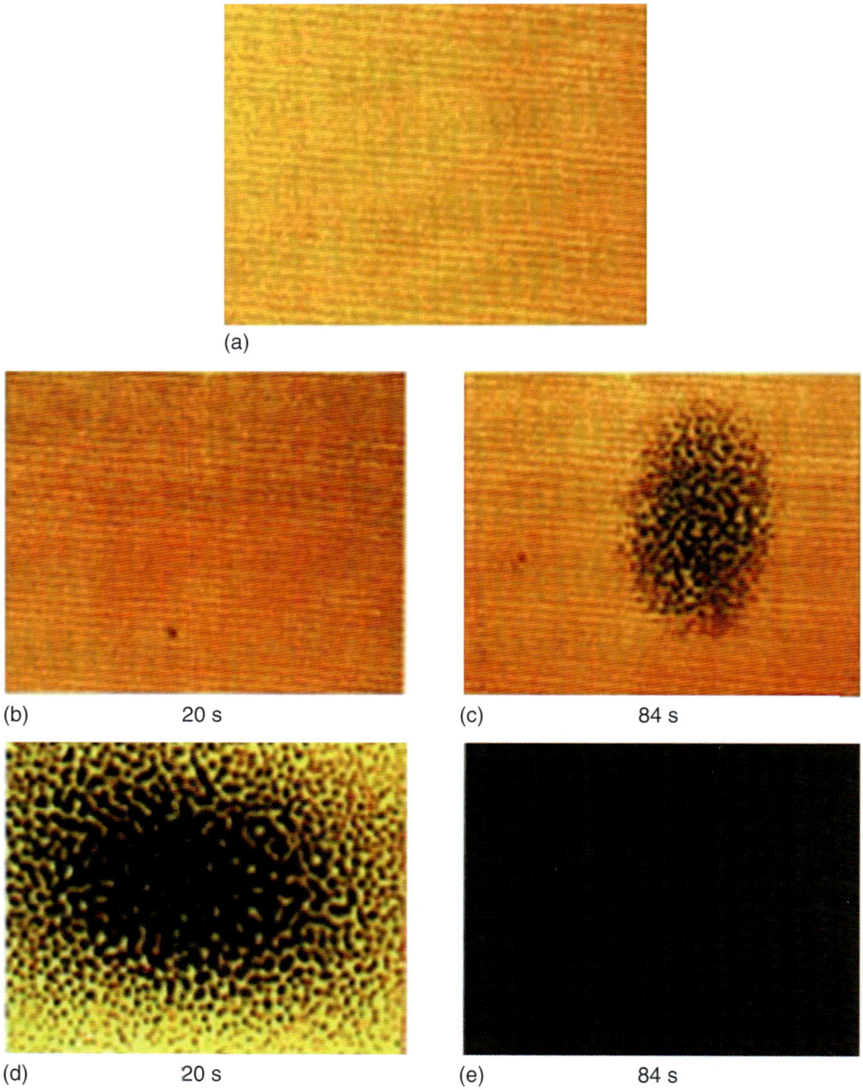

Figure 7.11 Optical textures observed during photochemical phase transition of the BMAB/5CB mixture ([BMAB] = 3.0 mol%). (a) Before irradiation, (b–e) after irradiation at 366 nm. Irradiation times are indicated in the figure. Source: Sung et al. 2002 [40]. Reproduced with permission of American Chemical Society.

7.3.1.2 Photoinduced Nematic to Smectic Phase Transition

On exposure to UV radiation, photoactive molecules based on azobenzene are known to induce a transition from the N phase (ordered) to the isotropic (disordered) phase, due to their change from the elongated *E* form to the bent *Z* form (Figure 7.12a). However, Prasad and Nair reported a novel example of a photoinduced disorder-to-order transition, which is contrary to the generally observed photoinduced order-to-disorder sequence. A photo-controlled

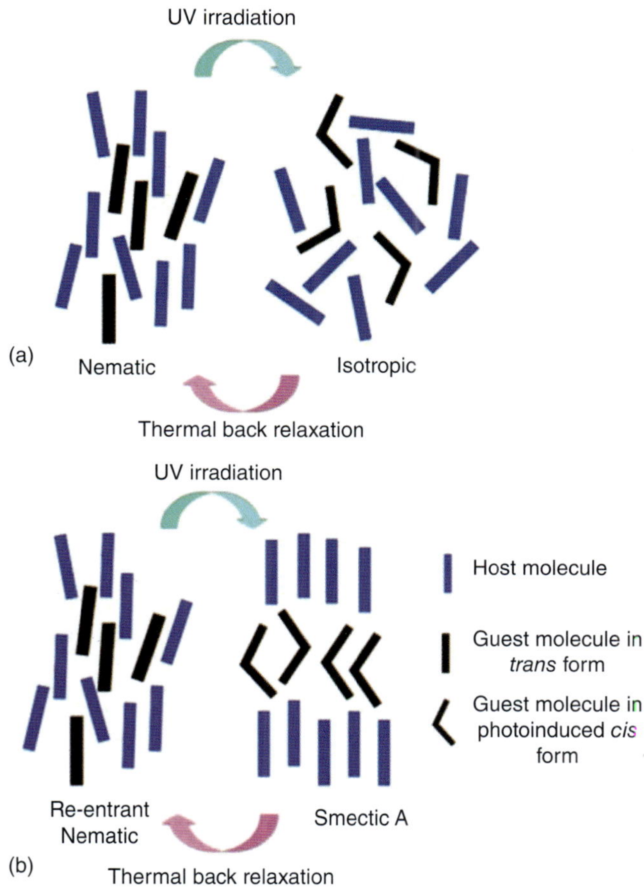

Figure 7.12 Schematic illustration of the photoinduced N to I phase transition due to *trans-to-cis* photoisomerization of the guest azo molecules (a); photoinduced re-entrant N to SmA phase transition due to *trans-to-cis* photoisomerization of the guest azo molecules (b). (As the *trans* form is energetically more stable, turning the radiation OFF results in the azo molecules converting back from the *cis-to-trans* form and the N phase is recovered by thermal back relaxation.)

nanophase segregation mechanism is utilized to explain this transition from a re-entrant nematic to a lamellar smectic A (SmA) phase (Figure 7.12b). They have observed that the speed of this transition is much slower than the nematic-to-isotropic transition, which occurs within 500 ns as compared to the 10 ns taken for nematic to isotropic transition [44].

The host material used for the measurements was a 24 wt% mixture of 4-cyanophenyl-4′-octyloxybenzoate in 4-octyloxy-4′-cyanobiphenyl. The photo-active guest compound was *p*-(*p*-ethoxyphenylazo)phenyl hexanoate doped at a low concentration of 2.9 wt% to the host mixture. The transition temperatures of the resultant guest–host mixture are isotropic 80.6 °C, nematic 48.8 °C, and smectic A 34.8 °C re-entrant nematic (I-N-SmA-N_{re}). The notable feature of this mixture is the existence of the N phase both above and below the SmA phase; the

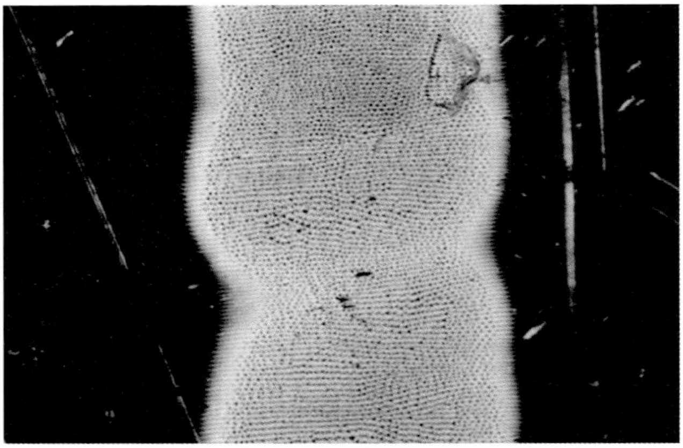

Figure 7.13 Photomicrograph images obtained by irradiating the sample in the N_{re} phase through a mask. The middle portion shows the photoinduced focal conic texture characteristic of the Sm A phase, while the darker regions on the left- and right-hand sides are in the N_{re} phase. Source: Prasad and Nair 2001 [44]. Reproduced with permission of John Wiley and Sons.

lower temperature N phase is termed re-entrant because of its reappearance on the temperature scale. When irradiated in the SmA and high-temperature N phases, the material transforms into N and I phases respectively (ordered structure to disordered structure). This fact is in agreement with the results in the literature that suggest that UV irradiation can lead to a melting of the phase, or in other words, a transition to a less ordered state [45, 46]. But when the experiment was done in the N_{re} phase, they observed a novel aspect, i.e. the transition of N_{re} phase to the SmA phase, i.e. to a more ordered state (Figure 7.13). This photoinduced effect, particularly in the N_{re} phase, can be explained with the help of photocontrolled nanophase segregation mechanism. The principle behind this nanophase segregation (see Figure 7.12b) is that when the UV radiation is absent the azo molecules are in their *trans* form, with a rod-like shape and hence can be easily accommodated into the smectic layers. However, in the photoisomerized *cis* form, these molecules assume a bent shape, and thus get expelled from the layers and occupy the region between the layers. This shows up as an increase in the overall layer spacing value, as noted from XRD studies. It may be mentioned that in the UV irradiated state, no additional reflections, which may be expected to originate from the phase separated regions, have been seen. This is perhaps because of the low concentration of the photochromic molecules used.

7.3.1.3 Photoinduced Phase Transition in Bent-Core Systems

In 1975, Meyer et al. showed that the tilted Sm LC phases composed of chiral molecules exhibit a spontaneous polarization [47]. However, in 1996, Niori et al. reported a ferroelectrically switchable B_2 phase exhibited by an achiral system with bent-core or banana-shaped molecules [48]. The origin of the spontaneous polarization (P_s) is attributed to a combination of factors such as (i) the highly

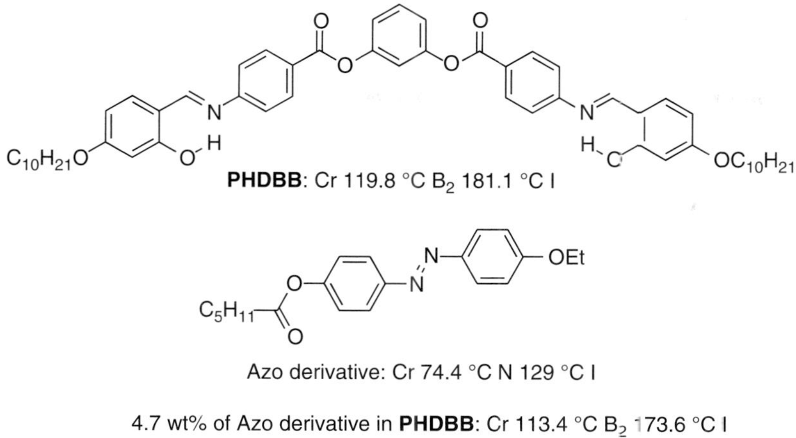

Figure 7.14 Molecular structures of the guest UV-active azo dopant and the banana-shaped host compound **PHDBB**. The transition temperatures, obtained from the DSC traces on cooling, for the azo dopant, **PHDBB** and for 4.7 wt% mixture of azo dopant in **PHDBB**, are also given. Source: Matsui et al. 2000 [50]. Reproduced with permission of AIP Publishing LLC.

polar character of the molecules; (ii) their arrangement in layers with the dipoles pointing along a common direction within the layers; and (iii) a tilt of the molecules with respect to the layer normal. These factors give rise to a chiral layer symmetry although the molecules themselves are achiral. The B_2 phase is known to be antiferroelectric in nature. The doping of azobenzene derivatives in such B_2 phase materials has shown changes in the magnitude of spontaneous polarization [49]. The P_s values were measured before UV irradiation, after irradiation, and after switching OFF the radiation. UV radiation caused the decrease in the P_s values, and while switching OFF the radiation reversed and recovered the original P_s value. This can be explained as follows. The non-photoactive bent-core molecules (PHDDB) (Figure 7.14) and the photoactive azo molecule coexist in a layer of B_2 phase (Figure 7.15). Schematic arrangements of these molecules are shown later. The length of the PHDBB is more than the double of the azo molecule in its *trans* form. In the absence of UV radiation, the azobenzene derivative will be in its *trans* form with a rod-like shape and will be arranged as shown in Figure 7.15a. When such molecules are arranged in pairs in a layer, they support the polar character of the smectic layer. On photoisomerization, they assume a bent shape (*cis* form) and due to the nanophase segregation they will be expelled and will occupy the region between the smectic layers (Figure 7.15b). This leads to an apparent increase in the volume of the system and thus to a reduction in the dipole moment/volume. Consequently, the measured P_s was found to be lower than the one obtained before UV illumination. X-ray diffraction studies confirmed the increase in the layer spacing by 0.05 nm upon UV illumination. But this volume change of 1% is not sufficient to decrease the P_s to such an extent (Figure 7.15c). As an alternative, they proposed a mechanism

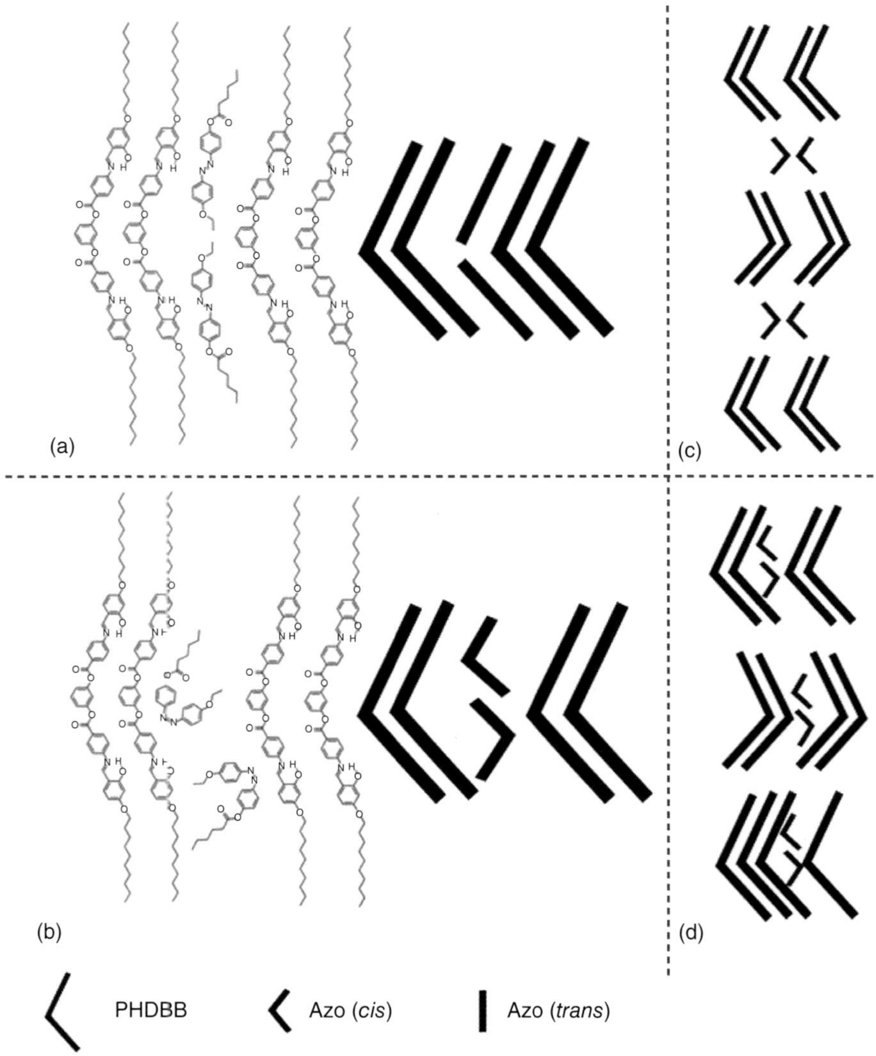

Figure 7.15 Schematic diagram illustrating (a) the disposition of the PHDBB and azo molecules before UV illumination. A pair of azo molecules that are in the *trans* conformation can stack in the fashion shown, as they have a much shorter length compared to the PHDBB molecule. (b) The bent conformation of the azo molecules in the *cis* form induced by the UV radiation. The pictures (c) and (d) show two possible mechanisms for the observed decrease in P_s values upon UV illumination. All the illustrations show the view in the polar plane, i.e. the plane containing the twofold symmetry axis and the layer normal. Source: Adapted from Prasad et al. 2004 [35] and Nair et al. 2001 [49].

where the azo molecules remain within the smectic layer (Figure 7.15d). Unlike for the host PHDDB molecules, the polar axis of the bent azo molecules within one layer could point in either direction compared to the polar axis of the smectic layer. This antiferroelectric nature of such an arrangement leads to a decrease in the P_s values.

7.3.1.4 Photoinduced Phase Transitions Involving Smectic and Chiral Phases

The photoisomerization of photoactive LCs decreases their clearing as well as mesophase–mesophase transition temperatures due to light-induced disordering of the system [32, 34–36, 51]. In some cases, even the intensity of the POM lamp source is sufficient to bring about such phase transitions [52–54]. Examples of photoactive compounds [53, 55] showing such effects and exhibiting mesophases SmA and SmC*, TGB, and N* and BP are given in Figure 7.16. The shifts in the mesophase–mesophase transition temperatures are found to be dependent on the illumination intensity and the sample thickness. This is obvious because the rate of conversion from *trans* to *cis* isomers depends on the penetration length of the incident light. Moreover, the photo-modulation of the spontaneous polarization of the SmC* phase has been established for some of these

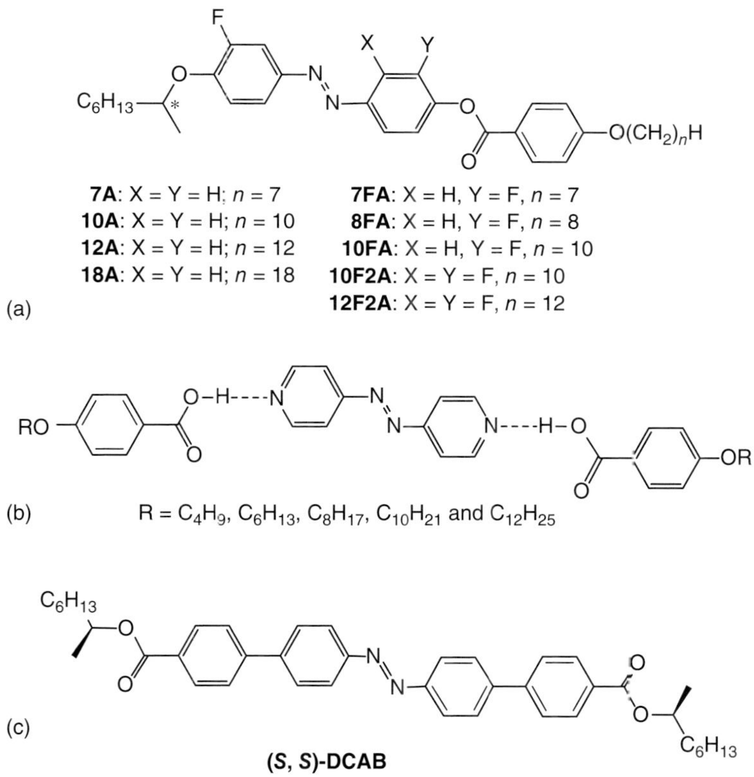

Figure 7.16 (a) General molecular structure of some optically active azobenzene-based monomeric LCs employed to study light-induced effect on (i) LC phase sequences involving SmA, SmC*, TGB, BP, and N* phases and (ii) spontaneous polarization of the SmC* phase. Source: Adapted from Joly et al. 1999 [54] and Boussoualem et al. 2009 [55]. (b) General molecular structure of supramolecular LC motifs formed by 4,4-bipyridine and alkoxybenzoic acids through H-bond association exhibiting photoinduced SmA–Cr transition. Source: Mallia et al. 1999 [56]. Reproduced with permission of American Chemical Society. (c) Molecular structure of an optically active azobenzene-based atypical LC exhibiting reversible photoinduced SmQ_L–SmQ_H and SmQ_H–I transitions. Source: Yamamoto et al. 2008 [57]. Reproduced with permission of Taylor and Francis.

compounds [52]. Further, in one of such compounds (**18A** in Figure 7.16a), a continuous decrease in SmC* to TGBA (TGB phase having SmA blocks) the phase transition temperature was observed with increase in the UV light intensity [54].

It is worth noting here that the photoinduced SmA–Cr phase transition has been observed in supramolecular complexes formed by interacting 4-*n*-alkoxybenzoic acids with the bifunctional acceptor 4,4′-bipyridine via hydrogen bonding (Figure 7.16b) [33, 42]. Hence, a transition from the less ordered state to a high ordered state occurs upon light irradiation, which is, in fact, contrary to that generally observed in other photoinduced phase transitions where photoirradiation triggers a reduction in the ordering of the system.

It is interesting to note that the photoinduced generation of BPs or the interconversion between different BPs or BPI phase transition has been observed in some mixtures consisting of different photoactive nematic hosts and photoinsensitive chiral components [58]. For example, when irradiating the N* phase of one of the mixtures with UV light (λ = 365 nm) light at ambient temperature, BPII was formed while heating; and BPI and BPII were stabilized during the cooling cycle. This was the first report on photoinduced generation of BPs. In another mixture, the photoinduced interconversion of BPs has been observed; that is, when BPI is exposed to UV light for a period of two, three, and four minutes, transitions to BPII, BPIII, and the isotropic phase occurs, respectively. Further, they have shown the modulation of the BP selective reflection with low levels of UV and visible light [58] in this study. The observed phenomena have been ascribed to the changes in refractive indices and helical pitch. In this context, it is worth mentioning that a photoinduced deformation in the BP structure has been established in chiral LC mixtures consisting of an N* LC, chiral dopant (**S811**), and azobenzene compound namely, 4-dodecyloxy-4′-methoxyazobenzene [59, 60]. Irradiating the BP phase with the 473 nm light, led to the deformation in the cubic unit cell with a concurrent shift in the photonic band gap (PBG).

Photoinduced effects on a smectic Q (SmQ) phase have been reported in a chiral azobenzene derivative (***S,S*)-DCAB** (Figure 7.16c). It is interesting to note that the SmQ phase with a chiral 3D architecture can be modulated with the help of light and such materials can be utilized in the construction of photocontrollable switching devices. This compound exhibits two SmQ phases, where the first one is a low temperature one (SmQ_L), while the second one is a high temperature one (SmQ_H). The SmQ_L on UV irradiation transforms to the SmQ_H phase; while the SmQ_H phase transforms to the isotropic phase on UV illumination. On switching off the UV light reverse transitions to the earlier phases occur. The photoisomerization of ***(S,S)*-DCAB** is responsible for these phase transitions [57].

Several azobenzene containing chiral dimers (**D4–D21**, Figure 7.17), where the azobenzene is connected to cholesterol moiety through odd/even spacers have shown photoinduced phase transitions [41–43, 61]. The SmA–I (Figure 7.18a–c) and N*–I transitions were noted on irradiation of these materials with UV light (λ = 365 nm) [63]. A blue shift was noticed in the selective reflection wavelength of the N* phase, when it is exposed to UV light; however, the magnitude of the

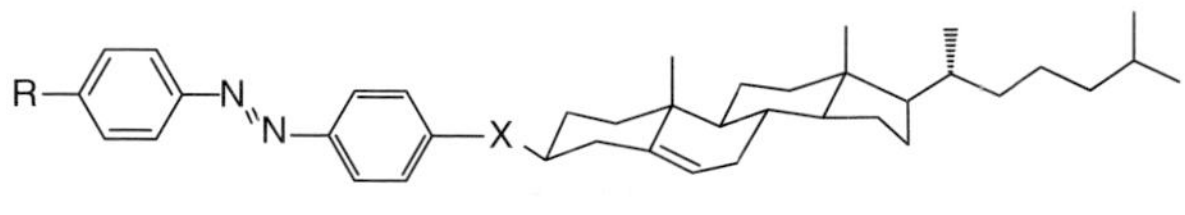

R	X	Azobenzene derivatives
n-C_mH_{2m+1}	$OOC(CH_2)_8$-$(C{=}C)_2$-$(CH_2)_8COO$	**D-4** (m = 4), **D-5** (m = 5), **D-6** (m = 12), **D-7** (m = 14) and **D-8** (m = 16)
H	$OOC(CH_2)_8$-$(C{=}C)_2$-$(CH_2)_8COO$	**D-9** (m = 9)
H	$OOC(CH_2)_nCOO$	**D-10** (n = 20), **D-11** (n = 6), **D-12** (n = 7), **D-13** (n = 8), **D-14** (n = 9), **D-15** (n = 10), **D-16** (n = 11), **D-17** (n = 12), **D-18** (n = 13), **D-19** (n = 14)
n-C_7H_{15}	$OOC(CH_2)_nCOO$	**D-20** (n = 7) and **D-21** (n = 8)
CH_3	OCOO	**1-Me**
CH_2CH_3	OCOO	**1-Et**

Figure 7.17 General molecular structure of chiral azobenzene-based LC dimers exhibiting various reversible photoinduced transitions [41–43, 61, 62].

shift was found to be dependent on the irradiation energy, temperature, and the length of alkyl tail of the azobenzene segment. The authors exploited this phenomenon in a rewritable full color image recording in photon mode [61] (Figure 7.18d). The shift in the selective reflection wavelength is attributed to *cis*-isomers, which have lower absorption maxima. These photosensitive dimers with a dicholesteryl ester have also been employed as dopants to obtain a stable glassy N* state [41]. In these materials, irradiating the SmA phase with UV light (λ = 366 nm) leads to the N* phase and with continued illumination the N* phase collapses to the isotropic phase; thus, SmA–N* and N*–I transitions have been observed [42]. In another dimer, exhibiting SmC*, SmA, and N* phases, a direct photoinduced SmC*–I transition occurs without passing through other two phases [43]. The percentage of *cis* isomer required for SmC*–I, SmA–N*, and SmA–I transition were estimated with the help of HPLC to be 30%, 6%, and 19%, respectively.

Li et al. synthesized two cholesterol appended azobenzene derivatives (**1-Me** and **1-Et**, Figure 7.17 exhibiting cholesteric phase [62]. The compound **1-Et** was used as a chiral dopant in conventional nematic LC **5CB** (achiral N–I transition at 35 °C); at the composition of 25% **1-Et** in 5CB, the mixture is homogeneously mixed and undergoes an N*–I transition temperature at 39.9 °C. UV irradiation at 365 nm for 10 seconds showed the photoinduced isothermal N*–I transition. Removal of UV light immediately leads to the reverse process of N* domain formation from the isotropic phase appearing as droplet nucleation followed by coalescence. The reversion to the original polygonal fingerprint texture was reached within approximately two hours at room temperature in the dark (Figure 7.19).

Zhou et al. showed the photoinduced reversible lattice deformation and transformation of a simple cubic soft lattice, i.e. blue phase II, in a chiral LC system to a helical superstructure, i.e. cholesteric LC (Figure 7.20) [63]. This is attained by doping an achiral photoresponsive molecular switch-functionalized polyhedral oligomeric silsesquioxane nanocage (**Azo-POSS**) into a chiral LC

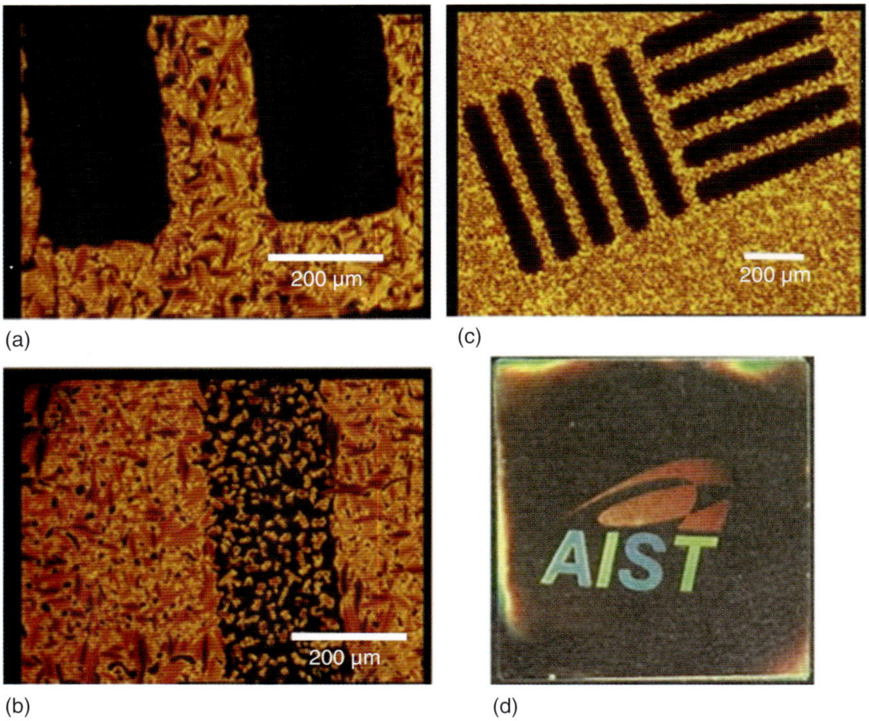

Figure 7.18 Optical textures of the compound **D-6** at 98 °C observed after the UV irradiation at 366 nm through a photomask with transparent lines of 200 μm width. (a) Just after the irradiation. The dark area was irradiated. (b) After several tens of seconds, color picture recorded using the laser scanning system in the thin film of the mixture of dicholesteryl-10,12-docosadiynoate and **D-6** (5 wt%). (c) Optical textures of the compound **D-6** at room temperature observed after UV irradiation at 366 nm through a photomask with transparent lines of 65 μm width at 98 °C followed by quenching by dipping the sample into ice-water. (d) Color picture recorded using the laser scanning system in the thin film of the mixture of dicholesteryl-10,12-docosadiynoate and **D-6** (5 wt%). Source: Tamaoki et al. 2003 [61]. Reproduced with permission of American Chemical Society.

host. Furthermore, a well-defined conglomerate micro-pattern composed of simple cubic soft lattice and helical superstructure, which is challenging to fabricate in organic and inorganic crystalline materials, is produced using photomasking technology.

A photoinduced reversible SmC*–N* phase transition under the influence of an applied electric field has been reported in mixtures consisting of an achiral smectic C (SmC) liquid crystal (4-hexyloxyphenyl-4-octyloxybenzoate, HOPOOB), chiral dopant [4-(methyl-heptyloxycarbonyl)phenyl 4-hexyloxybenzoate: **S811**], and azo- benzene derivative (4-diethyl-aminoazobenzene) as a dopant in 3–7 wt% concentration (Figure 7.21) [50]. The SmC* phase formed upon cooling the N* phase held under the influence of a DC electric field shows uniform unidirectional layer alignment in the entire region of the rubbed cell. Upon reversing the polarity of the electric field, the molecules tilt with respect to rubbing direction; further, the photoisomerization of this state gives rise to an N*

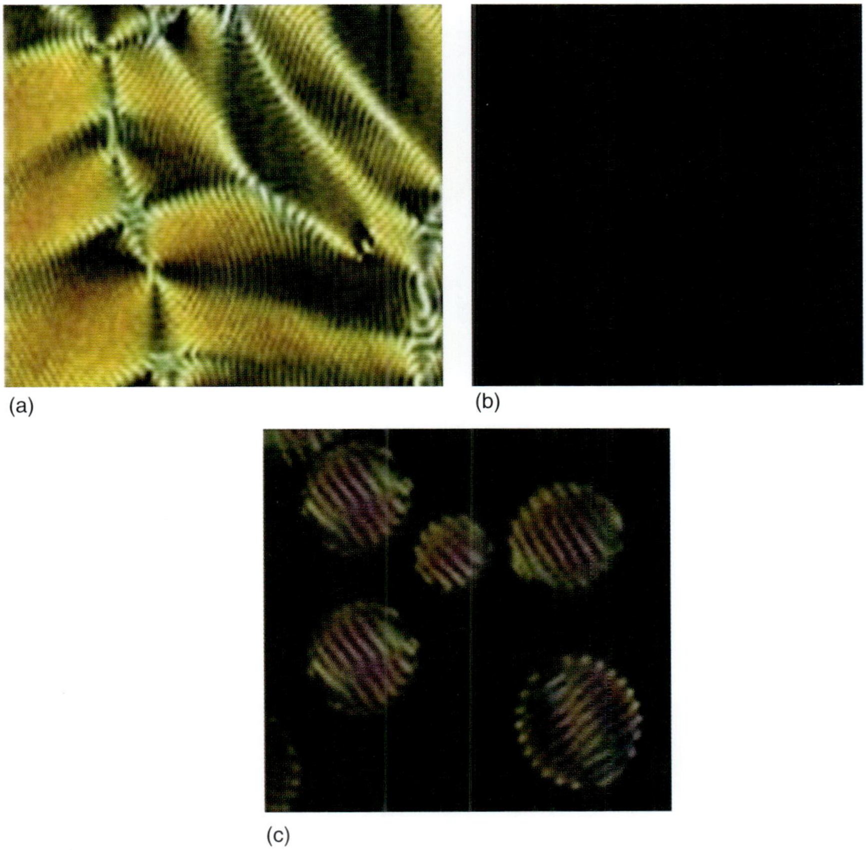

Figure 7.19 Crossed polarized optical texture micrograph of the mixture of 25% **1-Et** in 5CB on cooling at 38.9 °C: (a) before UV irradiation; (b) after UV irradiation for 10 seconds; (c) 20 seconds after removal of UV light at the isotropic phase. Source: Li et al. 2005 [62]. Reproduced with permission of American Chemical Society.

phase where the molecules tend to orient along the rubbing direction and in the absence of UV light, the N*–SmC* phase transition occurs. This light-assisted control of smectic phase alignment/switching, holds promise in fabricating LC display, optical memory, and grating devices.

Supramolecularly halogen-bonded LCs (Figure 7.22) were successfully fabricated using azapyridine derivatives with molecular iodine or bromine. The existence of halogen–base interactions was confirmed by XPS, Raman spectroscopy, and SAXS results. Among these complexes, the supramolecular LC compounds with bromine were first prepared with high stability, indicating that the N···Br interaction is strong enough to form ordered mesophases in this system. The iodine-bonded LCs showed reversibly photoinduced phase transition upon UV irradiation, similarly to other photoresponsive LC materials based on azobenzene derivatives reported earlier. However, such a photoinduced phase transition was not observed in brominated compounds [64].

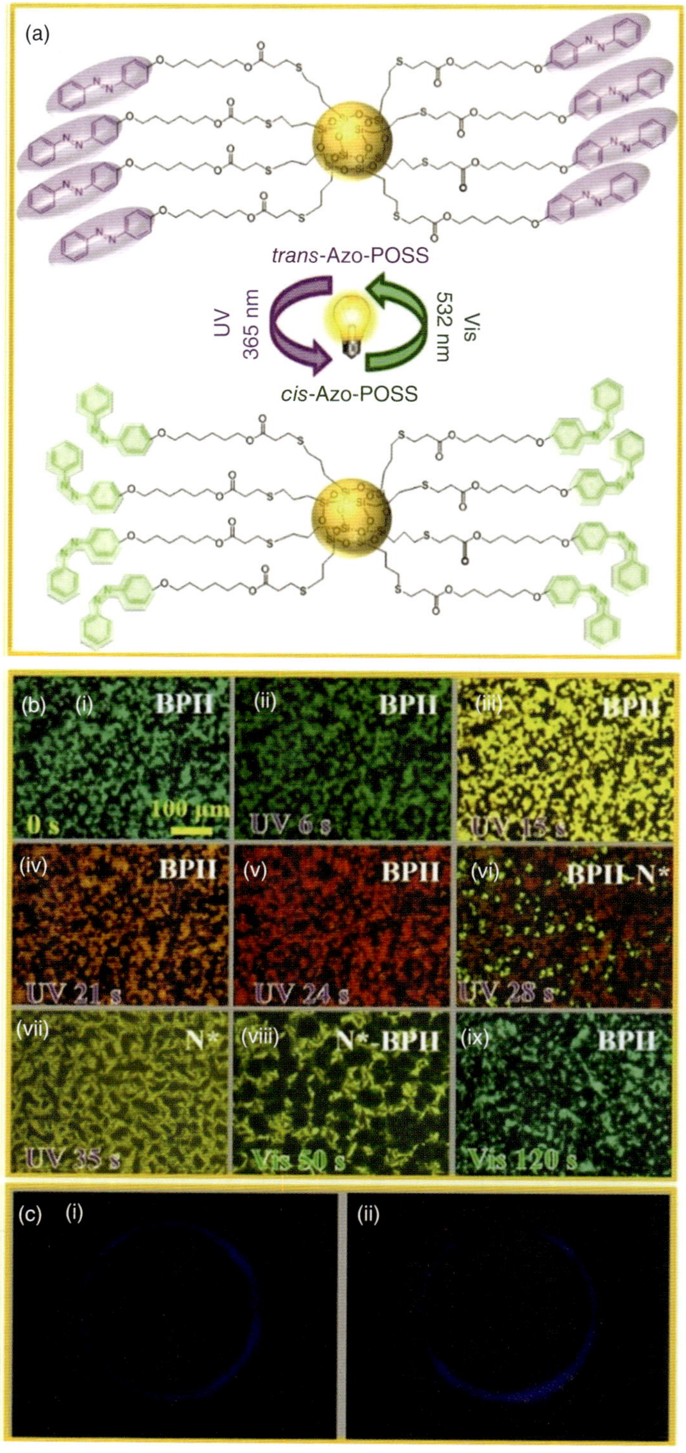
(a)
trans-Azo-POSS
UV
365 nm
Vis
532 nm
cis-Azo-POSS
(b) (i) BPII
0 s
100 μm
(ii) BPII
UV 6 s
(iii) BPII
UV 15 s
(iv) BPII
UV 21 s
(v) BPII
UV 24 s
(vi) BPII-N*
UV 28 s
(vii) N*
UV 35 s
(viii) N*-BPII
Vis 50 s
(ix) BPII
Vis 120 s
(c) (i)
(ii)

Figure 7.20 Light-driven reversible transformation between self-organized BPII soft sc lattice and a helical superstructure. (a) Schematically depicted photoisomerization of Azo-POSS from the *trans*-conformation to the *cis*-conformation by the irradiation of 365 nm UV light; a following irradiation with 532 nm visible light recovers such *cis*-conformation to *trans*-conformation. (b) The deformation of self-organized BPII sc lattice during UV-irradiation, showing the changes of reflection color from the original (i) bluish green, passing through (ii) the green (6-s duration), (iii) yellow (15-s duration), (iv) orange (21-s duration), to (v) the red (24-s duration). A further UV-irradiation led to (vi, vii) the transition from such sc BPII to N*; (viii, ix) the N* structure gradually reconstructed to BPII sc lattice with a following irradiation of 532 nm laser for about 120 seconds. Herein, the concentration of Azo-POSS was controlled at 2.0 wt%. The output light intensities of UV and laser were 1.0 mW cm^{-2} and 10.0 mW cm^{-2}, respectively. Scale bar: 100 μm. (c) Kössel diagrams at (i) the original state, i.e. corresponding to the texture shown in Figure b-i, and (ii) the reconstructed BPII sc lattices, i.e. corresponding to the texture shown in Figure b-ix, respectively, displayed a similar circular ring pattern. Source: Zhou et al. 2018 [63]. Reproduced with permission of John Wiley and Sons.

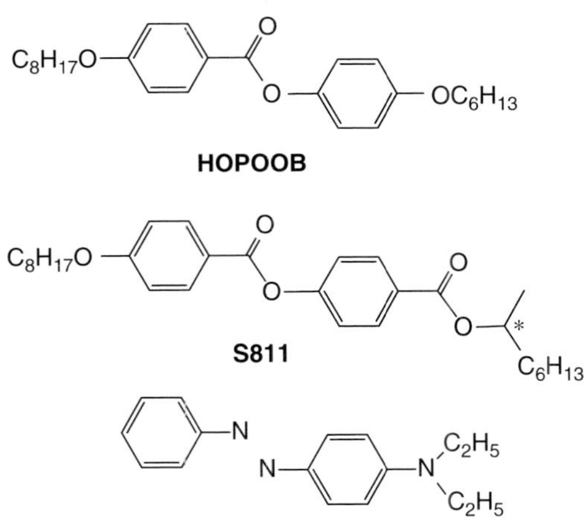

Figure 7.21 Molecular structures of a liquid crystal (**HOPOOB**), chiral dopant (**S811**), and azo dye used to demonstrate reversible SmC*–N* phase transition under the influence of an applied electric field. Source: Matsui et al. 2000 [50]. Reproduced with permission of AIP Publishing LLC.

Tamaoki and coworker recently reported asymmetric chiral dimers containing cholesterol and azobenzene moieties connected through flexible alkylenedioxy bridges (Figure 7.23a). These dimers acted as photon-mode chiral switches for reversible tuning of self-assembled helical superstructures. The chiral switches were doped into nematic liquid crystals, resulting in an N* phase. Under irradiation with UV light, photoisomerization of the azobenzene units led to unprecedented switching of the cholesteric pitch and helical twisting power (HTP, β), with a higher HTP found in the *cis*-rich state (bent form) than in the *trans*-state (rod form). They attributed this behavior to the elongated cybotactic smectic clusters disrupting the helical orientation of the molecules in the N* phase (Figure 7.23b). Their reversible decay and reassembly were evidenced upon

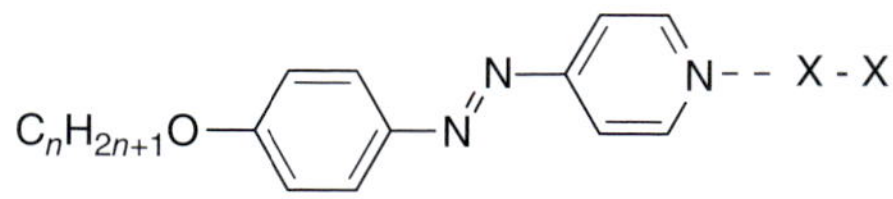

X = I or Br, *n* = 6, 8, 10, 12, 14

*n*I	Thermal behavior		*n*Br	Thermal behavior	
6I	———		**6Br**	Cr 75.1	SmA 108.8 I
8I	———		**8Br**	Cr 100.3	SmA 116.0 I
10I	Cr 123.6	SmA 135.5 I	**10Br**	Cr 100.3	SmA 107.5 I
12I	Cr 1119.6	SmA 135.5 I	**12Br**	Cr 70.5	SmA 158.3 I
14I	Cr 117.2	SmA 133.8 I	**14Br**	Cr 94.8	SmA 107.3 I

Figure 7.22 Structures and thermal behaviors of supramolecular halogen-bonded LCs. Source: Chen et al. 2014 [64]. Reproduced with permission of Royal Society of Chemistry.

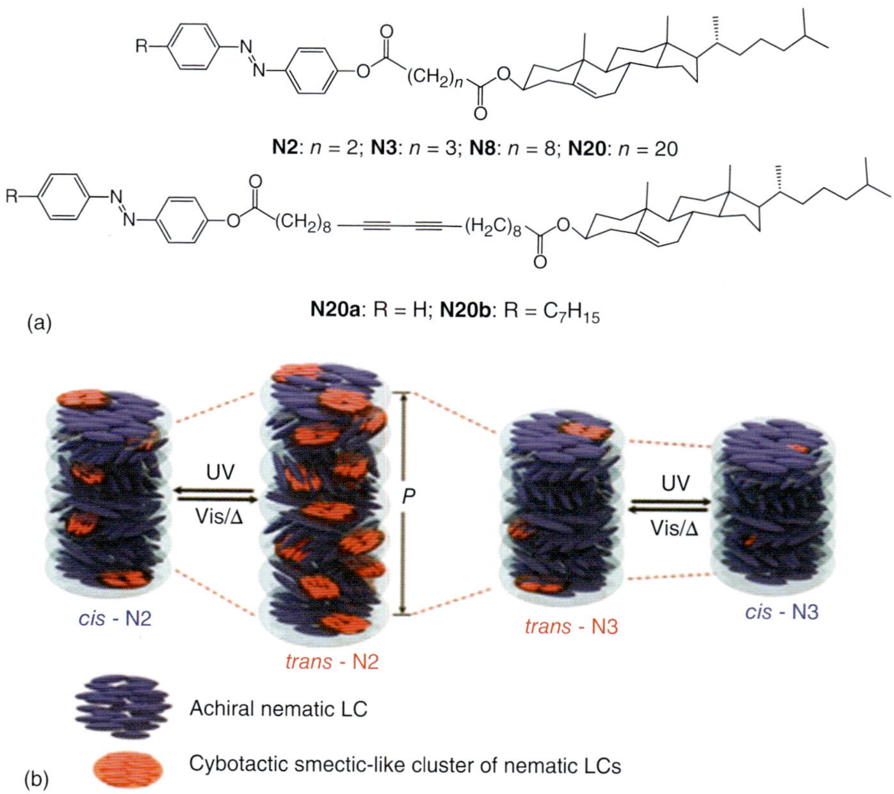

Figure 7.23 Molecular structures of chiral dimers (a); cooperative effect of *trans–cis* photoisomerization and even (**N2**)-/odd (**N3**)-numbered methylene linkers of the dopant on the smectic clusters, resulting in changes in helical pitch lengths and HTPs (b) Source: Kim and Tamaoki 2016 [65]. Reproduced with permission of American Chemical Society.

sequential irradiation with UV and visible light, respectively. In addition to the photoisomerization of the azobenzene units, the odd/even parity of the alkylenedioxy linkers of the dimeric dopants also had a dramatic effect on the transitions of the cybotactic smectic domains. On the basis of the large rotational reorganization of the cholesteric helix and HTP switching ($\Delta\beta/\beta_{ini}$ of up to 50%), they could control the macroscopic rotational motion of micron-sized glass rods upon irradiating the surface of a cholesteric liquid crystal film featuring a polygonal fingerprint texture using UV and visible light [65].

7.3.1.5 Photoinduced Phase Transitions Involving Columnar Phases

Most of the photoinduced phase transitions reported involved calamitic mesophases, while very few reports are available with reference to discotic LCs [66]. This is surprising despite the reports on many photoactive DLCs [67–70]. Recently, Norikane et al. reported photoinduced Col-isotropic phase transition in a macrocyclic compound containing three fused azobenzene mesogenic segments (Figure 7.24) [71]. Upon light irradiation, the room temperature Col phase of the compound transforms into isotropic liquid where nonplanar (bent) molecules forms due to their *E*–*Z* isomerization destabilizing the stacking of the molecules in columns; this process is reversible due to thermal back relaxation and quite sensitive to even small fractions of *Z*-isomer. Tanaka et al. reported triphenylene derivative with six 4-tetradecyloxyazobenzene units connected through propylene-ester linkages (**1-C14**). This exhibited enantiotropic SmA and Col_r mesophases due to the interchange of their shape from disk to rod-like in response to temperature. Interestingly, the photo-irradiation (λ = 365 nm) leads to the photoisomerization of azobenzene units from *trans* to *cis* forms. This results in isothermal phase transitions (Col_r–SmA–I) as in the thermotropic ones, but with a certain decrease of phase transition temperatures. This serves as a first example, where the isothermal LC phase transitions are controlled by

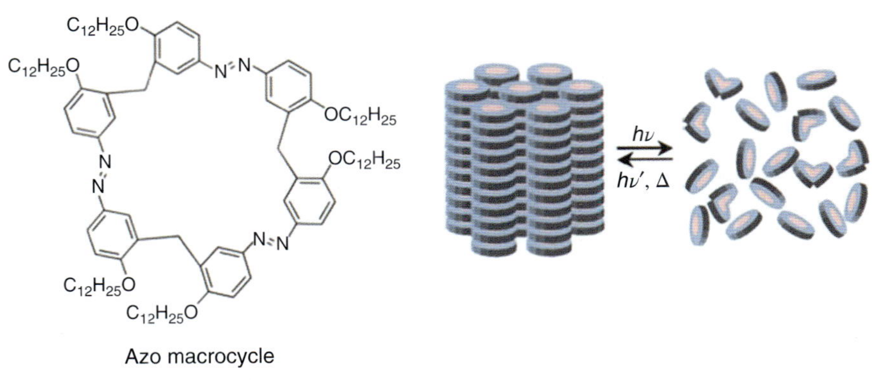

Figure 7.24 Schematic representation of self-assembly of photoactive discotic molecule into columnar LC structure which upon light irradiation transforms into isotropic phase due to *trans–cis* isomerism where the flat disks fold up leading to the destabilization of columnar organization. Note that when the thermal relaxation from *cis* to *trans* isomers occurs, the isotropic liquid state transforms into the Col phase. Source: Norikane et al. 2011 [71]. Reproduced with permission of Royal Society of Chemistry.

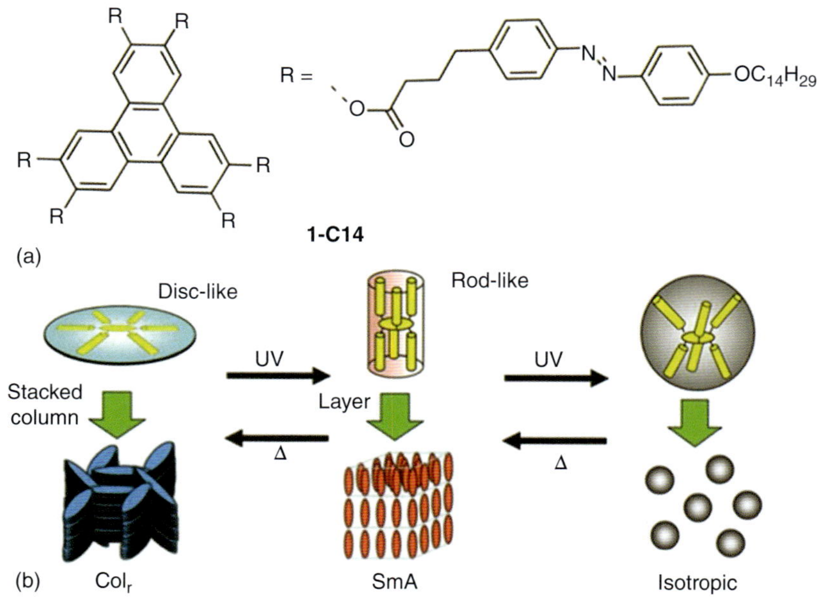

Figure 7.25 (a) Chemical structure of **1-C14**; (b) schematic illustration showing the isothermal molecular shape and LC phase changes depending on UV light intensity. Source: Tanaka et al. 2012 [72]. Reproduced with permission of Royal Society of Chemistry.

photo-irradiation among three phases (Col_r, SmA, and Iso), which are also accompanied with the shape change of molecules between rod-like and disk-like (Figure 7.25) [72].

Cheng and coworkers reported polycatenar LCs containing two 1,3,4-thiadiazole rings interconnected by an azobenzene central linkage. The compound with two tetradecyloxy chains exhibited a photoinduced transition from SmC to I, while the corresponding tetracatenar showed a transition from Col_h phase to isotropic phase. These transitions were found to be reversible upon visible light irradiation (Figure 7.26) [73].

7.3.2 Axially Chiral Azo Compounds

As discussed in the introduction section, when a chiral molecule is dissolved in an achiral nematic LC, its molecular chirality gets transferred to the nematic solvent, with the stabilization of N* LC capable of selective reflection according to Bragg's law. The wavelength λ of the selective reflection is defined by $\lambda = nP$, where P is the pitch length of the helical structure and n is the average refractive index of the LC material. The ability of a chiral dopant to twist the N LC phase is defined as its HTP "β," which is defined according to the equation: $\beta = (Pc)^{-1}$, where c is either the weight concentration or the molar fraction of the chiral dopant. The isomerization upon photoirradiation can control the HTP and the helical pitch of the cholesteric phases, providing the tool for optical display addressing. Presently, the chiral dopants are used at high concentrations (≥20%) and they have twisting powers of $\beta < 15\,\mu m^{-1}$. Further, the addition of chiral

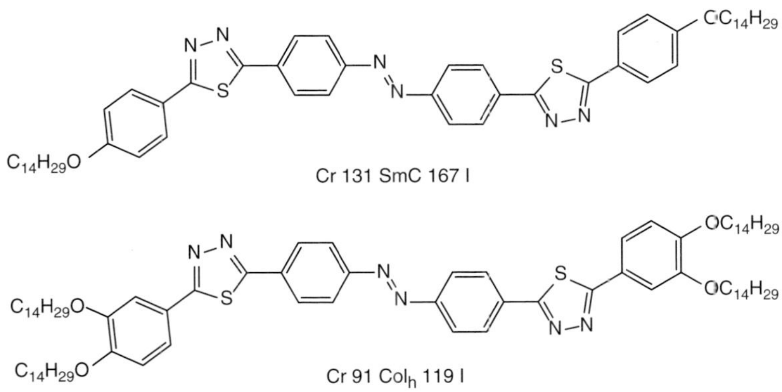

Figure 7.26 Molecular structures of compounds that exhibit photoinduced mesophase (SmC to I and Col_h to I) to isotropic transition. Source: Peng et al. 2017 [73]. Reproduced with permission of Royal Society of Chemistry.

dopant in such a high amount will alter or destabilize the LC phase. As a solution, scientists started developing of a chiral azo benzene derivative with high HTP and good solubility in a nematic host. The helical twisting power of azobenzene switches was considerably enhanced by incorporation of axially chiral binaphthyl moieties. Especially binaphthyldiamine-based switches **67**, **72**, and **73** (Figure 7.27), represent a powerful system for optical switching of the N* pitch, as the high HTP values allow for color generation and switching without added chiral codopants [74, 75]. Furthermore, UV irradiation of an LC film doped with switch **73** (β = +37 μm^{-1}) leads to a cholesteric phase of opposite helicity (β = −31 μm^{-1}). Owing to the molecular switch having two azo linkages, UV irradiation leads to reversible *trans–cis* isomerization of azo configurations producing two other isomers containing one or two *cis* configurations, respectively. The sequence of the photochemical switch of the three isomers is (*trans, trans*), (*trans, cis*), and (*cis, cis*). The reverse process from (*cis, cis*), (*trans, cis*),

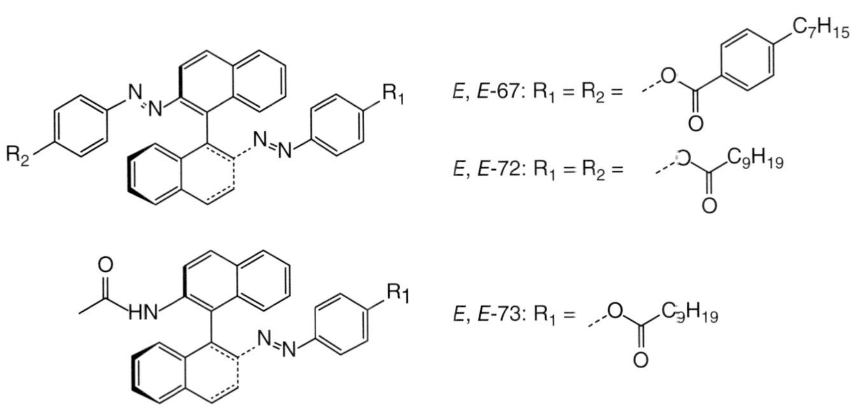

Figure 7.27 Molecular structures of axially chiral dopants. Source: Adapted from Pieraccini et al. 2003 [74] and Lin et al. 2013 [75].

and (*trans, trans*) can occur thermally or photochemically with visible light irradiation.

Li and coworkers reported the wide optical tuning of a PBG of a BP facilitated by an axially chiral azobenzene switch *E,E*-**67** (Figure 7.27) in a commercial LC mixture [75]. The resulting BP exhibited a light-induced unprecedented reversible PBG tuning over the visible region across red, green, and blue wavelengths accompanied by a unique, not previously seen light-induced isothermal phase transition from a BPII to a BPI phase, depending on irradiation time (Figure 7.28). The process of the light-induced phase transition from BPII to BPI were verified with reference to a Kossel diffraction diagram [76]. The photoresponsive temporal behavior as a function of composition is also reported.

7.3.3 Azoxybenzenes

Although it is known that azoxybenzene compounds form LCs with properties similar to azobenzene materials, such systems have not been extensively investigated, although they have been employed as the photoactive compounds in few experiments [58, 77]. Azoxybenzene compounds are stable and can be formulated into mixtures with wide LC phase ranges. It is therefore of interest to know whether the optical properties of azoxybenzene compounds differ from those of similar azobenzene materials, or whether their LC properties are affected contrarily due to *trans–cis* isomerization (Figure 7.29). The absorption spectra for the *trans* isomer of azoxybenzene compounds also have a strong absorption at similar wavelengths to azobenzene compounds, but lack the weak absorption peak around 440 nm. ZhK-440 (NIOPIK) is a mixture of two-thirds *p-n*-butyl-*p*-methoxyazoxybenzene and one-third *p-n*-butyl-*p*-heptanoylazoxybenzene. If the molecules are *trans* isomers, ZhK-440 possesses a nematic LC phase from 0 to 70 °C. The transition temperature from the N LC phase to the isotropic liquid phase decreases if molecules change from *trans* isomers to *cis* isomers, eventually reaching room temperature and probably eliminating the LC phase completely. Authors opine that azoxybenzene compounds should be just as useful in optical applications as are azobenzene compounds [78]. The same group also observed changes of the pitch (color) and optical anisotropy and also N*–I phase transition in response to light, when they doped ZhK-440 to a cholesteric LC [77].

7.3.4 Spiropyrans and Naphthopyrans

Spiropyrans are an interesting family of photochromic materials due to their unique properties such as excellent photofatigue resistance, strong photocoloration, and fast thermal relaxation. The colorless ring-closed spiro forms of spiropyran can be transformed into the colored ring-opened merocyanine form upon UV light irradiation. Its reverse process occurs thermally in the dark or photochemically by visible light irradiation (Figure 7.30). An exclusive feature of spiropyran is its significantly increased dipole moment after photoisomerization from the ring-closed spiropyran form to the ring-opened charge-separated zwitterionic merocyanine form. The physical and chemical properties of the

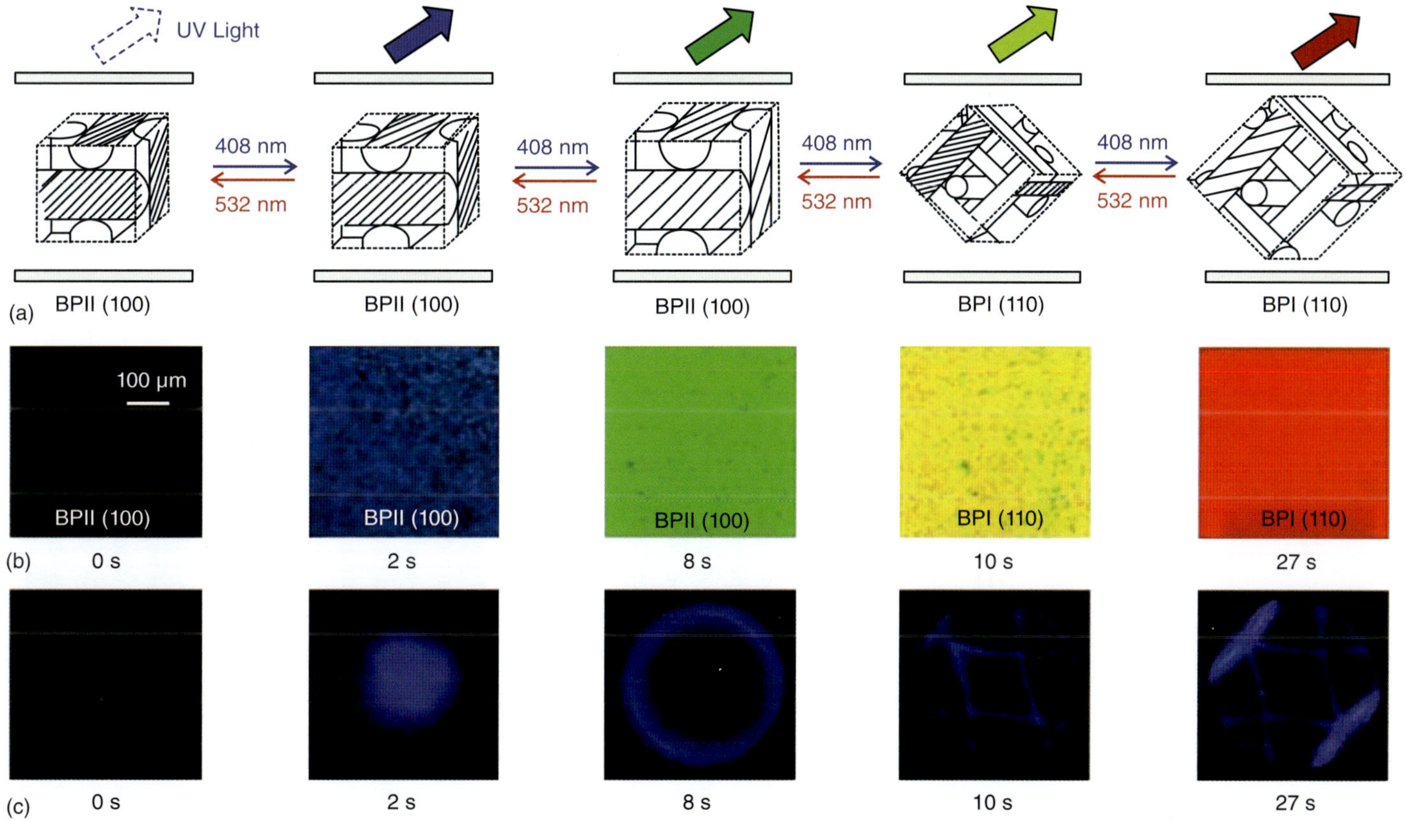

Figure 7.28 (a) Schematic illustration of the phototuning of the BP composed of 3.5% chiral switch 1 in commercially available 53.5% nematic liquid crystal E48, 29.5% S811, and 13.5% R811. (b) Reflection color images of 3.5% chiral switch *E,E*-**67** in commercially available 53.5% nematic liquid crystal, 29.5% S811, and 13.5% R811 in 7.5-μm thick planar cell with 408 nm light irradiation at 37 °C. (c) The corresponding Kossel diagrams. Source: Lin et al. 2013 [75]. Reproduced with permission of John Wiley and Sons.

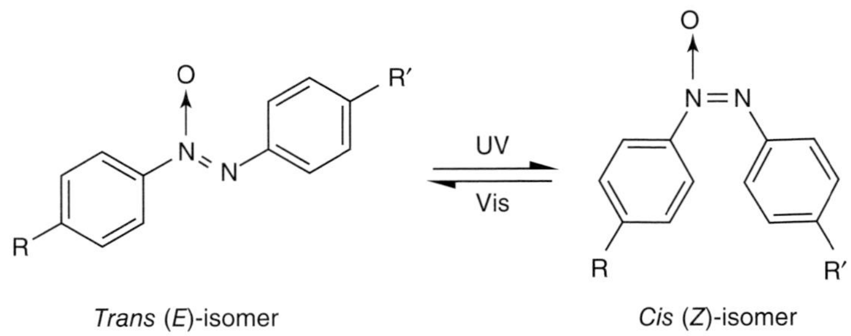

Figure 7.29 Photoisomerization of azoxybenzene derivatives.

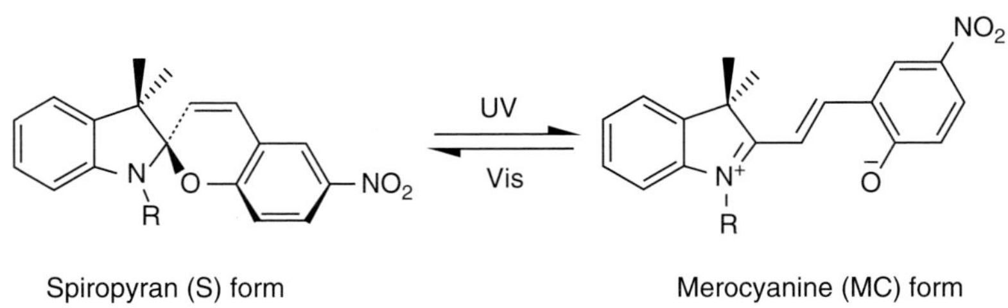

Figure 7.30 Photoisomerization of spiropyran derivatives.

two forms of spiropyran are specifically different; thus, the thermally reversible photochromic switching can be exploited for the construction of smart materials. Since the geometric shape changes upon photoirradiation are not drastic and the initial and final isomeric states are more or less elongated, spiropyrans have been used in the development of photochromic LC materials. The phase transition behavior of mixtures of LCs and photochromic spiropyran derivatives, induced by isomerization of the guest molecules, has been investigated. Two types of spiropyran derivatives were employed: normal spiropyrans that are stable in the closed form (spiropyran form, SPF) and open form (merocyanine form, MCF) stabilized derivatives. These spiropyran derivatives were doped into two LC compounds: **5CB** and 4-butoxyphenyl-4-ethylcyclohexanecarboxylate (**ECH 204**). SPF-stabilized derivatives exhibited photoisomerization behavior from SPF to MCF in LC media, which was accompanied by an isotropic to nematic phase transition of the mixtures. On the other hand, N–I phase transition of the mixtures was observed when the MCF-stabilized spiropyran doped in both LCs was photoisomerized from MCF to SPF. The result was interpreted in terms of shape anisotropy of the guest molecule associated with each isomer of the spiropyrans [79].

Kosa et al. reported a class of naphthopyran-based materials (Figure 7.31) that exhibit photoinduced conformational changes in molecular structure capable of yielding order-increasing phase transitions [80]. Appropriate functionalization of the naphthopyran molecules leads to an exceedingly large order parameter in the open form, which results in a clear to strongly absorbing dichroic state. The increase in order with light exposure has profound implications in optics, photonics, lasing, and displays and will merit further consideration for applications

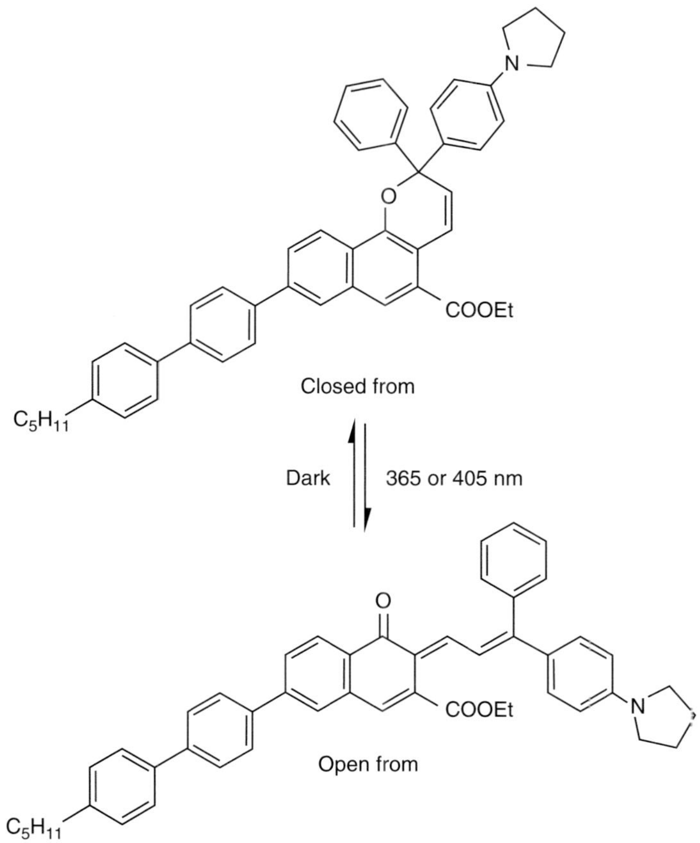

Figure 7.31 Photoinduced ring opening of methyl 8-(4′-pentylbiphenyl-4-yl)-2-phenyl-2-(4-fluorophenyl)-2*H*-naphtho[1,2-*b*]pyran-5-carboxylate upon exposure to ultraviolet light of wavelength 365 or 405 nm. Source: Kosa et al. 2012 [80]. Reproduced with permission of Springer Nature.

in solar energy harvesting. The large photoinduced isotropic to nematic transition switches the transmission through the cell from clear to strongly absorbing and dichroic (Figure 7.32). The clear to dichroic transition has been long pursued in ophthalmic applications, such as polarized variable transmission lenses.

7.3.5 Fulgides

Fulgide-based systems show thermal irreversibility and they are solely driven by light (Figure 7.33). The hexatriene group under UV irradiation undergoes cyclization to yield a cyclohexadiene unit in the compound, whereas the reverse process can be driven back by visible light irradiation. The research group of Schuster reported cholesteric LCs that do not show fatigue during light-driven reversible pitch modulation with the use of chiral fulgide **F1** as a dopant [81, 82]. The HTP values are different for the open form and closed form, which is the basis of light-driven pitch modulation in CLCs. Yokoyama and Sagisaka reported the N* pitch

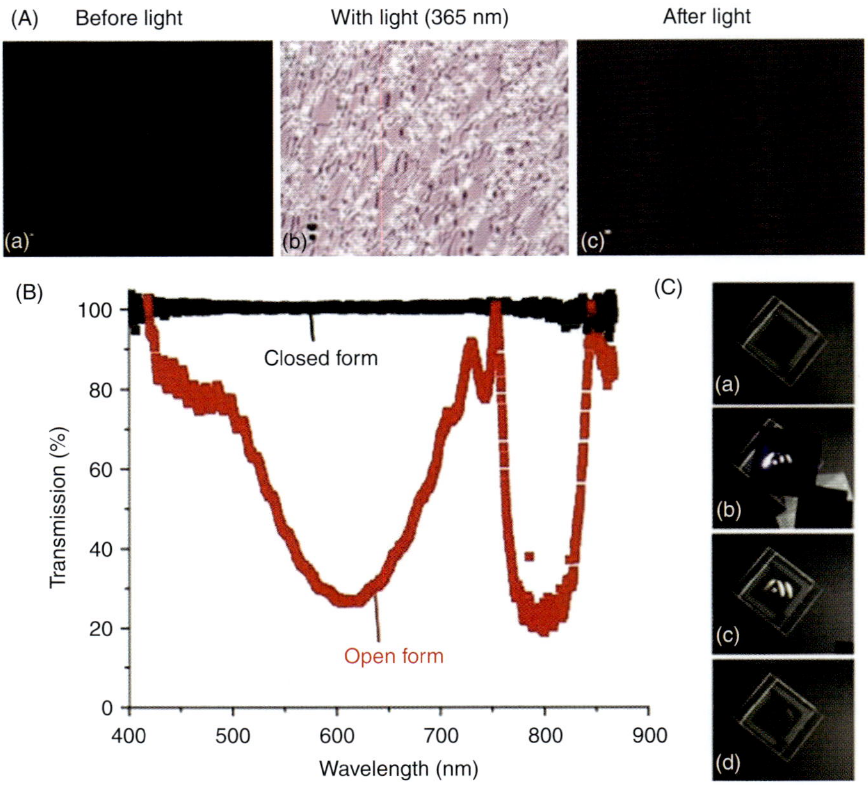

Figure 7.32 Phototropic phase transition of naphthopyran derivative (4.0 wt%) mixed with R1011 (6 wt%)/5CB. (A) Polarized optical micrographs of phototropic phase transitions with exposure to 365 nm ultraviolet light at 5 mW cm^{-2}: (a), before exposure; (b), during exposure; and (c), after exposure. (B) Transmission spectra of mixture before (black squares) and during (red squares) ultraviolet exposure. (C) Spatial patterning with a mask is demonstrated in a twisted nematic cell placed on a polarized light source viewed through a crossed analyzer: (a) before exposure; (b) during exposure. On removal of the light, the light is transmitted through the exposed area (c) before quickly vanishing (d). Source: Kosa et al. 2012 [80] Reproduced with permission of Macmillan Publishers Limited.

modulation by using compound **F2** as the light-driven chiral dopant [83]. Further, they demonstrated light-driven handedness inversion in a CLC by using a photoinsensitive codopant with compound **F2**. Similarly, compound **F3** has also shown the ability to optically reconfigure color-stable cholesteric LC films.

Owing to the advantage of thermal irreversibility of the dopant, long-lasting reflection colors at arbitrary spectral positions have been demonstrated [84].

7.3.6 Ketones

Unsaturated ketones undergo *trans–cis* photoisomerization. Accordingly, many such compounds have been used as chiral dopants for the induction of photoresponsive cholesteric LCs and to modulate their properties by photoirradiation.

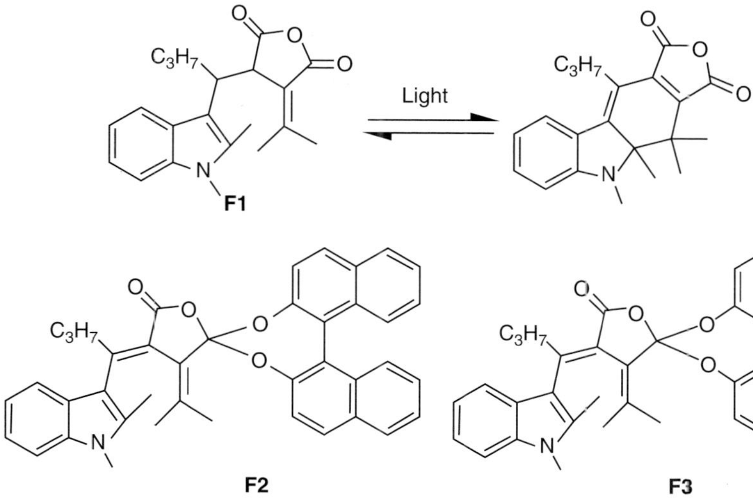

Figure 7.33 Molecular structures of fulgide derivatives.

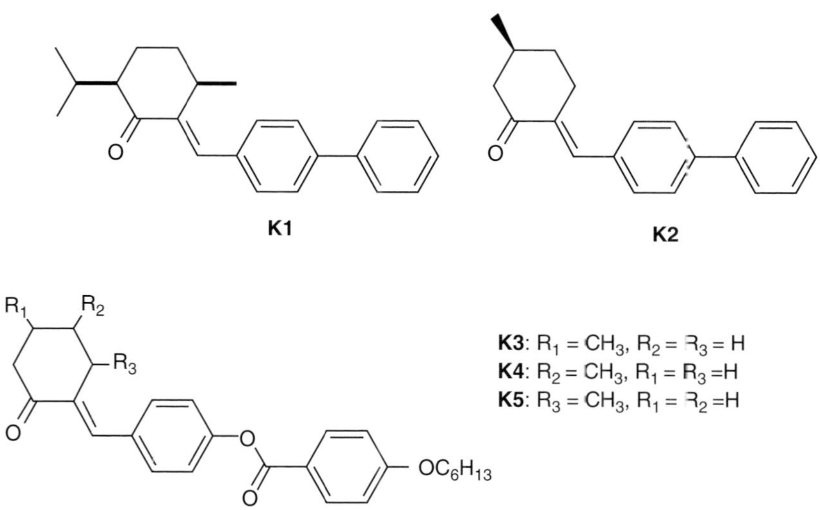

Figure 7.34 Chiral dopants **K1–K5** based on α,β-unsaturated ketones [85–89].

Yarmolenko et al. designed and synthesized menthone derivative **K1** (Figure 7.34) to study its behavior toward the induction of cholesteric LCs [85]. By linking the menthone moiety through an exocyclic C—C double bond, they rendered this chiral compound photoisomerizable. It was noticed that the *E* and *Z* isomers show sign inversion of their HTP values in nematic LC host 4-methoxybenzylidene-4′-butylaniline (MBBA). The *E* isomer induced a left-handed helix, while the *Z* isomer induced a right-handed helix in the LC matrix. This behavior

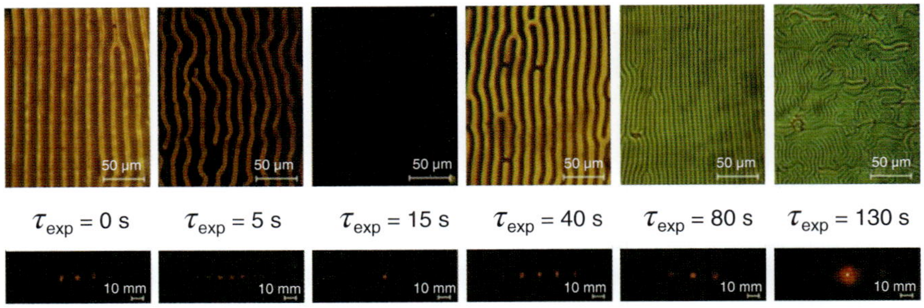

Figure 7.35 Diffraction gratings with an induced cholesteric LC exhibiting handedness inversion. The top row shows the formation of gratings in a homeotropic cell and the bottom row shows the corresponding diffraction patterns. (Gvozdovskyy et al. 2012 [87] Reproduced from with permission of Optical Society of America.)

is attributed to the dramatic change in the molecular shape of the dopant upon irradiation with UV light (λ = 313 nm). Recently, Gvozdovskyy et al. used this dopant in combination with the photoinsensitive dopant **R811** and observed light-driven handedness inversion of the induced cholesteric LC fabricated from the nematic liquid crystal MLC6884 (Merck) [86, 87]. The observed helicity inversion is due to the interplay of the opposite HTP signs of the two chiral dopants in the nematic host. By taking advantage of this property, they demonstrated optically controlled diffraction gratings and beam steering (Figure 7.35). By using CLC samples capable of handedness inversion, it was observed that a light beam could be steered in one or two directions in a single photoirradiation step. Krivoshey et al. studied light-driven handedness inversion in CLCs containing the chiral unsaturated ketone **K2** [88]. This dopant induced a left-handed helix in 5CB and, upon irradiation with UV light (λ = 313 nm), the CLC underwent helix inversion, thereby resulting in a right-handed helical superstructure.

Lub et al. designed and synthesized chiral dopants **K3–K5** (Figure 7.34) based on methyl-cyclohexanones and studied the effect of the position of the stereogenic center (i.e. methyl group) with respect to the carbonyl group on the CLC induction behavior in nematic host E7 [89]. They observed that the compounds display a low HTP when the chiral center is moved away from the ketone group. However, these dopants were found to exhibit a change in the sign of the HTP upon photoisomerization with UV irradiation from the *E* form to the *Z* form in the E7 nematic host. Thus, these dopants enable handedness inversion of the induced CLCs. Although these and few other examples mentioning handedness inversion of helical mesophases do not exactly fall under photoinduced phase transition, such examples may enlighten the reader about the future possibilities. For more such examples, the reader could refer to a review article [90].

7.3.7 Diarylethenes

Dithienyl cyclopentene or diarylethene derivatives as chiral dopants were introduced by the Feringa group. They studied the induction of photoresponsive

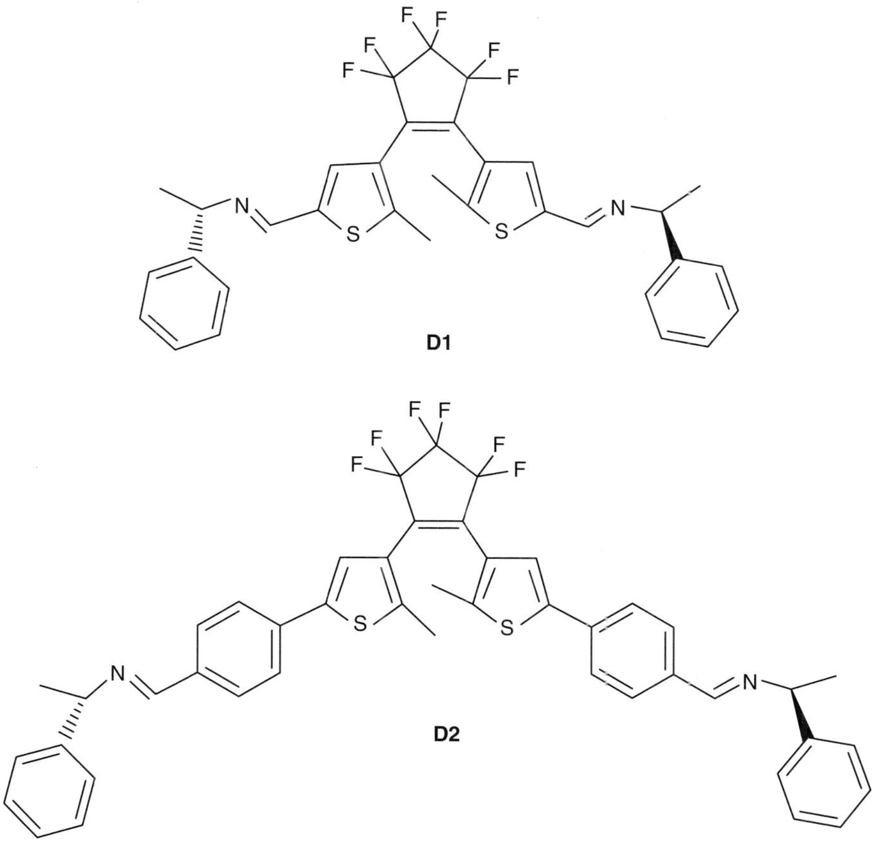

Figure 7.36 Molecular structures of diarylethene derivatives (**D1** and **D2**). Source: Adapted from Denekamp and Feringa 1998 [91] and van Leeuwen 2011 [92].

N* LCs by doping compound **D1** (Figure 7.36) into a nematic host. Light-driven modulation of these N* LCs was carried out [91, 92]. The photocyclization of the dopant leads to the pitch variation of the N* LC and even the N* to nematic phase transition has been reported. These events are a result of the different HTPs of the open and closed forms of the chiral dopant. Similarly, compound **D2** was used as a chiral light-driven molecular switch in the induction of N* phase. Reversible photocontrol over the pitch of the induced N* LC was demonstrated through irradiation with UV and visible light. It was found that usually the open form showed higher HTP than the closed form. Similar tansitions from chiral nematic to nematic was reported by the Irie group by utilizing chiral dithienylcyclopentene-based compounds as dopants [93–96]. Owing to good fatigue resistance of the dopant, CLC to the nematic reversible transition could be repeated many times. The drawbacks of these dopants were lower HTP values. To overcome this, Li and coworker synthesized compounds with a dithienyl cyclopentene moiety flanked by two axially chiral binaphthyl groups, which exhibited higher HTP values [37].

Akagi and coworkers reported a dynamic photoswitching of the helical inversion in the N* LC phase that contains photoresponsive axially chiral dopants [97]. These chiral dithienylethene derivatives had two axially chiral binaphthyl moieties. The dihedral angle of the binaphthyl rings changes via the photoisomerization between the open and closed forms of the dithienylethene moiety. The N* LCs induced by the dithienylethene derivatives that are used as chiral dopants exhibit reversible photoswitching behaviors, including a helical inversion in the N* LC and a phase transition between the N* LC and the nematic LC. These compounds are the first chiral dopants that induced a helical inversion in N* LC via the photoisomerization between open and closed forms of the dithienylethene moiety (Figure 7.37).

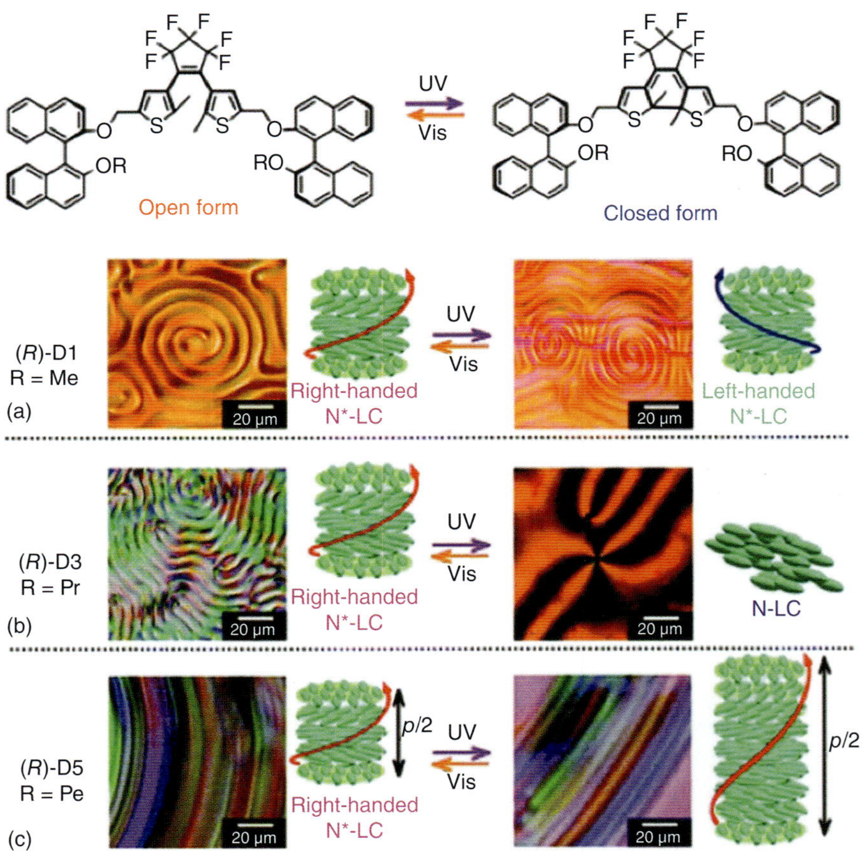

Figure 7.37 Structures of chiral binaphthyl derivatives bearing photoresponsive dithienylethene moieties and polarized optical micrographs (POMs) of photoresponsive N* LCs containing (a) (*R*)-D1, (b) (*R*)-D3, and (c) (*R*)-D5 (1.0 mol %) between the open form (left) and PSS (right) by irradiation of UV ($\lambda = 254$ nm, 16 W, handy lamp) and vis ($\lambda > 400$ nm, 100 W) light at 30 °C. Along the corresponding POMs are described schematic illustrations of reversible helical inversion between N* LCs with opposite screw senses, upon alternating irradiations of UV (left) and vis (right) light. Source: Hayasaka et al. 2012 [97]. Reproduced with permission of American Chemical Society.

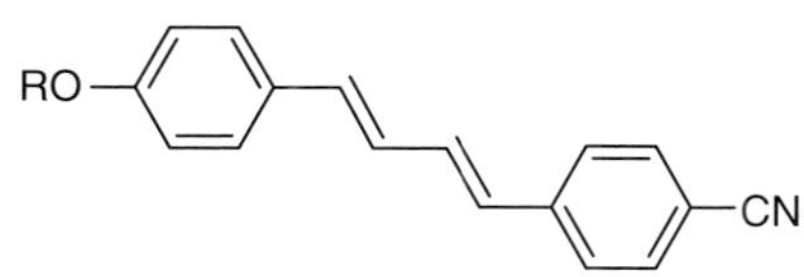

BC1: R = CH_3; **BC4**: R = C_4H_9; **BC8**: R = C_8H_{17}

(a) **BC10**: R = $C_{10}H_{21}$; **BC12**: R = $C_{12}H_{25}$; **BC16**: R = $C_{16}H_{33}$

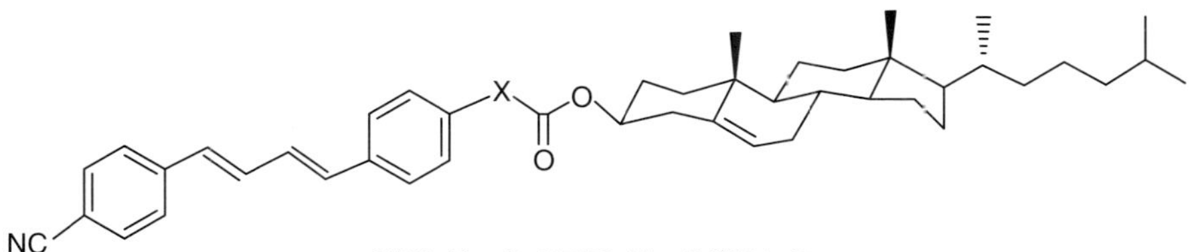

CBC: X = O; **CBC8**: X = $O(CH_2)_8O$

CBC11: X = $O(CH_2)_{11}O$; **CBC12**: X = $O(CH_2)_{12}O$

(b)

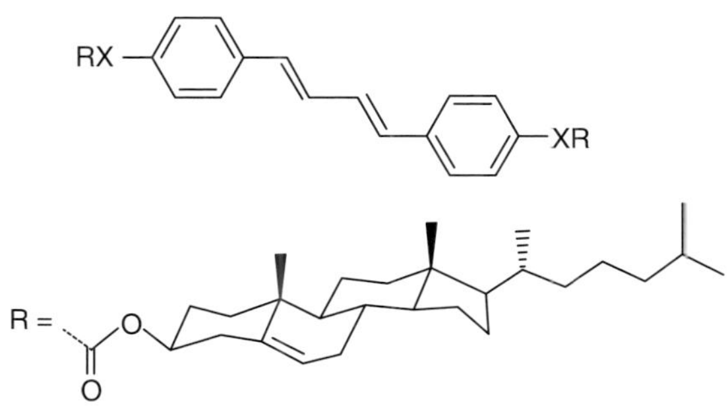

(c) **DCBC**: X = O; **DCBC8**: X= $O(CH_2)_8O$; **DCBC12**: X = $O(CH_2)_{12}O$

Figure 7.38 Structures of photoresponsive substituted diphenylbutadienes.

7.3.8 Butadienes

Diphenylbutadienes form an important class of photochromic materials. The *EE* isomers of alkoxy-cyano-substituted diphenylbutadienes (Figure 7.38a) exhibit N and Sm LC phases depending on the length of their terminal chains [98]. The methoxy-substituted derivative (**BC1**) exhibited an N phase. A nematic LC phase was also observed for **BC4**, whereas both nematic and smectic LC phases were observed for **BC8**. The **BC12** compound possessed only an SmA phase. The *EE* isomers of these derivatives undergo photoisomerization to yield thermally stable *EZ* and *ZE* isomers, and neither of these photoisomers possesses LC phases. The drastic difference in the phase transition characteristics between the all-*trans* isomer and the photoisomers makes these materials very suitable for studying photoinduced phase transitions. Photoirradiation with 360 nm light of the SmA phase of **BC12** at 120 °C resulted in the destruction of the LC nature and in the isothermal transition to the isotropic phase. Because of the thermal stability of the photoisomers, the isotropic phase thus formed does not revert to the LC phase even on heating. However, the photochemical recovery of the SmA phase was achieved.

Alkoxy cyano-substituted diphenylbutadienes were found to be good candidates for studying the molecular basis of photoinduced changes of cholesteric pitch lengths because they possess a rod-shaped structure and transform into thermally stable bent *Z* isomers upon light irradiation [99]. Doping the glass-forming N* LC, dicholesteryl-10,12-docasadiynediote, with 1-(alkoxyphenyly)-4-(cyanophenyl)buta-1*E*,3*E*-dienes (Figure 7.38b) resulted in a red shift in the wavelength of light reflected by the host CLC. The extent of the red shift was dependent on the concentration of the dopant, the temperature of the film, and also on the molecular length of the dopant. The photoisomerization of these dopants leads to a blue shift of the reflected light, allowing color tuning of the films over a large part of the visible region. The cholesteric pitch and hence the reflected light by these materials could be fixed in a glassy state by cooling them from their cholesteric temperatures to 0 °C. They reported that the dimesogenic liquid-crystalline molecules (Figure 7.38b) [100] possessed glass-forming properties (from the cholesterol moiety) and photochromic properties (from the butadiene moiety). Photoinduced *cis–trans* isomerization of the butadiene chromophore in these materials could be utilized to bring about an isothermal smectic to an N* phase transition. By photochemically controlling the *cis–trans* isomer ratio, they could tune the selective reflection color of the film over the entire visible region, and the color images thus generated could be freezed by converting them to glassy N* states by rapidly quenching the irradiated film to ~0 °C (Figure 7.39). Upon photoirradiation, the formation of the *cis* isomers

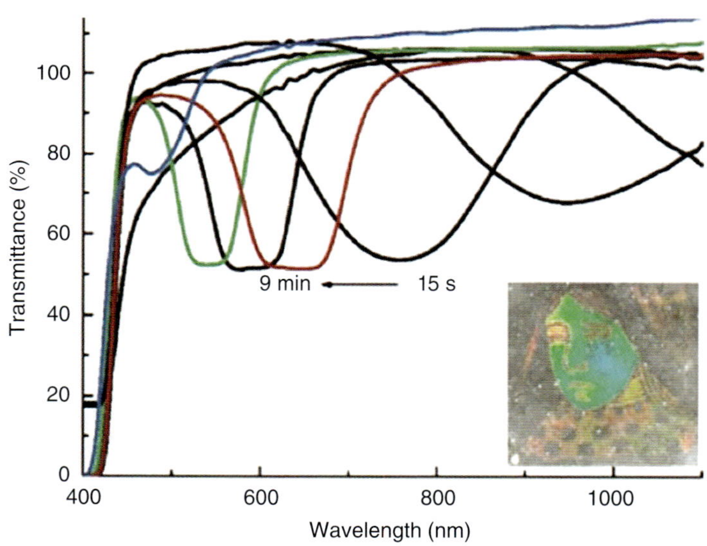

Figure 7.39 Changes in reflection band of the N* phase of **CBC12** developed on photoirradiation of its smectic film held at 105 °C. Inset shows an image stored in a cholesteric glass, obtained by energy modulated irradiation of a smectic film at 105 °C followed by temperature quenching to ~0 °C. Source: Abraham et al. 2006 [100]. Reproduced with permission of American Chemical Society.

initially destabilizes the SmA* phase, bringing about a reduction in the SmA*–N* phase transition temperature. With the increasing concentration of the *cis* isomers, the pitch of the N* phase decreases, leading to a blue shift in its reflection band. At high *cis* concentrations, an isothermal N*–I phase transition occurs. The isotropization temperature was lowered by around 90 °C, which is substantially larger than in any of the previous reports on photoresponsive LCs. Although the photogenerated *cis* isomers were thermally stable, they could be reverted photochemically. Thus, 266 nm photoirradiation of a film containing a mixture of the *cis* and *trans* isomers in the solid state resulted in nearly complete recovery of the all-*trans* isomer. Using this procedure, the images recorded using 366 nm light could be erased by irradiating them with 266 nm laser light. The pitch of the N* phase of the trimesogenic derivatives containing two cholesterol terminal moieties linked to a central diphenylbutadiene chromophore (Figure 7.38c) [101] could also be continuously varied thermally and photochemically to tune the color of the film over the entire visible region, and the color images thus generated could be stabilized by converting them to glassy N* states.

7.3.9 Near Infrared Light-Driven Phase Transition in Hybrid Materials

Recently, plasmonic gold nanorods (GNRs) have attracted increasing attention especially in biomedical applications due to their appealing "photothermal effect," i.e. converting NIR light into heat through a non-radiative relaxation process of longitudinal surface plasmon resonances (LSPR) [102]. It is also noteworthy in this context that the characteristic wavelength of the LSPR signal corresponding to NIR absorption can be expediently tuned on demand by suitably altering the aspect ratio of GNRs [103], and therefore GNRs have been proposed for use in a few triggering systems where UV or visible light is not favorable [104]. However, adequate and homogeneous dispersion of conventional GNRs into a complex LC medium with self-organized 3D photonic superstructures has always been a challenging task due to their poor miscibility in LCs and destabilization of the 3D lattices. Li and coworkers reported an NIR-light-responsive self-organized 3D photonic superstructure that was fabricated by incorporating new hydrophobic mesogen-functionalized GNRs (M-GNRs) into a BP LC medium composed of commercially available components (Figure 7.40a) [105]. The introduction of M-GNRs was found to be beneficial in stabilizing the cubic nanostructure. Importantly, the resultant 3D photonic nanostructures could be switched between body-centered cubic and simple cubic symmetry under irradiation using an 808 nm NIR laser due to the significant photothermal effect of M-GNRs. The reverse process occurs upon removal of the NIR laser irradiation. Furthermore, dynamic NIR-light-directed red, green, blue (RGB) reflections of the light-driven 3D soft photonic crystals were demonstrated for the first time (Figure 7.40b,c).

Further, Li and coworkers also reported mesogen-functionalized GNRs that were soluble in organic solvents and displayed the typical transverse surface plasmon resonance and LSPRs of GNRs (Figure 7.41a). The protected GNRs were homogeneously dispersible in a commercially available thermotropic nematic liquid crystal host **E7**. The homogeneous LC nanocomposites were found to

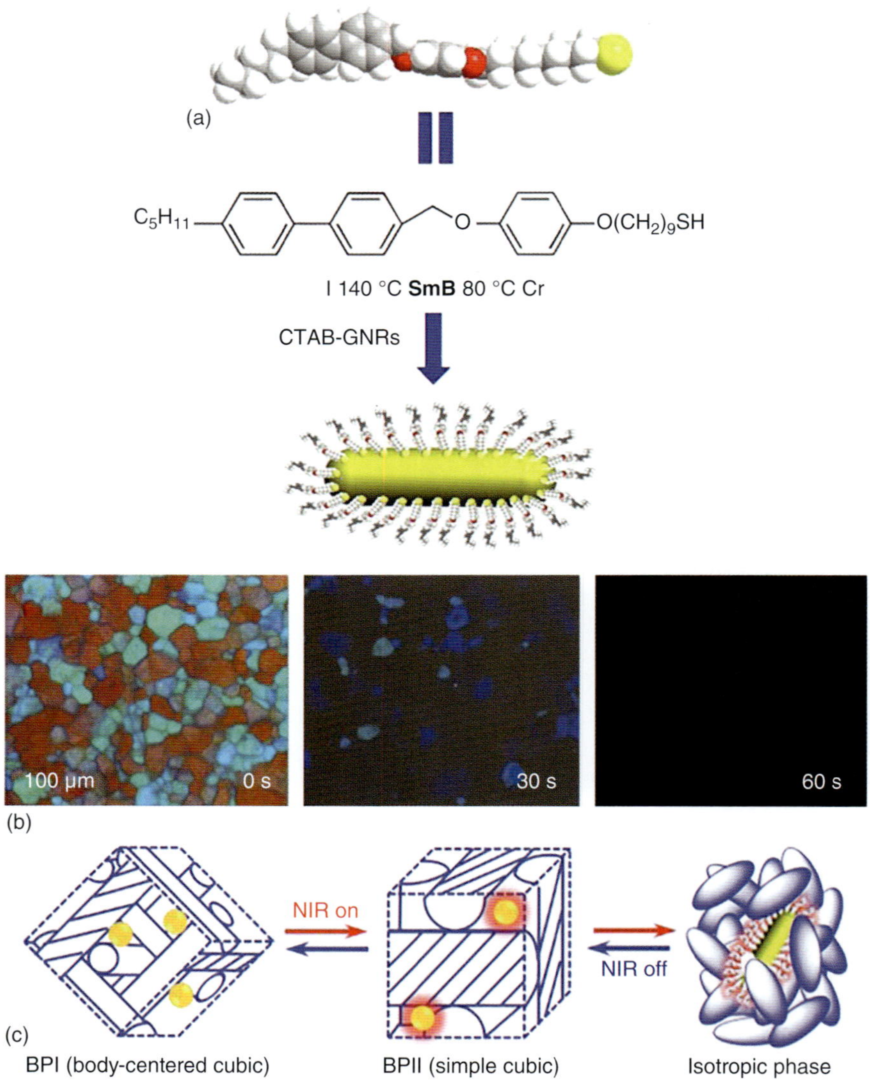

Figure 7.40 (a) Schematic representation of new mesomorphic surfactant 5 grafted onto the GNR by covalent Au–S linkage to form hydrophobic anisotropic plasmonic M-GNRs; (b) typical POM images of BPs doped with 0.3 wt% M-GNRs in 10 mm thick untreated cells upon irradiation with an 808 nm NIR laser (2 W); (c) schematic illustration of the NIR-light-tunable self-organized soft photonic superstructures. The yellow dots represent the cross-section of GNRs in the BP matrices. Source: Wang et al. 2015 [105]. Adapted with permission of Royal Society of Chemistry.

undergo a reversible N–I phase transition within a few seconds upon NIR laser irradiation, resulting from the photothermal effect of the embedded plasmonic GNRs (Figure 7.41b) [106]. The nanocomposites with higher concentration of GNRs exhibited faster phase transitions.

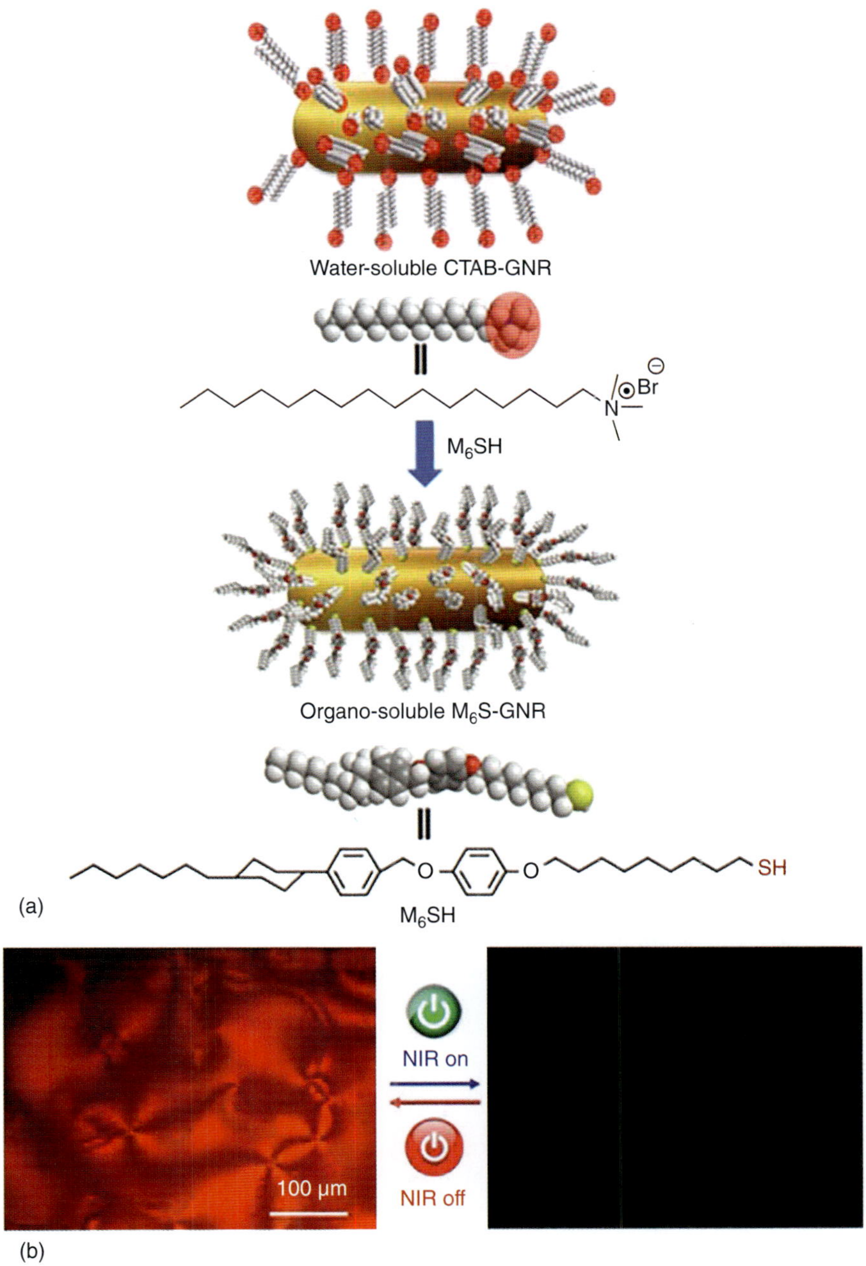

Figure 7.41 (a) Synthesis of the organo-soluble hybrid M6S-GNRs by the new hydrophobic mesogenic thiol M6SH grafted on the GNR via the covalent Au–S linkage starting from the as-prepared water-soluble CTAB-GNRs. (b) Phase transition of the 0.5 wt% M6S-GNR/E7 composite at $\Delta T = 3\,°C$ with NIR laser irradiation took approximately two seconds from the nematic to the isotropic phase whereas it took approximately one second to return from the isotropic to the nematic phase when the NIR light was turned off. Source: Gutierrez-Cuevas et al. 2015 [106]. Reproduced with permission of Royal Society of Chemistry.

7.4 Applications of Light-Driven Phase Transitions

7.4.1 Holography

Photoactive LCs, especially the polymer-based ones, exhibiting photoinduced N–I transitions have been explored from the viewpoint of their application in dynamic holography phenomenon. This holds great potential for real-time image processing, phase conjugate mirrors, and other optical communication applications [107]. Photoresponsive polymeric LC as a holographic recording medium has been discovered to have good thermal stability of the stored information, high resolution, and storage capability. Moreover, a variety of low molar mass LCs [108–111] and LC/polymer composites [112, 113] have also been studied extensively in the context of both static and dynamic holographic gratings. The pioneering work in this direction was done by Wendorff's group, with the production of holograms in LC copolymers containing azobenzene moieties and mesogenic groups [114–117].

Following this, Ikeda and Tsutsumi [118] established the high potentials of N–I transition in optical image storage application by achieving fast photochemical change (200 μs) both in low molar mass (**BMAB** and **8AB8**) and polymeric (**PA6AB2**) nematic LCs (Figure 7.15). In this report, the N phase of polymeric LCs, covered with a photomask, is subjected to photoinduced N–I transition. Here the light illuminated regions transform to the isotropic liquid phase, while the unexposed regions remain in the N phase. The sample was then taken to the glassy state that maintains the stored pattern, although a reverse isomerization (*cis–trans*) occurs, pointing to the fact that no phase transition takes place in the glassy environment. This feature has been ascribed to the fact that while the segmental motions of polymer chains are possible, no macroscopic ordering (necessary for the isotropic liquid state to return to the equilibrium N phase) is accomplished. Following this, different types of azobenzene–low molar mass LC composites [119, 120], azobenzene–polymeric LCs [121], polymeric LCs [122], and copolymers [123] have been synthesized and their performances were investigated.

7.4.2 Optical Storage Device

Some nematogens possess the dual frequency feature, and this can be utilized to create an optical storage device. The N phase of such materials exhibits a crossover frequency "f_c" at which the dielectric anisotropy (difference between the parallel and perpendicular values) changes its sign, from being positive for $f<f_c$ to negative for $f>f_c$. Hence, upon application of an electric field, the molecules would experience opposing dielectric torque when the field frequency is below and above the crossover frequency. In other words, above a certain magnitude of the electric field (Freedericksz threshold), the director prefers to be aligned along or perpendicular to the electric field in the two cases, allowing the orientation (homeotropic/planar) to be controlled by the field itself, instead of being controlled by surface forces. Such a switching of configuration can be achieved, by maintaining a constant frequency and by illuminating the sample with actinic light [124]. The principle behind such a device is the substantial increase in the dielectric relaxation frequency f_R (and thus indirectly in the crossover frequency f_c), induced by the radiation. These changes are connected with strong changes in the transmitted intensity of the LC cell. Thus this phenomenon can be exploited to fulfill a photonic addressable device in which the bright and dark states are achieved without allowing the material to transform to the isotropic phase, a highly desirable aspect particularly in view of enhancing the chemical stability of the LC substance employed.

7.4.3 Photocontrol in Liquid Crystal Displays

The characteristics of LCs such as dielectric constant, magnetic constant, viscosity, elasticity, and so on are hugely dependent on their alignment. Owing to their optical properties, LC materials can be used as actuators for light, which is the underpinning aspect for the modern LCD industry. When low molecular weight LC materials are placed in a cell and sandwiched between two transparent electrodes (usually ITO) coated with alignment layers, an LC cell is obtained. Upon application of an electric field with a driving voltage between the two electrodes, these LCs in the cell change their alignment directions from parallel (homogeneous) to perpendicular (homeotropic) with respect to the conductive substrates. Since LC molecules can change the polarization direction of light to their optic axes when the linearly polarized light (LPL) passes through them, and its transmittance through a pair of crossed polarizers, with the LC cell placed between them. A bright/dark contrast is produced at each electrically controlled pixel, and two-dimensional images can be displayed in the flat panel of LCDs. Usually, unidirectionally rubbed polyimide layer is used on the electrodes to attain the alignment for liquid crystals. Although it is widely used it has some drawbacks like dust contamination and production of static charge during the rubbing processing, which may lead to serious problems such as speckles and reduced display contrast of LCDs. Moreover, the rubbing method is limited to flat surfaces, which cannot be applied on curved substrates. Therefore, the noncontact method of photoalignment is considered to be one of the promising ways to solve the above-mentioned problems and can be easily used to assemble LCDs with high quality, wide viewing angles and large-area display.

Phototuning of N* LCs are expected to be very important, as these can act as active media for reflection-type displays, particularly if the wavelength of the reflected light is in the visible region. Such cholesteric LC materials have high potential for full-color LCDs with high quality and extremely low energy consumption. Further, the freezing of the cholesteric color by immobilizing the molecular ordering in a glassy state provides a way to achieve optical materials with a static function and for optical data storage [37]. Similarly, the ferroelectric LCs (FLCs), based on the helical SmC* phase, possess spontaneous polarization (P_s) and exhibit microsecond responses to the change in applied electric fields in a surface-stabilized state (flip of polarization). If a flip of polarization of surface-stabilized FLCs (SSFLCs) can be photoinduced in the presence of an applied electric field, a photoresponse of FLCs in the microsecond time region can be obtained. Here, the bias voltage remains constant before and after photoirradiation, but the threshold value for the flip of polarization is reduced by photoisomerization of the photoactive LCs. As a result, a flip of polarization is induced by photoirradiation, which leads to a change in alignment of the FLCs [125, 126]. A similar method was extended to antiferroelectric LCs too and already a device has been fabricated based on anti-FLCs [127].

7.4.4 Photocontrol of Mechanical Motion in Liquid Crystal Elastomers

Alteration of LC alignment by light not only induces an extreme change in refractive index, which is very difficult to achieve with amorphous materials, but also gives rise to some promising phenomena, e.g. photomechanical and photomobile effects in liquid crystalline elastomers (LCEs), molecular cooperative motion in low molecular weight LCs and LCPs, and supramolecular cooperative motion in LC block copolymers (LCBCs) [107].

Recently, the photoinduced N to I transformation has been also used in developing soft actuators derived from LCEs. This provides a possibility of coupling between orientational order and mechanical properties in LCEs, which was initially predicted by de Gennes [128, 129]. This concept finds potential application in the development of artificial muscles and mechanically tunable optical elements [130]. Experimentally, this was supported by a spontaneous contraction along the director axis of the nematic LCE films upon heating toward the N–I transition [131]. Notably, the deformation in the shape up to the extent of 300% has been achieved in nematic LCE films derived from side chain siloxane polymers [131]. Based on the concept of photomechanical effects of a polyamide-based monolayer [132], large and reversible light-driven deformation of LCEs has been demonstrated. The motion of LCE films was controlled by the gradient of the cross-linking density, polarization direction of actinic light, and the initial alignment of the LC molecules, etc. [133–135]. The occurrence of this photomechanical effect has been attributed to the decrease in the order parameter caused by *trans–cis* isomerization. An example, is the achievement of photoinduced three-dimensional deformation of the azobenzene-based nematic LCEs swollen in suitable solvents or heated above their T_g in air by Ikeda's group [136–138]. They successfully developed a single film capable of repeatedly bending precisely along the guided direction [107, 137, 138]. Similarly, a number of investigations have been reported to show that the LCEs containing azobenzene moiety, in response to the illuminated light, exhibit extraordinary changes in shape and volume where the light energy directly gets transformed into mechanical energy. Photoswitchable ferroelectric LCEs have also been realized as alternate ways to control the shape changes through optical (by photoisomerization) as well as electric (Piezo effect) means in addition to the existing thermal path [139].

7.5 Summary and Perspective

Over the years, application of several physical stimuli such as temperature, concentration of the solvent, pressure, and light of a specific wavelength (actinic light) has been demonstrated and documented in driving LC phase transitions.

The ability of light to alter/stabilize a particular LC phase through photoisomerization and consequent change in the molecular shape of the constituent molecules, provides an interesting tool to investigate condensed matter from a new dimension. Nature has been utilizing such photoinduced effects to perfection, as in the case of retinal protein rhodopsin, which enables vision. Scientists have been mimicking nature with the utilization of photosensitive organic molecules, which has resulted in a number of photoactive materials, for example, azobenzene or its derivatives. These photoactive materials are being utilized either in pure form or as a guest component in guest–host mixtures, to investigate their response to the light stimulus provided by the photoresponsive part and its effect on the whole system. In LC systems, a distinct variety of photoinduced phase transitions have been observed involving achiral as well as chiral materials, nematic, smectic, columnar, and blue phases. Thus, after a brief overview of LC systems, a detailed account of these photostimulated but isothermal transformations provided in this chapter illustrates their importance in both basic research and applied sciences.

Considering the fact that phase transition involving different mesophases or mesophase to isotropic phase leads to enormous changes in their optical properties, the usage of light to achieve such goals continues to be a captivating field for exploration. Since azobenzene systems are easily accessible, most of the studies are based on azobenzene-derived LCs or dopants. In addition, several other systems such as diarylethenes, azoxybenzenes, diphenylbutadienes, spiropyrans, fulgides, and several axially chiral azo compounds have also been discussed. In recent years, the effects of parameters such as molecular conformation, surface fields, electric field, elevated pressure, anisotropy reducing component, and magnitude of orientational order on the photoinduced N–I transition have been well studied to give a new dimension to their technological potential. Similarly, dynamic self-assembly of a layered phase induced and stabilized by light of a particular wavelength, tilt susceptibility of chiral smectic system, photoferroelectric effects in an antiferroelectric LC, etc. emphasize the fact that actinic light can be used as a new tool to study phase transitions and the associated critical phenomena that could also bring about effects that are not seen in equilibrium situations. Most importantly, two applications originating from the photoisomerization, i.e. optical storage and photomechanical effect, have been well corroborated. In addition, these studies will motivate several non-display applications of photoactive LC materials. With progress in the design of new photochromic molecules and photoinduced phase transitions, it can be expected that new technological advances and/or breakthroughs would occur in LC science in the future.

References

1 Bisoyi, H.K. and Li, Q. (2014). Liquid crystals. In: *Kirk-Othmer Encyclopedia of Chemical Technology*. Wiley.

2 Goodby, J.W., Collings, P.J., Kato, T. et al. (eds.) (2014). *Handbook of Liquid Crystals: Fundamentals*. Weinheim: Wiley-VCH.

3 Li, Q. (ed.) (2014). *Nanoscience with Liquid Crystals: From Self-Organized Nanostructures to Applications*. Heidelberg: Springer-Verlag.
4 Li, Q. (ed.) (2013). *Intelligent Stimuli Responsive Materials: From Well-Defined Nanostructures to Applications*. Wiley.
5 Kato, T., Mizoshita, N., and Kishimoto, K. (2006). Functional liquid crystalline assemblies: self-organized soft materials. *Angew. Chem. Int. Ed.* 45: 38.
6 Tschierske, C. (2007). Liquid crystal engineering – new complex mesophase structures and their relation to polymer morphologies, nanoscale patterning and crystal engineering. *Chem. Soc. Rev.* 36: 1930.
7 Bahadur, B. (ed.) (1992). *Liquid Crystals Applications and Uses*. Singapore: World Scientific.
8 Li, Q. (ed.) (2012). *Liquid Crystals Beyond Displays: Physics, Chemistry, and Applications*. Hoboken, NJ: Wiley.
9 Chandrasekhar, S., Sadashiva, B.K., and Suresh, K.A. (1977). Liquid crystals of disc-like molecules. *Pramana* 9: 471.
10 Kitzerow, H.S. and Bahr, C. (2001). *Chirality in Liquid Crystals*. New York: Springer-Verlag.
11 Wang, Y., Singh, G., Agra-Kooijman, D.M. et al. (2015). Room temperature heliconical twist-bend nematic liquid crystal. *CrystEngCommun* 17: 2778.
12 Wang, Y., Yoon, H., Bisoyi, H.K. et al. (2012). Hybrid rod-like and bent-core liquid crystal dimers exhibiting biaxial smectic-A and nematic phases. *J. Mater. Chem.* 22: 20363.
13 Rameshbabu, K., Urbas, A., and Li, Q. (2011). Synthesis and characterization of thermally irreversible photochromic cholesteric liquid crystals. *J. Phys. Chem. B* 115: 3409.
14 Li, L., Kang, S., Harden, J. et al. (2008). Nature inspired light-harvesting liquid crystalline porphyrins for organic photovoltaics. *Liq. Cryst.* 35: 233.
15 Reddy, R.A. and Tschierske, C. (2006). Bent-core liquid crystals: polar order, superstructural chirality and spontaneous desymmetrisation in soft matter systems. *J. Mater. Chem.* 16: 907.
16 Takezoe, H. and Takanishi, Y. (2006). Bent-core liquid crystals: their mysterious and attractive world. *Jpn. J. Appl. Phys.* 45: 597.
17 Dierking, I. (2003). *Textures of Liquid Crystals*. Weinheim: Wiley-VCH.
18 Tamaoki, N. (2001). Cholesteric liquid crystals for colour information technology. *Adv. Mater.* 13: 1135.
19 Gray, G.W. and Goodby, J.W. (1984). *Smectic Liquid Crystals*. Glasgow: Leonard Hill.
20 Brown, G.H. (1979). *Advances in Liquid Crystals* (ed. R. Schaetzing and J.D. Lister). Academic Press.
21 Garoff, S. and Meyer, R.B. (1978). Reply to "behavior of electric susceptibility and electroclinic coefficient near the chiral smectic A–C* transition". *Phys. Rev. A*. 18: 2739.
22 Helfrich, W. and Oh, C.S. (1971). Optically active smectic liquid crystal. *Mol. Cryst. Liq. Cryst.* 14: 289.
23 Lagerwall, S.T. (1999). *Ferroelectric and Antiferroelectric Liquid Crystals*. Wiley-VCH.

24 Gray, G.W., Hird, M., Lacey, D., and Toyne, K.J. (1989). The synthesis and transition temperatures of some 4,4″-dialkyl- and 4,4″-alkoxyalkyl-1,1′ :4′,1″-terphenyls with 2,3- or 2′,3′-difluoro substituents and of their biphenyl analogues. *J. Chem. Soc., Faraday Trans. 2* 2041.

25 Sun, G.X., Chen, G.B., Hong, T., and Xu, S.Y. (2003). Synthesis and physical properties of novel liquid crystals containing 2,3-difluorophenyl and 1,3-dioxane units. *J. Mater. Chem.* 13: 742.

26 Yelamaggad, C.V., Shanker, G., Hiremath, U.S., and Prasad, S.K. (2008). Cholesterol-based nonsymmetric liquid crystal dimers: an overview. *J. Mater. Chem.* 18: 2927.

27 Kuczynski, W. and Stegemeyer, H. (1995). Twist grain boundary phases in binary mixtures. *Mol. Cryst. Liq. Cryst.* 260: 377.

28 Renn, S.R. and Lubensky, T.C. (1988). Abrikosov dislocation lattice in a model of the cholesteric-to-smectic-A transition. *Phys. Rev. A.* 38: 2132.

29 Goodby, J.W., Waugh, M.A., Stein, S.M. et al. (1989). Characterization of a new helical smectic liquid crystal. *Nature* 337: 449.

30 Nguyen, H.T., Bouchta, A., Navailles, L. et al. (1992). TGBA and TGBC phases in some chiral tolan derivatives. *J. Phys. II* 2: 1889.

31 IUPAC Gold Book (1997). *IUPAC Compendium of Chemical Terminology (the "Gold Book")*, 2, Compiled by A.D. McNaught and A. Wilkinson. e. Oxford: Blackwell Scientific Publications. XML on-line corrected version: http://goldbook.iupac.org (2006) created by M. Nic, J. Jirat, and B. Kosata; updates compiled by A. Jenkins. ISBN: 0-9678550-9-8. doi: 10.1351/goldbook.

32 Ikeda, T. (2003). Photomodulation of liquid crystal orientations for photonic applications. *J. Mater. Chem.* 13: 2037.

33 Das, S. and Davis, R. (2003). Photoresponsive liquid crystals. *Proc. Indian Natl. Sci. Acad.* 69A: 109.

34 Yu, Y. and Ikeda, T. (2004). Alignment modulation of azobenzene-containing liquid crystal systems by photochemical reactions. *J. Photochem. Photobiol., C: Photochem. Rev.* 5: 247.

35 Prasad, S.K., Nair, G.G., Sandhya, K.L., and Shankar Rao, D.S. (2004). Photoinduced phase transitions in liquid crystalline systems. *Curr. Sci.* 86: 815.

36 Prasad, S.K., Nair, G.G., Hegde, G. et al. (2005). Photoinduced effects in nematic liquid crystals. *Phase Transitions* 78: 443.

37 Bisoyi, H.K. and Li, Q. (2016). Light-driven liquid crystalline materials: from photo-induced phase transitions and property modulations to applications. *Chem. Rev.* 116: 15089.

38 Prasad, S.K., Sandhya, K.L., Nair, G.G. et al. (2002). Opto-dielectric effect on a nematic liquid crystal doped with a photoactive azo mesogen. *J. Appl. Phys.* 87: 2084.

39 Tazuke, S., Kurihara, S., and Ikeda, T. (1987). Amplified image recording in liquid crystal media by means of photochemically triggered phase transition. *Chem. Lett.* 16: 911.

40 Sung, J.-H., Hirano, S., Tsutsumi, O. et al. (2002). Dynamics of photochemical phase transition of guest/host liquid crystals with an azobenzene derivative as a photoresponsive chromophore. *Chem. Mater.* 14: 385.

41 Mallia, V.A. and Tamaoki, N. (2003). Photoresponsive vitrifiable chiral dimesogens: photo-thermal modulation of microscopic disordering in helical superstructure and glass-forming properties. *J. Mater. Chem.* 13: 219.
42 Mallia, V.A. and Tamaoki, N. (2003). Photochemically driven smectic–cholesteric phase transition in an inherently photoactive dimesogen. *Chem. Mater.* 15: 3237.
43 Mallia, V.A. and Tamaoki, N. (2004). Photoactive dimesogen having different pathways of light driven phase transitions at different temperatures. *Chem. Commun.* 22: 2538.
44 Prasad, S.K. and Nair, G.G. (2001). Effects of photo-controlled nanophase segregation in a re-entrant nematic liquid crystal. *Adv. Mater.* 13: 40.
45 Tamai, N. and Miyasaka, H. (2000). Ultrafast dynamics of photochromic systems. *Chem. Rev.* 100: 1875.
46 Truong, T.V., Chen, C.Y., Tabiryan, N.V., and Shen, Y.R. (2007). Phase transition induced by two-photon absorption in azobenzene liquid crystals. *J. Opt. Soc. Am. B* 24: 2623.
47 Meyer, R.B., Liebert, L., Strzelecki, L., and Keller, P. (1975). Ferroelectric liquid crystals. *J. Phys. Lett.* 36: 69–71.
48 Niori, T., Sekine, T., Watanabe, J. et al. (1996). Distinct ferroelectric smectic liquid crystals consisting of banana shaped achiral molecules. *J. Mater. Chem.* 6: 1231.
49 Nair, G.G., Prasad, S.K., Hiremath, U.S., and Yelamaggad, C.V. (2001). Effect of light on the polarization of a banana-shaped achiral compound doped with a photoactive azobenzene material. *J. Appl. Phys.* 90: 48.
50 Matsui, T., Nakayama, K., Ozaki, M., and Yoshino, K. (2000). Photoinduced layer alignment control in ferroelectric liquid crystal with N*–C* phase transition doped with photo-chromic dye. *Appl. Phys. Lett.* 76: 1228.
51 Das, S. and Davis, R. (2003). Photoresponsive liquid crystals. *Proc. Indian Natl. Sci. Acad. A* 69: 109.
52 Garcia-Amoros, J., Szymczyka, A., and Velasco, D. (2009). Nematic-to-isotropic photo-induced phase transition in azobenzene-doped low-molar liquid crystals. *Phys. Chem. Chem. Phys.* 11: 4244.
53 Joly, G., Anakkar, A., Ismaili, M. et al. (2002). Chiral azobenzene liquid crystals under illumination: thickness influence and spontaneous polarization variations. *Ferroelectrics* 277: 67.
54 Joly, G., Anakkar, A., and Nguyen, H.T. (1999). Light induced shifts of ferroelectric mesophase transitions. *Liq. Cryst.* 26: 1251.
55 Boussoualem, Y., Ismaili, M., Buisine, J.M. et al. (2009). Dielectric and electro-optical properties of a photosensitive liquid crystal. *Liq. Cryst.* 36: 899.
56 Mallia, V.A., George, M., and Das, S. (1999). Photochemical phase transition in hydrogen-bonded liquid crystals. *Chem. Mater.* 11: 207.
57 Yamamoto, T., Nishiyama, I., and Yokoyama, H. (2008). Photo-control of chirality-induced liquid-crystalline 3D structures. *Ferroelectrics* 365: 39.
58 Chanishvili, A., Chilaya, G., Petriashvili, G., and Collings, P.J. (2005). *Trans–cis* isomerization and the blue phases. *Phys. Rev. E* 71: 051705.
59 Liu, H.Y., Wang, C.T., Hsu, C.Y., and Lin, T.H. (2010). Optically tuneable blue phase photonic band gaps. *Appl. Phys. Lett.* 96: 121103.

60 Liu, H.Y., Wang, C.T., Hsu, C.Y., and Lin, T.H. (2011). Photonic bandgaps controllable blue phase liquid crystal. *Proc. SPIE* 7955: 795506.
61 Tamaoki, N., Aoki, Y., Moriyama, M., and Kidowaki, M. (2003). Photochemical phase transition and molecular realignment of glass-forming liquid crystals containing cholesterol/azobenzene dimesogenic compounds. *Chem. Mater.* 15: 719.
62 Li, Q., Li, L., Kim, J. et al. (2005). Reversible photoresponsive chiral liquid crystals containing a cholesteryl moiety and azobenzene linker. *Chem. Mater.* 17: 6018.
63 Zhou, K., Bisoyi, H.K., Jin, J.-Q. et al. (2018). Light-driven reversible transformation between self-organized simple cubic lattice and a helical superstructure enabled by a molecular switch functionalized nanocage. *Adv. Mater.* 30: 1800237.
64 Chen, Y., Yu, H., Zhang, L. et al. (2014). Photoresponsive liquid crystals based on halogen bonding of azopyridines. *Chem. Commun.* 50: 9647.
65 Kim, Y. and Tamaoki, N. (2016). Asymmetric dimers of chiral azobenzene dopants exhibiting unusual helical twisting power upon photoswitching in cholesteric liquid crystals. *ACS Appl. Mater. Interfaces* 8: 4918.
66 Kumar, S. (2010). *Chemistry of Discotic Liquid Crystals: From Monomers to Polymers*. Boca Raton, FL: CRC Press.
67 Shimizu, Y., Kurobe, A., Monobe, H. et al. (2003). Calamitic and discotic mesophases formed by kinetically controlled rod–disc alternation of molecular shape in a triphenylene-azobenzene mesogenic system. *Chem. Commun.* 14: 1676.
68 Yagai, S., Nakajima, T., Kishikawa, K. et al. (2005). Hierarchical organization of photoresponsive hydrogen-bonded rosettes. *J. Am. Chem. Soc.* 127: 11134.
69 Pecinovsky, C.S., Hatakeyama, E.S., and Gin, D.L. (2008). Polymerizable photochromic macrocyclic metallomesogens: design of supramolecular polymers with responsive nanopores. *Adv. Mater.* 20: 174.
70 Westphal, E., Bechtold, I.H., and Gallardo, H. (2010). Synthesis and optical/thermal behavior of new azo photoisomerizable discotic liquid crystals. *Macromolecules* 43: 1319.
71 Norikane, Y., Hirai, Y., and Yoshida, M. (2011). Photoinduced isothermal phase transitions of liquid-crystalline macrocyclic azobenzenes. *Chem. Commun.* 47: 1770.
72 Tanaka, D., Ishiguro, H., Shimizu, Y., and Uchida, K. (2012). Thermal and photoinduced liquid crystalline phase transitions with a rod–disc alternative change in the molecular shape. *J. Mater. Chem.* 22: 25065.
73 Peng, X., Gao, H., Xiao, Y. et al. (2017). Synthesis and self-assembly of photoresponsive and luminescent polycatenar liquid crystals incorporating an azobenzene unit interconnecting two 1,3,4-thiadiazoles. *New J. Chem.* 41: 2004.
74 Pieraccini, S., Masiero, S., Spada, G.P., and Gottarelli, G. (2003). A new axially-chiral photochemical switch. *Chem. Commun.* 0: 598.
75 Lin, T.-H., Li, Y., Wang, C.-T. et al. (2013). Red, green and blue reflections enabled in an optically tunable self-organized 3D cubic nanostructured thin film. *Adv. Mater.* 25: 5050.

76 Liu, H.-Y., Wang, C.-T., Hsu, C.-Y., and Lin, T.-H. (2011). Pinning effect on the photonic bandgaps of blue-phase liquid crystal. *Appl. Opt.* 50: 1606.
77 Chanishvili, A., Chilaya, G., Petriashvili, G., and Sikharulidze, D. (2004). Light induced effects in cholesteric mixtures with a photosensitive nematic host. *Mol. Cryst. Liq. Cryst.* 409: 209.
78 Aronzon, D., Levy, E.P., Collings, P.J. et al. (2007). *Trans–cis* isomerization of an azoxybenzene liquid crystal. *Liq. Cryst.* 34: 707.
79 Kurihara, S., Ikeda, T., Tazuke, S., and Seto, J. (1991). Isothermal phase transition of liquid crystals induced by photoisomerization of doped spiropyrans. *J. Chem. Soc., Faraday Trans.* 87: 3251.
80 Kosa, T., Sukhomlinova, L., Su, L. et al. (2012). Light-induced liquid crystallinity. *Nature* 485: 347.
81 Janicki, S.Z. and Schuster, G.B. (1995). A liquid crystal opto-optical switch: nondestructive information retrieval based on a photochromic fulgide as trigger. *J. Am. Chem. Soc.* 117: 8524.
82 Zhang, M. and Schuster, G.B. (1992). Photoracemization of optically active 1,1′-binaphthyl derivatives: light-initiated conversion of cholesteric to compensated nematic liquid crystals. *J. Phys. Chem.* 96: 3063.
83 Yokoyama, Y. and Sagisaka, T. (1997). Reversible control of pitch of induced cholesteric liquid crystal by optically active photochromic fulgide derivatives. *Chem. Lett.* 26: 687.
84 White, T.J., Zhao, A.D., Cazzell, S.A. et al. (2012). Optically reconfigurable color change in chiral nematic liquid crystals based on indolylfulgide chiral dopants. *J. Mater. Chem.* 22: 5751.
85 Yarmolenko, S.N., Kutulya, L.A., Vashchenko, V.V., and Chepeleva, L.V. (1994). Photosensitive chiral dopants with high twisting power. *Liq. Cryst.* 16: 877.
86 Gvozdovskyy, I., Yaroshchuk, O., and Serbina, M. (2011). Photoinduced nematic – cholesteric structural transitions in liquid crystal cells with homeotropic anchoring. *Mol. Cryst. Liq. Cryst.* 546: 202.
87 Gvozdovskyy, I., Yaroshchuk, O., Serbina, M., and Yamaguchi, R. (2012). Photoinduced helical inversion in cholesteric liquid crystal cells with homeotropic anchoring. *Opt. Express* 20: 3499.
88 Krivoshey, A.I., Shkolnikova, N.I., Chepeleva, L.V. et al. (2004). Photoinduced cholesteric helix inversion in some systems containing some chiral α,β-unsaturated ketones. *Funct. Mater.* 11: 76.
89 Lub, J., Hoeve, W.T., Nijssen, W.P.M. et al. (2002). The effect of the position of the methyl group on the helical twisting powers of aldol condensation products of methyl cyclohexanones. *Liq. Cryst.* 29: 995.
90 Bisoyi, H.K. and Li, Q. (2016). Light-directed dynamic chirality inversion in functional self-organized helical superstructures. *Angew. Chem. Int. Ed.* 55: 2994.
91 Denekamp, C. and Feringa, B.L. (1998). Optically active diarylethenes for multimode photoswitching between liquid-crystalline phases. *Adv. Mater.* 10: 1080.
92 van Leeuwen, T., Pijper, T.C., Areephong, J. et al. (2011). Reversible photochemical control of cholesteric liquid crystals with a diamine-based diarylethene chiroptical switch. *J. Mater. Chem.* 21: 3142.

93 Yamaguchi, T., Inagawa, T., Nakazumi, H. et al. (2001). Photoinduced pitch changes in chiral nematic liquid crystals formed by doping with chiral diarylethene. *J. Mater. Chem.* 11: 2453.

94 Yamaguchi, T., Inagawa, T., Nakazumi, H. et al. (2000). Phase transition of a liquid crystal induced by chiral photochromic dopants. *Mol. Cryst. Liq. Cryst. Sci. Technol., Sect. A* 345: 287.

95 Uchida, K., Kawai, Y., Shimizu, Y. et al. (2000). An optically active diarylethene having cholesterol units: a dopant for photo-switching of liquid crystal phases. *Chem. Lett.* 29: 654.

96 Yamaguchi, T., Inagawa, T., Nakazumi, H. et al. (2000). Photoswitching of helical twisting power of a chiral diarylethene dopant: pitch change in a chiral nematic liquid crystal. *Chem. Mater.* 12: 869.

97 Hayasaka, H., Miyashita, T., Nakayama, M. et al. (2012). Dynamic photoswitching of helical inversion in liquid crystals containing photoresponsive axially chiral dopants. *J. Am. Chem. Soc.* 134: 3758.

98 Das, S., Varghese, S., and Kumar, N.S.S. (2010). Butadiene-based photoresponsive soft materials. *Langmuir* 26: 1598.

99 Davis, R., Mallia, V.A., Das, S., and Tamaoki, N. (2004). Butadienes as novel photochromes for color tuning of cholesteric glasses: influence of microscopic molecular reorganization within the helical superstructure. *Adv. Funct. Mater.* 14: 743.

100 Abraham, S., Mallia, V.A., Ratheesh, K.V. et al. (2006). Reversible thermal and photochemical switching of liquid crystalline phases and luminescence in diphenylbutadiene-based mesogenic dimers. *J. Am. Chem. Soc.* 128: 7692.

101 Vijayaraghavan, R.K., Abraham, S., Akiyama, H. et al. (2008). Photoresponsive glass-forming butadiene-based chiral liquid crystals with circularly polarized photoluminescence. *Adv. Funct. Mater.* 18: 2510.

102 Chen, H., Shao, L., and Wang, J. (2013). Gold nanorods and their plasmonic properties. *Chem. Soc. Rev.* 42: 2679.

103 Nikoobakht, B. and El-Sayed, M.A. (2003). Preparation and growth mechanism of gold nanorods (NRs) using seed-mediated growth method. *Chem. Mater.* 15: 1957.

104 Shanmugam, V., Selvakumar, S., and Yeh, C.-S. (2014). Near infrared light-responsive nanomaterials in cancer therapeutics. *Chem. Soc. Rev.* 43: 6254.

105 Wang, L., Gutierrez-Cuevas, K.G., Bisoyi, H.K. et al. (2015). NIR light-directing self-organized 3D photonic superstructures loaded with anisotropic plasmonic hybrid nanorods. *Chem. Commun.* 51: 15039.

106 Gutierrez-Cuevas, K.G., Wang, L., Xue, C. et al. (2015). Near infrared light-driven liquid crystal phase transition enabled by hydrophobic mesogen grafted plasmonic gold nanorods. *Chem. Commun.* 51: 9845.

107 Yu, H. and Ikeda, T. (2011). Photocontrollable liquid-crystalline actuators, photocontrollable liquid-crystalline actuators. *Adv. Mater.* 23: 2149–2180.

108 Ortler, R., Brauchle, C., Miller, A., and Riepl, G. (1989). Reversible holographic-optical data storage in cholesteric liquid-crystalline siloxanes. *Macromol. Chem. Rapid Commun.* 10: 189.

109 Wiederrecht, G.P., Yoon, B.A., and Wasielewski, M.R. (1995). High photorefractive gain in nematic liquid crystals doped with electron donor and acceptor molecules. *Science* 270: 1794.

110 Wiederrecht, G.P., Yoon, B.A., Svec, W.A., and Wasielewski, M.R. (1997). Photorefractivity in nematic liquid crystals containing electron donor–acceptor molecules that undergo intramolecular charge separation. *J. Am. Chem. Soc.* 119: 3358.
111 Sasaki, T., Katsuragi, A., and Ohno, K. (2002). Spontaneous polarization vector reorientation photorefractive effect in dye-doped ferroelectric liquid crystals. *J. Phys. Chem. B* 106: 2520.
112 Kurihara, S., Ikeda, T., and Tazuke, S. (1990). Photochemically induced isothermal phase transition in liquid crystals. Effect of interaction of photoresponsive molecules with matrix mesogens. *Mol. Cryst. Liq. Cryst.* 178: 117.
113 Ruslim, C. and Ichimura, K. (1998). *Z*-Isomers of azobenzenes highly compatible with liquid crystals. *Chem. Lett.* 27: 789.
114 Eich, M., Wendorff, J.H., Reck, B., and Ringsdorf, H. (1987). Reversible digital and holographic optical storage in polymeric liquid crystals. *Macromol. Chem. Rapid Commun.* 8: 59.
115 Eichand, M. and Wendorff, J.H. (1987). Erasable holograms in polymeric liquid crystals. *Macromol. Chem. Rapid Commun.* 8: 467.
116 Wendorff, J.H. and Eich, M. (1989). Nonlinear optical phenomenon in liquid crystalline side chain polymers. *Mol. Cryst. Liq. Cryst.* 169: 133.
117 Anderle, K. and Wendorff, J.H. (1994). Holographic recording, using liquid crystalline side chain polymers. *Mol. Cryst. Liq. Cryst.* 243: 51.
118 Ikeda, T. and Tsutsumi, O. (1995). Optical switching and image storage by means of azobenzene liquid-crystal films. *Science* 268: 1873.
119 Kurihara, S., Ikeda, T., Sasaki, T. et al. (1990). Time-resolved observation of isothermal phase transition of liquid crystals triggered by photochemical reaction of dopant. *J. Chem. Soc., Chem. Commun.* (24): 1751.
120 Kurihara, S., Ikeda, T., Sasaki, T. et al. (1991). Time-resolved observation of isothermal transition of liquid crystals induced by photoisomerization of azobenzene dopant. *Mol. Cryst. Liq. Cryst.* 195: 251.
121 Ikeda, T., Sasaki, T., and Kim, H.B. (1991). Intrinsic response of polymer liquid crystals in photochemical phase transition. *J. Phys. Chem.* 95: 509.
122 Tsutsumi, O., Kitsunai, T., Kanazawa, A. et al. (1998). Photochemical phase transition behavior of polymer azobenzene liquid crystals with electron-donating and accepting substituents at the 4,4′-positions. *Macromolecules* 31: 355.
123 Sasaki, T., Ikeda, T., and Ichimura, K. (1992). Time-resolved observation of photochemical phase transition in polymer liquid crystals. *Macromolecules* 25: 3807.
124 Hegde, G., Nair, G.G., Prasad, S.K., and Yelamaggad, C.V. (2005). A photodriven dual-frequency addressable optical device. *J. Appl. Phys.* 97: 093105.
125 Ikeda, T., Sasaki, T., and Ichimura, K. (1993). Photochemical switching of polarization in ferroelectric liquid-crystal films. *Nature* 361: 428.
126 Sasaki, T., Ikeda, T., and Ichimura, K. (1994). Photochemical control of properties of ferroelectric liquid crystals: photochemical flip of polarization. *J. Am. Chem. Soc.* 116: 625.

127 Shirota, K. and Yamaguchi, I. (1997). Optical switching of antiferroelectric liquid crystal with azo-dye using photochemically induced SmC_A*–SmC* phase transition. *Jpn. J. Appl. Phys.* 36: L1035.
128 de Gennes, P.G. (1969). Possibilities offered by polymer crosslinking in the presence of a liquid crystal. *Phys. Lett. A* 28: 725.
129 de Gennes, P.G., Hebert, M., and Kant, R. (1997). Artificial muscles based on nematic gels. *Macromol. Symp.* 113: 39.
130 Finkelmann, H., Kim, S.T., Munoz, A. et al. (2001). Tunable mirrorless lasing in cholesteric liquid crystalline elastomers. *Adv. Mater.* 13: 1069.
131 Clarke, S.M., Hotta, A., Taj-bakhsh, A.R., and Terentjev, E.M. (2001). Effect of cross-linker geometry on equilibrium thermal and mechanical properties of nematic elastomers. *Phys. Rev. E* 64: 061702.
132 Blair, H.S., Pague, H.I., and Riordan, J.E. (1980). Photoresponsive effects in azo polymers. *Polymer* 21: 1195.
133 Finkelmann, H., Nishikawa, E., Pereira, G.G., and Warner, M. (2001). A new opto-mechanical effect in solids. *Phys. Rev. Lett.* 87: 015501.
134 Yu, Y., Nakano, M., and Ikeda, T. (2003). Directed bending of a polymer film by light. *Nature* 425: 145.
135 van Oosten, C.L., Corbett, D., Davies, D. et al. (2008). Bending dynamics and directionality reversal in liquid crystal network photo-actuators. *Macromolecules* 41: 8592.
136 Ikeda, T., Nakano, M., Yu, Y. et al. (2003). Anisotropic bending and unbending behavior of azobenzene liquid-crystalline gels by light exposure. *Adv. Mater.* 15: 201.
137 Yu, Y. and Ikeda, T. (2006). Soft actuators based on liquid-crystalline elastomers. *Angew. Chem. Int. Ed.* 45: 5416.
138 Yu, Y., Nakano, M., Shishido, A. et al. (2004). Effect of cross-linking density on photoinduced bending behavior of oriented liquid-crystalline network films containing azobenzene. *Chem. Mater.* 16: 1637.
139 Beyer, P. and Zentel, R. (2005). Photoswitchable smectic liquid-crystalline elastomers. *Macromol. Rapid Commun.* 26: 874.

8

Photomechanical Soft Nanocomposites: Synergies Between Soft Matrix and Energy Conversion Additives

Jing Hu[1], *Shudeng Ma*[1], *Haifeng Yu*[1], *and Quan Li*[2]

[1] *Peking University, Department of Material Science and Engineering, College of Engineering and Key Laboratory of Polymer Chemistry and Physics of Ministry of Education, No. 5 Yiheyuan Road, Beijing, 100871, China*
[2] *Kent State University, Liquid Crystal Institute and Chemical Physics Interdisciplinary Program, 1425 Lefton Esplanade, Kent, OH 44242, USA*

8.1 Introduction

Photomechanical materials can directly transform light energy into mechanical movements that provide mechanical devices with many advantages, such as non-contact, remote control, and spatiotemporal resolution. In the past decades, a variety of photoresponsive materials have been developed, and the system based on liquid crystalline polymers (LCPs) is one of the most fascinating ones from the viewpoint of applications as soft photoactuators [1–8]. However, a simple-component material system sometimes develops into a bottleneck, such as difficulty in the realization of multiwavelength responsiveness or the flexibility of fabrication. Recently, nanocomposite materials, which combine the superiority of both organic and inorganic materials, have shown more and more potential in photomechanical researches. It is necessary to summarize the recent progresses made with reference to photomechanical soft nanocomposites. In this chapter, we will mainly discuss their design strategy, fabrication, operating principle, and potential applications.

Being one of the outstanding candidates for stimuli-responsive materials, some nanocomposites not only respond to light, but also to other stimuli, such as moisture [9–11], electric field [12, 13], magnetic field [14], temperature [15–18], and chemical energy [19], and may also simultaneously respond to multi-stimuli [20–25]. Such photomechanical soft nanocomposites are usually made up of soft matrices and functional additives, which utilize the synergistic effect in each component and magnify the microcosmic molecular changes into macrocosmic device motions. The photo-driven actuators originating from these soft nanocomposites have broader materials choices while performing multiple deformations under different light-irradiation conditions.

To interpret the data from researches made on photomechanical soft nanocomposites in a systematic way, their classification from different perspectives is

Photoactive Functional Soft Materials: Preparation, Properties, and Applications, First Edition.
Edited by Quan Li.

Table 8.1 Classification of photomechanical soft nanocomposites.

Photomechanical soft nanocomposites	
1. Assembly methods	
1.1 Homogeneously dispersed functional additives in soft matrix	
1.1.1 Soft matrix [26–30] Cross-linked liquid crystalline polymers (CLCPs) [31–33] Shape-memory polymers (SMPs) [24, 34] Hydrogels [35, 36]	1.1.2 Functional additives [37–39] Carbon materials [29, 40] Noble metal [41, 42] Photoresponsive dyes [31, 43] Up-conversion nanoparticles [44, 45]
1.2 Asymmetric assembly [46–48]	
Bilayer or multilayer films [49–51]	
2. Working principles 2.1 Photothermal actuator [38, 39] Surface plasma resonance (SPR) [42] Energy band transitions [52] 2.2 Photochemical actuator Photoisomerization [53–56] Photodimerization [57, 58] 2.3 Coexistence [59]	3. Actuating conditions 3.1 Light wavelength Ultraviolet [4] Visible [50] Infrared [40] 3.2 Light polarization [24, 60] 3.3 Light intensity Attenuated [61] Uniform [62]

summarized, as shown in Table 8.1. From the perspective of fabricating a nanocomposite actuator, they can be mainly divided into two classes: homogeneously dispersed functional additives in soft matrix [26–45] and asymmetric assembly [46–51]; however, there must be a soft matrix to act as a structural foundation of the material system and functional additives to respond to the specific light-stimuli. Making sense of the working principle is essential; we can also classify the nanocomposite materials into photothermal actuators [38–42, 52] and photochemical actuators [53–58]. In some cases, the same system can be actuated by both the photoconversion mechanisms [30, 34, 59]. Moreover, if considering the nanocomposite actuators through their responsive light parameters, they can be sorted according to the differences of light wavelength, light polarization, light intensity distribution, light interference, and so on [60–63]. In this chapter, the second classification method is adopted to introduce the photomechanical soft nanocomposites, in the order of the nanocomposites actuated by photothermal effect, than the photochemical ones.

8.2 Photomechanical Nanocomposites Based on Photothermal Effect

Photothermal effect refers to the phenomenon of directly converting light energy into thermal energy. Several kinds of materials possess outstanding photothermal conversion efficiency, and the most commonly investigated photothermal conversion agents are noble metal nanostructures, such as gold (Au) nanoparticles, nanorods, nanoshells, hollow nanospheres and nanocages, as well as carbon materials including graphene, graphene oxide (GO), and carbon nanotubes (CNTs) [23, 64–69]. In addition, organic compounds such as indocyanine green (ICG) dye and polyaniline, as well as copper chalcogenide semiconductors including CuS and $Cu_{2-x}Se$ nanocrystals, are also reported for their excellent photothermal performances [70]. Among these photothermal conversion agents, metal nanostructures exhibit close to the highest conversion efficiency due to their intensive near-infrared (NIR) absorption and tunable surface plasmon resonance (SPR) [42, 70–74]. These photothermal materials have been widely used in solar energy harvesting, cancer photothermal therapy, drug delivery, and biomedical imaging. Recently, researchers of photomechanical materials have become conscious of the advantages offered by photothermal agents, compositing them with soft matrix, which finally produces many interesting works.

8.2.1 Design Strategy

Generally, photomechanical soft nanocomposites based on photothermal effects utilize the thermal energy generated by the photothermal agents, heating the soft matrix and causing changes in the volume of the material system. To actuate the whole system by light, there must be a connecting bridge between the microcosmic change and macrocosmic deformation. It is well known that the elevating temperature can bring about the phase transition of crystals or liquid crystals (LCs) [75], the movement of polymer chain segments [76, 77], and the changes in volume of macromolecules or hydrogels [78]. A rational design strategy is essential to magnify these changes into mechanical deformation.

Figure 8.1 summarizes the design strategies employed for fabricating soft nanocomposite photoactuators based on the photothermal effect. Two mainstream design methods are demonstrated as homodispersed single-layer film and asymmetric bilayer or multilayer film. The former has been mainly developed from liquid crystalline elastomers (LCEs), and the latter has been under development recently. The Noble Prize winner de Gennes first theoretically predicted that, similar to muscles, the nematic LCPs would deform in response to environmental stimuli through calculation [1, 2]. Then, the German scientist, Finkelmann, successfully fabricated monodomain-aligned LCEs and observed a large deformation when inducing the LC to isotropic phase transition for the first time in the laboratory. Later, the groups of Ikeda and Keller further independently developed this.

Photoactuating LCEs usually rely on the photoisomerization of azobenzene moieties upon irradiation of actinic light [79, 80]. It is well known that an LCE simultaneously possesses the order of an LC and the elasticity of an elastomer [4], enabling it to exhibit reversible deformation enabling diverse applications as functional materials. When the ordering of LCs is disrupted, the LCE will deform along the longitudinal axis of the LC molecules due to the molecular cooperative motion, which is one of the most fascinating properties of LCs. Excluding the exertion of polarizing light, we can find that the two key points of LCEs, which

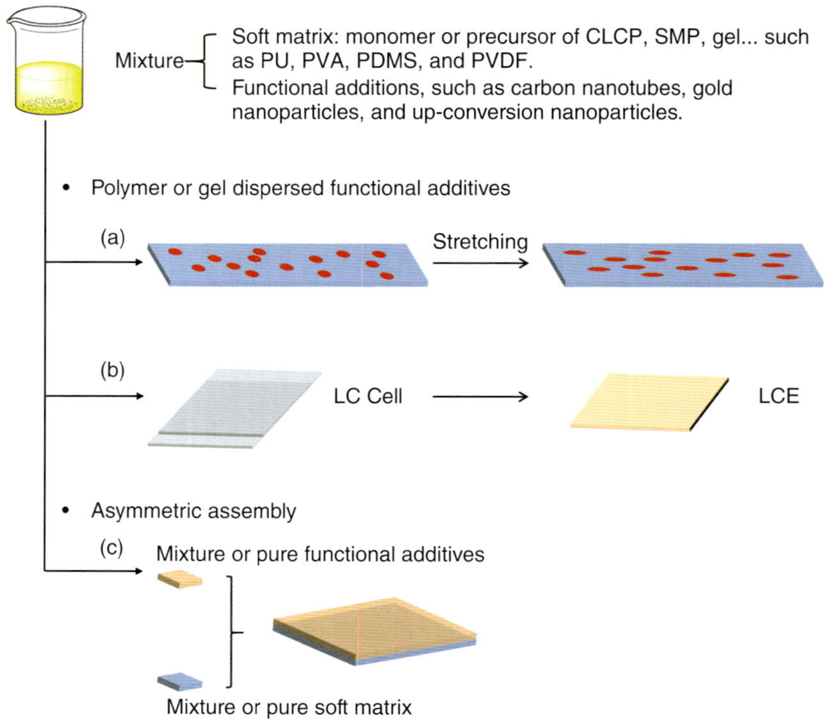

Figure 8.1 Summary of fabrication methods of the photomechanical soft nanocomposites. (a, b) Two routes of orienting the LC molecules in the substrate by stretching or using LC cell, respectively. (c) The structure of asymmetric assembled films.

are very important for their actuating behaviors, are namely the orientation of the LCs and cross-linking. Both guarantee the coherent and coordinating deformation of LCE-based materials. Then, the question as to how to get the deformation of LCEs can be simplified by focusing on how to disrupt the orientation of LCs. In addition to the photochemical reactions of azobenzene and cinnamate derivatives, the elevating temperature can also cause the disorder of LCs by inducing phase transition. By introducing the photothermal agents into the LCEs, the obtained nanocomposites can be easily photo-manipulated.

However, we should notice that the LC property is not a prerequisite to designing the photomechanical soft nanocomposites. The nanocomposites relying on sol–gel transition [35] or shape-memory effect (SME) [38] can also perform photomechanical motion with the aid of photothermal agents, as shown in Figure 8.1. Firstly, the monomers or the precursors of the soft matrix, such as cross-linked LCPs (CLCPs), shape-memory polymers (SMPs), and gels, are mixed homogeneously with the photothermal agents, which are then used to fabricate freestanding films through different routes. When the composite film contains LCs, two common ways are often applied to orient the mesogenic molecules, namely mechanical stretching or pre-alignment by the LC cell, as shown in Figure 8.1a,b.

Table 8.2 Summary of properties of some materials in flexible electronics [20, 83, 84].

Materials	Young's modulus (Gpa)	CTE ($\times 10^{-6}$/K)	CHE (%RH^{-1})
Paper	2–6	4–16	$3–19 \times 10^{-4}$
PEDOT: PSS	1.8–2.3	Not given	4.7×10^{-4}
PET	2.55	38–59.4	$5–85 \times 10^{-6}$
PI	2.5–8	17–28	Moisture absorbance <0.68%
BOPP	1.1	137	Moisture absorbance <0.03%
PDMS	$0.4–3.5 \times 10^{-3}$	266–310	Not given
PDMS (Ecoflex)	$6.9–8.3 \times 10^{-5}$	279–284	Not given
Graphene	1000	-8	Not given
GO	208	−67	$2.7–9 \times 10^{-2}$
Carbon nanotube	1000	−1.5	Not given

The homogeneously dispersed single-layer film can contract, expand, bend, and twist or perform other mechanical activities due to the deformation gradient along the film thickness caused by the nonuniform light illumination and molar extinction coefficient of chromophore additives [81].

The second strategy for fabricating soft nanocomposite photoactuators is asymmetric fabrication, as shown in Figure 8.1c. In this way, the nanocomposite materials usually have two layers with a large numerically differential coefficient of thermal expansion (CTE). The general structure of asymmetric fabrication is bilayer or bilamellar. It is worth noting that either layer can be either polymer or gel containing homogeneously dispersed functional additives (Figure 8.1a,b) or just a slice of a single component film. The nanocomposite actuating complies with the well-known principle of bimetallic mechanism [82], and the photothermal agent acts as the part of a nanoheater.

Table 8.2 shows the Young's modulus, CTE, and coefficient of hygroscopic expansion (CHE) of some often applied materials in flexible electronics. It is convenient to design a photomechanical device by choosing two materials with a largely different CTE, such as poly (dimethylsiloxane) (PDMS) and graphene to fabricate a nanocomposite film [50]. As given in Table 8.1, the CTE of GO and CNT is negative, which offers a large amount of photo-energy converted heat and shrinks them, endowing the bilayer films with a larger light-manipulated deformation. Therefore, carbon photothermal agents are widely used in designing the bilayer structure.

8.2.2 Fabrication

In the following sections, several concrete examples are introduced as fabrication prototypes of photomechanical soft nanocomposites based on photothermal effect.

8.2.2.1 Homogeneous Single-Layer Films

GO-5CB/PVA and GO/PVDF-HFP Recently, Yu's group carried out several photo-actuating researches on GO/polymer nanocomposite materials [27, 29]. GO is usually prepared by the modified Hummers method [85, 86], which can be ultrasonically homo-dispersed in deionized water or organic solvents such as dimethylformamide (DMF) and dimethyl sulfoxide (DMSO). With easy modification, GO often possesses relatively better dispersibility in solvents than do CNTs or gold nanoparticles. In the fabrication process, GO was firstly dispersed in the solution of a soft matrix, such as a polymer-dispersed LC system like 5CB/polyvinyl alcohol (PVA), or an SMP system like poly(vinylidenefluoride-hexafluoro propylene) (PVDF-HFP). Then the homogeneous mixture of GO and polymer in solution was coated on a substrate (e.g. glass slide), forming a film by volatilization (Figure 8.2a,c). To acquire the photoactuating feature, the GO-5CB/PVA film should undergo mechanical stretching to orient the LC molecules. It is well known that 5CB is a typically nematic LC with an LC-to-isotropic phase transition temperature slightly higher than room temperature. As a result, the heat originated from the photothermal effect of GO can induce the phase transition of 5CB, enabling the film to deform quickly (Figure 8.2b). As for the GO/PVDF-HFP system, the heat energy stimulates the reversible crystal-to-crystal phase transition of PVDF-HFP and facilitates the nanocomposites with a specific shape in a particular range of temperature.

Organic Near-Infrared (NIR) Dye/LCEs In 2016, Yang's group reported the design and synthesis of one kind of LC-like NIR absorbing dyes YHD796 [37]. This organic photothermal agent was inspired by Smith's croconaine dye [87], as compared with the ICG dye. In addition, they noticed the potential of the thiophene-croconaine dye to become the rigid mesogenic core for building calamitic LC molecular structures. By utilizing the like-dissolve-like theory, they mixed the synthesized NIR dye with LCEs via the LC cell fabrication method as shown in Figure 8.1b to ensure the stability and efficiency of the photothermal agent. The LC mixture and the NIR dye were simultaneously injected into the LC cell, which were then cured under illumination of one UV light. The freestanding film was obtained by dissolving the glass substrates with hydrofluoric acid (HF).

AuNPs/Semicrystalline Polymers (SCPs) and AuNRs/LCEs Gold nanoparticles (AuNPs) or gold nanorods (AuNRs) have a very high molar extinction coefficient and photothermal conversional efficiency thanks to their inherent SPR feature [41, 42]. However, they are generally immiscible with an organic matrix and easily aggregate, losing their SPR properties. In the nanocomposite actuating system, the well-dispersed gold nanostructures should be grafted on the polymers with the assistance of a thiol ligand. Zhao's group successfully fabricated optically controllable LCPs-based actuators by surface functionalization of AuNPs with thiol-substituted poly(ethylene-*co*-vinyl acetate) (EVA) [41]. Zhang's group [42] stabilized the AuNRs through a three-step sequential thiol-click approach and embedded the AuNRs into the LCEs. More interestingly, the nanocomposites

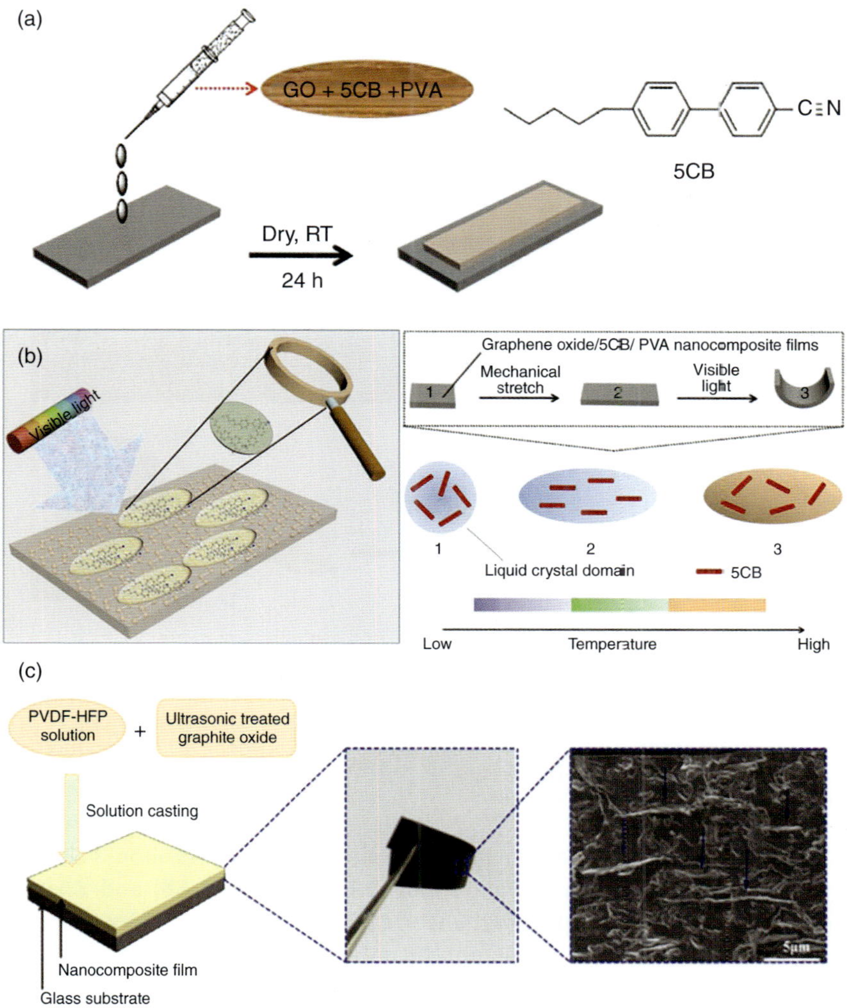

Figure 8.2 Fabrication and properties of the homogeneous single layer films. (a) Nanocomposites of GO-5CB/PVA. (b) The structure and the working principles of the GO-5CB/PVA nanocomposite film. Source: (a,b) Yu et al. 2014 [29]. Copyright 2014. Reproduced with permission of Royal Society of Chemistry. (c) Nanocomposites of GO/PVDF-HFP. Source: Yu and Yu 2015 [27]. Copyright 2015. Reproduced with permission of American Chemical Society.

consisting of gold nanostructures were found to deform and perform self-propelling motions [35].

8.2.2.2 Asymmetric Assembled Films

Bilayer Films with Single Component in Either Layer

SWCNTs/Polycarbonate (PC) and GO/BOPP

Owing to the immiscible feature between organic and inorganic components, the inevitable microphase separation is a challenge to the fabrication of composite material systems [20, 88]. This problem has been elegantly overcome in the bilayer structure with only single component in either layer. However, the

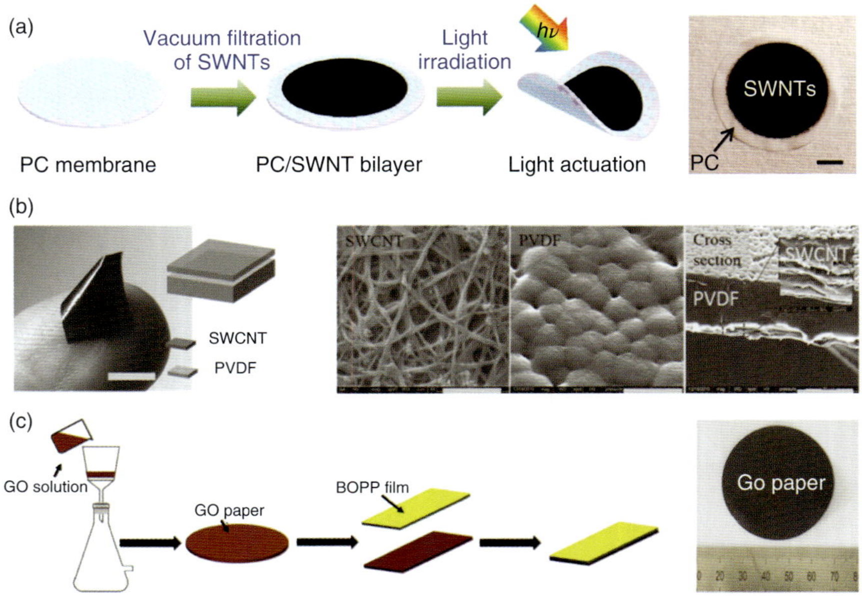

Figure 8.3 Schematic diagrams of the fabrication methods of bilayer films. (a) Nanocomposites of SWCNTs/PC. Source: Zhang et al. 2014 [88]. Copyright 2014. Reproduced with permission of Macmillan Publishers. (b) Nanocomposites of SWCNTs/PVDF. Source: Tai et al. 2016 [84]. Copyright 2016. Reproduced with permission of Wiley-VCH. (c) Nanocomposites of GO/BOPP. Source: Chen et al. 2017 [20]. Copyright 2017. Reproduced with permission of Royal Society of Chemistry.

interfacial interaction between the inorganic and organic layers should be seriously considered. As shown in Figure 8.3a,c, Javey's group [88] and Zhang's group [20] independently reported the fabrication of a bilayer film by adopting an easy and economical way of vacuum filtration. To obtain single-walled carbon nanotubes (SWCNTs)/PC film, the commercially available SWCNT powder was first ultrasonically treated to remove impurities and the residual metal catalyst particles, which was then deposited on a polycarbonate (PC) membrane by vacuum filtration to form a bilayer film. In the fabrication of a GO/BOPP bilayer film, the GO sheet was firstly obtained by vacuum filtration, which was then attached to the biaxial oriented polypropylene (BOPP) film pre-coated with pressure-sensitive adhesive of acrylic ester.

SWCNT/PVDF

As shown in Figure 8.3b, Yang's group reported another method for fabricating bilayer films by directly spin coating on a suitable and separable substrate [84]. When fabricating the SWCNT/PVDF bilayer film, they firstly spin coated one homogeneous PVDF solution on a glass slide and then dried the PVDF layer, which was then treated with plasma to make it hydrophilic. The SWCNT layer was deposited on the PVDF layer by the second spin coating. The bilayer film could be easily peeled from the glass substrate due to the high surface energy of the plasma-treated PVDF.

Multicomponents in one of the Layers

Graphene-chitosan/PE and RGO-CNT/PDMS

Xie's group successfully fabricated the graphene-chitosan/PE bilayer film by choosing chitosan and polyethylene (PE) as the matrix for the sake of material transparency [50, 89]. The GO was synthesized from purchased graphite oxide according to the modified Hummers method, and then the reduced GO (RGO) was obtained through a typical refluxing method [90]. The PE film was chemically treated to become hydrophilic, enabling the homogeneous mixture of chitosan and graphene to be easily coated on the film surface. After drying, the nanocomposite film was peeled from the glass substrate, as shown in Figure 8.4a. Similarly, Chen's group utilized PDMS as the polymer matrix due to its large CTE discrepancy from the RGO-CNT film, as given in Table 8.2 [50]. The RGO-CNT film was prepared by casting their mixed suspension on substrate followed

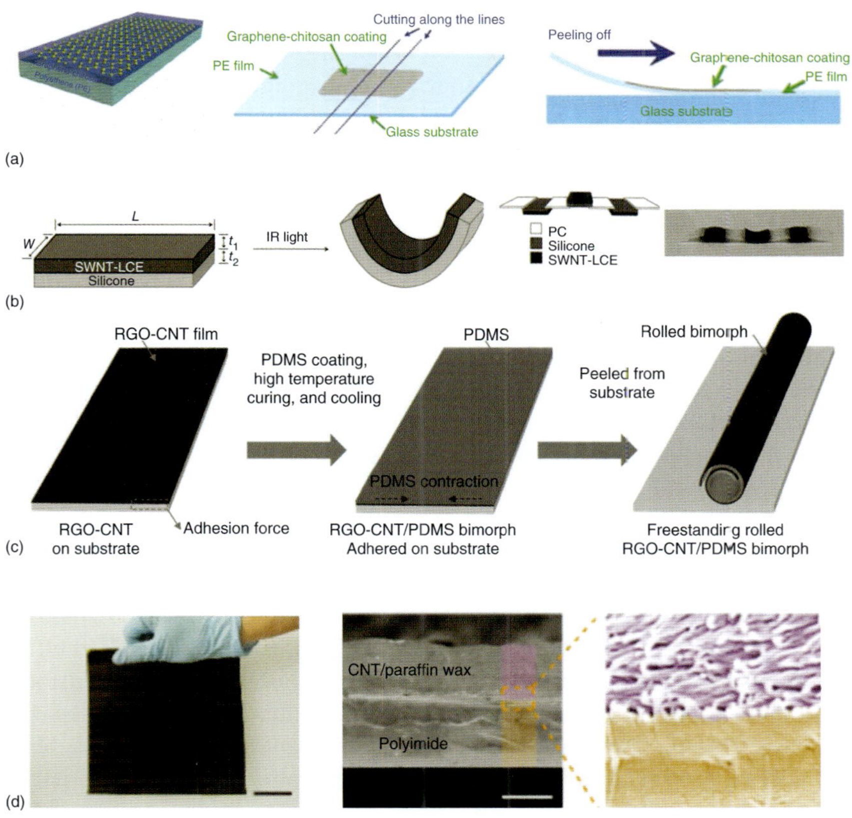

Figure 8.4 Schematic diagrams of the fabrication methods of asymmetric assembled films with multicomponents in one layer. (a) Nanocomposites of Graphene-chitosan/PE. Source: Wu et al. 2011 [89]. Copyright 2011. Reproduced with permission of Royal Society of Chemistry. (b) Nanocomposites of SWCNT-LCEs/silicone. Source: Kohlmeyer et al. 2013 [91]. Copyright 2013. Reproduced with permission of Wiley-VCH. (c) Nanocomposites of RGO-CNT/PDMS. Source: Hu et al. 2015 [50]. Copyright 2015. Reproduced with permission of Wiley-VCH. (d) Nanocomposites of CNT-Paraffin wax/PI. Source: Deng et al. 2016 [15]. Copyright 2015. Reproduced with permission of American Chemical Society.

by evaporating the solvent. Then the viscous PDMS mixture was coated on the RGO-CNT film and cured at a high temperature (Figure 8.4c).

SWCNT-LCE/Silicone and CNT-Paraffin Wax/PI

Since CNTs easily agglomerate when mixing with most organic liquids, it is a big challenge to get well-dispersed and uniform CNTs in matrix [15, 91]. Researchers have developed several methods such as shear-mixing and ultrasonication to avoid the aggregation [92, 93]. One of the elegant ways is to introduce a dispersing agent, which can interact with both the CNT and polymer matrix, making the CNT to homo-disperse in matrix [40]. For instance, poly(*p*-phenyleneethynylene) (PPE)-SWCNTs were synthesized and mixed with LCE monomers, and then the mixture was poured into molds and made to undergo photo-polymerization [91]. The SWCNT-LCE nanocomposites were further composited with silicone rubber slice to fabricate SWCNT-LCE/silicone bilayer film (Figure 8.4b). Another example is CNT-Paraffin wax/PI reported by Peng's group. They firstly deposited the paraffin wax on the famous Kapton (polyimide) film by heating and spin coating [15]. Then the CNT sheet synthesized by chemical vapor deposition was stacked on the bilayer film. Finally, the nanocomposite was reheated to make the CNT merge in the wax to obtain CNT-Paraffin wax/PI bilayer film, as shown in Figure 8.4d.

8.2.3 Properties of Photothermal Actuators

8.2.3.1 Characterization

After the photothermal nanocomposite actuators are fabricated, it is necessary to characterize their surface and interface morphology, degree of orientation, mechanical performances, photoresponsive properties, etc. For measurement of the surface and interface morphology, the most commonly utilized methods are scanning electron microscope (SEM) (Figure 8.2c, Figure 8.3b, and Figure 8.4d) and transmission electron microscope (TEM) [42]. When the composites have birefringent crystal or LC phases, polarizing optical microscope (POM) comes in handy [29]. To test the mesogenic or photothermal agents' orientation degree of LCEs and polymer-dispersed LCs or other ordered systems, the small-angle X-ray scattering (SAXS), two-dimensional (2D) SAXS [15, 39] and polarized ultraviolet–visible absorption spectroscopy (UV–Vis) or polarized Fourier transform infrared spectroscopy (FTIR) techniques are often useful. The dynamic mechanical analysis (DMA) or dynamic mechanical thermal analysis (DMTA) has been widely applied for characterization of the mechanical properties of the nanocomposite films, especially for the SMP-based polymer composite films [21]. Owing to the existence of photothermal agents, it is very important to know the *in-situ* photothermal conversion efficiency in the composites with the assistance of an infrared thermal imager or a thermometer [15, 88]. Finally, the photoresponsive properties, or in other words, the photomechanical performances of the nanocomposites can be studied under different light-irradiation conditions. The main parameters are responsive velocity, deformation displacement, repeatability, and stability.

8.2.3.2 Properties

Photothermal nanocomposite materials mainly respond to NIR and visible light, while the reports on the UV-responsive photothermal actuators [38, 39] are fewer, since the photothermal agents usually have intensive absorption in the NIR region. In practice, visible light or NIR light has an advantage over UV light for its safety and innocuity to humans [37, 40]. Besides, NIR light possesses biological tissue penetration ability and, therefore, exhibits excellent biocompatibility and broad utility in biomedical applications [37]. However, visible light is easily available and cheap.

Generally, the photomechanical nanocomposites based on photothermal effect often respond to visible and NIR light, while those responsive to UV light are often based on photochemical effect. This is decided by the functional additives' response to the wavelength of light. Obviously, if the photothermal agents and photochemical agents are simultaneously integrated into the same matrix, the nanocomposites may realize the NIR–Vis–UV full-range light responsiveness. Multi-wavelength-responsive nanocomposite actuators can be fabricated by a similar method [30, 34, 59]. Thus, we can acquire different shape changes under photoirradiation of different light wavelengths [34, 59]. Since some photothermal agents are sensitive to the wavelength of light, there are some reports about wavelength-selective actuators [88, 91]. One such report mentions that the researchers can modulate the responsive wavelength band of the actuators by choosing CNTs with different chirality distributions [88].

Some photomechanical nanocomposites not only respond to multi-wavelength light, but also to other stimuli, such as temperature [15] and moisture [20, 84]. The deformation of photothermal nanocomposites fabricated by asymmetric bilayer structures relies on the thermal expansion discrepancy between the two layers. As a result, the nanocomposites themselves can directly deform according to environmental temperature changes [15]. In addition to the CTE differential between the two layers, researchers find that the CHE differential of materials also can be used to prepare selective actuating (Table 8.2). For instance, the bilayer actuators made up of GO/BOPP [20] or SWCNT/PVDF [34] can deform by light illumination or by changing environmental moisture content.

The mechanical performance of the photomechanical nanocomposites is also an important parameter to evaluate from the viewpoint of applications. The introduction of carbon nanomaterials such as CNTs or metal nanoparticles into the soft matrix may reinforce the Young's modulus of the nanocomposites to some extent [94]. However, the ductility and durability of the nanocomposites may be influenced by the microphase separation between organic and inorganic matters. This is another challenge in the fabrication of photomechanical nanocomposite materials.

The photoinduced deformation velocity and degree are the basic indexes to evaluate photomechanical nanocomposite materials. The responsive and recovery velocity of photothermal actuators are often decided by several factors, such as the photothermal conversion efficiency of the photothermal agents, the thermal conductivity of each component [40], the CTE discrepancy between the two layers, the modulus of the matrix, and other related elements. Owing to the time cost of the photothermal conversion, the responsive time of the photothermal actuators may range from several milliseconds [35] to few tens of seconds [21]. The deformation extent is generally decided by the photoinduced driving force, the modulus of the matrix, and so on. Usually, photothermal actuators with the maximum bending of 90° can easily realize the functions such as switches [41] and robotic arms [15]. The bending direction of the azobenzene-LCEs is usually phototropic while the bilayer system can easily realize the phototropic and photophobic bending in one film, which originates from its asymmetric structure. To improve the properties of photothermal actuators, researchers may change and modify the photothermal agents or choose a suitable soft matrix. To make the actuators carry out specific work, researchers can design and combine the nanocomposites in a specific way [50, 91].

8.3 Photomechanical Nanocomposites Based on Photochemical Effect

Photochemically induced deformation essentially relies on the microscopic photosensitive chromophores in the composite materials. Over the past decades, much attention has been paid to photochemical strategies such as photoisomerization of spiropyran [95, 96] or azobenzene derivatives [4, 97, 98], photoinduced dimerization of diarylethenes [99], and cycloaddition of anthracene derivatives [100]. As for soft actuators, azobenzene and its derivatives are the most widely utilized photoresponsive moieties for the great changes in molecular conformation and convenient attachment into the polymer chain or hydrogel system. In addition, host–guest inclusion complexes whose interactions break under appropriate wavelength of light irradiation were introduced, especially in photodeformable SMP-based polymer composites. Unlike the photothermal part (see Section 8.2) which is classified according to the structure of nanocomposites, we introduce this part based on the different functions of the additives.

8.3.1 Photodeformable Supramolecular Systems

SMPs or hydrogels are among the most intensively studied stimuli-responsive materials. Generally, the shape-transform "switches" are based on thermally induced effects such as glass transition of polymers or phase transition of LCs. Photoresponsive supramolecular systems were integrated into shape-memory materials to achieve low temperature response and remote and precise control.

8.3.1.1 Reversible Metal–Ligand Coordination

Organic ligands such as amidogen, pyridine, triphenylphosphine, and others can be readily attached to polymer chains [101–103]. When metal ions like Zn^{2+}, Cd^{2+}, and Co^{3+} are integrated into the polymeric system, the metal–ligand coordination can aid cross-linking that reinforces the materials and fixes the shape. Through breaking of these bonds, the nanocomposite materials can acquire the shape-memory capability due to the release of residual stresses.

Rowan and coworker studied end-capped polybutadiene with pyridine groups to form metallo-supramolecular polymers when mixed with the metal salts in solution [39]. Then the cast films were photo-crosslinked through a thiol-ene reaction with tetra-functional thiol to form a permanent shape. Upon irradiation of UV light, the metal–ligand complexes absorbed the light energy and converted it into heat. The localized heating disassembled the coordinated complexes and softened the polybutadiene chain, making the polymers deformable. When removing the UV light, the interaction between the metal ions and the pyridine groups rearranged to fix the temporary shape. Subsequent exposure to continuous UV light released the films to the original shape. Thus, the shape-memory performance was obtained in this way.

Note that the above-mentioned disintegration still relied on photothermal effect, and the photoacid generator (PAG) provided a pure photochemical approach to trigger the shape-memory effect. Yang and coworkers synthesized a slightly cross-linked poly(-acrylamide-*co*-*N*-vinylimidazole) hydrogel [104]. They forced the hydrogels into predesigned shapes and then immersed the hydrogels into solution containing metal ions and diphenyliodonium nitrate (that served as the PAG) without removal of the load. As a result, the imidazole groups interacted with the metal ions and held the temporary shape. The PAG generated protons under UV light irradiation lowering the pH value and further leading to the disassembly of the metal–ligand complexes, which ultimately brought the films back to the original permanent shape (Figure 8.5a).

It is worth pointing out that the PAG strategy can also be applied to polyelectrolyte hydrogels. Aoyagi and coworkers integrated nitrobenzaldehyde as the PAG into bilayer hydrogels composed of a polyacid layer and a polybase layer [105]. Under irradiation of UV light, the PAG accomplished a fast proton release, reducing the pH value, which enables the polyacid layer to contract and makes the polybase layer expand together, leading to a bending deformation, as shown in Figure 8.5b. More appealingly, with short and localized UV irradiation, the gels returned to their original shape after the removal of the light source due to the diffusion of protons and the increase of pH within the whole gels.

8.3.1.2 Interaction Between Cyclodextrin and Azobenzene Derivatives

Cyclodextrin (CD) refers to a series of ring-like sugar molecules that have been extensively utilized in the construction of various supramolecular systems because of the existence of distinctive cavities [106]. Many kinds of stimuli-responsive inclusive guests have been developed including photoactive azobenzene groups. Generally, *trans*-azobenzene has superior complementarity with CD molecules whereas *cis*-azobenzene has a low association with CD molecules because of the bending configuration [107]. Thus, remote light irradiation can switch on or off the interaction between CD and azobenzene derivatives.

Recently, Wu and coworkers reported supramolecular hydrogels based on tetra-*ortho*-methoxy-substituted azobenzene and β-CD (7-membered sugar ring molecule) [108], showing red-light responsive properties. The two components were mixed in dark conditions to fabricate the supramolecular gel. When the obtained gel was exposed to red light (625 nm, 60 mW cm^{-2}) after

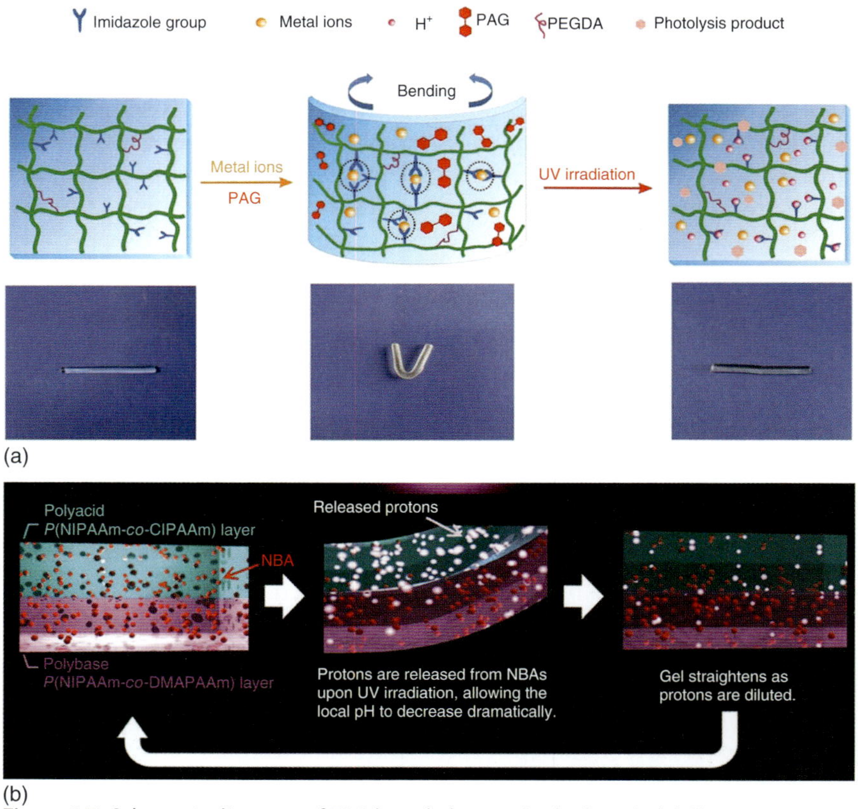

Figure 8.5 Schematic diagrams of PAG-based photoactive hydrogels. (a) Shape-memory effect of hydrogels. Source: Feng et al. 2015 [104]. Copyright 2015. Reproduced with permission of Royal Society of Chemistry. (b) Reversible bending of bilayer gels. Source: Techawanitchai et al. 2012 [105]. Copyright 2012. Reproduced with permission of Royal Society of Chemistry.

30 minutes, gel-to-sol transition occurred because of the triggered disassembly between *cis*-azobenzene and β-CD, as shown in Figure 8.6. The blue-light irradiation or heating can trigger the *cis-to-trans* isomerization of azobenzene, leading to the reverse sol-to-gel transition.

Furthermore, much effort has been made to induce macroscopic deformation in hydrogels beyond the mere gel-to-sol transition. Takashima et al. reported an excellent approach by slight covalent cross-linking of polymer chains containing azobenzene and α-CD (6-membered sugar ring molecule) groups [109]. It was found that the formation of host–guest inclusion complexes altered the total cross-linking density, thus endowing the hydrogel with expansion–contraction ability (Figure 8.7a). The α-CD-azobenzene gel immersed in water expanded approximately 24% in weight under 365 nm UV irradiation and shrank almost to the original state upon exposure to 430 nm visible light. Furthermore, they achieved the away-from-light bending motion in hydrogel strips with several millimeters thickness (Figure 8.7b), because of the poor penetrability of UV light in polymers containing azobenzene.

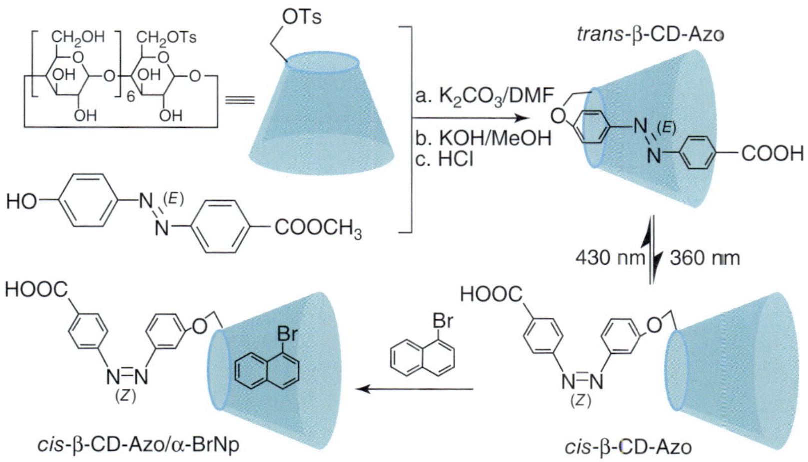

Figure 8.6 Photoactuating of CD-azobenzene supramolecular hydrogels. Illustration of inclusion between β-CD and azobenzene. Source: Wang et al. 2015 [108]. http://pubs.rsc.org/-/content/articlehtml/2015/sm/c5sm01888a. Licensed Under CC BY 3.0.

Li and coworkers designed triple shape-memory hydrogels by integrating the above-mentioned light-responsive azobenzene-CD inclusion complexes with pH-responsive dansyl-aggregations [110]. The initial flat film was deformed into an arc-shape under UV light exposure and then the temporary shape was fixed by visible light irradiation. Once UV light irradiated the deformed film again, the azobenzene–CD inclusion complexes separated and the film returned to the permanent flat shape in 60 minutes. Thus, the light-controlled reversible shape-memory was obtained.

8.3.2 Liquid Crystalline Polymer Nanocomposites

LCs have been intensively investigated ever since their discovery in the late nineteenth century and have been applied in various aspects such as display [111], detection [112], biomedicine [113], and as actuators [48]. LCP materials perform well in stimuli-responsive actuators due to the assembly of mesogens and the order change they undergo when exposed to external stimulus such as light [30], electric field [114], temperature [115], and solvent [116]. The difficulty lies in fixing the orientation of the mesogenic molecules. Finkelmann and coworker provided a two-step cross-linking technique in which polymers were stretched under a slightly cross-linked state and then the alignment was fixed by the second cross-linking reaction [117]. Besides, another method was employed by using pre-assembled LC cells to align the mesogenic monomer capillary filled inside and usually a subsequent photo-polymerization to fix the alignment by chemical cross-linking [48, 118]. As both the approaches are complicated and time-consuming, recent works have aimed to promote several new methods to achieve the orientation through the composite strategy.

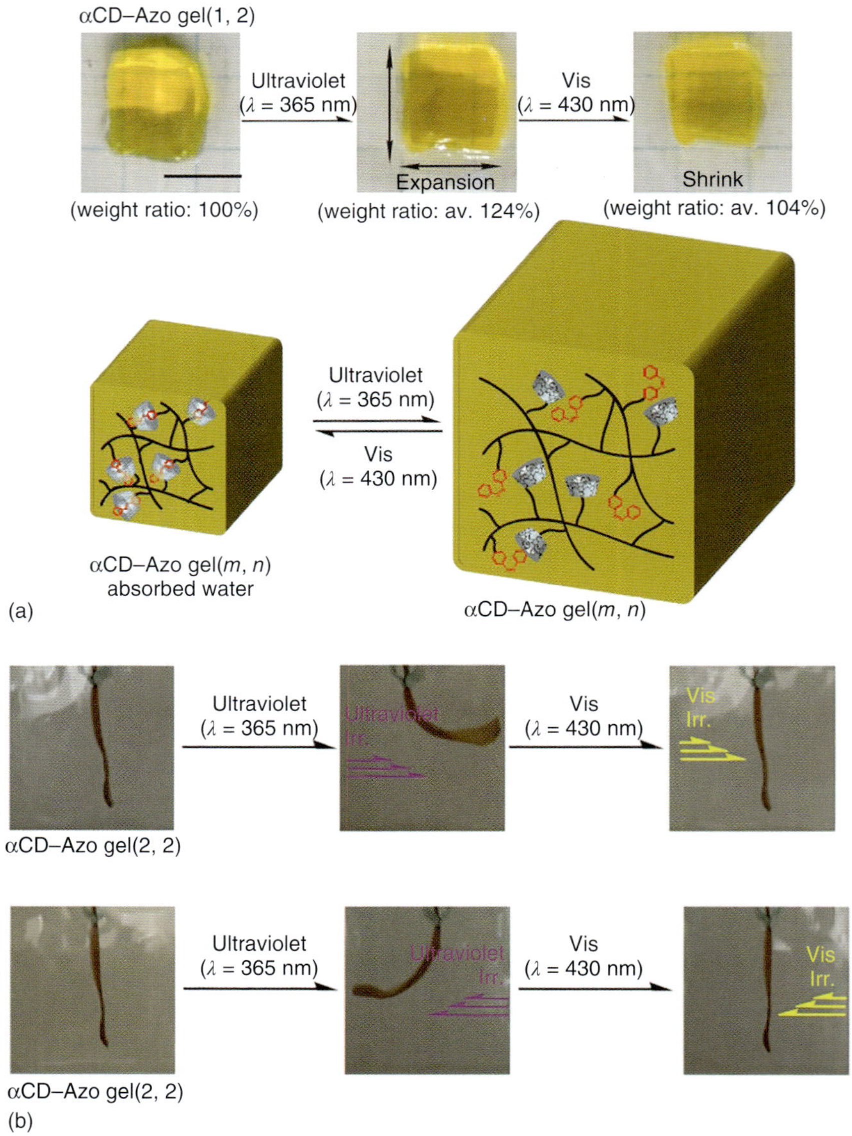

Figure 8.7 Photo actuating of CD-azobenzene supramolecular hydrogels. (a) and (b) Light-induced expansion–contraction and bending. Source: Takashima et al. 2012 [109]. Copyright 2012. Reproduced with permission of Nature Publishing Group.

8.3.2.1 Aligned Carbon Nanotube

CNTs have been widely utilized for their mechanical, electrical, and photothermal properties [119]. However, taking their intrinsic one-dimensional (1D) nanostructure into account, CNTs can be applied directly for the alignment of mesogenic molecules instead of employing a polyimide layer [26]. After injection of LC monomers containing azobenzene mesogens, cross-linkers, and photoinitiators and the following scheduled polymerization, light-responsive

CLCPs were obtained with CNT sheets integrated in both surfaces of the film. It was confirmed that the azobenzene mesogens were aligned along the orientation of the CNT sheets through POM and polarized UV–Vis absorption spectroscopy. Under irradiation of UV light, the *trans-to-cis* isomerization of azobenzenes caused a reduction of the orientational order, leading to a contraction of the irradiated surface layer and eventually inducing a bending motion toward the light source. The bent films reversibly returned to their flat state by exposure to 530 nm visible light. It is worth noting here that the surface CNT sheets reinforced the polymeric film and endowed it with additional electrical conductivity.

Furthermore, Peng and coworkers mixed aligned CNT arrays with the same CLCP [33], but in this condition, polymer chains were attached along the parallel CNTs and the mesogens were aligned perpendicularly in the voids. Surprisingly, this CLCP/CNT composite strip exhibited photodeformable reversibility when UV irradiation was applied alternately on both sides of the strip. They indicated that during the recovery of the strip shape induced by UV light, azobenzene mesogens at the firstly irradiated side experienced *cis-to-trans* isomerization due to the strong π–π interaction between azobenzene molecules and the aligned CNTs.

8.3.2.2 Polymer-Dispersed Hybrid Film

Researches on dispersing photoresponsive microparticles in supporting polymers resulted in providing solutions for the flexible regulation of integral mechanical properties by simply replacing parts of the substrate. This method elegantly provides another advantage that simply stretching the composite membrane can orient the particles inside the composite films. Yu et al. first reported such a kind of photoresponsive polymer-dispersed liquid crystalline polymer (PDLCP) films [26], as shown in Figure 8.8. They prepared wrinkled microparticles with an azobenzene-containing side chain LCP, which were homogeneously dispersed in polyvinyl alcohol (PVA) aqueous solution. After casting the mixture on clean glass substrates and drying for one day, the PVA-based PDLCP composite films were mechanically stretched to orient the azobenzene mesogens. During the evaporation of water, the LCP microparticles gradually concentrated on the upper layer of the films due to the low density of the polymer compared to that of water. Thus, upon irradiation of UV light, the upper layer underwent more contraction resulting in the bending motion toward the light source, as shown in Figure 8.8.

Then the same group developed a similar method by introducing both a photoresponsive azobenzene compound (5CAZ) and a low clearing-point LC (5CB, T_{NI} = 35 °C) together with GO as the nanoscale heating source into the PVA substrate [30], as shown in Figure 8.9. In addition to the UV light, GO can absorb visible and NIR light and converts these into thermal energy, inducing the nematic–isotropic phase transition of 5CB. The thermally induced changes in volume had the same effect as that of the *trans-to-cis* isomerization of 5CAZ. In addition, the responsive speed of photoactuating was much faster than that in the PVA-based PDLCP system (Figure 8.8) since the phase transition occurred in low-viscous small molecules. To obtain reversible actuating properties, Zhou et al.

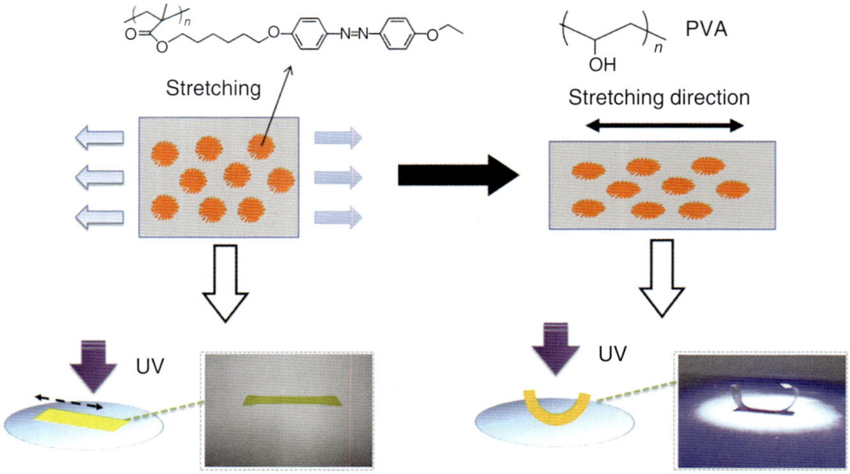

Figure 8.8 Schematic illustration of the mechanical stretching process of the PVA-based PDLCP hybrid films and the distinct UV response of stretched films. Source: Yu et al. 2011 [26]. Reproduced with permission of American Chemical Society.

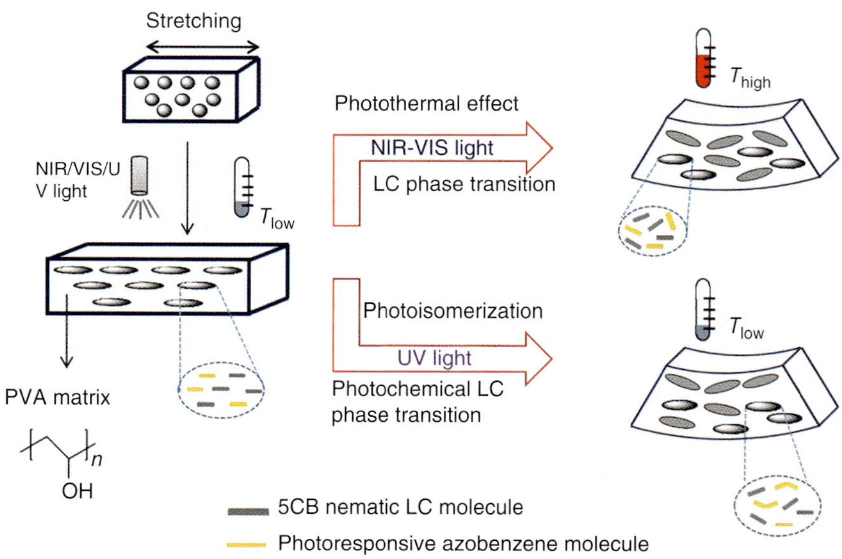

Figure 8.9 NIR–Vis–UV light response of the hybrid films though photothermal and photoisomerization effects. Source: Cheng et al. 2015 [30]. Copyright 2015. Reproduced with permission of American Chemical Society.

replaced the PVA with SMP polyurethane. After UV-light irradiation, the bent PDLC films recovered to the initial flat shape when being heated to 85 °C coinciding with the shape-memory temperature of the pure polyurethane [34, 120].

Interestingly, the stretched nanostructure in polymer films could also orient the mesogens filled inside. Ryabchun et al. utilized stretched porous PE membranes, which anchored the orientation of LCs along the direction of stretching

[28]. They infiltrated polymerizable LC monomers containing an azobenzene mesogen into the porous PE films and obtained a highly cross-linked polymer network after polymerization. Such a bent film induced by asymmetric UV irradiation can be entirely recovered in about two minutes after removal of the light source. The reversibility is dependent on the cross-linked structure and the mechanical property of the PE films.

Besides soft materials, photoinduced deformation was also observed in molecular crystals [121–124]. However, in these systems, actuators were always fragile and it was difficult to control size. Chen and coworker took advantage of the polymer-dispersed strategy to solve these problems [125]. They added molecular crystals of 2-hydroxynaphthylidene-1′-naphthylamine (HNAN) into a PVDF–HFP co-polymer. Upon exposure to UV light, the rod-like nanocrystals showed contraction along the long axis and expansion along the short axis, which reversed when the irradiation of UV light was stopped. They mixed the HNAN and PVDF–HFP in DMF to get a homogeneously dispersed solution and then cast the mixture onto a glass slide. Upon heating to remove DMF, the nanocrystals concentrated in the lower layer and the long axis tended to align along the membrane surface for decreasing the surface free energy during the fabrication process. Finally, the composite film was obtained after removal of the solvent. Interestingly, these microscopic structures determined the bending toward the lower surface regardless of which side was illuminated.

8.3.2.3 Bilayer Composite Film

Bilayer composite film is often utilized to transform the volume change of the LCP film into bending motion or to provide a tough supporting material for acquiring excellent mechanical properties. Current approaches required prefabrication of aligned mesogenic membranes and combination of two layers subsequently through adhesion [55, 126]. To simplify the method further, two steps were integrated by combining a commercially available soft film with photoactive materials [56]. As shown in Figure 8.10, as one typical example, low-density polyethylene (LDPE) was chosen as substrate, which was first polished carefully with silicon carbide sandpaper. The films were immersed into a solution containing small azobenzene molecules to obtain a photosensitive crystal layer and then put into a 40 °C oven for annealing. The microgrooves aligned the azobenzene crystals and simultaneously enhanced the interaction between the two layers. The bilayer films bent to the azobenzene layer upon UV irradiation due to the contraction of this layer along with the *trans-to-cis* isomerization.

8.3.3 Incorporation of Upconversion Nanophosphors

To date, the majority of the reported photoresponsive actuators undergo mechanical deformation upon UV light irradiation, which originates from the intrinsic property of integrated photosensitive molecules. However, UV light has a relatively short wavelength leading to a high energy density and a low penetrability, which is harmful to human health, thus blocking potential biologic applications. Large efforts have been made to develop photomechanical systems

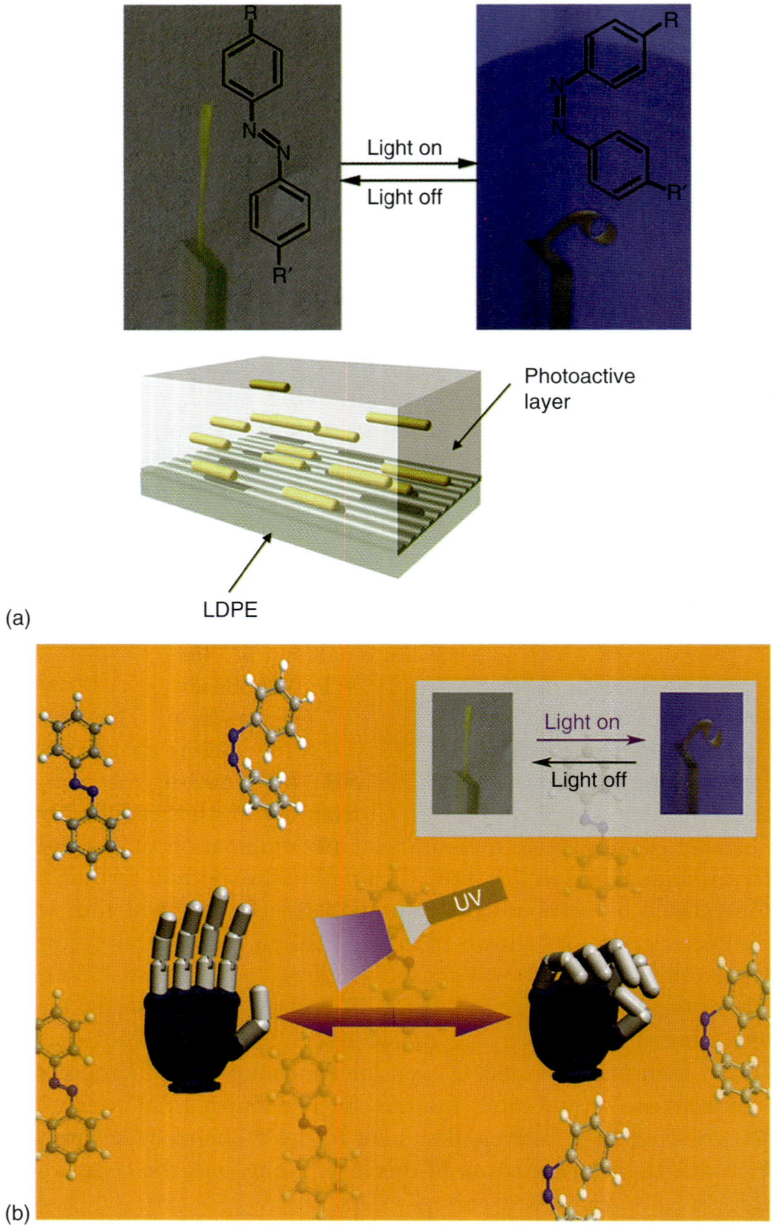

Figure 8.10 (a) Bending behavior toward azo layer and schematic diagrams of molecular distribution. (b) Imitation of grasping utilizing bilayer films. Source: Liu et al. 2015 [56]. Copyright 2015. Reproduced with permission of Wiley-VCH.

triggered by visible or NIR light such as incorporation of photothermal carbon nanomaterials [29, 127, 128] and modification of the substituent [129, 130].

Recently, a novel approach was reported on embedding upconversion nanoparticles into CLCPs. Originally, upconversion nanophosphors (UCNP) composed of NaYF4: Yb, Tm was chosen with emission peaks that matched the absorption band of azotolane groups and then the UCNP was coated on the surface of the homogeneously aligned CLCP film [44]. Upon irradiation with a 980 nm laser, the composite film bent toward the light source and returned to the flat state upon removal of the laser due to the rigid UCNP layer. However, this system required a high irradiation density of 15 W cm^{-2} that caused an overheating effect. Thus, a low-power excited triplet-triplet annihilation-based upconversion system was introduced in the composite system [45]. These composite films were fabricated by adhering upconversion particles containing polyurethane film to aligned CLCP film. Owing to the excellent penetration of the 635 nm excitation light source, the composite film can response to the light source blocked by a piece of pork, indicating its potential in biological applications.

8.4 Applications

Soft matters have always been expected to be promising alternatives to artificial muscle [117]. When they are incorporated with photic driving, various applications such as in actuators, switches [88], oscillators [131], generators [53], and tumblers [27] have been successfully obtained. As shown in Figure 8.11, a biomimetic actuator that simulated the blooming and fading of flowers was realized by taking advantage of triple shape-memory hybrid films with azobenzene and GO dispersed in polyurethane matrix [34]. Upon irradiation of UV light, the prestretched nanocomposite films bent toward the light source, which mimicked the blooming of flowers. When exposed to successive NIR irradiation, the GO served as a nanoscale heating source and triggered the shape-memory behavior of SMP polyurethane. As a result, the films reverted to the original short and flat state as if the flowers faded.

The most basic application of photomechanical soft nanocomposite materials is in actuator fabrication, but the challenge mainly lies in how to achieve

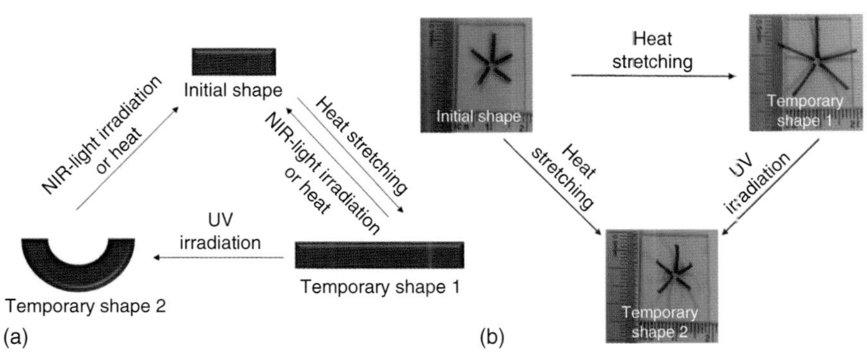

Figure 8.11 (a) Triple shape-memory effect of polyurethane-based hybrid films. (b) Schematic diagrams of biomimetic actuator that simulated the blooming and fading of flowers. Source: Zhou et al. 2016 [34]. Copyright 2016. Reproduced with permission of Royal Society of Chemistry.

more complicated motion than the bending deformation. Anisotropic orientation was utilized to realize complicated motion in LC systems and recent researches even achieved 2D and 3D deformation through introduction of topological defects [132, 133]. However, it is rarely reported in other soft matters. Recently, an anisotropic photothermal actuator was fabricated, which was inspired by hygroscopic seed dispersal in plants [15]. Aligned CNTs were embedded in paraffin wax on a polyimide substrate, and when cut into strips with the long axis parallel or perpendicular to the CNT orientation, the obtained strips showed opposite bending direction under irradiation of visible light. Peng and coworkers also designed a telescopic arm consisting of a pre-shaped helical strip that served as elevator, which underwent untwisting upon light irradiation, and six curving strips that served as claw, as shown in Figure 8.12A.

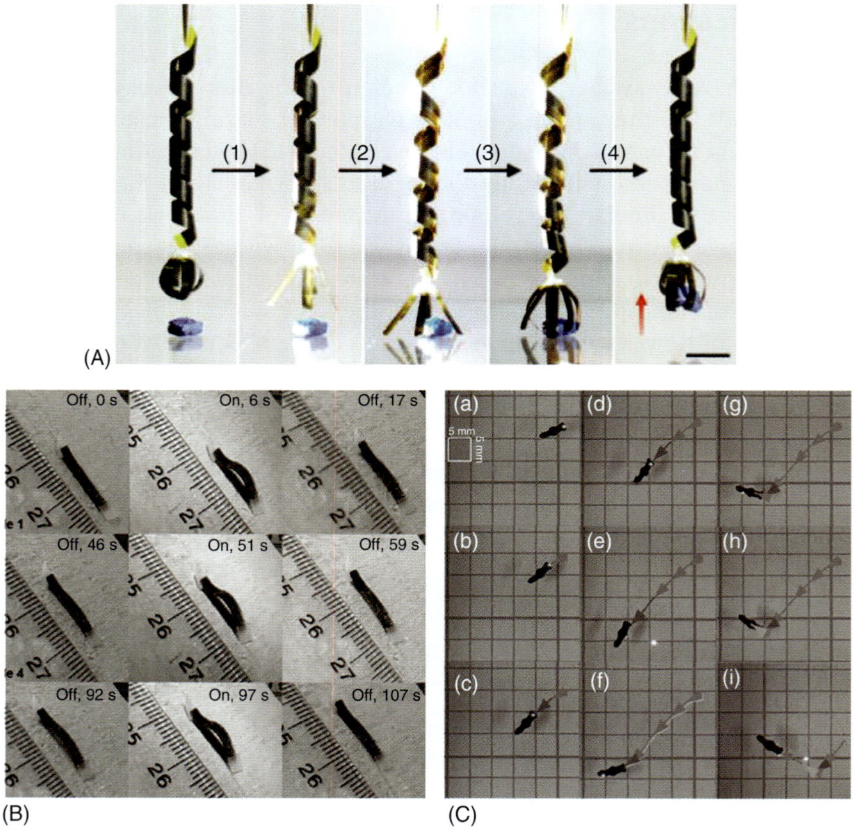

Figure 8.12 (A) Process of the artificial arm picking up an object. Source: Deng et al. 2016 [15]. Copyright 2016. Reproduced with permission of American Chemical Society. (B) The inchworm walker climbing a slope in response to on and off cycles of NIR light. Source: Kohlmeyer et al. 2013 [91]. Copyright 2013. Reproduced with permission of Wiley-VCH. (C) An artificial microfish. By adjusting the position of irradiation spot, the artificial microfish can move ahead or turn around. Source: Jiang et al. 2014 [134]. Copyright 2014. Reproduced with permission of Wiley-VCH.

In addition, actuators with novel designed structures have been developed into more functionalized devices. For example, an IR-light- driven actuator was reported based on bilayer composite materials composed of CNT-doped LCE layer and silicone layer [91]. The actuator can climb a slope in the loop of bending and reverting with the help of two PC films in different shapes (Figure 8.12B). Similarly, Jiang et al. prepared a PDMS platform with graphene nanoplatelets composited in one side [134]. They fabricated a fish-like swimmer that can move forward, backward, and sideward on water (Figure 8.12C).

As the global energy crisis is becoming increasingly serious, photovoltaic materials have been intensively studied as novel energy sources producing electricity energy from light energy. More efforts have also been taken for the same purpose in photomechanical soft nanocomposite materials. As shown in Figure 8.13, an optical pendulum generator was designed with copper coils in the front of a bilayer LCP/LDPE photoactive film [53], as shown in Figure 8.13. Being placed in a magnetic field, the device generated a voltage pulse when the light was turned on and a damped oscillate voltage when the light was turned off according to Faraday's law of electromagnetic induction. Interestingly, continuous electrical output was also achieved by self-shielding the light exposure realizing alternating photoinduced motions of bending and unbending. The fabricated optical pendulum generator showed advantages of simplicity, low cost and stability for coupling the photomechanical energy.

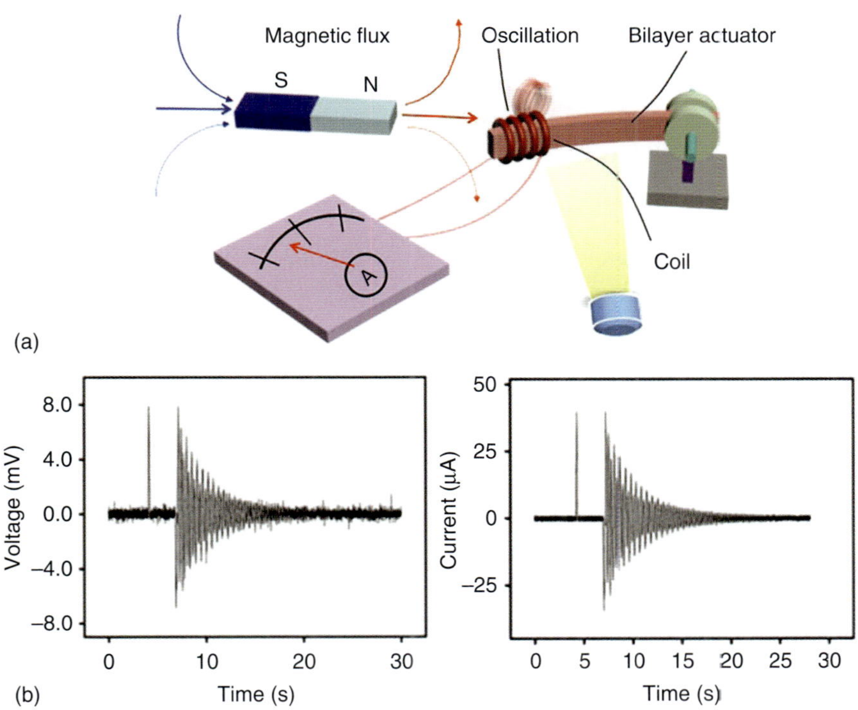

Figure 8.13 (a) Schematic illustration of optical pendulum generator. (b) Light-induced electric pulse signal and the successive damped oscillate. Source: Tang et al. 2015 [53]. Copyright 2015. Reproduced with permission of American Chemical Society.

Peng and coworkers proposed a new strategy by integrating photomechanical actuation and electrostatic effect. They prepared a 1D yarn containing two electrodes. One electrode was aligned CNTs coated with polytetrafluoroethylene (PTFE) and the other was the afore-mentioned CNT/paraffin wax/PI bilayer photoactuator [135]. The yarn was then treated by plasma to produce negative charge on the PTFE surface and then positive charge accumulated on CNT parts in both electrodes due to electrostatic attraction. When light was turned on, the photoinduced bending of yarn changed the distance between the PTFE surface and two CNT electrodes resulting in an electron flow. When the light was turned off, the recovery of distance led to an opposite voltage.

Recently, photoresponsive lamina and flexible GO/PVDF-HFP nanocomposite films were prepared by one simple solution-casting method. Fast, stable, and reversible photomechanical behavior of the nanocomposite films upon irradiation with visible light was observed based on the photothermal effect of GO and the shape-memory effect of the polymer matrix. According to the principle of equilibrium apparatus, light-powered tumbler movement was achieved in these films by imitating the structure of one wobbly man [27], as shown in Figure 8.14. Once the tumbler movement started, the oscillation continued for several hours as long as the visible light was irradiated. It is expected that more complicated movement can be achieved in these photomechanical nanocomposite films for full-plastic devices of advanced functions driven by visible light.

There are still many excellent works that are not mentioned in this chapter [136, 137], but to be honest, all these applications would be far away from being practical. They have shortcomings in mechanical properties, stability, repeatability, manufacturing, cost, and others. However, these shortcomings cannot overshadow the prospect of photomechanical actuators and that is the reason why this aspect attracts so many outstanding researchers.

Figure 8.14 Schematic of working principles: a wobbly man (a) and the light-powered tumbler based on PVDF-HFP/GO nanocomposites (b). Source: Yu and Yu 2015 [27]. Copyright 2015. Reproduced with permission of American Chemical Society.

8.5 Summary and Perspectives

In this chapter, we systematically introduced the fabrication, properties, and applications of photomechanical soft nanocomposite materials from different perspectives. The nanocomposites are on the basis of traditionally studied polymer matrices, such as LCPs, CLCPs, SMPs, hydrogels, supramolecular polymers, and other commercially available polymer films. By doping or asymmetric assembling the functional additives with soft matrices, the fabricated nanocomposites show amazing photomechanical properties. Besides, the nanocomposites usually possess more designability and easier fabrication features than the pure polymer matrices. The nanocomposites perform photodriven motions due to the photothermal or photochemical effect of the functional additives and the synergies between the additives and matrices, indicating their large potential in the photoactuating field.

However, the existing defects in the nanocomposites still could not be ignored. The interface between the composite components, organic and inorganic materials, makes the phases or layer separation of nanocomposites to probably occur, which may intensely influence the service life and application range of the nanocomposites. The requirements of durability, stability, and adaptability to various working conditions are still challengeable for novel composite photoactuators. Other challenges not only for photomechanical nanocomposites but also for other systems are how to enhance their driving force and mechanical strength, and how to utilize their various deformations in real life.

References

1 Küpfer, J. and Finkelmann, H. (1991). Nematic liquid single crystal elastomers. *Macromol. Rapid Commun.* 12: 717.
2 De Gennes, P.G., Hébert, M., and Kant, R. (1997). *Artificial Muscles Based on Nematic Gels.* Hüthig & Wepf Verlag.
3 Finkelmann, H., Nishikawa, E., Pereira, G.G., and Warner, M. (2001). A new opto-mechanical effect in solids. *Phys. Rev. Lett.* 87: 015501.
4 Ikeda, T., Mamiya, J.I., and Yu, Y. (2007). Photomechanics of liquid-crystalline elastomers and other polymers. *Angew. Chem. Int. Ed.* 46: 506.
5 Barrett, C.J., Mamiya, J.I., Yager, K.G., and Ikeda, T. (2007). Photo-mechanical effects in azobenzene-containing soft materials. *Soft Matter* 3: 1249.
6 Bisoyi, H.K. and Li, Q. (2016). Light-driven liquid crystalline materials: from photo-induced phase transitions and property modulations to applications. *Chem. Rev.* 116: 15089.
7 Li, Q. (ed.) (2013). *Intelligent Stimuli-Responsive Materials: From Well-Defined Nanostructures to Applications.* Hoboken, NJ: Wiley.
8 Li, Q. (ed.) (2014). *Nanoscience with Liquid Crystals: From Self-Organized Nanostructures to Applications.* Heidelberg: Springer-Verlag.

9 de Haan, L.T., Verjans, J.M., Broer, D.J. et al. (2014). Humidity-responsive liquid crystalline polymer actuators with an asymmetry in the molecular trigger that bend, fold, and curl. *J. Am. Chem. Soc.* 136: 10585.

10 Zeng, S., Li, R., Freire, S.G. et al. (2017). Moisture-responsive wrinkling surfaces with tunable dynamics. *Adv. Mater.* 29: 1700828.

11 Gong, J., Lin, H., Dunlop, J.W., and Yuan, J. (2017). Hierarchically arranged helical fiber actuators derived from commercial cloth. *Adv. Mater.* 29: 1605103.

12 Hu, Y., Lan, T., Wu, G. et al. (2014). A spongy graphene based bimorph actuator with ultra-large displacement towards biomimetic application. *Nanoscale* 6: 12703.

13 Chambers, M., Finkelmann, H., Remškar, M. et al. (2009). Liquid crystal elastomer-nanoparticle systems for actuation. *J. Mater. Chem.* 19: 1524.

14 Schmauch, M.M., Mishra, S.R., Evans, B.A. et al. (2017). Chained iron microparticles for directionally controlled actuation of soft robots. *ACS Appl. Mater. Interfaces* 9: 11895.

15 Deng, J., Li, J., Chen, P. et al. (2016). Tunable photothermal actuators based on a pre-programmed aligned nanostructure. *J. Am. Chem. Soc.* 138: 225.

16 Yang, R. and Zhao, Y. (2017). Non-uniform optical inscription of actuation domains in a liquid crystal polymer of uniaxial orientation: an approach to complex and programmable shape changes. *Angew. Chem. Int. Ed.* 56: 14202.

17 Wang, L., Gutierrez-Cuevas, K.G., Bisoyi, H.K. et al. (2015). NIR light-directing self-organized 3D photonic superstructures loaded with mesogen-functionalized anisotropic plasmonic nanorods. *Chem. Commun.* 51: 15039.

18 Yang, Z., Wang, Q., and Wang, T. (2016). Dual-triggered and thermally reconfigurable shape memory graphene-vitrimer composites. *ACS Appl. Mater. Interfaces* 8: 21691.

19 Maeda, S., Hara, Y., Sakai, T. et al. (2007). Self-walking gel. *Adv. Mater.* 19: 3480.

20 Chen, L., Weng, M., Zhou, P. et al. (2017). Multi-responsive actuators based on a graphene oxide composite: intelligent robot and bioinspired applications. *Nanoscale* 9: 9825.

21 Li, Y., Rios, O., Keum, J.K. et al. (2016). Photoresponsive liquid crystalline epoxy networks with shape memory behavior and dynamic ester bonds. *ACS Appl. Mater. Interfaces* 8: 15750.

22 Ji, Y., Marshall, J.E., and Terentjev, E.M. (2012). Nanoparticle-liquid crystalline elastomer composites. *Polymers* 4: 316.

23 Wang, L., Bisoyi, H.K., Zheng, Z. et al. (2017). Self-organized chiral liquid crystalline superstructures for adaptive windows enabled by mesogen-functionalized graphene. *Mater. Today* 20: 230.

24 Wang, L. and Li, Q. (2016). Stimuli-directing self-organized 3D liquid crystalline nanostructures: from materials design to photonic applications. *Adv. Funct. Mater.* 26: 10.

25 Wang, L. and Li, Q. (2018). Photochromism into nanosystems: towards lighting up the future nanoworld. *Chem. Soc. Rev.* 47: 1044.

26 Yu, H., Dong, C., Zhou, W. et al. (2011). Wrinkled liquid-crystalline microparticle-enhanced photoresponse of PDLC-like films by coupling with mechanical stretching. *Small* 7: 3039.

27 Yu, L. and Yu, H. (2015). Light-powered tumbler movement of graphene oxide/polymer nanocomposites. *ACS Appl. Mater. Interfaces* 7: 3834.
28 Ryabchun, A., Bobrovsky, A., Stumpe, J., and Shibaev, V. (2012). Novel generation of liquid crystalline photo-actuators based on stretched porous polyethylene films. *Macromol. Rapid Commun.* 33: 991.
29 Yu, L., Cheng, Z., Dong, Z. et al. (2014). Photomechanical response of polymer-dispersed liquid crystals/graphene oxide nanocomposites. *J. Mater. Chem. C* 2: 8501.
30 Cheng, Z., Wang, T., Li, X. et al. (2015). NIR–Vis–UV light-responsive actuator films of polymer-dispersed liquid crystal/graphene oxide nanocomposites. *ACS Appl. Mater. Interfaces* 7: 27494.
31 Camacho-Lopez, M., Finkelmann, H., Palffy-Muhoray, P., and Shelley, M. (2004). Fast liquid-crystal elastomer swims into the dark. *Nat. Mater.* 3: 307.
32 Wang, W., Sun, X., Wu, W. et al. (2012). Photoinduced deformation of crosslinked liquid-crystalline polymer film oriented by a highly aligned carbon nanotube sheet. *Angew. Chem. Int. Ed.* 51: 4644.
33 Sun, X., Wang, W., Qiu, L. et al. (2012). Unusual reversible photomechanical actuation in polymer/nanotube composites. *Angew. Chem. Int. Ed.* 51: 8520.
34 Zhou, L., Liu, Q., Lv, X. et al. (2016). Photoinduced triple shape memory polyurethane enabled by doping with azobenzene and GO. *J. Mater. Chem. C* 4: 9993.
35 Mourran, A., Zhang, H., Vinokur, R., and Möller, M. (2017). Soft microrobots employing nonequilibrium actuation via plasmonic heating. *Adv. Mater.* 29: 1604825.
36 Zhou, Y., Hauser, A.W., Bende, N.P. et al. (2016). Waveguiding microactuators based on a photothermally responsive nanocomposite hydrogel. *Adv. Funct. Mater.* 26: 5447.
37 Guo, L., Liu, M., Sayed, S.M. et al. (2016). A calamitic mesogenic near-infrared absorbing croconaine dye/liquid crystalline elastomer composite. *Chem. Sci.* 7: 4400.
38 Wu, Y., Hu, J., Zhang, C. et al. (2015). A facile approach to fabricate a UV/heat dual-responsive triple shape memory polymer. *J. Mater. Chem. A* 3: 97.
39 Kumpfer, J.R. and Rowan, S.J. (2011). Thermo-, photo-, and chemo-responsive shape-memory properties from photo-cross-linked metallo-supramolecular polymers. *J. Am. Chem. Soc.* 133: 12866.
40 Marshall, J.E., Ji, Y., Torras, N. et al. (2012). Carbon-nanotube sensitized nematic elastomer composites for IR-visible photo-actuation. *Soft Matter* 8: 1570.
41 Ge, F., Lu, X., Xiang, J. et al. (2017). An optical actuator based on gold-nanoparticle-containing temperature-memory semicrystalline polymers. *Angew. Chem. Int. Ed.* 56: 6126.
42 Yang, H., Liu, J.J., Wang, Z.F. et al. (2015). Near-infrared-responsive gold nanorod/liquid crystalline elastomer composites prepared by sequential thiol-click chemistry. *Chem. Commun.* 51: 12126.
43 Qin, C., Feng, Y., Luo, W. et al. (2015). A supramolecular assembly of cross-linked azobenzene/polymers for a high-performance light-driven actuator. *J. Mater. Chem. A* 3: 16453.

44 Wu, W., Yao, L., Yang, T. et al. (2011). NIR-light-induced deformation of cross-linked liquid-crystal polymers using upconversion nanophosphors. *J. Am. Chem. Soc.* 133: 15810.
45 Jiang, Z., Xu, M., Li, F., and Yu, Y. (2013). Red-light-controllable liquid-crystal soft actuators via low-power excited upconversion based on triplet-triplet annihilation. *J. Am. Chem. Soc.* 135: 16446.
46 Castagna, R., Nucara, L., Simoni, F. et al. (2017). An unconventional approach to photomobile composite polymer films. *Adv. Mater.* 29: 1604800.
47 Hosono, N., Kajitani, T., Fukushima, T. et al. (2010). Large-area three-dimensional molecular ordering of a polymer brush by one-step processing. *Science* 330: 808.
48 Yamada, M., Kondo, M., Miyasato, R. et al. (2008). Photomobile polymer materials-various three-dimensional movements. *J. Mater. Chem.* 19: 60.
49 Tang, R., Sang, W., Wu, Y. et al. (2017). Multi-wavelength light drivable oscillatory actuator on graphene-based bilayer film. *Macromol. Mater. Eng.* 302: 1600384.
50 Hu, Y., Wu, G., Lan, T. et al. (2015). A graphene-based bimorph structure for design of high performance photoactuators. *Adv. Mater.* 27: 7867.
51 Naka, Y., Mamiya, J.I., Shishido, A. et al. (2011). Direct fabrication of photomobile polymer materials with an adhesive-free bilayer structure by electron-beam irradiation. *J. Mater. Chem.* 21: 1681.
52 Tian, Q., Tang, M., Sun, Y. et al. (2011). Hydrophilic flower-like CuS superstructures as an efficient 980 nm laser-driven photothermal agent for ablation of cancer cells. *Adv. Mater.* 23: 3542.
53 Tang, R., Liu, Z., Xu, D. et al. (2015). Optical pendulum generator based on photomechanical liquid-crystalline actuators. *ACS Appl. Mater. Interfaces* 7: 8393.
54 Zhong, H.Y., Chen, L., Yang, R. et al. (2017). Azobenzene-containing liquid crystalline polyester with π–π interactions: diverse thermo-and photo-responsive behaviours. *J. Mater. Chem. C* 5: 3306.
55 Wen, H., Zhang, W., Weng, Y., and Hu, Z. (2014). Photomechanical bending of linear azobenzene polymer. *RSC Adv.* 4: 11776.
56 Liu, Z., Tang, R., Xu, D. et al. (2015). Precise actuation of bilayer photomechanical films coated with molecular azobenzene chromophores. *Macromol. Rapid Commun.* 36: 1171.
57 Commins, P., Desta, I.T., Karothu, D.P. et al. (2016). Crystals on the move: mechanical effects in dynamic solids. *Chem. Commun.* 52: 13941.
58 Hu, Y., Li, Z., Lan, T., and Chen, W. (2016). Photoactuators for direct optical-to-mechanical energy conversion: from nanocomponent assembly to macroscopic deformation. *Adv. Mater.* 28: 10548.
59 Wang, M., Lin, B.P., and Yang, H. (2016). A plant tendril mimic soft actuator with phototunable bending and chiral twisting motion modes. *Nat. Commun.* 7: 13981.
60 Yu, Y., Nakano, M., and Ikeda, T. (2003). Photomechanics: directed bending of a polymer film by light. *Nature* 425: 145.
61 Lv, J.A., Liu, Y., Wei, J. et al. (2016). Photocontrol of fluid slugs in liquid crystal polymer microactuators. *Nature* 537: 179.

62 Gelebart, A.H., Vantomme, G., Meijer, E.W., and Broer, D.J. (2017). Mastering the photothermal effect in liquid crystal networks: a general approach for self-sustained mechanical oscillators. *Adv. Mater.* 29: 1606712.
63 Camargo, C.J., Campanella, H., Marshall, J.E. et al. (2012). Batch fabrication of optical actuators using nanotube-elastomer composites towards refreshable braille displays. *J. Micromech. Microeng.* 22: 075009.
64 Liu, Z., Robinson, J.T., Tabakman, S.M. et al. (2011). Carbon materials for drug delivery & cancer therapy. *Mater. Today* 14: 316.
65 Markovic, Z.M., Harhaji-Trajkovic, L.M., Todorovic-Markovic, B.M. et al. (2011). In vitro comparison of the photothermal anticancer activity of graphene nanoparticles and carbon nanotubes. *Biomaterials* 32: 1121.
66 Yang, K., Zhang, S., Zhang, G. et al. (2010). Graphene in mice: ultrahigh in vivo tumor uptake and efficient photothermal therapy. *Nano Lett.* 10: 3318.
67 Yuan, Z., Xiao, X., Li, J. et al. (2018). Self-assembled graphene-based architectures and their applications. *Adv. Sci.* 5: 1700626.
68 Li, Q. (ed.) (2016). *Nanomaterials for Sustainable Energy*. Heidelberg: Springer-Verlag.
69 Li, Q. (ed.) (2015). *Anisotropic Nanomaterials: Preparation, Properties, and Applications*. Heidelberg: Springer-Verlag.
70 Tian, Q., Jiang, F., Zou, R. et al. (2011). Hydrophilic Cu_9S_5 nanocrystals: a photothermal agent with a 25.7% heat conversion efficiency for photothermal ablation of cancer cells in vivo. *ACS Nano* 5: 9761.
71 Xue, C. and Li, Q. (2015). Gold nanorods. In: *Encyclopedia of Surface and Colloid Science*, 3e. Taylor & Francis.
72 Gutierrez-Cuevas, K.G., Wang, L., Xue, C. et al. (2015). Near infrared light-driven liquid crystal phase transition enabled by hydrophobic mesogen grafted plasmonic gold nanorods. *Chem. Commun.* 51: 9845.
73 Xue, C., Birel, O., Gao, M. et al. (2012). Perylene monolayer-protected gold nanorods: unique optical, electronic properties and self-assemblies. *J. Phys. Chem. C* 116: 10396.
74 Xue, C., Xue, Y., Dai, L. et al. (2013). Size and shape dependent fluorescence quenching of gold nanoparticles on perylene dye. *Adv. Opt. Mater.* 1: 581.
75 Bisoyi, H.K. and Li, Q. (2014). *Liquid Crystals. Kirk-Othmer Encyclopedia of Chemical Technology*. New York: Wiley.
76 Pei, Z., Yang, Y., Chen, Q. et al. (2013). Mouldable liquid-crystalline elastomer actuators with exchangeable covalent bonds. *Nat. Mater.* 13: 3812.
77 Zheng, N., Fang, Z., Zou, W. et al. (2016). Thermoset shape-memory polyurethane with intrinsic plasticity enabled by *transcarbamoylation. Angew. Chem. Int. Ed.* 55: 11421.
78 Turnbull, D. and Cohen, M.H. (1961). Free-volume model of the amorphous phase: glass transition. *J. Chem. Phys.* 34: 120.
79 Ikeda, T., Nakano, M., Yu, Y. et al. (2003). Anisotropic bending and unbending behavior of azobenzene liquid-crystalline gels by light exposure. *Adv. Mater.* 15: 201.
80 Li, M.H., Keller, P., Li, B. et al. (2003). Light-driven side-on nematic elastomer actuators. *Adv. Mater.* 15: 569.

81 Kondo, M., Yu, Y., and Ikeda, T. (2006). How does the initial alignment of mesogens affect the photoinduced bending behavior of liquid-crystalline elastomers? *Angew. Chem. Int. Ed.* 45: 1378.

82 Timoshenko, S. (1925). Analysis of bi-metal thermostats. *J. Opt. Soc. Am.* 11: 233.

83 Amjadi, M. and Sitti, M. (2016). High-performance multiresponsive paper actuators. *ACS Nano* 10: 10202.

84 Tai, Y., Lubineau, G., and Yang, Z. (2016). Light-activated rapid-response polyvinylidene-fluoride-based flexible films. *Adv. Mater.* 28: 4665.

85 Hummers, W.S. Jr. and Offeman, R.E. (1958). Preparation of graphitic oxide. *J. Am. Chem. Soc.* 80: 1339.

86 Kovtyukhova, N.I., Ollivier, P.J., Martin, B.R. et al. (1999). Layer-by-layer assembly of ultrathin composite films from micron-sized graphite oxide sheets and polycations. *Chem. Mater.* 11: 771.

87 Spence, G.T., Hartland, G.V., and Smith, B.D. (2013). Activated photothermal heating using croconaine dyes. *Chem. Sci.* 4: 4240.

88 Zhang, X., Yu, Z., Wang, C. et al. (2014). Photoactuators and motors based on carbon nanotubes with selective chirality distributions. *Nat. Commun.* 5: 2983.

89 Wu, C., Feng, J., Peng, L. et al. (2011). Large-area graphene realizing ultrasensitive photothermal actuator with high transparency: new prototype robotic motions under infrared-light stimuli. *J. Mater. Chem.* 21: 18584.

90 Wang, G., Yang, J., Park, J. et al. (2008). Facile synthesis and characterization of graphene nanosheets. *J. Phys. Chem. C* 112: 8192.

91 Kohlmeyer, R.R. and Chen, J. (2013). Wavelength-selective, IR light-driven hinges based on liquid crystalline elastomer composites. *Angew. Chem. Int. Ed.* 52: 9234.

92 Koshio, A., Yudasaka, M., Zhang, M., and Iijima, S. (2001). A simple way to chemically react single-wall carbon nanotubes with organic materials using ultrasonication. *Nano Lett.* 1: 361.

93 Paredes, J.I. and Burghard, M. (2004). Dispersions of individual single-walled carbon nanotubes of high length. *Langmuir* 20: 5149.

94 Coleman, J.N., Khan, U., Blau, W.J., and Gun'ko, Y.K. (2006). Small but strong: a review of the mechanical properties of carbon nanotube-polymer composites. *Carbon* 44: 1624.

95 Whelan, J., Abdallah, D., Wojtyk, J., and Buncel, E. (2010). Micro-environmental fine-tuning of electronic and kinetic properties of photochromic dyes. *J. Mater. Chem.* 20: 5727.

96 Nakahara, Y., Okazaki, Y., and Kimura, K. (2012). Self-assembling phenomena of benzo-18-crown-6/crowned-spirobenzopyran copolymers in aqueous solution and controlled release of entrapped organic dyes using external stimuli. *Soft Matter* 8: 3192.

97 Van Oosten, C.L., Bastiaansen, C.W., and Broer, D.J. (2009). Printed artificial cilia from liquid-crystal network actuators modularly driven by light. *Nat. Mater.* 8: 677.

98 Ahn, S.K., Ware, T.H., Lee, K.M. et al. (2016). Photoinduced topographical feature development in blueprinted azobenzene-functionalized liquid crystalline elastomers. *Adv. Funct. Mater.* 26: 5819.

99 der Molen, S.J. and ávan Wees, B.J. (2006). Uni- and bi-directional light-induced switching of diarylethenes on gold nanoparticles. *Chem. Commun.* 34: 3597.

100 Shi, J. and Tang, C.W. (2002). Anthracene derivatives for stable blue-emitting organic electroluminescence devices. *Appl. Phys. Lett.* 80: 3201.

101 Vermonden, T., van Steenbergen, M.J., Besseling, N.A. et al. (2004). Linear rheology of water-soluble reversible neodymium(III) coordination polymers. *J. Am. Chem. Soc.* 126: 15802.

102 Kim, H.J., Lee, J.H., and Lee, M. (2005). Stimuli-responsive gels from reversible coordination polymers. *Angew. Chem.* 117: 5960.

103 Nair, K.P., Breedveld, V., and Weck, M. (2011). Multiresponsive reversible polymer networks based on hydrogen bonding and metal coordination. *Macromolecules.* 44: 3346.

104 Feng, W., Zhou, W., Zhang, S. et al. (2015). UV-controlled shape memory hydrogels triggered by photoacid generator. *RSC Adv.* 5: 81784.

105 Techawanitchai, P., Ebara, M., Idota, N. et al. (2012). Photo-switchable control of pH-responsive actuators via pH jump reaction. *Soft Matter* 8: 2844.

106 Cao, J., Ma, X., Min, M. et al. (2014). INHIBIT logic operations based on light-driven β-cyclodextrin pseudo[1]rotaxane with room temperature phosphorescence addresses. *Chem. Commun.* 50: 3224.

107 Ma, X., Cao, J., Wang, Q., and Tian, H. (2011). Photocontrolled reversible room temperature phosphorescence (RTP) encoding β-cyclodextrin pseudorotaxane. *Chem. Commun.* 47: 3559.

108 Wang, D., Wagner, M., Butt, H.J., and Wu, S. (2015). Supramolecular hydrogels constructed by red-light-responsive host-guest interactions for photo-controlled protein release in deep tissue. *Soft Matter* 11: 7656.

109 Takashima, Y., Hatanaka, S., Otsubo, M. et al. (2012). Expansion-contraction of photoresponsive artificial muscle regulated by host-guest interactions. *Nat. Commun.* 3: 1270.

110 Xiao, Y.Y., Gong, X.L., Kang, Y. et al. (2016). Light-, pH- and thermal-responsive hydrogels with the triple-shape memory effect. *Chem. Commun.* 52: 10609.

111 Kokawa, S., Nishihara, M., and Sato, Y. (1995). U.S. Patent and Trademark Office. US Patent 5,467,208. Washington, DC.

112 Schadt, M., Seiberle, H., and Schuster, A. (1996). Optical patterning of multi-domain liquid-crystal displays with wide viewing angles. *Nature* 381: 212.

113 Woltman, S.J., Jay, G.D., and Crawford, G.P. (2007). Liquid-crystal materials find a new order in biomedical applications. *Nat. Mater.* 6: 929.

114 Xie, X., Qu, L., Zhou, C. et al. (2010). An asymmetrically surface-modified graphene film electrochemical actuator. *ACS Nano* 4: 6050.

115 Miaudet, P., Derré, A., Maugey, M. et al. (2007). Shape and temperature memory of nanocomposites with broadened glass transition. *Science* 318: 1294.

116 Lee, W.E., Jin, Y.J., Park, L.S., and Kwak, G. (2012). Fluorescent actuator based on microporous conjugated polymer with intramolecular stack structure. *Adv. Mater.* 24: 5604.

117 Wermter, H. and Finkelmann, H. (2001). Liquid crystalline elastomers as artificial muscles. *e-Polymers* 1: 111.

118 Harris, K.D., Bastiaansen, C.W., and Broer, D.J. (2006). A glassy bending-mode polymeric actuator which deforms in response to solvent polarity. *Macromol. Rapid Commun.* 27: 1323.

119 Jorio, A., Kauppinen, E., and Hassanien, A. *Carbon-Nanotube Metrology*, 2007. Berlin, Heidelberg: Springer-Verlag.

120 Pang, X., Xu, B., Qing, X. et al. (2018). Photo-induced bending behavior of post-crosslinked liquid crystalline polymer/polyurethane blend films. *Macromol. Rapid Commun.* 39: 1700237.

121 Kobatake, S., Takami, S., Muto, H. et al. (2007). Rapid and reversible shape changes of molecular crystals on photoirradiation. *Nature* 446: 778.

122 Morimoto, M. and Irie, M. (2010). A diarylethene cocrystal that converts light into mechanical work. *J. Am. Chem. Soc.* 132: 14172.

123 Zhu, L., Al-Kaysi, R.O., and Bardeen, C.J. (2011). Reversible photoinduced twisting of molecular crystal microribbons. *J. Am. Chem. Soc.* 133: 12569.

124 Koshima, H., Takechi, K., Uchimoto, H. et al. (2011). Photomechanical bending of salicylideneaniline crystals. *Chem. Commun.* 47: 11423.

125 Lan, T. and Chen, W. (2013). Hybrid nanoscale organic molecular crystals assembly as a photon-controlled actuator. *Angew. Chem. Int. Ed.* 52: 6496.

126 Lu, X., Guo, S., Tong, X. et al. (2017). Tunable photocontrolled motions using stored strain energy in malleable azobenzene liquid crystalline polymer actuators. *Adv. Mater.* 29: 1606467.

127 Yang, L., Setyowati, K., Li, A. et al. (2008). Reversible infrared actuation of carbon nanotube-liquid crystalline elastomer nanocomposites. *Adv. Mater.* 20: 2271.

128 Yang, Y., Zhan, W., Peng, R. et al. (2015). Graphene-enabled superior and tunable photomechanical actuation in liquid crystalline elastomer nanocomposites. *Adv. Mater.* 27: 6376.

129 Bushuyev, O.S., Tomberg, A., Friščić, T., and Barrett, C.J. (2013). Shaping crystals with light: crystal-to-crystal isomerization and photomechanical effect in fluorinated azobenzenes. *J. Am. Chem. Soc.* 135: 12556.

130 Liu, Y., Wu, W., Wei, J., and Yu, Y. (2016). Visible light responsive liquid crystal polymers containing reactive moieties with good processability. *ACS Appl. Mater. Interfaces* 9: 782.

131 Kumar, K., Knie, C., Bléger, D. et al. (2016). A chaotic self-oscillating sunlight-driven polymer actuator. *Nat. Commun.* 7: 11975.

132 Ware, T.H., McConney, M.E., Wie, J.J. et al. (2015). Voxelated liquid crystal elastomers. *Science* 347: 982.

133 McConney, M.E., Martinez, A., Tondiglia, V.P. et al. (2013). Topography from topology: photoinduced surface features generated in liquid crystal polymer networks. *Adv. Mater.* 25: 5880.

134 Jiang, W., Niu, D., Liu, H. et al. (2014). Photoresponsive soft-robotic platform: biomimetic fabrication and remote actuation. *Adv. Funct. Mater.* 24: 7598.

135 Yu, X., Pan, J., Deng, J. et al. (2016). A novel photoelectric conversion yarn by integrating photomechanical actuation and the electrostatic effect. *Adv. Mater.* 28: 10744.

136 Głowacki, E., Horovitz, K., Tang, C.W., and Marshall, K.L. (2010). Photoswitchable gas permeation membranes based on liquid crystals. *Adv. Funct. Mater.* 20: 2778.

137 Wang, E., Desai, M.S., and Lee, S.W. (2013). Light-controlled graphene-elastin composite hydrogel actuators. *Nano Lett.* 13: 2826.

9

Photoresponsive Polyolefins

Shaji Varghese, John R. Severn, and Albertus P. H. J. Schenning

Eindhoven University of Technology, Stimuli Responsive Functional Materials and Devices, Department of Chemical Engineering and Chemistry, Het Kranenveld, 5600 MB, Eindhoven, the Netherlands

9.1 Introduction

Nature is unmatched in its esthetics and precisely controlled complexity of biochemical structures of life and the way they function. Nature's sophisticated functional hierarchical structures are mostly made from biopolymers like polypeptides and polysaccharides, which, over time, have evolved to respond to various adverse external stimuli at levels ranging from molecular to macroscopic. The responses include change of color, shape, porosity, and stiffness [1, 2]. To cite a few, Cuttlefishes have mastered the art of color change for purposes of camouflage or as a way of warning against potential predators [3]. *Mimosa pudica* is a tropical plant that undergoes change in leaf orientation by closing and opening its foliage with response to touch and light [4]. The Venus flytrap, an insectivorous plant, can close its leaves to catch insects that touch its trigger hairs [5]. A vital organ in the human body, the heart, undergoes constant beats at the rate of 60–100 bpm, triggered by electrical impulses (Figure 9.1) [8]. Most of the man-made polymers on the other hand are robust, but lack the stimuli-responsive functional properties seen in their natural counterparts. Many of these synthetic polymers are extensively used in our day-to-day life, but modern nanotechnology and biomedical applications, for example, require functionally active smart polymers.

The most commonly used synthetic commodity polymers are polyolefins. But, just as most known synthetic polymers, functional properties of polyolefins are static. The challenge is to develop a commercially viable stimuli-responsive polymer material, based on commodity polymers such as polyolefins. If we can make it respond to external stimuli, like light, such photoresponsive materials will have a wide range of applications [9, 10] in optical storage, sensors, actuators, artificial muscle, and soft-robotics, etc. This chapter will discuss briefly the photoresponsive polymeric systems in general and the need for photoresponsive commodity polymers, followed by recent developments of photoresponsive

Photoactive Functional Soft Materials: Preparation, Properties, and Applications, First Edition.
Edited by Quan Li.

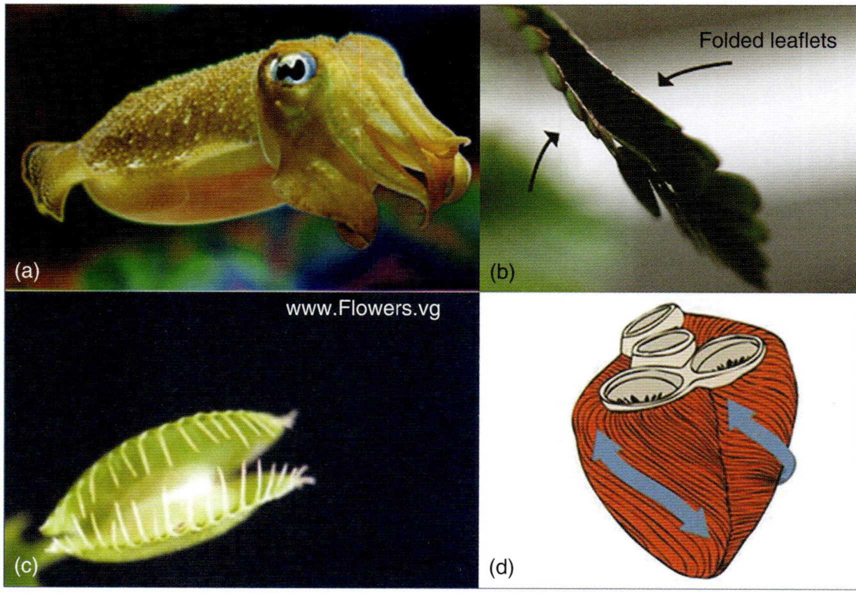

Figure 9.1 Stimuli-responsive systems found in Nature. Pictures of (a) a Cuttlefish; (b) a *Mimosa pudica*. Source: Su et al. 2015 [6]. Copyright 2014. Reproduced with permission from Wiley-VCH Verlag GmbH & Co. (c) A Venus flytrap. Source: (a,c) Zhai 2013 [1]. Copyright 2013. Reproduced with permission from Royal Society of Chemistry. and (d) Arrangement of fibers in the human heart. Source: Roche et al. 2014 [7]. Copyright 2013. Reproduced with permission from Wiley-VCH Verlag GmbH & Co.

polyolefins with suitable examples. Furthermore, the challenges and future directions of advancing the same are also discussed.

9.2 Photoresponsive Polymers

Photo-activation with a specific wavelength is one of the most commonly adopted external triggers for changing the properties of stimuli-responsive materials and also for the development of photoactuators and other micro-devices as it is very convenient to apply remotely, cheap, and easy to control intensity and wavelength. Such photo-activation of polymeric materials generally requires a photoresponsive chromophore incorporated in them. Azobenzene, spiropyran, and dithienylethene are examples of molecules that can reversibly change their chemical structure upon light excitation and are also the most commonly used light-responsive chromophores in stimuli (photo)-responsive polymers and materials. For example, azobenzene undergoes light-induced reversible isomerization with large geometrical changes between its thermally stable *trans* form with length of 0.9 nm and a metastable *cis* form with length of only 0.55 nm [11]. When this molecule is incorporated into a special polymeric network, it induces wavelength-controlled macroscopic deformation. Such photoresponsive materials can convert light energy

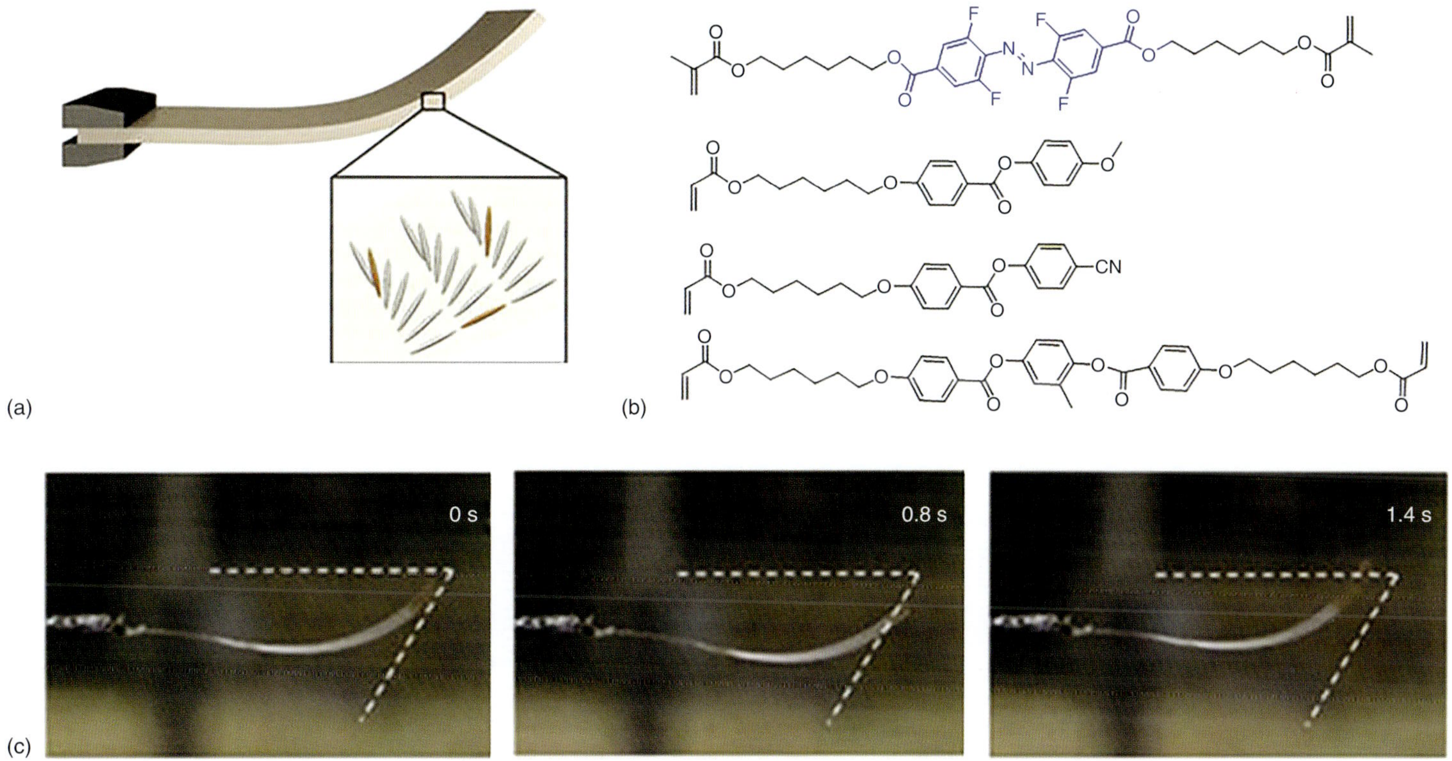

Figure 9.2 Sunlight-driven oscillatory motion of an LC polymer film. (a) Schematic of the LC splay aligned film (gray) containing fluoro-azobenzene molecules (orange). (b) Chemical structures of monomers and cross-linkers used for the preparation of LC network films. (c) Series of snapshots showing splay-oriented (homeotropic side on top) film oscillations. Source: Kumar et al. 2016 [18]. Reproduced with permission from Springer Nature.

into mechanical energy and demonstrate reversible three-dimensional shape changes, photomechanical actuators, etc. and a variety of such responsive systems have been reported during the past decade [12–14]. However, most of them are based on photoresponsive chromophore-embedded hydrogels and liquid crystalline networks (LCNs) [15, 16]. LCNs combine the property of polymer networks and the properties of liquid crystals (LC) with prominent anisotropic expansion. Azobenzene- related LCNs are very popular photoresponsive materials that undergo wavelength-controlled reversible shape changes due to their cooperative molecular effect and have found wide range of applications in nonlinear optics, holography, actuators, soft-robotics, etc. This wide exploration of azobenzene based LCNs is due to the greater control of relative degree of order or phase associated with macroscopic deformation by light of appropriate wavelength [17]. The response process is caused by a small reversible change of the local molecular order. In case of a uniaxial molecular orientation, this leads to shrinkage in the direction along the molecular director while expansion will occur in all other directions. This disorder can be induced by molecular switches such as azobenzenes. Altering the chemical structure by the derivatization of the azobenzene core was studied in order to use different wavelengths of light ranging from UV to visible light; these modified structures also influence the photochemical properties of LCNs. One recent example of such an advanced LCN-based photomechanical actuator was developed by Kumar et al. [18] to demonstrate the ability of utilizing sun light, the most abundantly available source of energy, to create a self-sustained oscillatory motion (Figure 9.2). This system used an engineered azobenzene moiety by ortho-substitution, which absorbs the light from the visible region of the electromagnetic radiation. This ortho-fluoro azobenzene polymer actuator has the potential for directly converting the solar energy into mechanical work, in addition to creating autonomous oscillatory function.

9.3 Need for Non-liquid Crystalline or Commodity Polymeric Materials

Tremendous progress has been made in the fabrication of photoresponsive hydrogels and LCNs with the development of next generation micro-devices that are capable of mimicking natural movements [19, 20]. However, the photomechanical work output or light energy conversion efficiency is not high. One possible reason for this may be due to the low elastic modulus and low strength [21]. For the future, for commercial development of light-controlled micro-devices and soft-robotics greater photomechanical force output with minimal expense is required as the mechanical work output is an important performance tool for the polymer actuators or micro-devices. The work output of natural muscle is a common efficiency indicator to compare the performance of these devices; for example, mammalian skeleton muscle generally produces 0.35 MPa stress and 40% strain with the work density of 40 $kJ\,m^{-3}$ and power to mass ratio of 284 $W\,kg^{-1}$ at its highest performance [22, 23]. Figure 9.3 demonstrates the work efficiency of various responsive materials including LC materials, shape-memory alloy and

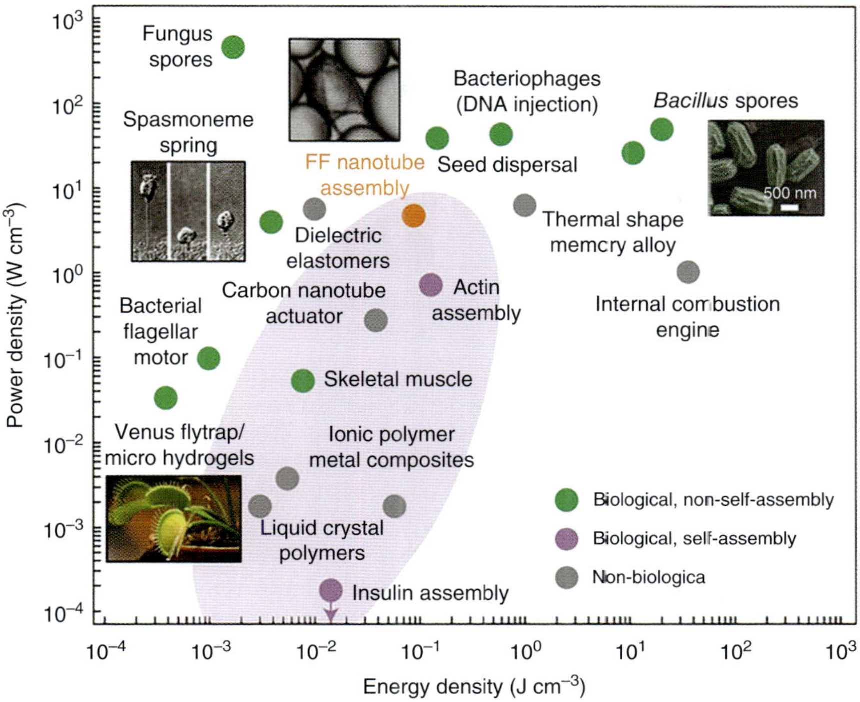

Figure 9.3 Comparison of energy (J cm^{-3}) and power (W cm^{-3}) densities generated by biological and responsive synthetic systems. Source: Levin et al. 2016 [24]. Copyright 2016. Reprinted with permission from Nature Publishing Group.

biological systems (e.g. natural muscles), which clearly shows the need for new materials for the responsive applications with maximum work output [24]. To improve the work efficiency, dielectric elastomers have been demonstrated with the potential of creating artificial muscles [25]. An alternative strategy that is both cost-effective and direct, is to use responsive flexible semi-crystalline commodity plastic materials [26].

High photo-mechanical work output can be obtained by photoresponsive semi-crystalline polymeric materials such as polyolefins, because of their flexibility and high Young's modulus. Since mechanical energy is directly related to the Young's modulus, using such polymeric systems is one way to improve work output significantly [25]. Additionally, these polymers are also available in large quantities at a cheaper price. However, it should be noted that the strains of these materials are low, which can be overcome by careful device design. Research in this direction requires renewed effort although the first reported photoresponsive material is based on such a category of polymer. In 1966, Merian observed that a semi-crystalline polymer fiber, nylon dyed with an azobenzene derivative shrank upon photoirradiation [27]. Although the observed contraction was very small (0.1%), his research laid the foundation for azobenzene-based photoresponsive materials. The photoinduced contraction in this case may have been a combined effect of *trans–cis* isomerization and photothermal effect or just the latter. This work was followed by Eisenbach

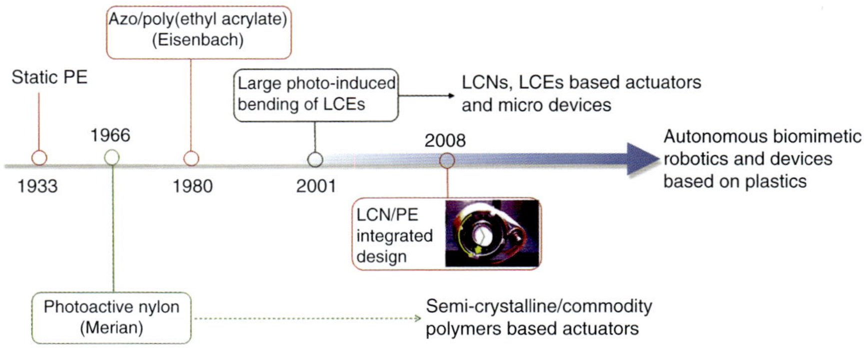

Figure 9.4 Timeline showing the evolution of static plastic to responsive plastic objects with the possibility of creating light-driven soft-robotics and micro-devices with autonomous control of complex movements. Inserted image of light-driven plastic motor. Source: Yamada et al. 2008 [29]. Copyright 2008. Reproduced with permission from Wiley-VCH Verlag GmbH & Co.

in 1980 who demonstrated improved responsive nature when azobenzene-functionalized poly(ethyl acrylate) was used. The improvement, however, was not very remarkable as the photoinduced contraction reached only ~0.25% [28]. The plastic objects evolved from static to next generation complex responsive materials (Figure 9.4) during the last decade after the discovery of large photoinduced bending of liquid crystalline elastomers (LCEs) in 2001 [17, 30]. The development in this direction includes the semi-crystalline-based polymers as well [31].

9.4 Polyolefins

Polyolefins can range from ultra-rigid thermosets to high-performance elastomers via all conceivable thermoplastic and elastoplastic materials in between. As a result they account for more than 55% of the global commodity plastic materials demand (Figure 9.5). Yet, if one looks at their chemical composition, polyolefins are surprisingly limited: polyethylene (PE), polypropylene (PP), a few copolymers of ethylene with propene or another α-olefin, and little else [33]. This success has come from a deep understanding of the synthesis and processing of these polymers at a multiple-length scale, from chain topology to crystalline morphology and orientation. The combination the know-how of this structure property, cost of the materials, and base monomer means that a polyolefin solution often finds itself as the preferred option, from a price/performance balance. Consequently, polyolefins are employed in a wide range of applications such as infrastructure (water and gas piping and energy transmission); advanced packaging, light and reliable packaging that reduce transport emissions while increasing the self-life of perishable goods; and automotive application replacing metal with light material that allows for lighter automobiles and niche areas such as biomedical material and high-strength ultra-high molecular weight PE (UHMW-PE) fibers. Highly oriented UHMW-PE fibers are

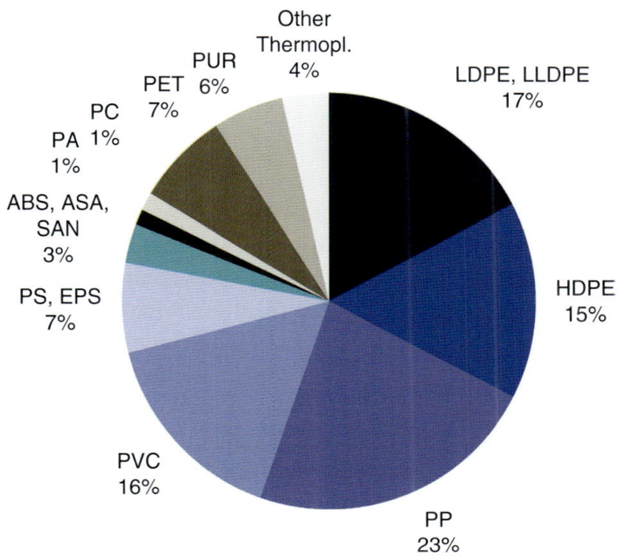

Figure 9.5 World plastic materials demand by resin types 2015 [32] (Plastics Europe Market Research Group (PEMRG) 2015). Low-density PE (LDPE), high-density PE (HDPE), polypropylene (PP), polyvinyl chloride (PVC), polystyrene (PS), polyethylene terephthalate (PET), and polyurethane (PUR). Source: PlasticsEurope Market Research Group (PEMRG)/ Consultic Marketing & Industrieberatung GmbH.

special types of linear polyolefins containing very long chain PE with an average molecular weight over 1000 kg mol^{-1} and crystallinity of more than 90%, which makes them impressive polymers with highly remarkable mechanical properties such as high Young's modulus (>200 GPa), high abrasion resistance, high impact strength, and low coefficient of friction. Owing to these unique qualities of UHMW-PE fibers, they are highly suitable for high-performance applications such as protective clothing and gloves, fishing lines, marine ropes, anti-ballistics, prosthetic implants, surgical sutures, etc. [34]. Dyneema®, is one of the commercial brand of highly aligned UHMW-PE fiber from DSM and is already marketing such high-performance products [35]. The intriguing question we can ask here, is, can we make these static material into functional systems that can mimic nature? In other words, is it possible to make a responsive polyolefin for advanced applications? To answer this question, we highlight some of the impressive research done in that direction.

9.5 Photoresponsive Polyolefins

Polyolefins have good mechanical properties, but the functional properties of their products remain static to light. In order to make photoresponsive polyolefin-based micro-devices, researchers combined well-known responsive properties of LC materials with polyolefins. One notable earlier approach was to make composite bilayer film of LCs and PE or PP. PE layer in these responsive films act as the low-cost, flexible base or supportive layer to provide good mechanical strength. Some of the work based on such bilayer composite layer films is summarized further.

9.5.1 Bilayer Actuators

Ikeda's light-driven plastic motor device is one of the first examples that used PE bilayer layer films based designs. This plastic motor is based on a composite film consisting of a light-responsive, azobenzene-based LCN layer laminated on unstretched low-density PE (LDPE) sheet [29]. The resultant bilayer sheet is connected to both ends to form a ring, which in turn is attached to a homemade pulley system as a belt in such a way that the azobenzenes are aligned in the ring direction (Figure 9.6) facing the light source. When the prepared system is irradiated by UV light (365 nm, 240 mW cm^{-2}) from the top right and by visible light (>500 nm, 120 mW cm^{-2}) from top left simultaneously, the plastic belt is rotated, which in turn drives the two pulleys in a counterclockwise direction at room temperature. This is the first plastic device that was reported to be able to convert the light energy into mechanical work in the form of continuous rotation in one direction. They have explained the mechanism of the light-controlled, unidirectional movement as follows: a local contraction force was generated along the alignment direction of the azobenzene mesogen upon UV irradiation due to the *trans* to *cis* isomerization of azobenzene, which is parallel to the long axis of the ring. This contraction force acted on the right pulley to rotate in the counterclockwise direction. At the same time, irradiation by visible light produced a local expansion force due to the *cis* to *trans* isomerization, which caused a counterclockwise rotation of the left pulley. These continuous contraction and

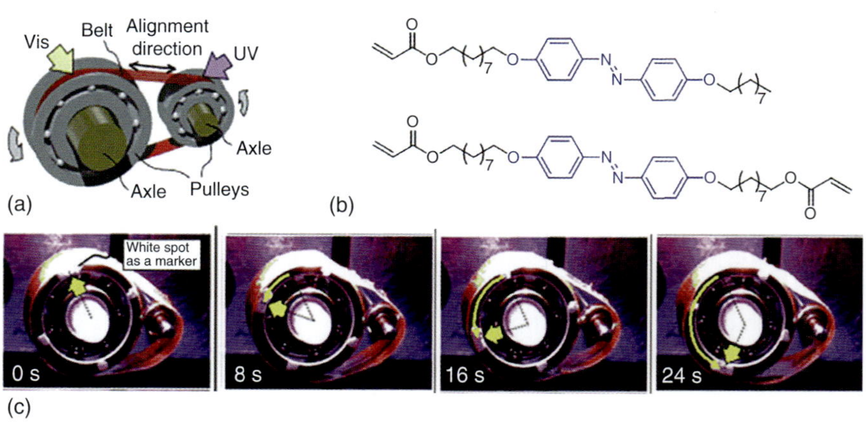

Figure 9.6 (a) Schematic diagram of the light-driven plastic motor system. (b) Chemical structures of the azobenzene mono- and diacrylates used for the preparation of LCN films. (c) Series of images showing the rotation of the light-activated plastic motor with LCN laminated film induced by simultaneous UV and visible light irradiation at distinct position of the film at room temperature. This led to the continuous motion of the motor. Source: Yamada et al. 2008 [29]. Copyright 2008. Reproduced with permission from Wiley-VCH Verlag GmbH & Co.

expansion forces generated at different parts along the long axis of the belt, resulting in the rotation of the pulleys with the ring was unidirectional. Although the PE layer did not contribute directly to the photoresponsiveness, it provided better mechanical strength and flexibility to the system.

The bilayer composition of a flexible PE (or PP) film and an LCN was also used to demonstrate light-driven micro-devices such as: (i) a simple device that moved like a "robotic arm"; (ii) a device that can walk in one direction like an "inch-worm"; (iii) a soft micro-robot that can pick, lift, move, and place objects; and (iv) a plastic wheel with tunable rolling (or moving) direction and speed. The detailed design and working principle of such photoresponsive devices are described further.

Ikeda and coworkers have also developed a "robotic arm" based on a laminated film, which consists of photoresponsive azobenzene LCN and a flexible unstretched PE layer [36]. As shown in Figure 9.7, the laminated film contains LCN layers at two places with the azobenzenes aligned along the long axis. This system shows sequential and flexible motion upon continuous irradiation of the

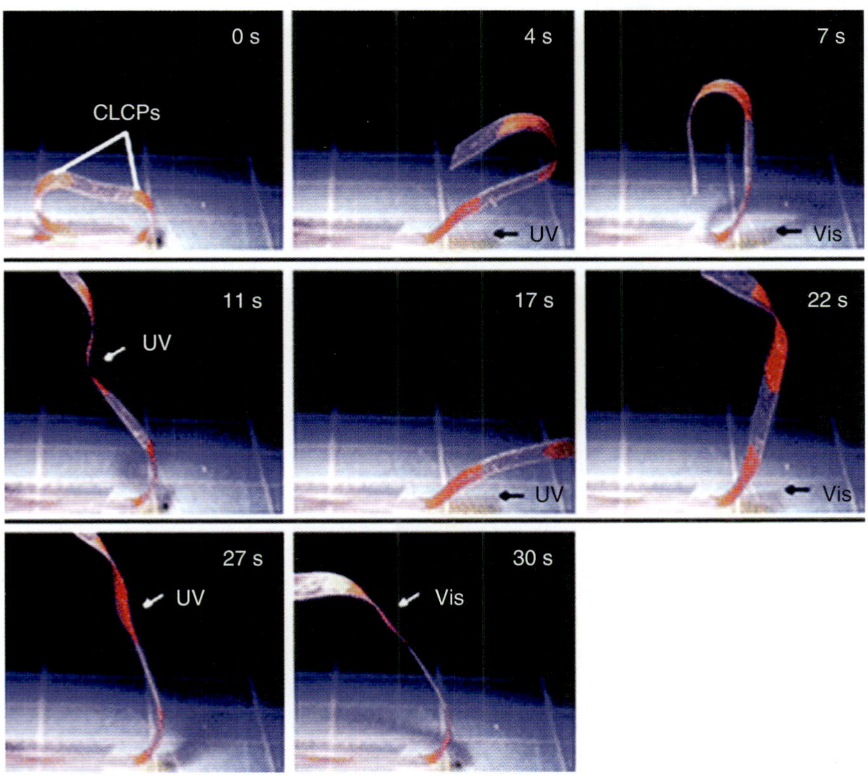

Figure 9.7 Light-driven robotic arms: series of images showing time profiles of the flexible robotic arm motion of the LCN laminated film induced by irradiation with UV (366 nm, 240 mW cm^{-2}) and visible light (>540 nm, 120 mW cm^{-2}) at room temperature. Arrows indicate the direction of photoirradiation. Thickness of the PE layers is 50 μm and that of the LCN layers is 16 μm. Source: Yamada et al. 2009 [36]. Copyright 2008. Reprinted with permission from Royal Society of Chemistry.

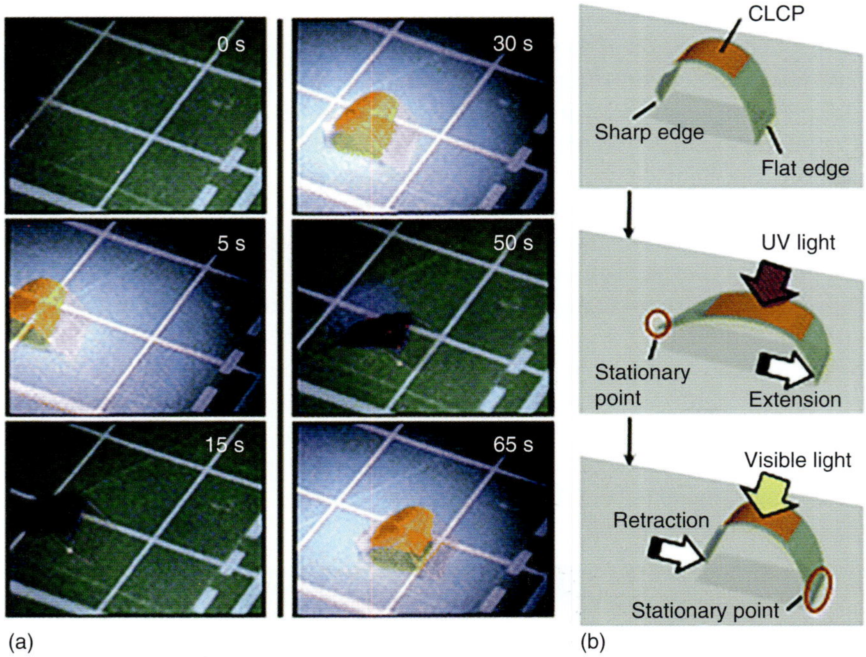

Figure 9.8 (a) Series of images showing time profiles of the light-induced inchworm walk of the LCN laminated film by alternate irradiation with UV (366 nm, 240 mW cm^{-2}) and visible light (>540 nm, 120 mW cm^{-2}) at room temperature. The film moved on the plate with 1 cm × 1 cm grid. (b) Schematic illustrations showing a plausible mechanism of the photoinduced inchworm walk of the LCN laminated film. Upon exposure to UV light, the film extends forward because the sharp edge acts as a stationary point, and the film retracts from the rear side by irradiation with visible light because the flat edge acts as a stationary point. Thickness of the PE layer is 50 μm and that of the LCN layer is 18 μm. Source: Yamada et al. 2009 [36]. Copyright 2008. Reprinted with permission from Royal Society of Chemistry.

UV and visible light. A bent laminated film is based on a similar bilayer design but the LCN layer at the central part shows lateral movement by alternate irradiation with UV and visible light, which mimics the walk of an inchworm [36]. The plastic inchworm had an asymmetric end-shape design to make such photoresponsive directional movement (Figure 9.8). Upon alternate irradiation with UV and visible light at room temperature, the sharp and flat edges of the plastic inchworm acted as the stationary points, respectively. As a result, the inchworm walked unidirectionally.

In addition to the controlled thermal back isomerization, the main advantage of azobenzene-based photoactive material is that the substitution pattern can result in difference in absorption and, therefore, a modified azobenzene can be used for visible light or even direct sunlight as the energy source. Yu and coworkers designed and synthesized a new LCN/PE bilayer composite film that contains longer conjugated azotolane structure in the side chain (Figure 9.9b) as the photo-reversible chromophore in the LCN layer, transfers visible light directly into mechanical energy [37]. Using this approach they have demonstrated a visible-light-driven soft micro-robot, which was manipulated to pick, lift, move,

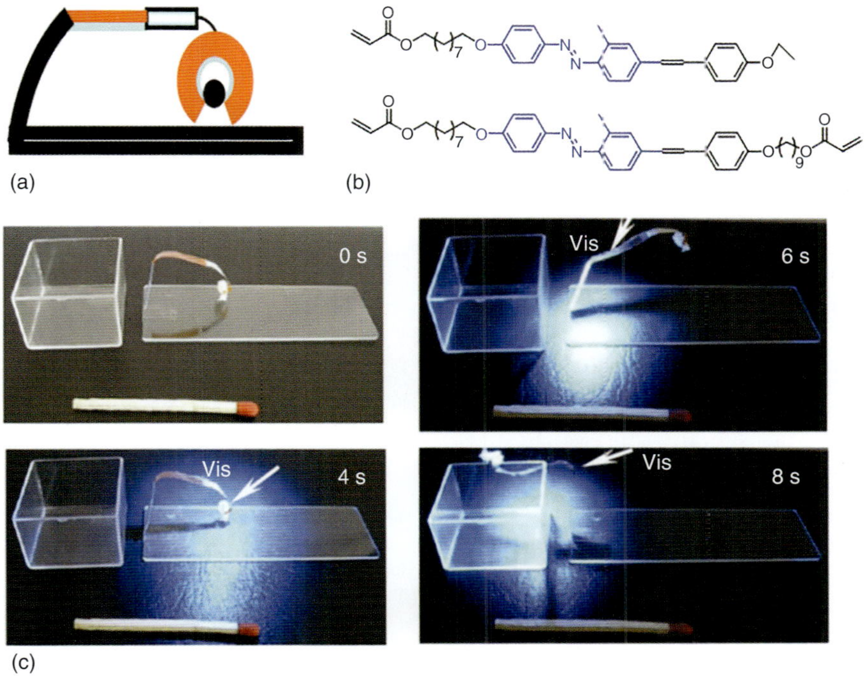

Figure 9.9 (a) Schematic illustrations of the states of the micro-robot during the process of manipulating the object. (b) Chemical structures of monomer and cross-linkers used for the preparation of photoresponsive LC layer. (c) Images showing the micro-robot that moves an object through its head to a container by irradiation with visible light (470 nm, 30 mW cm^{-2}). Length of the match in the pictures: 50 mm. Thickness of PE and LCN films: 12 μm. Object weight: 1 mg. White arrows denote the parts irradiated with visible light. Source: Cheng et al. 2010 [37]. Copyright 2010. Reprinted with permission from Royal Society of Chemistry.

and place milligram-scale objects by light irradiation at different part as shown in Figure 9.9. Opening and closing motion of the hand part of this plastic robot was achieved simply by switching the light on and off instead of changing the wavelength of the light used. This is because the bilayer curvature can be reverted to its original flat shape by turning off the light.

Lu et al. fabricated a light-driven wheel based on a laminated bilayer composite film composed of azobenzene-containing LCE layer and a transparent PP layer of thickness around 45 μm [38]. Carefully laminated architectural design by controlled setting of the LCEs film as the outer or inner layer of the bilayer wheels enabled the wheels to roll away from and toward the light source, respectively on a flat surface (Figure 9.10). They have explained the mechanism of the movement as the nonuniform bending of the LCE layers, which can create a curvature of the laminated film by incident UV light. This asymmetric deformation of the wheel is considered to be the driving force of the rotatory movement. The speed of the rotatory movement can also be controlled by adjusting the strain energy stored in the film. For example, film with more energy stored in it moves faster. To achieve this, the wheels were fabricated with LCEs films stretched to either 200% or 100% strain before lamination. They have also demonstrated that the speed of

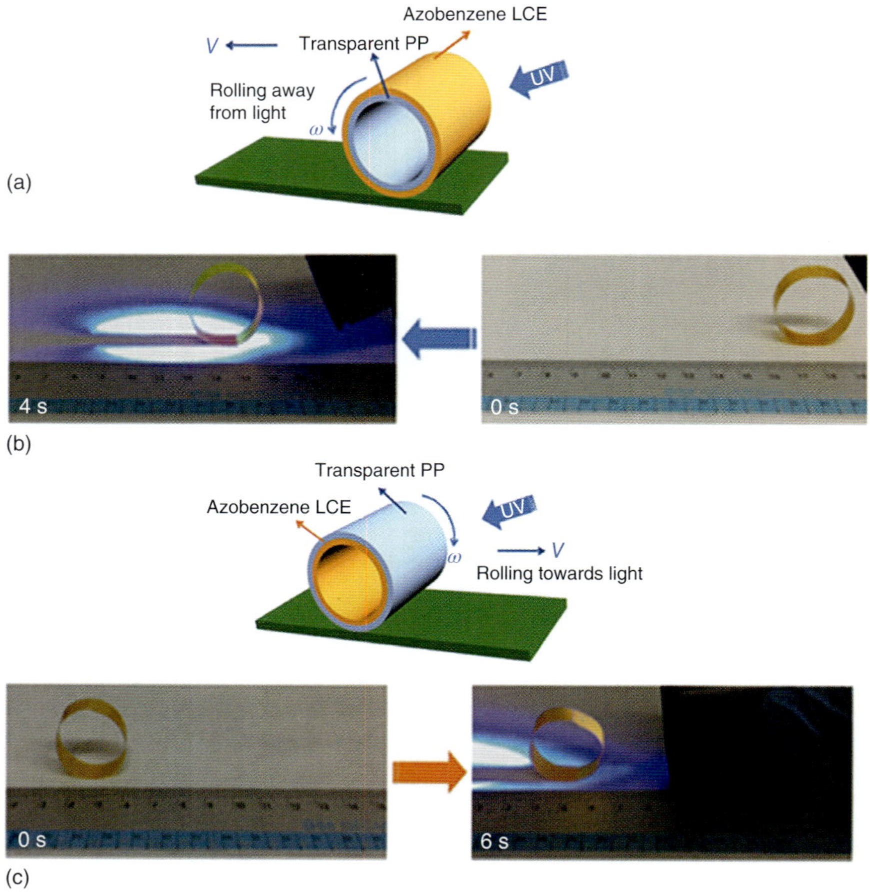

Figure 9.10 Light-driven plastic wheel. (a) Schematic illustration of a direction controlled light driven-plastic wheel system. (b) Images showing light-induced direction profiles of the movement of the wheel fabricated with a laminated film of cross-linked azobenzene LCEs on the outside and a PP substrate on the inside [38]. (c) Schematic illustration and images showing the light-induced opposite direction profiles of the movement of the wheel fabricated with a laminated film of azobenzene LCEs now at the inside and the PP substrate at the outside. Source: Lu et al. 2017 [38]. Copyright 2017. Reproduced with permission from Wiley-VCH Verlag GmbH & Co.

rotatory movement can be controlled by other parameters such as intensity of incident light or the radius of the wheel.

9.5.2 Single-Layer Actuators

The above examples of bilayer composite design have already shown the use of polyolefins as a supportive layer for the photoresponsive micro-device design due to its flexibility and modulus. However, the laminated composite actuators with an adhesive layer may suffer low life time due to the delamination of the two layers after long exposures to the light source [39]. Direct incorporation of a

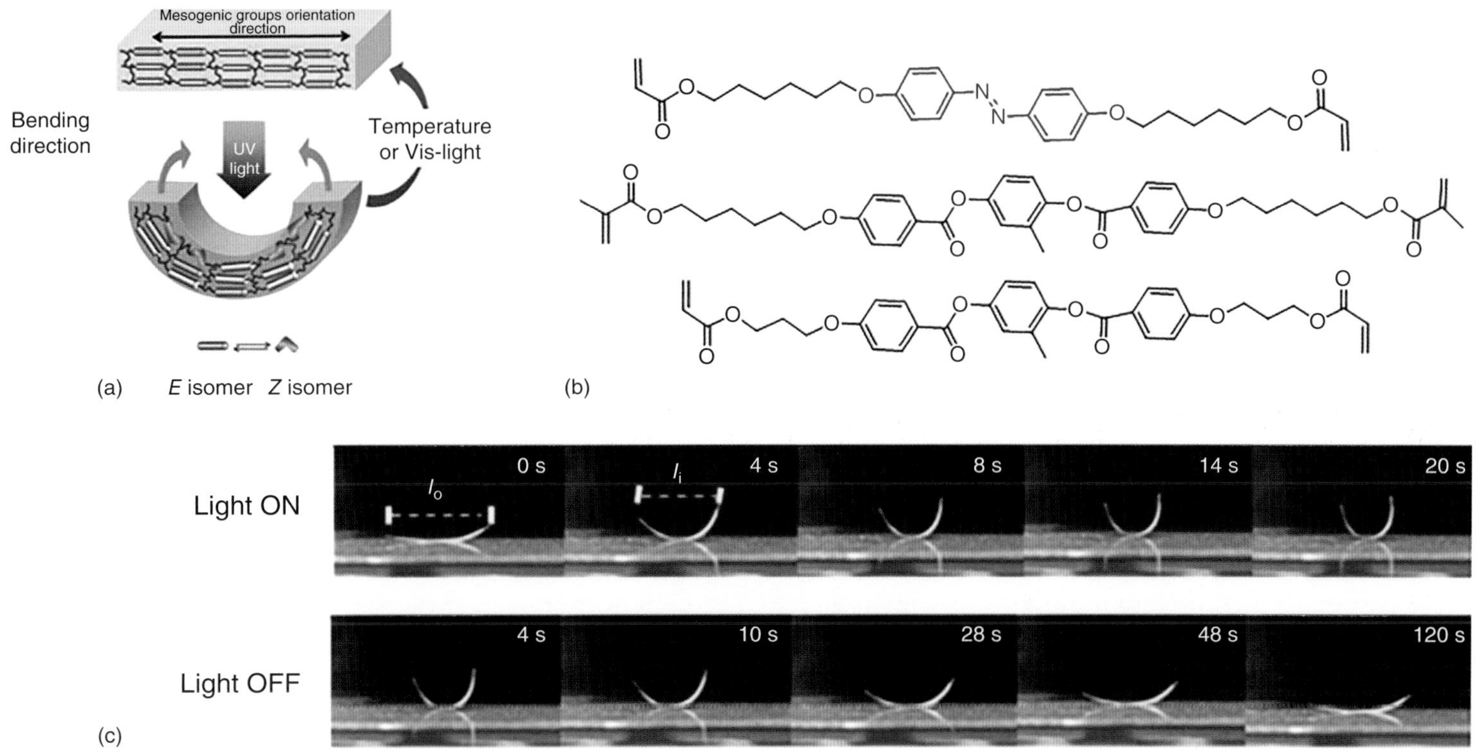

Figure 9.11 (a) Schematic illustration of the reversible bending of the PE-based LC polymer actuator with uniaxial molecular organization. (b) Chemical structures of monomers and cross-linkers used for the preparation of LC/PE monolayer film. (c) Bending behavior of azobenzene LCNs/PE composite film (1 × 10 mm × 12 μm) by controlling incident UV light irradiation (364 nm, 27.4 mW cm^{-2}). Source: Ryabchun et al. 2012 [40]. Copyright 2012. Reproduced with permission from Wiley-VCH Verlag GmbH & Co.

photoresponsive chromophore or materials in the polyolefin matrix is a very straightforward strategy to make a cost-effective (mass-production) responsive material. Commercial availability, low cost, flexibility, and high Young's modulus of the polyolefins make it the best host polymer matrix for the large-scale production of light-responsive actuators or soft-robotics based on monolayers. Some of the attempts in this direction are mentioned further.

Ryabchun et al. have reported such an approach for the first time. They have used the stretched porous PE as the matrix for azobenzene-based LCNs to create photo-actuators. These plastic films were capable of reversibly deforming with response to light irradiation [40]. High deformation values of these new PE composites film and its reversibility were demonstrated (Figure 9.11). These plastic photoactuators combined several advantages such as lack of the need for aligning coatings and complex cell preparation process, good mechanical properties, and relatively low-cost fabrication. They have also studied the kinetic and thermodynamic features of the process of bending of the LCNs/PE monolayer composite actuators.

9.6 Photo Patterning

Apart from photoactuators based on polyolefins, photopattering, or surface micro-patterning of PE films have wide range of applications including in optical recording, optical memory devices, sensors, biofouling, and tissue engineering [41]. Several methods were developed for micro-patterning of polymers like photolithography, holography, micro- and nanoprinting and inkjet printing, etc. Bobrovsky et al. [42, 43] reported a photopatternable polymer composite film based on stretched porous PE and LC. The resultant film was analyzed by SEM (Figure 9.12a) in order to confirm the porous aligned nature of the LC/PE composite matrix as showed in Figure 9.12b. The chemical structures of LC monomers and cross-linkers are showed in Figure 9.12c. This film was then irradiated with UV light through a mask, Figure 9.12d–f shows a polarized optical microscopy image of such optical photo-recording. This pattern is quite stable over time, due to the slow *cis–trans* back isomerization. This slow back isomerization is mainly due to the controlled substitution pattern of the azobenzene unit.

Uchida et al. extended such a photopattering approach into a useful three bits eight state recording system by incorporating three different photoresponsive diarylethene derivatives into a polyolefin matrix (poly(cyclopentene) polymer) [44]. When irradiated with UV light, the three different diarylethene derivatives changed to their closed-ring structure with the formation of yellow, red, and blue colors. They were able to selectively bleach the color by applying the specific wavelength of UV light that matched the closed ring form of different derivatives. Since the spectra of the closed forms of the diarylethene derivatives are in the IR region in addition to UV–Vis, IR light of a specific wavelength could be used to read the eight states. Figure 9.13 shows the schematic of the recording on photochromic polymer film.

Surface micro-patterning of uniaxially drawn high-density PE (HDPE) films using interference holography with a nanosecond pulsed laser was demonstrated

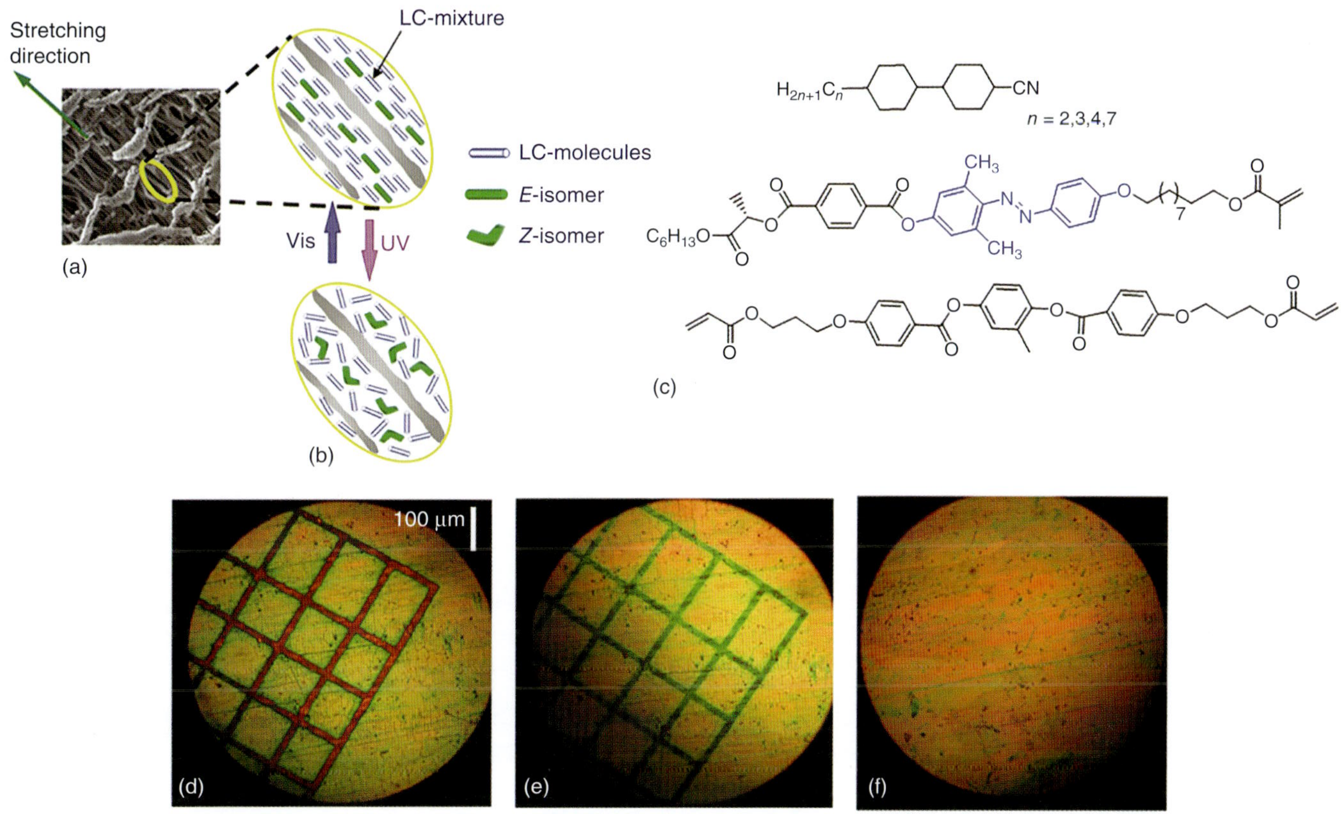

Figure 9.12 (a) SEM image showing the microporous structure of the oriented porous PE film. (b) Schematic representation of the orientation of the LC molecules along the PE fibrils and the photo-induced phase transition initiated by *trans*(*E*)–*cis*(*Z*) isomerization of the azobenzene unit. (c) Chemical structures of monomers and cross-linkers used for the preparation of LC/PE monolayer film. (d) Polarizing optical microscopy images of the LC/PE composite film (c) after 20 minutes of UV irradiation through the mask. (e) After 14 days at room temperature. (f) After 15 minutes of visible light irradiation (436 nm). Source: Bobrovsky et al. 2014 [43]. Copyright 2014. Reprinted with permission from Royal Society of Chemistry.

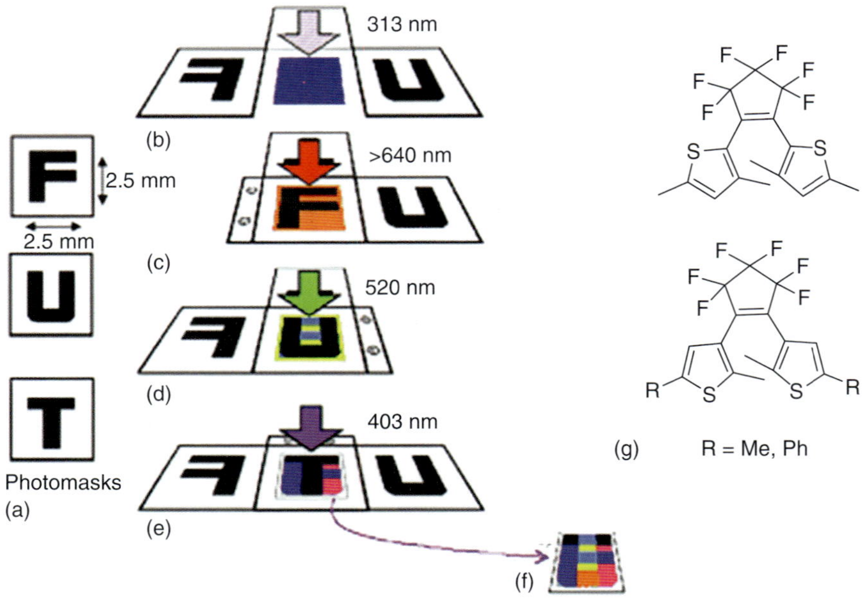

Figure 9.13 Schematic illustration of recording process by using masks F, U, and T (a) on the photochromic film based on polyolefin with three bits eight states. (b–e) represents the stepwise light irradiation process and (f) represents the recorded film. (g) Chemical structures of diarylethene derivatives used for the preparation photochromic polymer film. Source: Uchida et al. 2005 [44]. Copyright 2004. Reproduced with permission from Wiley-VCH Verlag GmbH & Co.

recently by Shen et al. [41]. They used UV absorber, 2-(2*H*-benzotriazole-2-yl)-4,6,-diterpentylphenol (BZT) incorporated oriented HDPE film to generate a micro-pattern. Upon illumination with an interference pattern, the BZT molecule induces transition from light to heat creating a height relief grating. This topography was found to be dependent on the concentration of the BZT used and also on the light energy dose applied. The detailed schematic diagram of the experimental setup for the interference is shown in Figure 9.14A. The generated surface patterns were analyzed by AFM technique. Figure 9.14B (a and b) shows the AFM images and the height profile of the oriented HDPE films containing 5 wt% BZT after illumination with interference pattern ($\Lambda_o = 9.5\,\mu m$) at an energy dose of $120\,mJ\,cm^{-2}$ and at various angles (α) between the grating direction and the film drawing direction. Additionally, this aligned and chain-extended PE films with the surface relief gratings are also evaluated as strain sensors in tensile deformation. By this approach, strains below 10% are monitored accurately in a transmission mode.

Polyolefins-based photoresponsive systems are not limited only to the above-mentioned applications. Despite the fact that these are the old polymers, incorporated functions offered us exciting new perspectives, which cope with the current society demands. The following section will discuss some of the challenges, recent developments, and future directions.

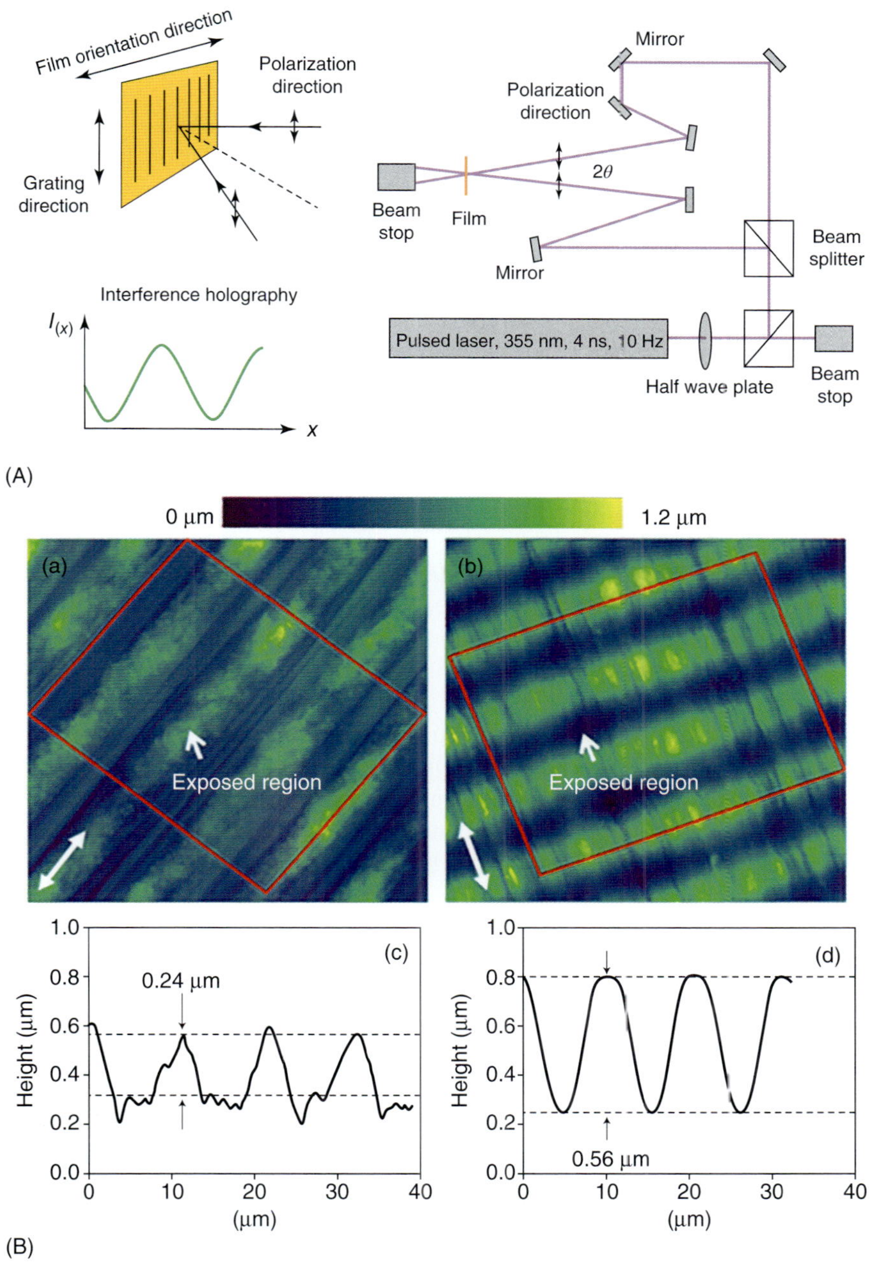

Figure 9.14 (A) Schematic diagram of the experimental setup used for the interference holography. The angle between two laser beams is 2θ. (B) AFM images of oriented and chain-extended PE film containing 5 wt% BZT after illumination. Source: Shen et al. 2017 [41]. Copyright 2017. Reprinted with permission from American Chemical Society.

9.7 Challenges for Photoresponsive Polyolefins and Future Directions

One of the major challenges to be addressed during the development of polyolefin-based photoresponsive materials is photo-oxidative degradation. It should be noted that UV radiation is the greatest contributor for deep deterioration of ultimate mechanical properties and degradation of polyolefins in the presence of oxygen [45]. This can be addressed by choosing an appropriate photoresponsive unit that can also act as the stabilizer, in addition to maintaining the photoresponsiveness [46, 47]. This approach will prevent the auto-oxidation cycle of the photodegradation of responsive polyolefins and will increase the lifetime.

Another challenge is the incorporation of the photoresponsive chromophore to the polyolefin, due to its immiscibility with the polymer matrix as polyolefins are highly nonpolar. But this can be addressed by careful design and chemical modification of the photoresponsive molecular unit while maintaining the thermal stability and photoisomerization. Compatibility with biological systems for the development of integrated bionics can also be considered. UHMW-PE already showed utility in human hip/knee transplants in the static forms [48]. Recently, advanced medical textiles based on medical grade UHMW-PE fibers have also been developed for various implantable devices. These developments shed the hope for the easy human–machine interface of responsive-polyolefins. One of the immediate practical medical applications we can consider are photoresponsive intravenous needles of such highly oriented polymer fibers, which can be activated by near-infrared (NIR) light inside the body. If successful, they can be promising tools with minimal invasive medical application [49, 50].

Other semi-crystalline, amorphous commodity polymers (Figure 10.5) also can be used instead of contemporary light-responsive polymers in order to design devices that can do greater mechanical work. For example, azobenzene-functionalized polyimides have shown large deformation and higher photogenerated force output by White and coworkers [51, 52]. These polymers are amorphous to semi-crystalline and exhibit high glass transition temperature (T_g) and high storage moduli with good thermal stability. Photoinduced bending behavior with a bending angle ranging from 0° to 90° in 30 minutes has been demonstrated for one such polymer. This research already indicates that a careful rational design and polymer modification of polyolefins with an azobenzene unit can be used to create photoresponsive polymeric materials. The above work additionally signals that the alignment of the photoresponsive chromophore in the polymer matrix is one of the key factors to be considered while developing the photoresponsive semi-crystalline or commodity polymers. This can be achieved either by synthesizing the polymer with photoresponsive unit incorporated in the main chain [53] or creating a composite polymer material where the photoresponsive chromophore is aligned within the host polymer matrix as a guest.

The responsive nature of semi-crystalline polymers need not only be limited to photoresponsiveness; we can also consider other external stimuli such as humidity, pH, temperature, electricity, mechanical stress, etc. The amorphous and the crystalline region of some of the highly oriented (e.g. Nylon, PE) fibers contract in length and expand in volume, respectively, in response to temperature [54, 55]. Based on this principle, temperature-responsive artificial muscles were developed utilizing such commercial polymers by Baughman and coworkers with the specific work output of $2000\,J\,kg^{-1}$ during contraction which is 20 times larger than that of natural muscle [55]. The same group earlier developed a guest-filled, twist-spun carbon nanotube yarn, which is not only temperature responsive, but also photothermal responsive [56]. Moreover, semi-crystalline polymers based on shape-memory response were studied for several interesting applications [57]. These include the strong polymer fiber, UHMW-PE [58]. In addition to temperature-responsive commodity polymers, humidity-responsive polyamides have also been demonstrated with the possibility of developing next-generation smart textiles [59]. Other responsive functions such as change of color can be obtained by incorporating chromophores such as spiropyran; the resultant system can be used for the development of sensors. Recently, responsive cholesteric liquid crystals having structural color have been attached to nylon to fabricate color changing textile and optical strain sensors [60]. Such developments and outlooks are in the direction of future commercial development of smart materials with interesting yet useful functions.

The responsive materials based on semi-crystalline (non-liquid crystalline) or commodity polymers have begun to grow in demand due to their commercial availability, easy fabrication process, flexibility, and high Young's modulus which are expected to generate greater work output in addition to mechanical stability.

9.8 Conclusions

This chapter has highlighted some of the recent developments in the area of photoresponsive polymeric systems and the possibility of generating future responsive functional materials out of semi-crystalline commodity polymers especially using polyolefins. Converting light energy to mechanical energy or motion utilizing photoresponsive polymeric materials have led to smart micro-devices that can do useful work. The work output most likely can be improved by utilizing high Young's modulus, semi-crystalline commodity polymers such as polyolefins with careful device design. Such photoresponsive systems are a promising yet inexpensive way of converting light into mechanical work. Additionally, this approach provides a tool to control the bulk polymer properties of polyolefins remotely. These developments foresee a future commercial development of responsive commodity polymers in fields such as soft-robotics and micro-devices with greater adaptability, sensors, motors and artificial muscles, etc.

References

1 Zhai, L. (2013). Stimuli-responsive polymer films. *Chem. Soc. Rev.* 42: 7148.
2 Ko, H. and Javey, A. (2017). Smart actuators and adhesives for reconfigurable matter. *Acc. Chem. Res.* 50: 691–702.

3 Hanlon, R. (2007). Cephalopod dynamic camouflage. *Curr. Biol.* 17: R400–R404.
4 Volkov, A.G., Foster, J.C., and Markin, V.S. (2010). Signal transduction in *Mimosa pudica*: biologically closed electrical circuits. *Plant Cell Environ.* .
5 Forterre, Y., Skotheim, J.M., Dumais, J., and Mahadevan, L. (2005). How the Venus flytrap snaps. *Nature* 433: 421.
6 Su, B., Gong, S., Ma, Z. et al. (2015). Mimosa-inspired design of a flexible pressure sensor with touch sensitivity. *Small* 11: 1886.
7 Roche, E.T., Wohlfarth, R., Overvelde, J.T.B. et al. (2014). A bioinspired soft actuated material. *Adv. Mater.* 26: 1200.
8 Jones, C.M., Baker-Groberg, S.M., Cianchetti, F.A. et al. (2014). Measurement science in the circulatory system. *Cell. Mol. Bioeng.* 7 (1).
9 Zhou, H., Xue, C., Weis, P. et al. (2016). Photoswitching of glass transition temperatures of azobenzene-containing polymers induces reversible solid-to-liquid transitions. *Nat. Chem.* 9: 145.
10 Kizilkan, E., Strueben, J., Staubitz, A., and Gorb, S.N. (2017). Bioinspired photocontrollable microstructured transport device. *Sci. Rob.* 2: eaak9454.
11 Hartley, G.S. (1937). The *cis*-form of azobenzene. *Nature* 140: 281–281.
12 Yu, Y., Nakano, M., and Ikeda, T. (2003). Directed bending of a polymer film by light: photomechanics. *Nature* 425: 145.
13 Ionov, L. (2015). Polymeric actuators. *Langmuir* 31: 5015.
14 Bushuyev, O.S., Aizawa, M., Shishido, A., and Barrett, C.J. (2018). Shape-shifting azo dye polymers: towards sunlight-driven molecular devices. *Macromol. Rapid Commun.* 39: 1700253.
15 Stumpel, J.E., Gil, E.R., Spoelstra, A.B. et al. (2015). Stimuli-responsive materials based on interpenetrating polymer liquid crystal hydrogels. *Adv. Funct. Mater.* 25: 3314–3320.
16 Gelebart, A.H., Vantomme, G., Meijer, E.W., and Broer, D.J. (2017). Mastering the photothermal effect in liquid crystal networks: a general approach for self-sustained mechanical oscillators. *Adv. Mater.* 29: 1606712.
17 White, T.J. (2018). Photomechanical effects in liquid crystalline polymer networks and elastomers. *J. Polym. Sci., Part B: Polym. Phys.* 56: 695.
18 Kumar, K., Knie, C., Bléger, D. et al. (2016). A chaotic self-oscillating sunlight-driven polymer actuator. *Nat. Commun.* 7: 11975.
19 Gelebart, A.H., Jan Mulder, D., Varga, M. et al. (2017). Making waves in a photoactive polymer film. *Nature* 546: 632.
20 Wani, O.M., Zeng, H., and Priimagi, A. (2017). A light-driven artificial flytrap. *Nat. Commun.* 8: 15546.
21 Liu, D. and Broer, D.J. (2014). Liquid crystal polymer networks: preparation, properties, and applications of films with patterned molecular alignment. *Langmuir* 30: 13499.
22 Hines, L., Petersen, K., Lum, G.Z., and Sitti, M. (2017). Soft actuators for small-scale robotics. *Adv. Mater.* 29: 1603483.
23 Madden, J.D.W., Vandesteeg, N.A., Anquetil, P.A. et al. (2004). Artificial muscle technology: physical principles and naval prospects. *IEEE J. Oceanic Eng.* 29: 706.
24 Levin, A., Michaels, T.C.T., Adler-Abramovich, L. et al. (2016). Elastic instability-mediated actuation by a supra-molecular polymer. *Nat. Phys.* 12: 926.

25 Harris, K.D., Cuypers, R., Scheibe, P. et al. (2005). Large amplitude light-induced motion in high elastic modulus polymer actuators. *J. Mater. Chem.* 15: 5043.

26 Haines, C.S., Lima, M.D., Li, N. et al. (2014). Artificial muscles from fishing line and sewing thread. *Science* 343: 868.

27 Merian, E. (1966). Steric factors influencing the dyeing of hydrophobic. Fibers. *Text. Res. J.* 36: 612.

28 Eisenbach, C.D. (1980). Isomerization of aromatic azo chromophores in poly(ethyl acrylate) networks and photomechanical effect. *Polymer* 21: 1175.

29 Yamada, M., Kondo, M., Mamiya, J. et al. (2008). Photomobile polymer materials: towards light-driven plastic motors. *Angew. Chem. Int. Ed.* 47: 4986.

30 Finkelmann, H., Nishikawa, E., Pereira, G.G., and Warner, M.A. (2001). New opto-mechanical effect in solids. *Phys. Rev. Lett.* 87: 015501.

31 Hiraoka, M., Nakamura, K., Arase, H. et al. (2016). Power-efficient low-temperature woven coiled fibre actuator for wearable applications. *Sci. Rep.* 6: 36358.

32 PlasticEurope Market Research Group (PEMRG)/Cunsultic Marketing and Industrieberatung GmbH. http://Www.Plasticseurope.Org/En (accessed 27 January 2018).

33 Whiteley, K.S., Heggs, T.G., Koch, H. et al. (2000). Ullmann's Encyclopedia of Industrial Chemistry. Weinheim: Wiley-VCH.

34 Peijs, T. (2018). 1.5 High performance polyethylene fibers. In: Comprehensive Composite Materials II, 86–126. Elsevier.

35 Dyneema. http://www.dsm.com/products/dyneema/en_GB/home.html (accessed 10 July 2018).

36 Yamada, M., Kondo, M., Miyasato, R. et al. (2009). Photomobile polymer materials – various three-dimensional movements. *J. Mater. Chem.* 19: 60.

37 Cheng, F., Yin, R., Zhang, Y. et al. (2010). Fully plastic microrobots which manipulate objects using only visible light. *Soft Matter* 6: 3447.

38 Lu, X., Guo, S., Tong, X. et al. (2017). Tunable Photocontrolled motions using stored strain energy in malleable azobenzene liquid crystalline polymer actuators. *Adv. Mater.* 29: 1606467.

39 Pang, X., Xu, B., Qing, X. et al. (2017). Photo-induced bending behavior of post-crosslinked liquid crystalline polymer/polyurethane blend films. *Macromol. Rapid Commun.* 1700237.

40 Ryabchun, A., Bobrovsky, A., Stumpe, J., and Shibaev, V. (2012) Novel generation of liquid crystalline photo-actuators based on stretched porous polyethylene films. *Macromol. Rapid Commun.* 33: 991.

41 Shen, L., Lafleur, S.S.D., Houben, S.J.A. et al. (2017). Surface micropatterning of uniaxially oriented polyethylene films using interference holography for strain sensors. *Langmuir* 33: 14592.

42 Bobrovsky, A., Shibaev, V., and Elyashevitch, G. (2008). Photopatternable fluorescent polymer composites based on stretched porous polyethylene and photopolymerizable liquid crystal mixture. *J. Mater. Chem.* 18: 691.

43 Bobrovsky, A., Shibaev, V., Cigl, M. et al. (2014). Photochromic LC-polymer composites containing azobenzene chromophores with thermally stable Z-isomers. *J. Mater. Chem. C* 2: 4482.

44 Uchida, K., Saito, M., Murakami, A. et al. (2005). Three bits eight states photochromic recording and nondestructive readout by using IR light. *Chem. Eur. J.* 11: 534.
45 Raab, M., Kotulák, L., Kolařík, J., and Pospíšil, J. (1982). The effect of ultraviolet light on the mechanical properties of polyethylene and polypropylene films. *J. Appl. Polym. Sci.* 27: 2457.
46 Yousif, E. and Haddad, R. (2013). Photodegradation and photostabilization of polymers, especially polystyrene: review. *SpringerPlus* 2: 398.
47 Zayat, M., Garcia-Parejo, P., and Levy, D. (2007). Preventing UV-light damage of light sensitive materials using a highly protective UV-absorbing coating. *Chem. Soc. Rev.* 36: 1270.
48 Blunn, G., Brach del Preva, E.M., Costa, L. et al. (2002). Ultra high molecular-weight polyethylene (UHMWPE) in total knee replacement: fabrication, sterilisation and wear. *J. Bone Joint Surg.* 84: 946.
49 Lendlein, A. (2002). Biodegradable, elastic shape-memory polymers for potential biomedical applications. *Science* 296: 1673.
50 Maitland, D.J., Metzger, M.F., Schumann, D. et al. (2002). Photothermal properties of shape memory polymer micro-actuators for treating stroke. *Lasers Surg. Med.* 30 (1).
51 Baczkowski, M.L., Wang, D.H., Lee, D.H. et al. (2017). Photomechanical deformation of azobenzene-functionalized polyimides synthesized with bulky substituents. *ACS Macro Lett.* 6: 1432.
52 Lee, K.M., Wang, D.H., Koerner, H. et al. (2012). Enhancement of photogenerated mechanical force in azobenzene-functionalized polyimides. *Angew. Chem.* 124: 4193.
53 Appiah, C., Woltersdorf, G., and Binder, W.H. (2017). Synthesis of Photoresponsive main-chain oligomers with azobenzene moieties via ADMET oligomerization and their micellization properties. *Polym. Chem.* 8: 2752.
54 Mirvakili, S.M. and Hunter, I.W. (2017). Artificial muscles: mechanisms, applications, and challenges. *Adv. Mater.* 1704407.
55 Haines, C.S., Li, N., Spinks, G.M. et al. (2016). New twist on artificial muscles. *Proc. Natl. Acad. Sci.* 113: 11709.
56 Lima, M.D., Li, N., Jung de Andrade, M. et al. (2012). Electrically, chemically, and photonically powered torsional and tensile actuation of hybrid carbon nanotube yarn muscles. *Science* 338: 928.
57 Leng, J., Lan, X., Liu, Y., and Du, S. (2011). Shape-memory polymers and their composites: stimulus methods and applications. *Prog. Mater Sci.* 56: 1077.
58 Kaloshkin, S., Maksimkin, A., Kaloshkina, M. et al. (2012). Shape memory behavior of ultra-high molecular weight polyethylene. *MRS Proc.* 1403.
59 Dai, M., Picot, O.T., Verjans, J.M.N. et al. (2013). Humidity-responsive bilayer actuators based on a liquid-crystalline polymer network. *ACS Appl. Mater. Interfaces* 5: 4945.
60 Picot, O.T., Dai, M., Billoti, E. et al. (2013). A real time optical strain sensor based on a cholesteric liquid crystal network. *RSC Adv.* 3: 18794.

10

A Photoresponsive Multi-Bilayered Film for a Tunable Photonic Crystal

Sunnam Kim and Seiji Kurihara

Kumamoto University, Department of Applied Chemistry and Biochemistry, 2-39-1 Kurokami, Kumamoto, 860-8555, Japan

10.1 Introduction

Over the last half-century, advances in science based on the evolution of semiconductor physics and transistor electronics have greatly influenced the development of our lives. The key to this was the ability to design crystalline materials that tailor the conduction properties of electrons to suit diverse applications. More recently, the optical communication revolution has shifted the frontier in high-speed, multiplex, and wide-bandwidth information processing from electrons to *photons*, a shift that demands a new kind of synthetic soft material to mold the gigabits [1–3]. To address this need, recent research has suggested a way to tailor the propagation of light much like that of electrons, via the creation of periodic optical media dubbed *photonic crystals* (PCs).

10.1.1 Photonic Crystals

Many colors found in nature cannot be explained simply by the phenomenon of absorption and reflection of light. The origins of such colors were not understood until the middle of the last century and it was only after electron microscopy had become an established technique, that the origins of these colors could be explained. These colors arise from the interaction of light with highly ordered structures, the sizes of which are in the order of the wavelength of visible light. The most prominent feature of such structures is that they present iridescences as a result of diffraction. Some natural examples are the multilayered structure of pearls, the flashing wings of several insects, and natural opals as shown in Figure 10.1.

Recently, there has been growing interest in developing bioinspired nanostructures that suppress defects, and control the symmetry and size of PCs [4–10]. PCs are composed of spatially ordered dielectric structures with their lattice parameter comparable to the wavelength of light. PCs are largely divided into one-dimensional (1D), two-dimensional (2D), or three-dimensional (3D) according to the dimensionality of the periodicity as shown in Figure 10.2. At specific

Photoactive Functional Soft Materials: Preparation, Properties, and Applications, First Edition.
Edited by Quan Li.

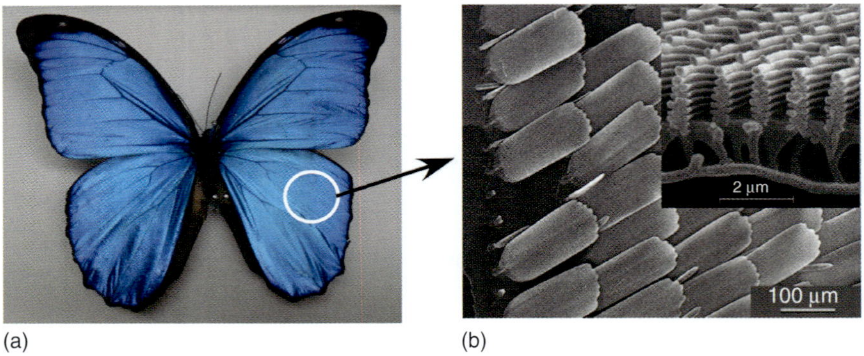

Figure 10.1 An example of photonic crystals in nature. Morpho butterfly (a) and SEM images of the wing of a morpho butterfly (b).

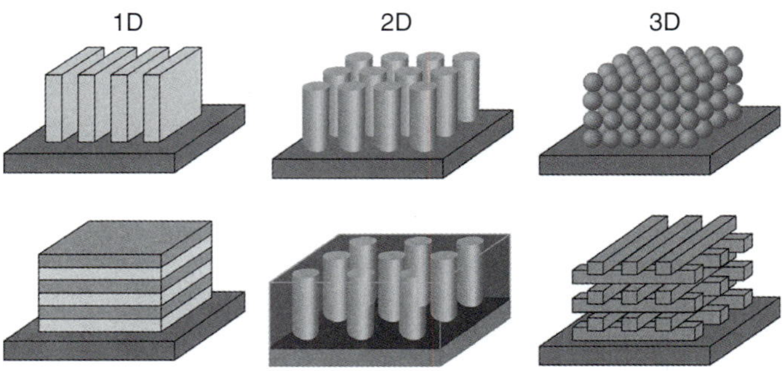

Figure 10.2 Examples of one-dimension (1D), two-dimension (2D), and three-dimensional (3D) photonic crystals.

wavelength ranges, the propagation of light can be suppressed to travel through PCs by strong interference. Thus, PCs can reflect the light coming onto the crystal. The forbidden wavelength ranges are called a photonic band gap (PBG), which performs similar behavior as the electronic band gap created by the periodic arrangement of atoms in a semiconductor [11, 12]. Based on the properties of PCs confining a light, increasing attention has been paid recently to manipulate a light for optical signal processing, optical integration technology, and color display by using PCs.

10.1.2 Tunable Photonic Crystals

Many potential applications of PCs require some unique capabilities for tuning the PBG through external stimuli. The PBG mainly depends on the angle of incidence, the lattice constants, and material composition of PCs. When the light passes in a PC of spherical particles as shown in Figure 10.3, the diffraction of light can be described by the Bragg–Snell law [3, 7, 13–20],

$$m\lambda = 2d\sqrt{{n_{\mathrm{eff}}}^2 - \sin^2\theta} \tag{10.1}$$

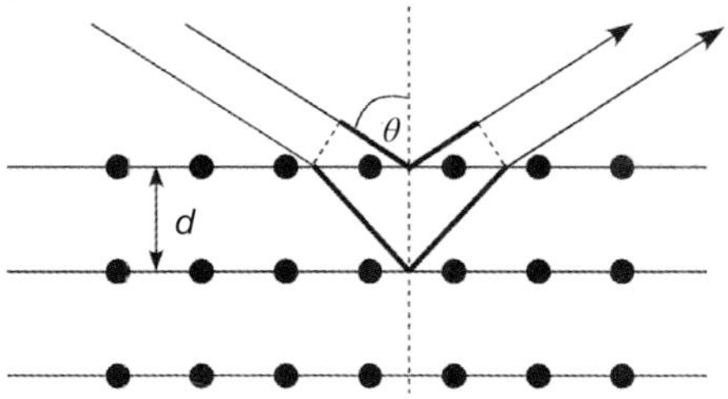

Figure 10.3 Schematic diagram of Bragg diffraction in a periodic structured material.

where, m is the order of diffraction, λ is the wavelength of light, d is the interplanar spacing, θ is the angle of incidence of light with respect to the normal, and n_{eff} is the effective refractive index which is related to the sum of the refractive indices of the two media, weighted by their volume fraction. Therefore, any variation in either spacing or refractive index results in a shift of the Bragg reflection peak.

Some practical schemes for tunable PCs that are based on the concept of changing the lattice constants have been reported. For example, control of the spatial structure has also been realized by taking advantage of the volume phase transition of hydrogels, where PBGs have been tuned by controlling the temperature, the pH, and the ionic state, or by using photoisomerization of azobenzene derivatives [21–24]. Figure 10.4 shows an example of the tuning of a PBG based on the concept of changing lattice constants by *trans–cis* photoisomerization of azobenzene molecules attached in a hydrogel network around a self-assembled crystalline colloidal array [22]. 3D periodic structure was constructed by self-assembly of colloidal particles due to long-range electrostatic repulsions between particles, and following the formation of the hydrogel network.

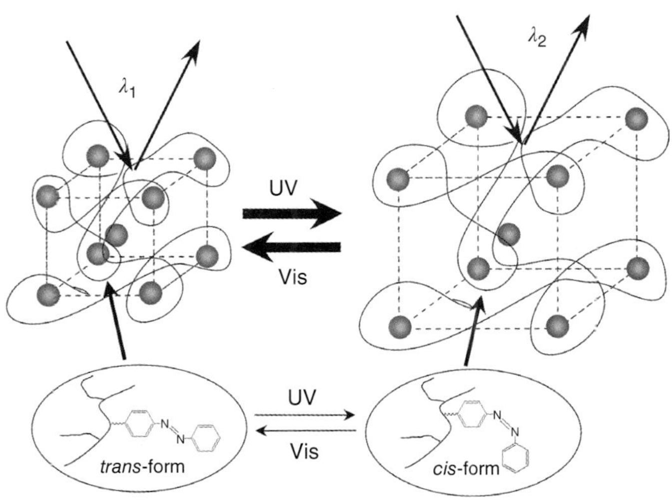

Figure 10.4 An example of the tuning of optical properties based on the changes in lattice constants of PCs. Source: Kamenjicki et al. 2003 [22]. Copyright 2003. Reprinted with permission from Wiley-VCH Verlag GmbH & Co.

Large shifts in PBGs have been demonstrated by controlling the lattice constants. However, these techniques have a disadvantage because of incomplete recovery and deformation when structural changes occur on the order of micrometer dimensions, resulting in difficulties for practical applications. On the other hand, PBG depends on the refractive index change as well as lattice constant according to Eq. (10.1). The photochemical PBG switching by refractive index change will be the main focus in this chapter.

10.2 Photo-Tunable 1D PCs

Several methods of fabricating such materials have been explored since the concept of PCs was proposed. The simplest example of a PC is the multilayer-stacked material. A multi-bilayered film, which is one of the multilayer-stacked materials, can be prepared by stacking of alternate hydrophilic and hydrophobic polymer layers. When a photoresponsive azobenzene polymer was employed for a hydrophobic layer, the refractive index of the multi-bilayered film was controlled by changing the molecular orientation of an azobenzene group, thermally and photochemically. In this section, we specifically describe the tuning of the PBG of multi-bilayered films depending on the refractive index change with molecular orientation of an azobenzene group attached to a polymer segment.

10.2.1 Photoresponsive Properties of Azobenzene Molecules

Since the wavelength and intensity of the reflection strongly depend on the changes in refractive indices of the materials, PBG switching is possible by using a liquid crystal (LC) exhibiting stimuli-responsive orientation behavior. Among various promising stimuli, light shows distinct and significant advantages due to its spatial, remote, and temporal controllability. It is also noteworthy that the light sources with different wavelengths, intensity, and polarization are readily accessible. Photocontrol of chemical and physical functions of various materials has been studied vigorously using the photochemical configurational change of azobenzene derivatives. The photochromism of azobenzene and its derivatives due to *trans–cis* isomerization has been widely investigated [17, 25–30]. The azobenzene compounds are well known to show reversible photoisomerization between the *trans* and *cis* forms (Figure 10.5). In particular, an irradiation with linearly polarized light results in re-orientation of the

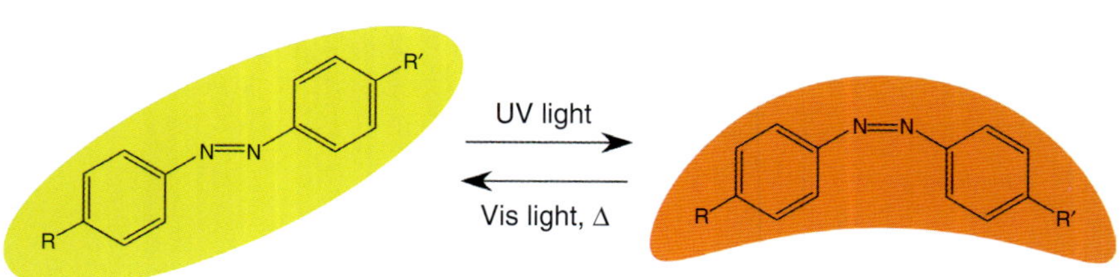

Figure 10.5 Photochromism of azobenzene molecules between *trans*- and *cis*-forms.

azobenzene groups perpendicular to the polarization axis of the light when the azobenzene molecules are in a viscous media, or are attached to polymer segments. A *trans*-azobenzene molecule has an anisotropic shape; consequently, the change in the molecular orientation causes the change in the optical anisotropy, the so-called *Weigert effect*. Therefore, the change in PBG will be achieved by controlling the molecular orientation of azobenzene molecules or other anisotropic molecules in PCs, resulting in the change in optical anisotropy.

10.2.1.1 Optical Anisotropy Based on Molecular Orientation

Molecular orientation of low molecular weight liquid crystals (LMWLCs) can be controlled easily: e.g. in-plane as well as out-of-plane molecular orientation (Figure 10.6) can be achieved statically by the use of alignment layers. In addition, application of external stimuli such as an electric field causes a dynamic change in the molecular orientation perpendicular or parallel to the direction of the external stimuli corresponding to the property of LMWLCs [32]. However, it is not easy to control the molecular orientation of polymer liquid crystals (PLCs) mainly because of their high viscosity. Therefore, the development of a facile technique to control the molecular orientation of PLCs uniformly, or the development of PLCs showing spontaneous molecular orientation, will open the door for new functions or new materials. There are some reports on polymers showing out-of-plane molecular orientation by annealing: side chain type PLCs, main

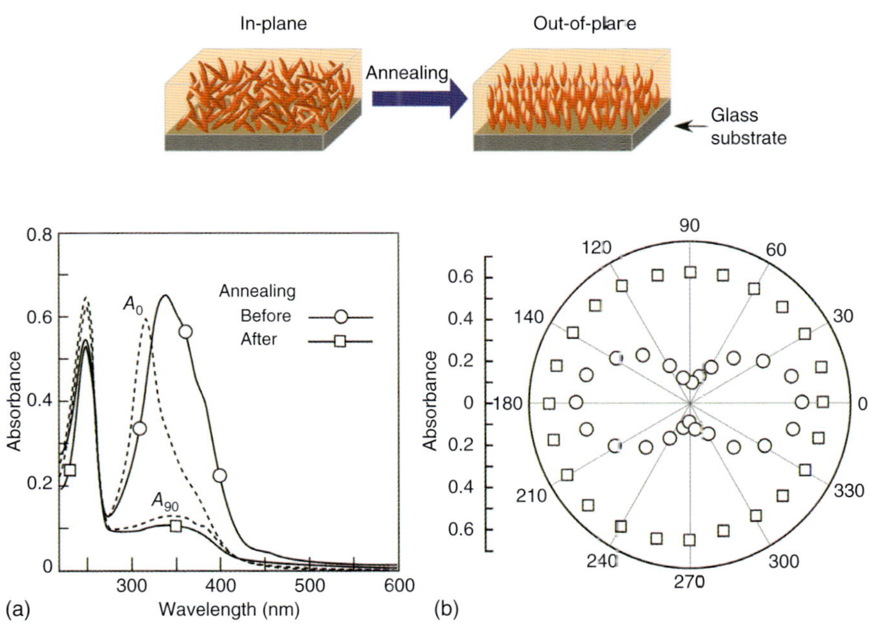

Figure 10.6 (a) Absorption spectra of the PA6Az1 film before and after annealing (solid lines) and change in absorption spectra at an incident angle of 45° to the normal of the film by rotating the polarization plane of the monitor light (dashed line, A_0 and A_{90} are spectra at the rotation angle of 0° and 90°, respectively). (b) Polar plots at 320 nm (circle) and 250 nm (square) of the PA6Az1 film annealed. Source: Moritsugu et al. 2011 [31]. Copyright 2011. Reprinted with permission from Wiley-VCH Verlag GmbH & Co.

chain type PLCs, and so on [33–39]. It is likely that the polymers possessing a smectic phase tend to show thermal out-of-plane molecular orientation by annealing at the smectic phase. Bobrovsky et al. have reported that a polyacrylate copolymer having both 4-ethoxy-4′-hexyloxyazobenzene and cholesterol groups as a side chain group showed spontaneous out-of-plane molecular orientation on a glass substrate by annealing at 75 °C, whereas less out-of-plane molecular orientation was observed for a polymethacrylate homopolymer having the same azobenzene side group (PA6Az2 in Scheme 10.1) [40]. In contrast, we found that a polyacrylate homopolymer consisting of 4-methoxy-4′-hexyloxyazobenzene (PA6Az1 in Scheme 10.1) aligned spontaneously in the out-of-plane orientation on a glass substrate by annealing at a smectic phase (80 °C) [31]. It is interesting to note that a very small difference in the terminal substituent has a large effect on the spontaneous molecular orientation behavior of the polymers. However, there are few systematic studies on the thermal out-of-plane molecular orientation of polyacrylates and polymethacrylates with azobenzene side groups.

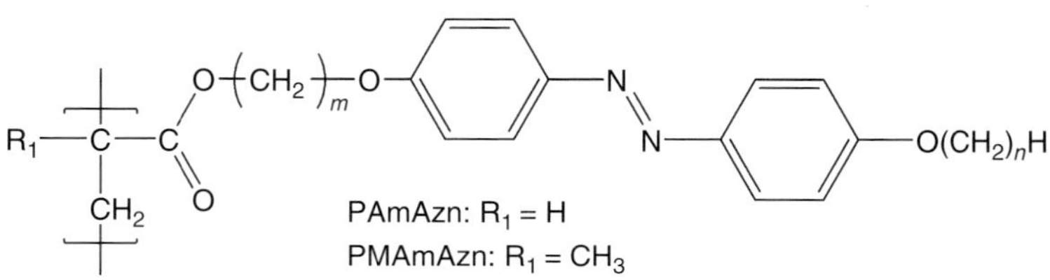

Scheme 10.1 Molecular structure of polyacrylates and polymethacrylates with azobenzene side groups.

Figure 10.6 shows the annealing effect on the molecular orientation of a PA6Az1 film. The film was prepared by spin coating a PA6Az1/cyclohexanone solution onto a glass substrate, and annealed at 80 °C. Before annealing, there were two strong peaks at c. 250 and 350 nm corresponding to φ–φ^* and π–π^* transitions, respectively. Upon annealing, the absorbance at 350 nm was significantly decreased, whereas no change in the absorbance at 250 nm was observed. It has been reported that the φ–φ^* transition is insensitive to the orientation; consequently, the change in the relative intensity of the π–π^* absorption band to the φ–φ^* absorption band ($A_{\pi-\pi^*}/A_{\varphi-\varphi^*}$) is related to the transformation from the in-plane orientation to the out-of-plane orientation. The film was rotated horizontally by 45°, and polarized absorption spectra of the PA6Az1 film at an incident angle of 45° to the normal of the film were measured by rotating the polarization plane of the monitor light (dotted line). As shown with the dotted line in Figure 10.6a, the π–π^* absorption band at the rotation angle of 0° (monitored with horizontally polarized light, curve A_0) was shifted to a shorter wavelength, whereas at 90° a weak π–π^* band was observed (curve A_{90}). In order to clarify the change in the molecular orientation by annealing, the absorbance at 250 and 320 nm was plotted as a function of the rotation angle in Figure 10.6b. A significant rotation angle dependence was observed for the absorbance at 320 nm. These results revealed that the annealing at 80 °C contributed not only to a transformation of the molecular orientation from the in-plane orientation to the out-of-plane orientation, but also to the formation of H-aggregates of the azobenzene groups. In order to discuss the structural effect on the spontaneous out-of-plane orientation of the polymers by annealing, the order parameter, S, of the out-of-plane orientation was evaluated using the following Eq. (10.2) [40]:

$$S = \frac{A_0 - A_{90}}{A_0 + 2A_{90}} \tag{10.2}$$

A_0 and A_{90} of the polymer films were measured after annealing at various temperatures for 10 minutes followed by cooling to ambient temperature. The order parameter was increased by elevating the annealing temperature up to a smectic phase, and then decreased at a higher temperature near the phase transition temperature from the smectic phase to the nematic phase. The highest order parameter of the out-of-plane orientation obtained in the temperature range studied for each polymer was measured as a function of methylene spacer length. In a series of PAmAz1, the polymers having methylene spacers in a range of 6–10 showed a higher order parameter [41]. In this range of spacer length, the order parameter of the polyacrylates (PAmAz1) is higher than that of the polymethacrylates (PMmAz1). This may be related to the rigidity of the main chain or steric hindrance associated with methyl groups attached to the main chain of the polymethacrylates.

10.2.1.2 Refractive Indices Depending on Molecular Orientation States

For normal direction to the films, refractive indices of out-of-plane orientation and random orientation rely on n_o and $n_{av,}$ respectively. Birefringence properties of azobenzene-containing polymers depend on their molecular structures. Refractive indices of PA6Az1 were measured by using a wedge cell with alignment treatment. A He–Ne laser light (633 nm) was used as a monitor light. For the refractive indices of PA6Az1 for ordinary and extraordinary lights of 633 nm, n_e and n_o, are estimated to be 1.71 and 1.48, respectively [42, 43]. Namely, the refractive index of the out-of-plane orientation of the PA6Az1 film should be 1.48. Assuming random orientation of the azo chromophores of PAmAz1, the average refractive index n_{av} is estimated to be 1.60.

Therefore, if PA6Az1 undergoes an out-of-plane orientation and photoisomerization in the multi-bilayered film consisting of PA6Az1 and polyvinyl alcohol (PVA), on–off switching of the reflection will be expected. Because, the refractive index of the PA6Az1 film must be controlled between 1.48 (n_o, out-of-plane state) and 1.64 (n_e, in-plane random state) by the annealing and following photoisomerization, and the refractive index of PVA is 1.49, which is nearly equal to the refractive index of the out-of-plane state of PA6Az1.

10.2.2 Fabrication of Multi-Bilayered Films

Multi-bilayered films with PVA and PA6Az1 were prepared by alternative spin coating of a PA6Az1/cyclohexanone solution and a PVA/water solution (Figure 10.7).

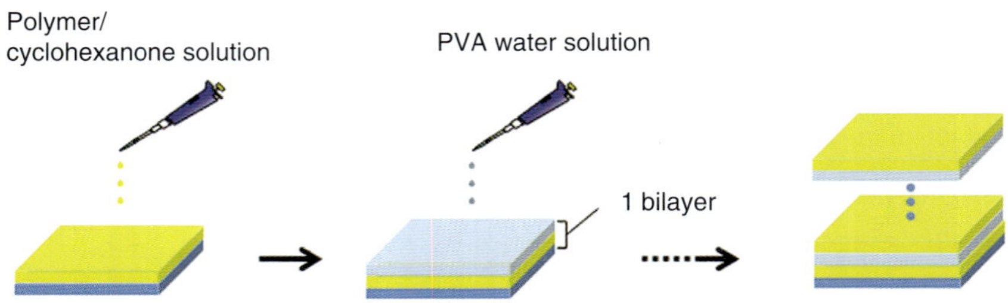

Figure 10.7 SEM image of cross-section of the PVA/PA6Az1 multi-bilayered film.

The wavelength of the reflected light from a periodic multi-bilayered material depends on the optical thickness of each layer and refractive index difference. The reflection wavelength is given by Eq. (10.3).

$$m\lambda = 2a\sqrt{{n_1}^2\frac{d_1}{a} + {n_2}^2\frac{d_2}{a}} \tag{10.3}$$

where a is the thickness of each bilayer, n_1, n_2 and d_1, d_2 are the refractive indices and the layer thicknesses of stacked materials 1 and 2 in each bilayer, respectively, and m is the diffraction order integer. In such multi-bilayered films, the wavelength of the reflection can be controlled by varying the thickness of each layer and/or the refractive index of the stacked materials. Namely, the multi-bilayered films reflect the light of specific wavelength which depends on the difference between the refractive indices of PA6Az1 (1.60, n_{av}) and PVA (1.49) layers, and the bilayer thickness.

10.2.2.1 Control of Reflection Wavelength

The wavelength of reflection can be easily controlled with film thickness. 20-bilayered film was prepared by spin coating of alternate hydrophilic and hydrophobic solutions. With different concentrations of the solution, the film

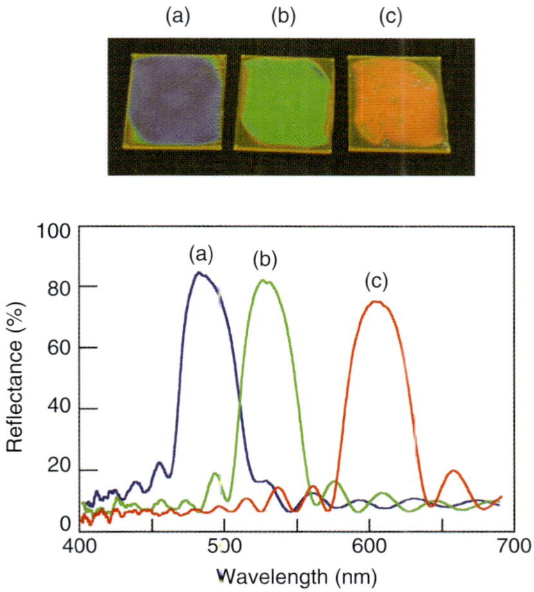

Figure 10.8 Reflection spectra of the PA6Az1/PVA multi-bilayered films with various PA6Az1 layer thicknesses, (a) 40 nm, (b) 55 nm, and (c) 80 nm. Source: Moritsugu et al. 2011 [31]. Copyright 2011. Reprinted with permission from Wiley-VCH Verlag GmbH & Co.

thickness can be controlled. When the thickness of the PVA layer is 120 nm, blue–green–red reflection can be achieved by changing the thickness of PA6Az1 layer as 40, 55, and 80 nm, respectively (Figure 10.8).

10.2.2.2 Control of Reflection Intensity

The reflectance, R, of the multi-bilayered films is given by the following equation for the normal incidence of light [44]:

$$R = \left[\frac{1-(n_H / n_L)^{2q}\left(n_H^2 / n_S\right)}{1+(n_H / n_L)^{2q}\left(n_H^2 / n_S\right)} \right]^2 \quad (10.4)$$

where n_H and n_L are the high- and low-refractive indices of the stacked materials in each bilayer, respectively, n_S is the refractive index of the substrate, and q is the number of bilayers. From Eq. (10.4), it can be observed that the reflectance is expected to increase by increasing the number of layers (Figure 10.9), and the reflectance of PA6Az1/PVA of multi-bilayered films measured with different layer numbers agreed well with the calculated value.

On the other hand, the reflectance becomes extremely low when n_H, n_L, and n_S are nearly equal, so that an on–off switching of the reflection will be achieved by controlling the difference in the refractive indices of the stacked materials between the equal state (n_H, n_L, and n_S) and the different state (n_H and n_L) [31, 41, 45, 46].

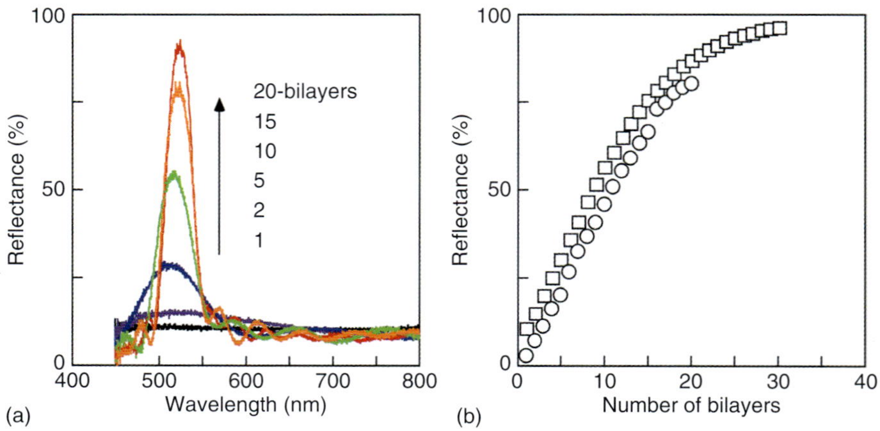

Figure 10.9 Changes in reflection spectra (a) and reflection intensity (b) of PA6Az1/PVA multi-bilayered films as a function of the bilayer number; (circle) observed, (square) calculated with Eq. (10.4). Source: Yagi et al. 2014 [41]. Copyright 2014. Reprinted with permission from Elsevier.

10.2.3 On–Off Switching of Reflection Based on Refractive Index Change

It is well known that azobenzene compounds show reversible photoisomerization between *trans*- and *cis*-forms, and there are some studies on the phase transition from a liquid crystalline phase to an isotropic phase relating to photoisomerization. For the multi-bilayered film composed of PA6Az1 and PVA, the reflection band disappeared by annealing at 80 °C for 60 seconds as can be seen in Figure 10.10a. The reflective state was visibly transformed to a transparent state as shown in the bottom part of the image given in Figure 10.10b. This result demonstrates that the annealing caused a transformation of the molecular orientation of the PA6Az1 layers from the in-plain orientation to the out-of-plane one even in the multi-bilayered film, and the small difference between the refractive indices of PA6Az1 (1.48, n_o) and PVA (1.49) contributed to the disappearance of the reflection band. The reflection band appeared again upon UV irradiation, because of the transformation from the out of-plane orientation to the photoinduced random state with photoisomerization from the *trans*- to *cis*-form (solid color line in Figure 10.10a and the upper part of the photographs in Figure 10.10b). The schematic diagram of the on–off switching of the reflection relative to the change in the molecular orientation is shown in Figure 10.11.

10.2.4 Improvement of Response Speed

10.2.4.1 Introduction of Biphenyl LC Group

The on–off switching of reflection was achieved by change in the PBG of the multi-bilayered film. However, long-time irradiation was required, because of higher absorbance in the UV region due to the azobenzene chromophores in the multi-bilayered films. To improve the light response efficiency across the entire

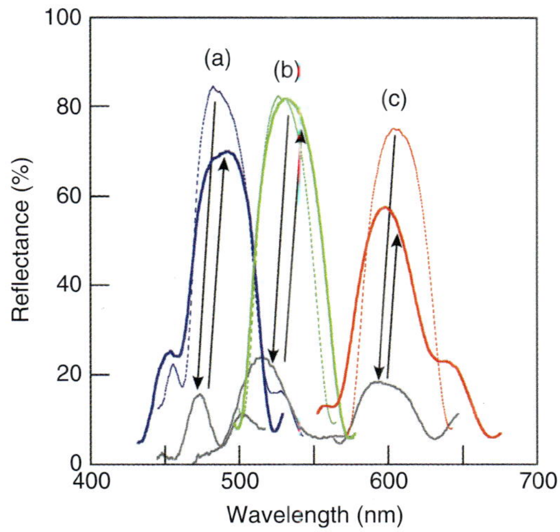

(a)

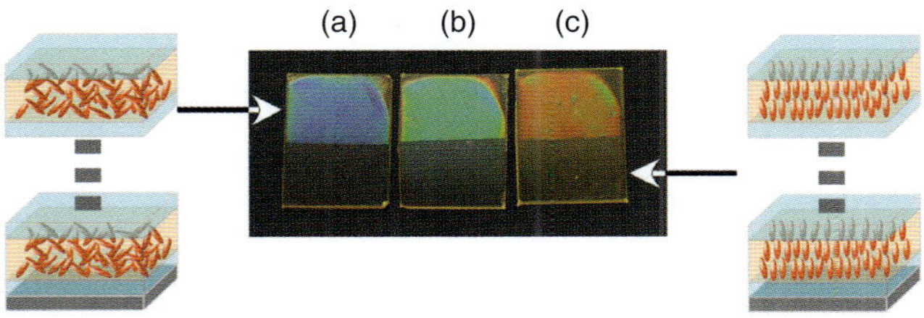

(b)

Figure 10.10 (A) Changes in reflection spectra of PA6Az1/PVA-20-bilayered films by annealing and following UV irradiation. Dotted lines for as-prepared: Gray lines for films annealed at 80 °C for 60 seconds: Solid lines for one irradiated with UV light (unpolarized, 365 nm, 40 mW cm^{-2}, 5 minutes) after annealing. (B) Bottom part for film annealed, upper part for one irradiated with UV light after annealing. Source: Moritsugu et al. 2011 [31]. Copyright 2011. Reprinted with permission from Wiley-VCH Verlag GmbH & Co.

films, the non-photoresponsive LC mesogen groups were employed because it is possible to dilute the azobenzene chromophore concentration and to help change the orientation states. So, the effect of introducing a biphenyl group as the non-photoresponsive LC mesogen in the hydrophobic azobenzene polymer (PA6Az1–PA6B1 (m:n) copolymers given in Scheme 10.2) was explored. If a UV transparent biphenyl group possesses cooperative orientation ability synchronized with motion of azobenzene group, the switching speed will be improved based on the improvement of the penetration property of UV light in the film due to the decrease in the azobenzene chromophore concentration (Figure 10.12) [46]. Irradiation time required for re-coloration was reduced to one-third by introduction of a biphenyl group more than 25% molar ratio. The patterns of the PA6Az1–PA6B1(m:n)/PVA-20-bilayered film through photomask was evidently clear even with low intensity of UV light (Figure 10.13). Furthermore, it was possible to write the number on the multi-bilayered film by scanning the UV laser at 130 mm s^{-1}. (Figure 10.14)

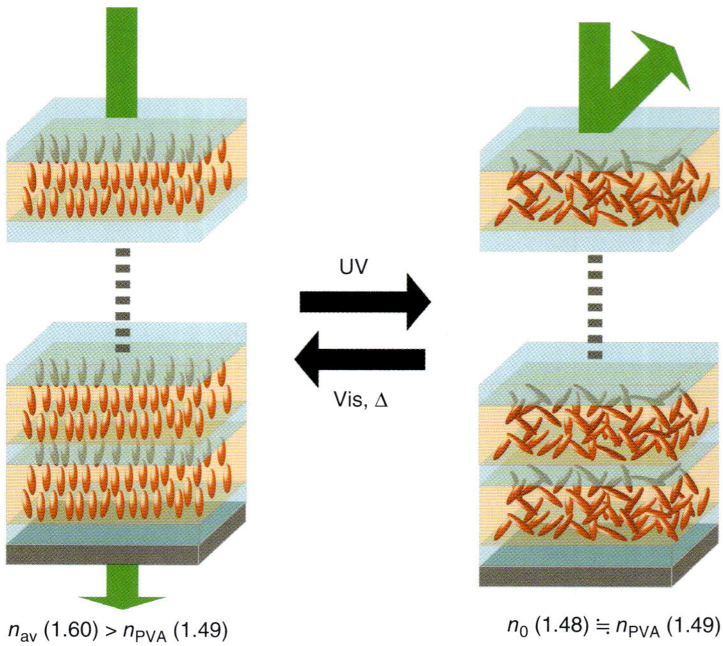

Figure 10.11 Schematic diagram of on–off switching of reflection of the multi-bilayered film.

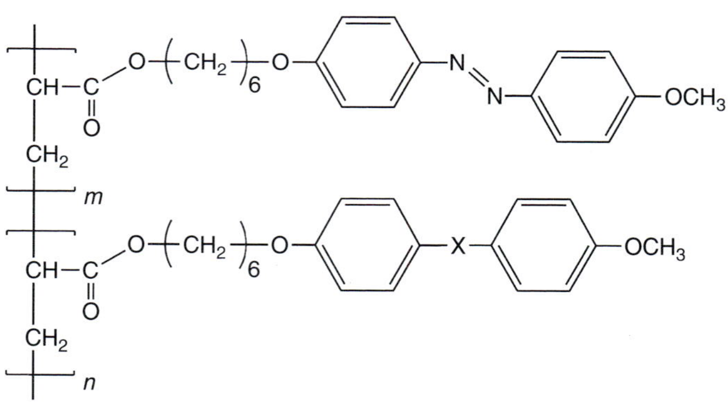

Scheme 10.2 Structure of copolymers PA6Az1–PA6B1 (m:n) consisting of azobenzene and biphenyl side groups.

10.2.4.2 Introduction of Longer Conjugated LC Groups

By using copolymers bearing azobenzene and biphenyl groups as the hydrophobic layer component, the light response property of the multi-bilayered films was improved. However, the reflection intensity deteriorated due to the low refractive index of the biphenyl group as reflection intensity relied on the refractive index difference of composite materials. In order to improve the reflection intensity, it is helpful to employ LC groups exhibiting high refractive indices as well as good cooperative orientation behavior. Since the dielectric constant increases with the long-conjugated system, molecules with the longer conjugated system

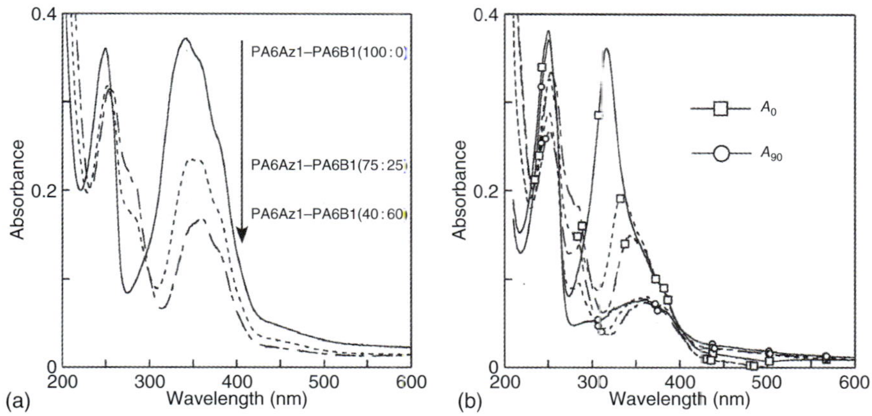

Figure 10.12 Changes in absorption spectra of PA6Az1–PA6B1(m:n) solid film (a) and polarized absorption spectra after thermal annealing (80 °C, 10 minutes) at an incident angle of 40° to the normal of the film by rotating the polarization plane of the monitored light (A_0 and A_{90}) (b). Source: Yagi et al. 2015 [46]. Copyright 2015. Reprinted with permission from Royal Society of Chemistry.

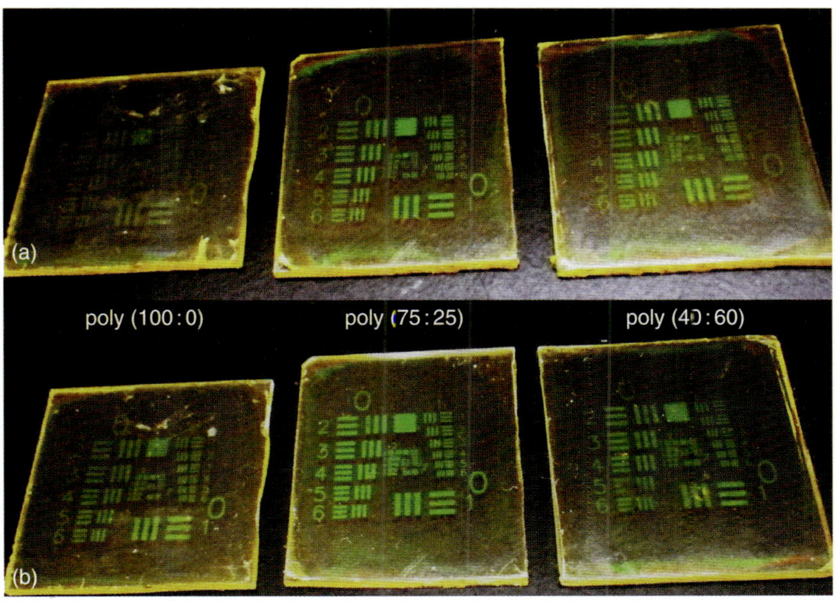

Figure 10.13 Photograph of PA6Az1–PA6B1 (m:n)/PVA-20-bilayered film after irradiation of different intensity UV light ((a) 5 mW cm^{-2} for 3 minutes and (b) 30 mW cm^{-2} for 3 minutes) with photomask. Source: Yagi et al. 2015 [46]. Copyright 2015. Reprinted with permission from Royal Society of Chemistry.

exhibit higher refractive indices. Introduction of a tolan group in a molecular system has been reported to increase the refractive index significantly with high conjugation along the molecular long axis, attributing to increase of average refractive index and birefringence [47–50]. Therefore, the introduction of LC

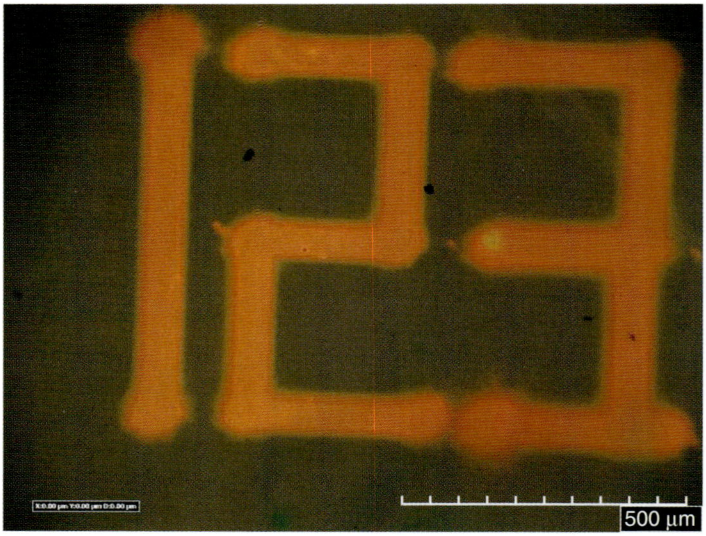

Figure 10.14 Photograph of PA6Az1–PMBP(75 : 25)/PVA-20-multi-bilayered film after annealing at 80 °C and irradiation with UV light by using UV laser (λ = 375 nm, 250 mW cm^{2}, scan speed was 130 mm s^{-1}). Source: Yagi et al. 2015 [46]. Copyright 2015. Reprinted with permission from Royal Society of Chemistry.

groups with long conjugation such as tolan, schiff base, and stilbene in the azobenzene polymers is expected to be effective to improve the photoinduced reflection switching behaviors for the multi-bilayered films composed of the azobenzene-containing copolymers and PVA.

LC groups with long-conjugated systems were employed for the copolymers of Az100 (PA6Az1), AzStb, AzTo, AzToMe, and AzSchi as shown in Scheme 10.3. For the thin films, the orientation states were controlled by irradiation of 365, 436, and 546 nm light, except for the copolymers with the Stb group. For the

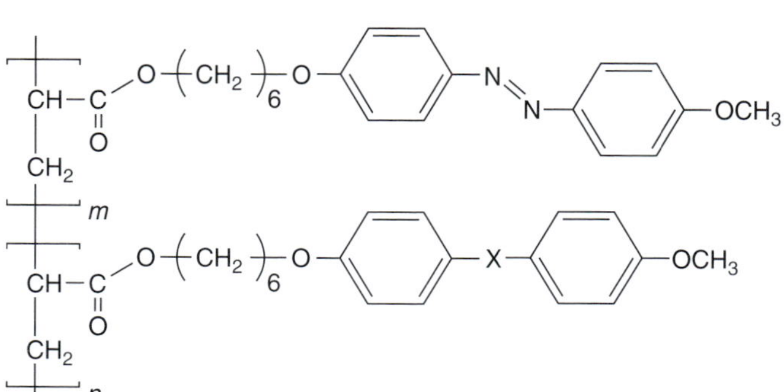

Az100 : R=H, X=N=N
AzStb : R=H, X=HC=CH; AzTo : R=H, X=C≡C;
AzToMe : R=CH_3, X=C≡C; AzSchi : R=H, X=CH=N

Scheme 10.3 Structures of azobenzene copolymers, Az100, AzStb, AzTo, AzToMe, and AzSchi.

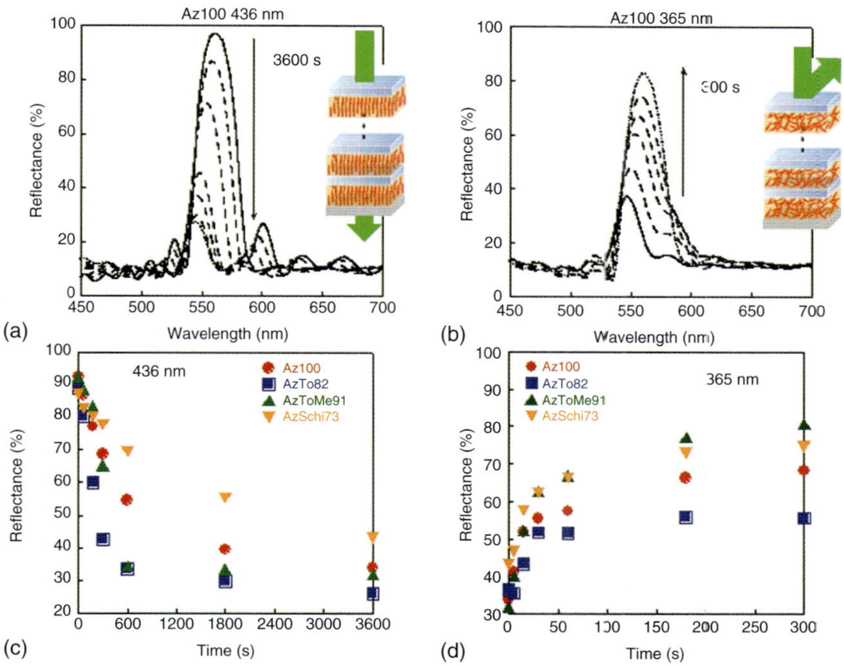

Figure 10.15 The reflection spectra (a, b) and time dependency (c, d) of reflection of Az100 after 436-nm light (a, c) and 365-nm light (b, d). The orientation states of films are illustrated in the figures. Source: Kim et al. 2017 [45]. Copyright 2017. Reprinted with permission from Royal Society of Chemistry.

multi-bilayered films comprising alternate azobenzene-containing layers and PVA layers, photoresponsive speed for on–off switching of the reflection intensity were compared, as shown in Figure 10.15. Four hundred and thirty six nanometers visible-light irradiation caused the decrease in the reflection intensity for all films due to the transformation to the out-of-plane orientation. For copolymers of AzTo82 and AzToMe91, the response speed of reflection was faster than that of Az100. The result indicates that the out-of-plane orientation behavior in the multi-bilayered films was improved by copolymerization with To and ToMe groups, possibly due to their lower T_g as compared to that of Az100. On the other hand, photo-response speed of reflection of AzSchi73 was slower than that of Az100 even with a lower T_g. It is attributed to the surface energy depending on the hydrophobicity of LC groups that are sandwiched by hydrophilic PVA layers.

In order to estimate the hydrophobicity of LC molecules, the dipole moments and partition coefficient (log P) were calculated using Spartan'14 (Wavefunction, Inc.) with the B3LYP/6-31G* method. For Az, Schi, and To groups, they were respectively estimated to be 0, 1.24, and 0 D for dipole moments; and 4.22, 3.65, and 3.72 for log P. Consequently, Schi group exhibiting higher polarity is relatively less hydrophobic. Therefore, AzSchi layers sandwiched by PVA layers are less preferable for the out-of-plane orientation compared with the other copolymers, based on higher surface energy. This is supported by the experimental results of thermally induced out-of-plane orientation behaviors. When AzTo73 and AzSchi73 were annealed at 70 °C, their S values for the out-of-plane orientation were respectively estimated to be 0.73 and 0.34. As expected, S values of AzSchi73 were lower than that of AzTo73.

On the other hand, for 365-nm UV light, the reflection for all films increased. The light response of AzToMe91 was faster than those of Az100 and other copolymers, indicating improvement of random orientation behavior, which was attributed to light penetration deep inside. On the other hand, AzTo82 had apparently poor response to UV light. Namely, random orientation was not sufficiently induced by accumulation of *cis* isomers of azobenzene groups. It is due to the strong intermolecular interaction between To groups as described in our previous reports [51]. It is also supported by the observation that the T_g of AzTo decreased lesser with increase of molar ratio of the mesogen group as compared to the T_g of AzToMe. In contrast, ToMe group is considered to be advantageous for random orientation because of steric hindrance of methyl group.

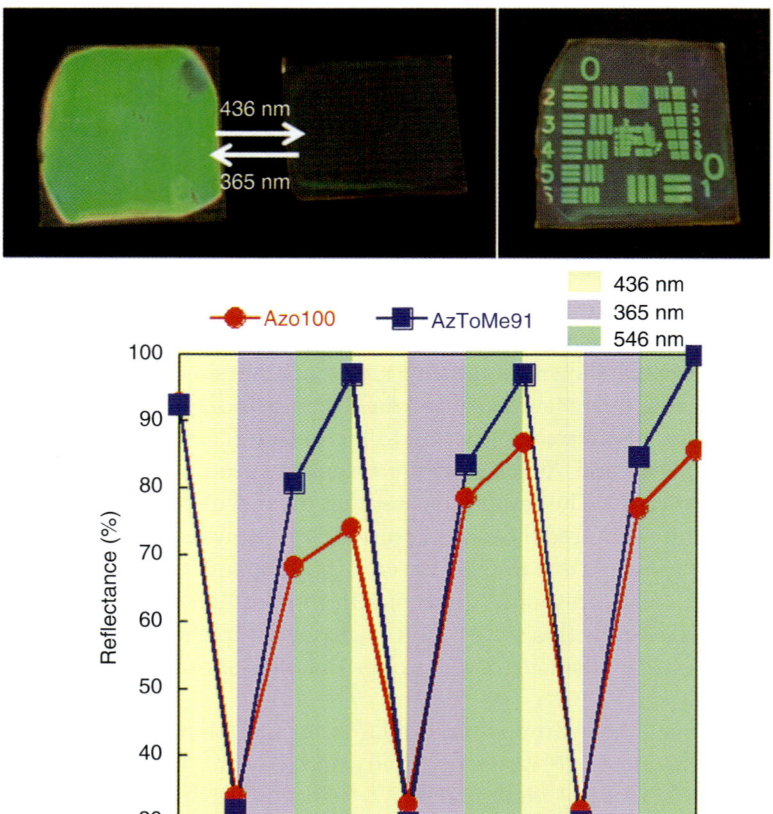

Figure 10.16 Photos of on–off state of reflection and writing patterns on the multi-bilayered film of AzToMe91. Repeated examination for reflection changes of the muli-bilayered films, Azo100 and AzToMe91, by light irradiation of 436 nm (150 mW cm^{-2}, 30 minutes), 365 nm (30 mW cm^{-2}, 10 minutes), and 546 nm (100 mW cm^{-2}, 10 minutes). Source: Kim et al. 2017 [45]. Copyright 2017.Reprinted with permission from Royal Society of Chemistry.

Reversible on–off switching of reflection was achieved by the sequence of light irradiation of 436-nm visible light (for the out-of-plane orientation state; reflection = OFF) and 365-nm UV light following 546-nm visible light (for the random orientation state; reflection = ON), as shown in Figure 10.16. For the photoinduced on–off switching, the reflection contrast of AzToMe91 was improved owing to better orientation behavior and higher refractive index.

10.3 Summary and Outlook

Development of bioinspired materials attracts growing interest, because nature is replete with hierarchically assembled hybrid materials where multi-scale structures provide unique properties and various functions. Therefore, mimicking nature allows us to develop ideal functional materials with high efficiency and sustainability. PCs observed in nature are based on the interaction of light with highly ordered structures.

Many potential applications of PCs require some capability for tuning PBG through external stimuli. Since the wavelength and intensity of PBG strongly depends on the changes in the refractive indices of the materials, PBG switching is possible by using a LC exhibiting stimuli-responsive orientation behavior. Among various promising stimuli, light shows distinct and significant advantages due to its spatial, remote, and temporal controllability. It is also noteworthy that the light sources with different wavelength, intensity, and polarization are readily accessible. Photocontrol of chemical and physical functions of various materials has been studied vigorously using photochemical configurational change of azobenzene derivatives.

This chapter described the on–off switching of reflection of 1D-PC. The 1D-PC was prepared by alternative spin coating of a hydrophobic polymer consisting of azobenzene groups and PVA as hydrophilic polymer, providing multi-bilayered films with precisely controlled layer thickness. The on–off switching was achieved by combining thermally spontaneous molecular orientation of azobenzene molecules and their photoisomerization between *trans*- and *cis*-forms, resulting in the change in PBG of the 1D-PC. To improve the light-response efficiency across the entire films, the molecular structure of azobenzene polymer was studied by introducing non-photoresponsive LC mesogen groups because of their UV transmittance, molecular orientation behavior, and birefringence characteristics. A reflection intensity higher than 90% was observed for the 1D-PC stacked with 20-bilyers, and contrast of reflection change was visually sufficient. The entire thickness of the 1D-PC is within a few microns, so they are light. Therefore, the photoresponsive 1D-PC will be promising candidates for applications to optical materials that need to be light, thin and save energy such as material replacing paper, self-standing reflective color display devices, and so on.

References

1 John, S. and Quang, T. (1997). Collective switching and inversion without fluctuation of two-level atoms in confined photonic systems. *Phys. Rev. Lett.* 78: 1888.

2 Wang, R. and John, S. (2004). Engineering the electromagnetic vacuum for controlling light with light in a photonic-band-gap microchip. *Phys. Rev. A* 70: 043805.
3 Yablonovitch, E. (1987). Inhibited spontaneous emission in solid-state physics and electronics. *Phys. Rev. Lett.* 58: 2059.
4 López, C., Vazquez, L., Meseguer, F. et al. (1997). Photonic crystal made by close packing SiO_2 submicron spheres. *Superlattices Microstruct.* 22: 399.
5 Joannopoulos, J.D., Johnson, S.G., Winn, J.N., and Meade, R.D. (1995). *Photonic Crystals: Molding the Flow of Light.* Princeton University Press.
6 Sukhorukov, A.A., Neshev, D.N., and Kivshar, Y.S. (2007). Shaping and control of polychromatic light in nonlinear photonic lattices. *Opt. Express* 15: 13058.
7 John, S. (1987). Strong localization of photons in certain disordered dielectric superlattices. *Phys. Rev. Lett.* 58: 2486.
8 Noda, S., Tomoda, K., Yamamoto, N., and Chutinan, A. (2000). Full three-dimensional photonic bandgap crystals at near-infrared wavelengths. *Science* 289: 604.
9 Akahane, Y., Asano, T., Song, B.-S., and Noda, S. (2003). High-*Q* photonic nanocavity in a two-dimensional photonic crystal. *Nature* 425: 944.
10 Braun, P.V., Rinne, S.A., and García-Santamaría, F. (2006). Introducing defects in 3D photonic crystals: state of the art. *Adv. Mater.* 18: 2665.
11 Yablonovitch, E. and Gmitter, T.J. (1989). Photonic band structure: the face-centered-cubic case. *Phys. Rev. Lett.* 63: 1950.
12 Yablonovitch, E., Gmitter, T.J., and Leung, K.M. (1991). Photonic band structure: the face-centered-cubic case employing nonspherical atoms. *Phys. Rev. Lett.* 67: 2295.
13 Kinoshita, S., Yoshioka, S., and Miyazaki, J. (2008). Physics of structural colors. *Rep. Prog. Phys.* 71: 076401.
14 Furumi, S., Yokoyama, S., Otomo, A., and Mashiko, S. (2006). Control of photonic bandgaps in chiral liquid crystals for distributed feedback effect. *Thin Solid Films* 499: 322.
15 Kubo, S., Gu, Z., Takahashi, K. et al. (2002). Control of the optical band structure of liquid crystal infiltrated inverse opal by a photoinduced nematic–isotropic phase transition. *J. Am. Chem. Soc.* 124: 10950.
16 Kang, D., Maclennan, J.E., Clark, N.A. et al. (2001). Electro-optic behavior of liquid-crystal-filled silica opal photonic crystals: effect of liquid-crystal alignment. *Phys. Rev. Lett.* 86: 4052.
17 Ozaki, M., Shimoda, Y., Kasano, M., and Yoshino, K. (2002). Electric field tuning of the stop band in a liquid-crystal-infiltrated polymer inverse opal. *Adv. Mater.* 14: 514.
18 Busch, K. and John, S. (1999). Liquid-crystal photonic-band-gap materials: the tunable electromagnetic vacuum. *Phys. Rev. Lett.* 83: 967.
19 Leonard, S., Mondia, J., van Driel, H. et al. (2000). Tunable two-dimensional photonic crystals using liquid crystal infiltration. *Phys. Rev. B* 61: R2389.
20 Shimoda, Y., Ozaki, M., and Yoshino, K. (2001). Electric field tuning of a stop band in a reflection spectrum of synthetic opal infiltrated with nematic liquid crystal. *Appl. Phys. Lett.* 79: 3627.

21 Xia, Y., Gates, B., Yin, Y., and Lu, Y. (2000). Monodispersed colloidal spheres: old materials with new applications. *Adv. Mater.* 12: 693.

22 Kamenjicki, M., Lednev, I.K., Mikhonin, A. et al. (2003). Photochemically controlled photonic crystals. *Adv. Funct. Mater.* 13: 774.

23 Kanai, T., Sawada, T., and Kitamura, K. (2003). Optical determination of the lattice constants of colloidal crystals without use of the refractive index. *Langmuir* 19: 1984.

24 Reese, C.E., Mikhonin, A.V., Kamenjicki, M. et al. (2004). Nanogel nanosecond photonic crystal optical switching. *J. Am. Chem. Soc.* 126: 1493.

25 Natansohn, A. and Rochon, P. (2002). Photoinduced motions in azo-containing polymers. *Chem. Rev.* 102: 4139.

26 Rochon, P., Gosselin, J., Natansohn, A., and Xie, S. (1992). Optically induced and erased birefringence and dichroism in azoaromatic polymers. *Appl. Phys. Lett.* 60: 4.

27 Yager, K.G. and Barrett, C.J. (2006). Novel photo-switching using azobenzene functional materials. *J. Photochem. Photobiol., A* 182: 250.

28 Barrett, C.J., Mamiya, J., Yager, K.G., and Ikeda, T. (2007). Photo-mechanical effects in azobenzene-containing soft materials. *Soft Matter* 3: 1249.

29 Mahimwalla, Z., Yager, K.G., Mamiya, J. et al. (2012). Azobenzene photomechanics: prospects and potential applications. *Polym. Bull.* 69: 967.

30 Rezig, H. and Vitrant, G. (2001). Feasibility of optically controlled integrated Mach–Zehnder device based on azo dye-doped PMMA thin films. *Opt. Commun.* 200: 261.

31 Moritsugu, M., Ishikawa, T., Kawata, T. et al. (2011). Thermal and photochemical control of molecular orientation of azo-functionalized polymer liquid crystals and application for photo-rewritable paper. *Macromol. Rapid Commun.* 32: 1546.

32 de Gennes, P.G. and Prost, J. (1995). *The Physics of Liquid Crystals*, International Series of Monographs on Physics No. 83. Oxford University Press.

33 Ujiie, S. and Iimura, K. (1992). Thermal properties and orientational behavior of a liquid-crystalline ion complex polymer. *Macromolecules* 25: 3174.

34 Date, R.W., Fawcett, A.H., Geue, T. et al. (1998). Self-ordering within thin films of poly(olefin sulfone)s. *Macromolecules* 31: 4935.

35 Sapich, B., Vix, A., Rabe, J. et al. (2006). Ordering and dewetting in spin-coated films of a liquid crystalline main chain polymer. *Thin Solid Films* 514: 165.

36 Rais, D., Zakrevskyy, Y., Stumpe, J. et al. (2008). Photoorientation of azobenzene side groups in a liquid-crystalline polybutadiene-based polymer. *Opt. Mater.* 30: 1335.

37 van der Wielen, M., Cohen Stuart, M.A., Fleer, G.J. et al. (1997). Order in thin films of side-chain liquid-crystalline polymers. *Langmuir* 13: 4762.

38 Wong, G.C., Commandeur, J., Fischer, H., and de Jeu, W.H. (1996). Orientational wetting in hybrid liquid crystalline block copolymers. *Phys. Rev. Lett.* 77: 5221.

39 Tian, Y., Watanabe, K., Kong, X. et al. (2002). Synthesis, nanostructures, and functionality of amphiphilic liquid crystalline block copolymers with azobenzene moieties. *Macromolecules* 35: 3739.

40 Bobrovsky, A., Boiko, N., Shibaev, V., and Stumpe, J. (2004). Comparative study of photoorientation phenomena in photosensitive azobenzene-containing homopolymers and copolymers. *J. Photochem. Photobiol., A* 163: 347.

41 Yagi, R., Katae, H., Kuwahara, Y. et al. (2014). On-off switching properties of one-dimensional photonic crystals consisting of azo-functionalized polymer liquid crystals having different methylene spacers and polyvinyl alcohol. *Polymer* 55: 1120.
42 Moritsugu, M., Kim, S., Ogata, T. et al. (2006). Photochemical switching behavior of azofunctionalized polymer liquid crystal/SiO_2 composite photonic crystal. *Appl. Phys. Lett.* 89: 153131.
43 Kurihara, S., Moritsugu, M., Kubo, S. et al. (2007). Photoswitching properties of photonic band gap material containing azo-polymer liquid crystal. *Eur. Polym. J.* 43: 4951.
44 Komikado, T., Inoue, A., Masuda, K. et al. (2007). Multi-layered mirrors fabricated by spin-coating organic polymers. *Thin Solid Films* 515: 3887.
45 Kim, S., Ishii, S., Yagi, R. et al. (2017). Photo-induced orientation behaviors of azobenzene liquid crystal copolymers for photonic crystals. *RSC Adv.* 7: 51978.
46 Yagi, R., Iwamoto, H., Kuwahara, Y. et al. (2015). On/off switching of structural color by using multi-bilayered films containing copolymers having azobenzene and biphenyl side groups. *RSC Adv.* 5: 84762.
47 Yoneyama, S., Yamamoto, T., Tsutsumi, O. et al. (2002). High-performance material for holographic gratings by means of a photoresponsive polymer liquid crystal containing a tolane moiety with high birefringence. *Macromolecules* 35: 8751.
48 Sarkar, S.D. and Choudhury, B. (2010). Study of refractive indices, density and order parameters of two nematogens and their eutectic mixture. *Acta Phys. Pol. A* 118: 665.
49 Ye, F., Orita, A., Yaruva, J. et al. (2004). High birefringent bisdiynes and hexatriynes based on double elimination of β-substituted sulfones. *Chem. Lett.* 33: 528.
50 Zakerhamidi, M., Ebrahimi, Z., Tajalli, H. et al. (2010). Refractive indices and order parameters of some tolane-based nematic liquid crystals. *J. Mol. Liq.* 157: 119.
51 Moritsugu, M., Shirota, T., Kubo, S. et al. (2009). Enhanced photochemical-shift of reflection band from an inverse opal film based on larger birefringent polymer liquid crystals: effect of tolane group on the photochemical shift behavior. *J. Polym. Sci., Part B: Polym. Phys.* 47: 1981.

11

Photoinduced Liquid Crystal Domain Engineering for Optical Field Control

Wei Hu, Peng Chen, and Yan-Qing Lu

Nanjing University, Department of Quantum Electronics and Optical Engineering, College of Engineering and Applied Sciences, 22 Hankou Road, Nanjing, Jiangsu, 210093, China

11.1 Introduction

Liquid crystals (LCs) have attracted considerable attention due to their pronounced high optical anisotropy/transparency over a wide electromagnetic spectrum, high sensibility to various external field stimuli (such as electric/magnetic field, light irradiation, and heat) and strong capability for matter and electromagnetic field interactions [1–4]. Thanks to their large birefringence and reconfigurable director (i.e. local optical axis) orientation, LCs have become one of the most attractive candidates for making optical devices requiring versatile functionalities. Particularly, liquid crystal display (LCD) holds a dominant position in the flat panel display market nowadays. Essentially, they are composed of a huge number of separate light intensity modulators in the visible range [3, 5]. In addition to amplitude, LCs can freely manipulate other spatial degrees of light, such as polarization and phase. Through specifical designing of the director distribution of LC domains, one could arbitrarily manipulate the optical field [6, 7].

Optical field control is a key requirement for optics and photonics. Structured light, also known as shaped light or tailored light, is a light beam that has customized intensity, polarization, and phase. Recently, specific beams including vortex beams [8, 9], vector beams (VBs) [10], and Airy beams [11] have been studied intensively due to their fantastic properties and diverse applications in optical trapping, laser processing, high-resolution imaging, etc. Generally, these beams can be directly generated from Gaussian beams by spatially manipulating the amplitude or phase of light. Such manipulations could be accomplished by lens [12, 13] or plates [14, 15] with specific curvature radius, photoresist masks [16–18], or metasurfaces [19–21] with delicately designed structures. However, their optical characteristics are static once the respective device is fabricated, which severely restricts corresponding applications. Thanks to the electro-optical tunability of LCs, structured LC domains make these devices dynamic and more efficient.

Photoactive Functional Soft Materials: Preparation, Properties, and Applications, First Edition.
Edited by Quan Li.

To date, several strategies have been developed for LC domain engineering. Typically, patterned electrodes [22–25] and polymer-dispersed LCs [26] were adopted to locally and poloidally guide the LC directors to form binary domains. However, their manufacture is complicated and the devices suffer from fixed structures. Dynamic LC domain generation could be accomplished by a commercial spatial light modulator (SLM) [27, 28]. As shown in Figure 11.1, an SLM consists of numerous discrete micro-size LC pixels driven separately. Nevertheless, the complex electrode matrix makes the SLM costly, optically inefficient, and limits the quality of the generated beams. All the above LC microstructures are realized through the polar angle control of LCs. Actually, their azimuthal angle could also be employed to form LC multi-domains. Micro-rubbing [29–31] and patterned rubbing [32, 33] are adopted to carry out the patterned alignment and subsequently guide the local directors to form the desired pattern. Compared to these mechanical methods, photoalignment is much more suitable for implementing accurate and high-resolution multi-domain LC alignments

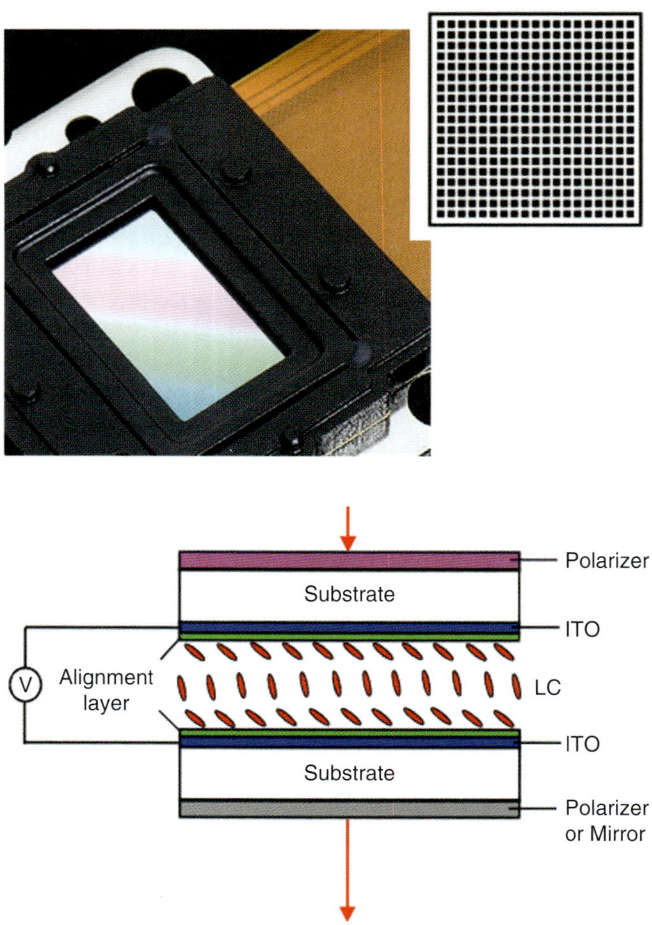

Figure 11.1 The photo and partial schematic diagram of an LC SLM.

[34, 35]. In the initial research, multistep photolithography [35, 36] and holography [37–40] are commonly utilized. More recently, several techniques exhibiting superior image-output flexibility are developed [41–44], facilitating the realization of complex LC orientations and subsequent optical characteristics. These techniques fully unlock a new degree of freedom to manipulate LC domains.

In this chapter, we concentrate on the recent progresses made in optical field control via photopatterned LCs over the past few years. A brief review on photoalignment technology is presented and some of the recently developed photopatterning techniques suitable for generating complicated LC domains are emphasized. Subsequently, fascinating microstructured LC domains with binary or space-variant orientations are demonstrated. Novel optical fields, including arbitrary optical vortices, VBs, and Airy beams, are identified and characterized. The photoinduced LC domain engineering drastically extends the capability of optical field control and affords a promising approach for arbitrarily tailoring the wavefront. It is likely to improvise traditional LC optics and photonics, and inspire promising innovations to this rapid developing field.

11.2 Photoalignment Technology and Photopatterning System

11.2.1 Photoalignment Technology

Since the first work on LC photoalignment in 1988 [45], tremendous improvements have been made over the past 30 years [46]. Compared to conventional contact LC alignment techniques such as rubbing, photoalignment is more promising for high-quality and high-resolution multi-domain LC alignment. Photoalignment avoids any mechanical damage, electrostatic charge, or dust contamination, and also possesses the ability for efficient LC alignment on curved or flexible substrates [46]. These years have witnessed a rapid development in LC photoalignment. The commercially available photoalignment agents are now widely adopted in information display industries.

LC photoalignment originates from the anisotropic intermolecular interactions between the alignment agent with a photoinduced order and adjacent LC molecules. The photoalignment agents can be divided into four types according to the following different mechanisms: (i) photochemical reversible *cis–trans* isomerization in azo-dye containing polymers [47–49], monolayers [50], and organic dye films [45, 51]; (ii) diffusion of azo-dye molecules due to their chromophore reorientation under the action of polarized light [52]; (iii) topochemical cross-linking in cinnamoyl side-chain polymers [34]; and (iv) photodegradation in polyimide materials [53, 54].

One photoaligning mechanism of azo-dye photoalignment agents is the reorientation of azo-dye molecules [52, 55]. The model of rotational diffusion is used to explain the azo-dye reorientation under the action of polarized UV light. Here, we focus on a typical photoalignment agent, i.e. sulfonic azo-dye SD1 [56] shown in Figure 11.2. The azo-dye molecules with transition dipole moments parallel to the direction of incident polarization will most probably get the excess energy, which

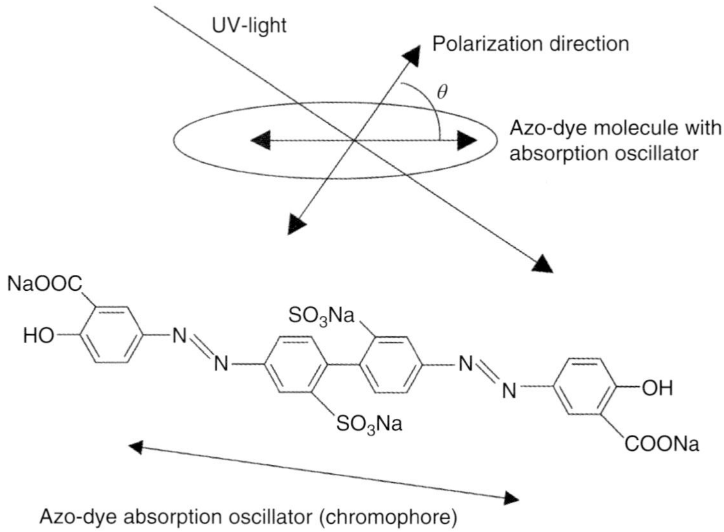

Figure 11.2 The sulfonic azo-dye SD. Under the action of polarized light, the SD molecules tend to reorient perpendicular to the polarization of the activated light. The absorption oscillator is parallel to the molecular axis. Source: Chigrinov et al. 2004 [52]. Copyright 2004. Reproduced with permission of American Physical Society.

results in their reorientation from the initial position. This results in an excess of chromophores with the absorption oscillator perpendicular to the recorded UV polarization. Hence, a thermodynamical equilibrium in the new oriented state will be established [52]. The orientation of SD1 will spread to adjacent LC molecules via intermolecular interactions and thus guide the local LC directors. The azo-dye layers supply sufficiently high anchoring energy to LCs. Additionally, SD1 is photo-rewritable and only the last written polarization information will be recorded. That will drastically facilitate the LC domain engineering.

11.2.2 Photopatterning System

During the past few years, several techniques suitable for flexible and high-resolution image output have been developed [41–43], facilitating the realization of complex LC domains that can present versatile optical functions. Figure 11.3 presents three photopatterning systems. As shown in Figure 11.3a, the first one enables arbitrary interferometric holographic recording via a modified Mach–Zehnder interferometer. The first polarizing beam splitter (PBS) separates the object (O) and reference (R) arms into orthogonal linear polarizations, which are then recombined by a second PBS. A quarter-wave retarder is utilized to transform the beams into orthogonal circular polarizations. The object is placed before the second PBS and interfered with the reference beam (usually a plane wave) to record its wavefront [6]. The second one directly writes the desired pattern by scanning a focused laser beam [43]. The laser direct writing system consists of a polarization rotator and a precise 2D translation system as depicted in Figure 11.3b.

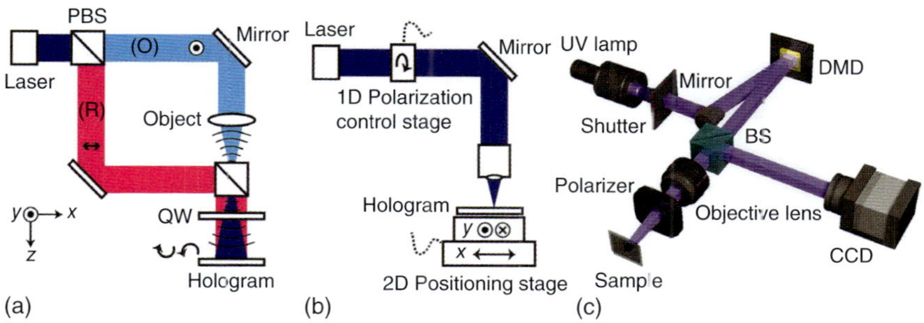

Figure 11.3 Schematic illustrations of photopatterning systems: (a) modified Mach–Zehnder interferometry, (b) laser direct writing. Source: (a,b) Kim et al. 2015 [6]. Copyright 2015. Reproduced with permission of Optical Society of America. and (c) DMD-based microlithography. Source: (c) Ji et al. 2016 [57]. Copyright 2016. Reproduced with permission of Nature Publishing Group.

Figure 11.3c exhibits the third one, namely a digital micro-mirror device (DMD) based dynamic microlithography system [41, 42]. The DMD consists of 1024×768 micro-mirrors, each of which can be independently tilted by an electrostatic force. These mirrors play the role of a dynamic mask. A uniform and collimated UV beam is reflected onto the DMD controlled by the computer. Subsequently, the bundle of light reflected by the "ON" state mirrors carries on the designed pattern. On subsequent focus by a 10× apo-chromatically corrected projection lens and polarized by a motorized rotating polarizer, the beam then projects onto an LC cell or glass substrate spin-coated with photoalignment films placed at the image plane. A charge-coupled device (CCD) is utilized to collect the light reflected by the sample in order to monitor the focusing process. The optical path between the sample and the DMD is conjugated to that of the sample to CCD. Owing to the excellent flexibility of image-output capability of the DMD, arbitrarily fine LC multi-domains can be conveniently obtained.

In addition to the above techniques, some others have also been introduced to accomplish high-resolution multi-domain LC alignments. For example, the plasmonic photoalignment technique enables patterning LC molecular directors with a diffraction-limited spatial resolution and single exposure [44]. The realization of well-fabricated LC domains is at the heart of all optical LC devices. These photoinduced LC domain engineering techniques make arbitrary tailoring of optical fields possible.

11.3 Binary LC Domains for Binary Optics

Through an LC cell with locally controlled azimuthal angles, the same linearly polarized light will experience different phase shifts due to the angle-dependent efficient refractive index. The predesigned LC director distribution results in a patterned phase profile. Therefore, arbitrary optical elements in binary optics can be obtained by properly designing alternative LC domains. Usually, a two-step photo-exposure process is performed [36, 58]. At first, the substrate or cell spin-coated with photoalignment layers is uniformly exposed with a linearly polarized UV light, and then is exposed again to receive the designed pattern with the polarizer rotating at a certain degree (for example, 90°). The exposed areas are realigned to be orthogonal to the first orientation. Therefore, the substrate(s) are synchronously imprinted with alternative 0° and 90° alignments consistent with the desired phase pattern.

The grating, a typical binary optical element, can be divided into two kinds: amplitude grating and phase grating. By introducing periodically orthogonal LC parallel alignment (PA) domains, binary phase grating is formed, while amplitude grating can be formed by alternative twist nematic (TN) domains. In initial researches, micro-rubbing [29–31], two-step rubbing [32, 33], and holographic recording method [59, 60] have been adopted to produce these gratings. In 2012, Hu et al. [36] introduced SD1 into the formation of an orthogonal PA LC grating (Figure 11.4). Different domains in adjacent regions are achieved by two-step photo-exposure to guide orientations of SD1 layers and further align the LC molecules. Both one-dimensional (1D) and two-dimensional (2D) switchable phase gratings have been demonstrated. Its polarization-independent diffraction is theoretically studied and experimentally tested. Such polarizer-free gratings show very high transmittance (~92%), diffraction efficiency (over 31%), and optical contrast (over 150). The alternate PA/TN gratings are also demonstrated in cells placed between two crossed polarizers [61, 62]. Both 1D- and 2D-tunable gratings are demonstrated. The work is based on the combination of phase and amplitude modulation. Some special LC materials, including dual-frequency

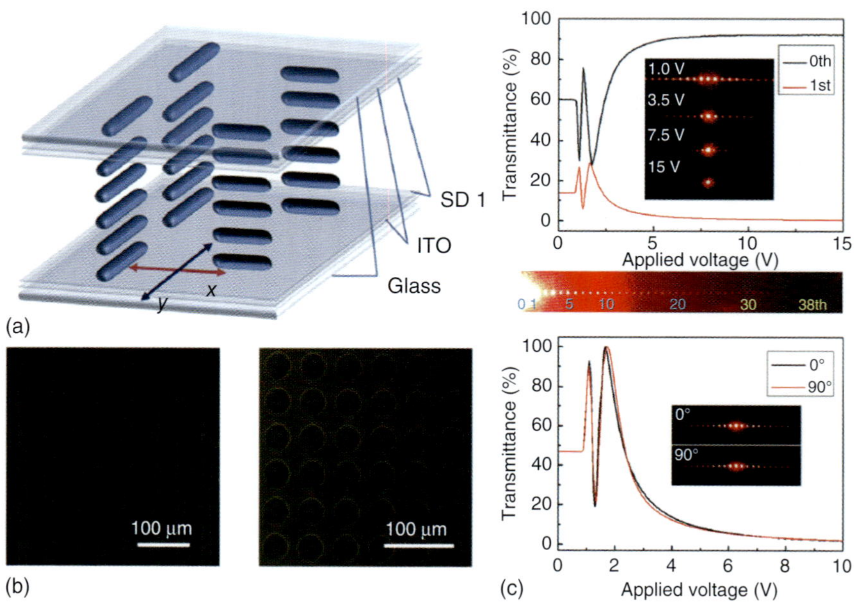

Figure 11.4 The schematic cell structure (a), micrograph (b), and diffraction properties (c) of binary LC gratings. Source: Hu et al. 2012 [36]. Copyright 2012. Reproduced with permission of American Institute of Physics.

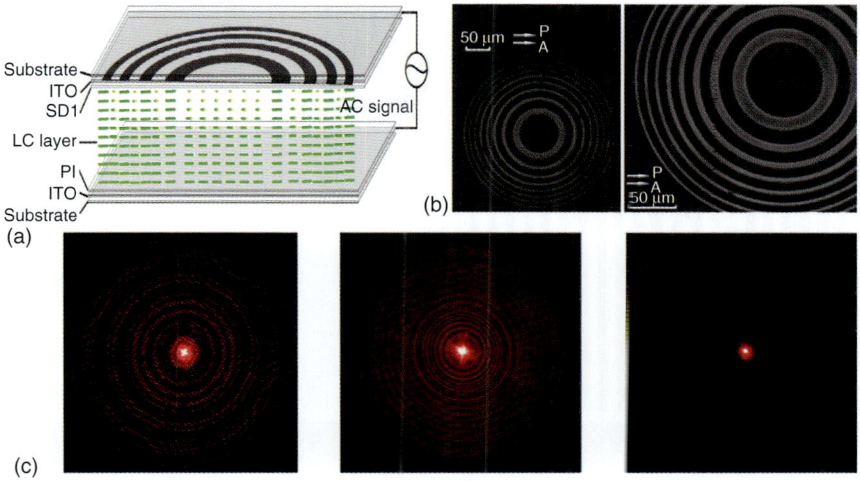

Figure 11.5 The schematic cell structure (a), micrograph (b), and diffraction patterns (c) of an LC Fresnel lens with alternative PA and TN domains. Source: Wang et al. 2013 [67]. Copyright 2013. Reproduced with permission of Optical Society of America.

LCs [63] and ferroelectric LCs [64], are introduced to improve the response of these gratings. Thanks to the rewritability of the photoalignment agent, the optical function of these elements can be dynamically changed [65, 66].

The electrically tunable and switchable lens is desired for wide applications. LC lenses have attracted extensive attention. Among them, LC Fresnel lens is promising due to its outstanding electro-optical performances. It can be realized by specifically aligning the LC directors to form a nonperiodic refractive index profile (Figure 11.5). Similar to those LC gratings, different alignment modes are adopted, including alternative PA and TN domains [67], single-side-patterned domains [68], and hybrid photo-aligned domains [69]. Different LC materials are also employed, including ferroelectric LCs [70] and dual frequency LCs [69], to improve the performance of these lenses.

Let us come back to grating. If a dislocation is introduced at the center, the so called "fork" grating is obtained. It is a computer-generated hologram of a Gaussian beam and an optical vortex (OV). OVs are characterized by helical phase fronts and donut-shaped intensity distributions. Their unique helical phase fronts result in an orbital angular momentum (OAM) of $m\hbar$ per photon [12]. More generally, they can be described by the so-called Laguerre–Gaussian (LG) mode with two specific indices: the topological charge m and the radial index p. The OAM of OVs adds a new degree of freedom to the manipulation of light, leading to applications such as optical tweezers, quantum informatics, and OAM based optical communications [8, 9, 71]. With fork grating, a Gaussian beam can be conveniently diffracted into a series of OVs.

LC fork gratings in various alignment modes with arbitrary topological charges are fabricated [58]. Figure 11.6 exhibits the micrograph and diffraction properties of a sample with $m = 1$. Thanks to the design of orthogonal alignment, the two perpendicular components derived from all incident polarization

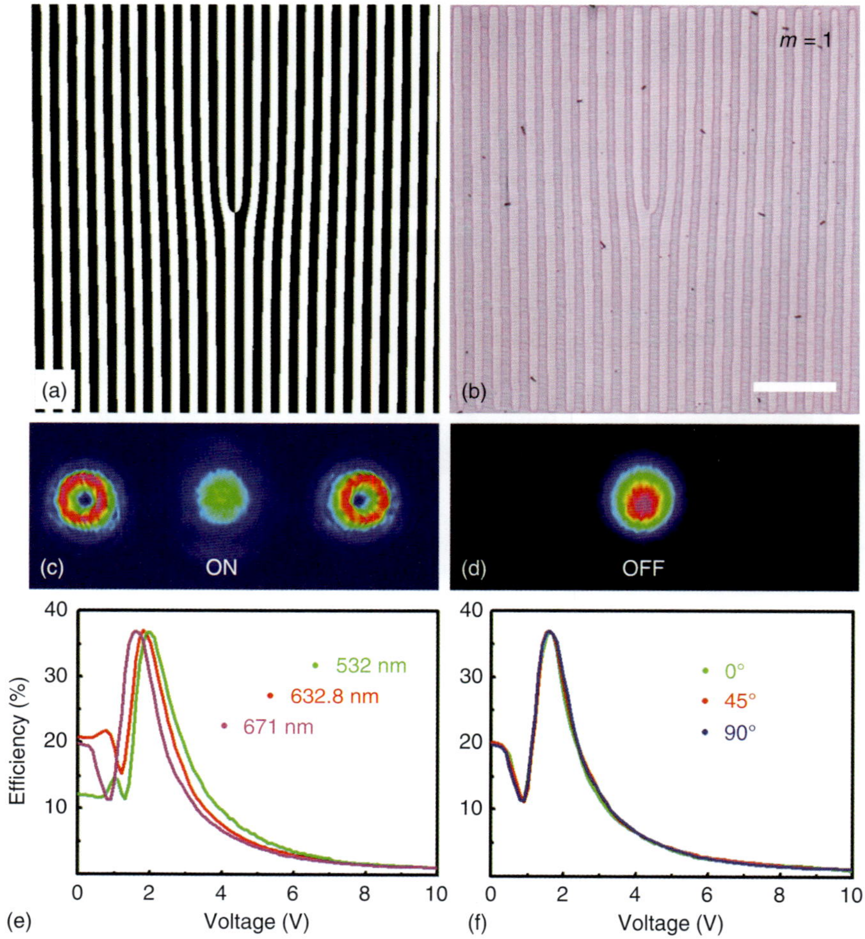

Figure 11.6 (a) A fork grating with $m = 1$. (b) Micrograph of an HAN LC fork grating. The scale bar is 100 μm. Corresponding diffraction patterns at (c) ON and (d) OFF state. V–η curves of the 1st order OV with $m = 1$ at (e) different incident wavelengths: 532, 632.8 and 671 nm; and (f) different incident polarization angles: 0°, 45°, and 90°. Source: Wei et al. 2014 [58]. Copyright 2013. Reproduced with permission of John Wiley and Sons.

experience the same refractive index change, resulting in an excellent polarization independency [36, 63]. A maximum efficiency of 37% (74% for total ±1st orders) is obtained (ON state) as shown in Figure 11.6c. Here, the obtained efficiency is close to the theoretical limitation, and is significantly improved compared to that generated by a commercial SLM. The dynamic switching between Gaussian mode and OVs can be easily accomplished through electrical tuning. Besides, the electrically tunable half-wave condition makes the device suitable to a broad wavelength range. Since SD1 is optically rewritable, the pattern could be erased and then rewritten at the optical isotropic state to obtain a new hologram. This means that the mode variation is also dynamic [58]. Ferroelectric LCs can be further introduced to accomplish a ferroelectric LC fork grating with the merit of ultra-fast response [72].

For conventional grating and fork grating with only two phase regions in each period, the efficiencies of higher orders are quite limited, restricting their applications. Dammann gratings (DGs) [73, 74] can satisfactorily solve this problem. Each period of a DG is composed of certain binary phase regions of different widths. By optimizing the number and values of phase transition points, the light energy can be distributed into desired orders with good uniformity and high efficiency [75]. Such DGs can be realized by photopatterning LCs as well [76, 77]. The concept and implementation of DGs can be introduced into various optical elements based on phase modulation, such as circular grating and fork grating. On one hand, different-order binary-phase LC circular DGs with two mutually orthogonal photoinduced domains in every two adjacent alignment domains are proposed for generating annular uniform-intensity optical fields (Figure 11.7a–c) [78, 80]. Such LC circular DGs exhibit merits of high efficiency, good uniformity, and outstanding optical/electrical tunability. On the other hand, by encoding DGs into fork gratings, the so-called Dammann vortex gratings (DVGs) are achieved. DVGs can provide equal-energy OV array with specific OAMs at designed diffraction orders [18, 81, 82], enabling parallel OAM detection that is a key requirement in OAM based optical communications. The LC DVG featured by alternatively orthogonal PA regions is presented [79]. Figure 11.7d exhibits the micrograph of a 5 × 5 LC DVG with m = 1 & 1. The diffraction pattern at 632.8 nm in Figure 11.7e shows a 2D donut-like OV array in high quality and good energy-distribution uniformity. Thanks to the electro-optical tunability of LCs, equivalent high efficiencies could be achieved

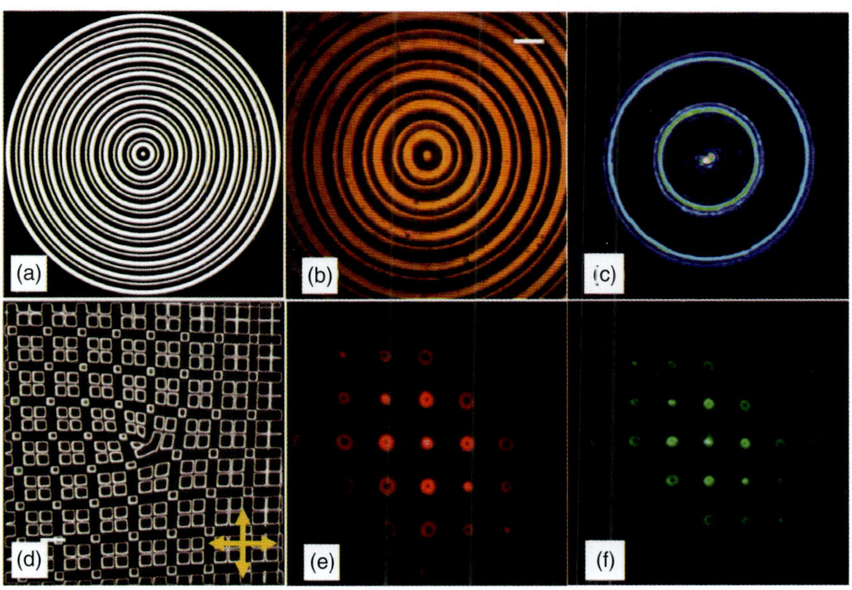

Figure 11.7 (a) Theoretical phase pattern, (b) micrograph, and (c) diffraction pattern of the LC circular DG. Source: (a–c) Luo et al. 2011 [78]. Copyright 2011. Reproduced with permission of Optical Society of America. (d) micrograph and diffraction patterns of the LC DVG at (e) 632.8 and (f) 532 nm. Source: (d–f) Chen et al. 2016 [79]. Copyright 2016. Reproduced with permission of American Physical Society.

for different wavelengths by slightly tuning the applied voltages. A case for 532 nm is shown in Figure 11.7f. These LC DVGs present excellent polarization independence, electrical switchability, and tunability, which are crucial characteristics for OAM generation, manipulation, and detection.

For the above discussed binary LC elements, the desired phase profiles arise from dynamic phases introduced by the in-plane optical path differences. Theoretically, the maximum diffraction efficiency for traditional binary optics is 40.5% for a single first order. To further improve the efficiency for optical field control, much higher phase orders between 0 and 2π should be adopted to maximally mimic a continuous phase modulation. These could be hardly realized via traditional SLMs which work based on dynamic phase modulation.

11.4 Space-Variant LC Domains for Geometric Phase Modulation

Unlike a dynamic phase, the geometric phase (also named Pancharatnam–Berry phase [83]) originates from the spin–orbit interaction of light, i.e. the coupling between the polarization and the trajectory of light [84, 85]. The geometric phase depends only on the geometrical characteristics and can be created from inhomogeneous anisotropic media, such as LCs [6, 86] and anisotropic scatters based metasurfaces [19, 87, 88]. Usually, its magnitude is proportional to the orientation angle of the effective local optical axis, and has a polarization-dependent sign. Arbitrary wavefront shaping can be realized by encoding the desired phase profile into the in-plane optical axis orientations. As it is very easy to manipulate the optical axis for geometric phase elements, a continuous geometric phase change can be conveniently achieved. Therefore, the input energy can be 100% diffracted into a single order.

Again, we choose grating as an example. Polarization grating (PG) is a geometric phase element characterized by periodically and spatially variant optical axes. It is usually fabricated with LCs. As shown in Figure 11.8, the LC directors are homogeneous along the z axis and obey the following equation in the x–y plane: $\alpha(x) = -\pi x/\Lambda$ [89], where Λ is the pitch of the polarization grating. The diffraction property of the PG can be obtained by Jones matrix calculation [4]. The PG has only three diffraction orders: ±1st orders and 0th order. The intensity distribution among the three orders depends on the phase retardation Γ and the incident polarization. By adjusting Γ, switching between 0th and ±1st orders can be realized. If $\Gamma = \pi$, the incident linearly polarized light will equally diffract to the ±1st orders. If the input light is circularly polarized, only one first order will be obtained. In this manner, theoretically we can predict a 100% diffraction efficiency for a single first order.

To accomplish such a space-variant LC director distribution, polarization holography is usually adopted, which employs two orthogonally circularly polarized beams [90–92]. The interferometric pattern is uniform in light intensity, but

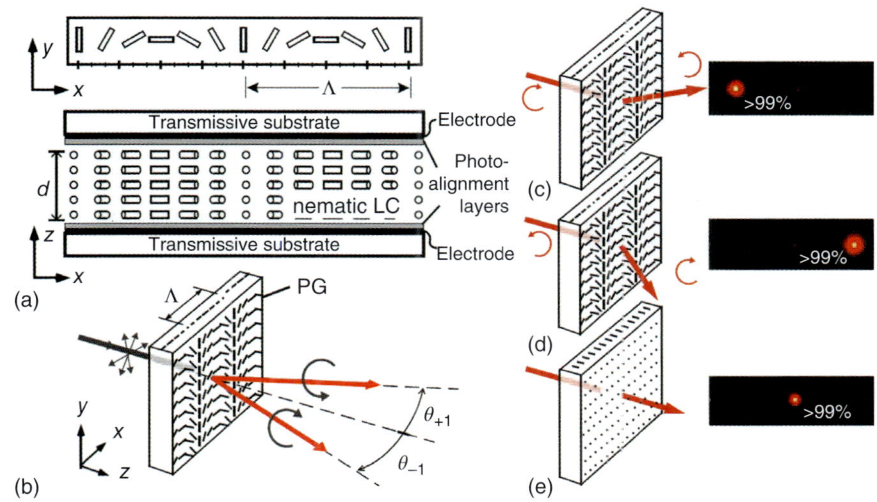

Figure 11.8 Schematic illustrations and diffraction properties of an LC PG [89].

has a spatially variant linear polarization obeying $\alpha(x) = -\pi x/\Lambda$. After recording such a polarization hologram onto the photoalignment films, the director distribution of the LC PG can be realized. Besides, the laser direct writing method and the DMD based microlithography can also be used [43, 93]. LC PGs can be applied in broad fields, including wide viewing angle displays, polarization beam steering, Stokes parameters measurement, and polarization image sensors [89]. Most recently, dual frequency LCs have been introduced to the design of PGs to realize fast-response and high-efficiency optical switch [93]; by integrating multidirectional LC PGs and multilayered functional LC polymer films, the high transmittance holographic polarizer is demonstrated with a transmittance efficiency up to 90% [94].

The focusing/defocusing function of traditional LC lens is independent of incident polarization, while geometric phase lens can locally modulate the incident polarization states and result in a circular-polarization-dependent continuous lens phase profile [95, 96]. The theoretical optical axes of such a lens are homogeneous along the z axis and obey $\alpha = \pi/\lambda\left(\sqrt{r^2 + f^2} - f\right)$ in x–y plane, where α is the azimuthal angle, r is the radius, $r^2 = x^2 + y^2$, f is the designed focal length, and λ indicates the free space wavelength. Through Jones matrix calculation, the property of the helicity-dependent focusing/defocusing can be theoretically predicted. As shown in Figure 11.9, when the phase retardation satisfies the half-wave condition $\Gamma = (2n + 1)\pi$ (n is an integer), for a right circularly polarized (RCP) incident beam, the output is transformed into left circularly polarization (LCP) with a spherical phase profile, leading to a concave spherical wavefront. Thus, a focused beam is obtained. On the contrary, for an LCP incident beam, the transformed RCP is defocused. Such helicity-determined lenses are promising in 3D displays, compact photonic spin filters [97], and multi-spot focusing engineering.

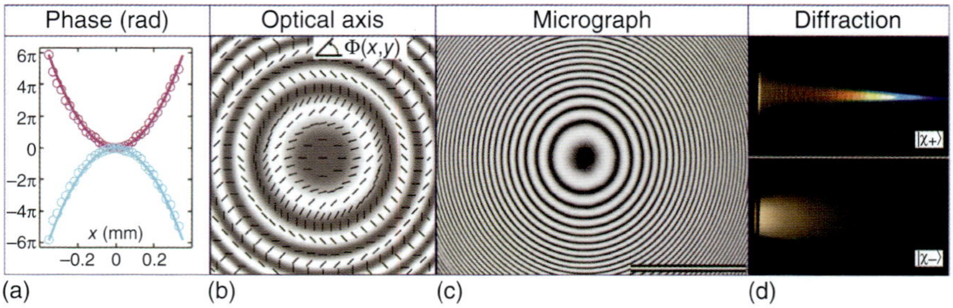

Figure 11.9 LC geometric phase lens. Source: Kim et al. 2015 [6]. Copyright 2015. Reproduced with permission of Optical Society of America.

Q-plate [86, 98–100] is another attractive geometric phase element, which has been intensively studied in OAM-lightened fields. Among various techniques for OV generation, the q-plate is especially promising due to the coupling of spin angular momentum (SAM) and OAM [86]. It not only plays an important role in classical optics [101, 102], but also enables many quantum-optical applications [103, 104]. As shown in Figure 11.10, the q-plate is a half-wave plate with space-variant optical axis in the transverse plane, which follows $\alpha(r, \varphi) = q\varphi + \alpha_0$, where r is the polar radius, φ is the azimuthal angle, and α_0 is the initial angle when $\varphi = 0$. The number q, which denotes the changing rate of the optical axis with respect to the azimuthal angle, depicts the specific geometry. The q-plates will convert circularly polarized light into an OV (topological charge $m = \pm 2q$). When a circularly polarized light beam with OAM states of m traverses the q-plate, an OAM variation of $\pm 2q\hbar$ is imposed. Herein, the sign depends on the input polarization, positive for LCP and negative for RCP. The output polarization is sign-inverted. The q-plate will convert linearly polarized light into a VB (polarization order $P = 2q$, depicts the degree of polarization symmetry) [99, 105]. Unlike OVs with phase singularity, VBs are featured by spatially variant polarization states and centric polarization singularities. They exhibit unique focusing properties thus enabling high-resolution imaging and optical manipulation [10].

Owing to the pronounced optical birefringence and controllable director distribution, LCs are the first choice for q-plate fabrication. Recent years have witnessed great progress in both the design and the fabrication of LC q-plates. The objective LC orientations can be obtained via self-assembled [106, 107] or extra-field-induced [108, 109] LC topological defects, and circular rubbing or photoaligning [99, 110]. In 2016, the concept of "*meta-q-plate*" was proposed, which enabled the arbitrary control of q and α_0 along r and φ [57]. The meta-q-plate supplies a point-to-point beam shaping method, thus drastically enhancing the capability of manipulation of geometric phases. By further integrating the spiral geometric phase of the q-plate with a PG, a geometric phase lens, or both of them, one can obtain a forked PG [111], a vortex lens [112], or a forked vortex lens [113]. These integrated geometric phase devices combine the features of polarization-dependent diffraction, OV/VB generation, and focusing and defocusing functions. Besides, vector vortex beams can be generated by stacking

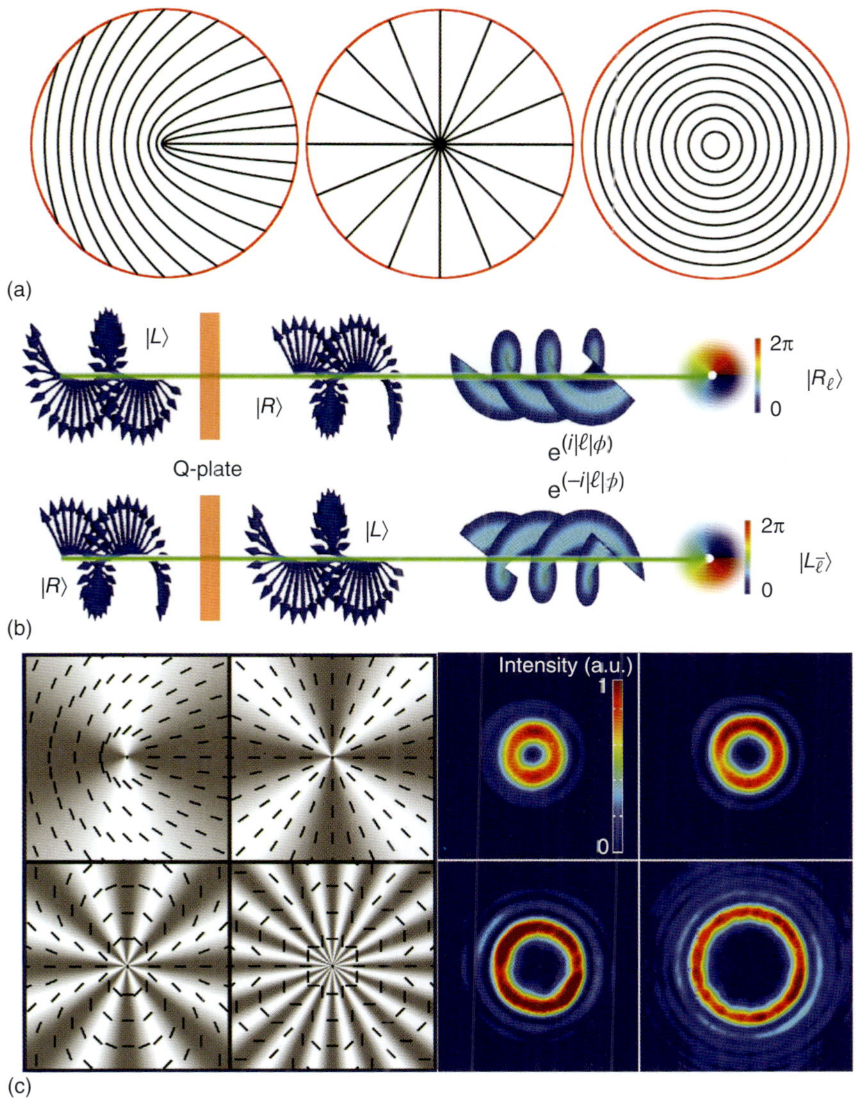

Figure 11.10 (a) Optical axis distributions of different LC q-plates. Source: Marrucci et al. 2006 [86]. Copyright 2006. Reproduced with permission of American Physical Society. (b) Spin–orbital coupling based on a q-plate. Source: Naidoo et al. 2016 [98]. Copyright 2016. Reproduced with permission of Nature Publishing Group. (c) Micrograph with local optical axes labeled and corresponding generated OVs. Source: Kim et al. 2015 [6]. Copyright 2015. Reproduced with permission of Optical Society of America.

different q-plates and waveplates [114]. Both the topological charge and polarization order of generated vector vortex beams can be controlled independently by stacking and reordering different optical films.

By encoding the phase profile of desired structured beams into the LC director orientations, arbitrary optical field shaping can be realized. Other examples include Airy beam [115] (Figure 11.11), Bessel beam [116], depolarized beam

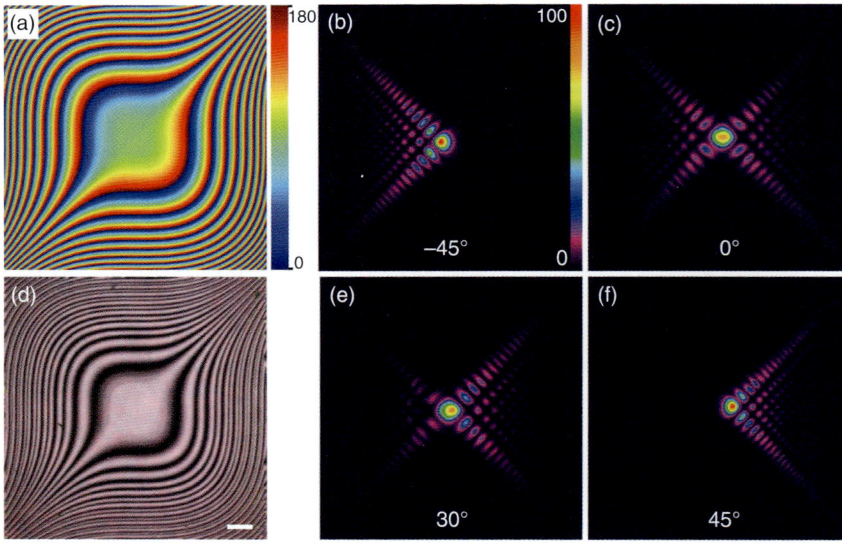

Figure 11.11 The LC polarization Airy mask and corresponding generated Airy beams. Source: Wei et al. 2015 [115]. Copyright 2015. Reproduced with permission of Nature Publishing Group.

[117], and even complex hologram [6]. Airy beam exhibits features of non-diffraction, transverse acceleration, and self-healing [11]. It has attracted considerable interest in broad fields including optical manipulation, micro-fabrication, and biology science. The geometric phase has been introduced to Airy mask to form an LC polarization Airy mask, in which the azimuthal angles of LC cubically vary along both x and y axes [115]. Via incident polarization control, switch between single- and dual-branch Airy beams is realized. Their non-diffraction and transverse acceleration features are verified. Furthermore, self-healing and transverse accelerating OVs are generated via modulating Gaussian beams through subsequent LC q-plate and polarization Airy mask [118].

All the above LC geometric phase elements are made of nematic LCs, which can be considered as a space-variant wave-plate. For such transmissive elements, half-wave conditions should be precisely satisfied to maximize the conversion efficiency. EO tuning is necessary for matching different wavelengths, restricting their applications in fields with wavelength-independency requirement.

Recently, the spin–orbit geometric phase was discovered for the light reflected off planar cholesteric liquid crystal (CLC) chiral superstructures [119–122] (Figure 11.12). CLC is a topical liquid crystalline phase, where the rod-like molecules self-assemble into a periodic helical superstructure, contributing to a natural 1D photonic crystal. Thanks to the circular-polarization selective Bragg reflection and the external-field tunablility [123], CLCs have attracted much attention in versatile applications beyond mere displays [124–128]. For light propagating along the CLC helix axis, a circular-polarization-selective Bragg reflection is exhibited over a wavelength range $n_o p \sim n_e p$, where p is the helical pitch, and n_o/n_e are the ordinary/extraordinary refractive indices, respectively

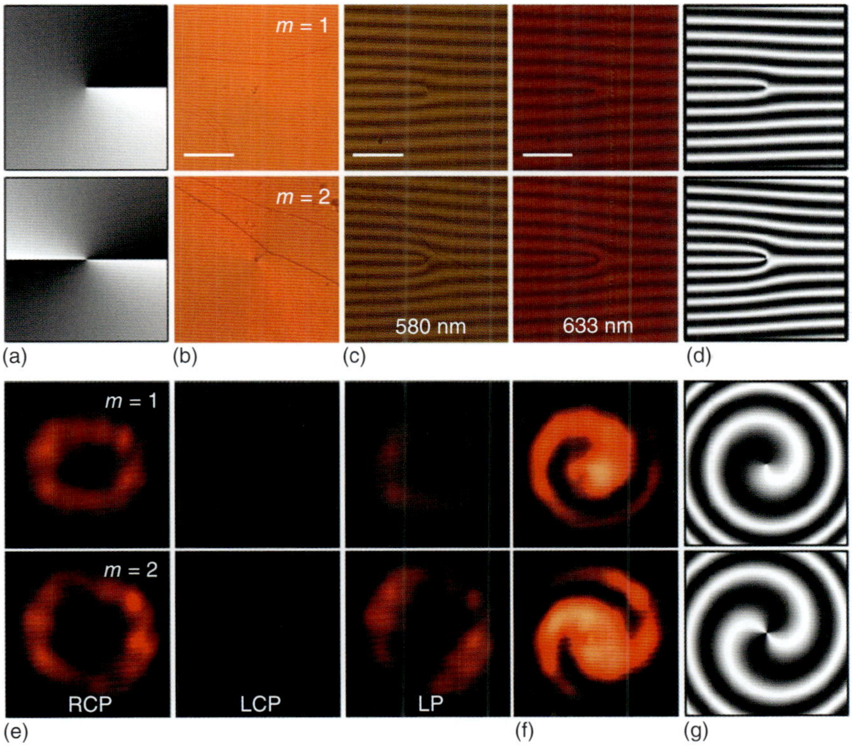

Figure 11.12 Broadband reflective OV generation from space-variant CLC domains. Source: Kobashi et al. 2016 [122]. Copyright 2016. Reproduced with permission of American Physical Society.

[4]. The circularly polarized light with the same handedness as the chiral helix of CLCs is reflected, while the orthogonal one is transmitted. The reflective geometric phase is attributed to the preservation of the polarization helicity from such circular Bragg reflection, which implies flipping of the optical SAM [119]. The magnitude of the reflected light phase is twice the local orientation angle of CLC planar alignment and exhibits a CLC handedness-dependent sign [120, 129]. Therefore, it supplies a new platform for broadband reflective geometric phase manipulation. Similar to those in nematic LCs, via preprogramming the CLC domains, planar reflective elements including light deflectors [121], lenses [121], and OV generators [122, 130] can be obtained.

11.5 Digitalized LC Domains for Digitalized Geometric Phase

Traditional geometric phases are usually limited to handling single-structured beams and are unsuitable for parallel manipulation. If this challenge could be overcome, the capability for programming geometric phases and tailoring structured beams will be drastically extended. For instance, the creation of higher order Poincaré (HOP) sphere beam arrays would inspire multiple optical tweezers [131], parallel laser fabrication, and higher efficiency-stimulated emission depletion microscopy [132]. The HOP sphere generalizes OVs and VBs, and vividly describes their higher order polarization states [133, 134]. Compared with traditional Poincaré sphere whose poles indicate two opposite SAM eigenstates, the poles of an HOP sphere represent orthogonally circularly polarized OVs. Two such eigenstates are featured by helical phase-fronts and OAMs. Each point on the HOP sphere is a linear superposition of them. Especially, the equator indicates VBs featured by space-variant linear polarizations. The HOP sphere incorporates SAM and OAM, thus extending to a higher dimension angular momentum and adding a new degree of freedom to the manipulation of light.

In 2016, the concept of digitalized geometric phase modulation was proposed and demonstrated by means of digitalizing LC domains [135]. As shown in Figure 11.13, traditional space-variant spiral geometric phase was digitalized by imprinting a particular binary phase referring to binary optics. With such digitalized geometric phase elements, HOP sphere beam lattices available for SAM and OAM parallel encoding are demonstrated for the first time. Through tuning the incident spin state, any point on the HOP sphere can be achieved, including special cases of VBs and OAM modes. Square as well as hexagonal lattices with different desired orders are realized in high efficiency and good energy uniformity. By electrically tuning, the phase retardation can reach π at different wavelengths, making the digitalized LC domains wavelength-tolerant. Besides, the electro-optical tunability of LCs makes it possible to dynamically switch between ON/OFF states.

The idea is not restricted to Dammann encoded spiral geometric phase, and can extend to other integration of binary phases and space-variant geometric phases, thus enabling tailoring of versatile optical beams. Actually, the medium is not limited to nematic LCs. Other natural or artificial birefringent materials could be adopted as well. As a special case, CLC chiral superstructures can be digitalized for advanced functionalities [136].

CLC chiral superstructure exhibits unique features of polychromatic and spin-determined phase modulation. By encoding a specifically designed binary pattern and emphasizing the spin-dependent phase invariance, an innovative CLC OV processor is demonstrated. As exhibited in Figure 11.14, up to 25 different OVs are extracted with equal efficiency over a wavelength range of 116 nm. The multiplexed OVs can be detected simultaneously without mode cross talk or distortion, permitting a polychromatic, large-capacity, and *in situ* way for parallel OV processing [136]. Such CLC domains exhibit versatile functionalities, and provide a satisfactory platform for OV-lightened and other cutting-edge territories, especially, the OAM-incorporated wavelength-division multiplexing (WDM), which is expected to dramatically expand the communication capacity [137, 138]. It is a vital step forward to extending the fundamental understanding and fantastic applications of ordered soft matters.

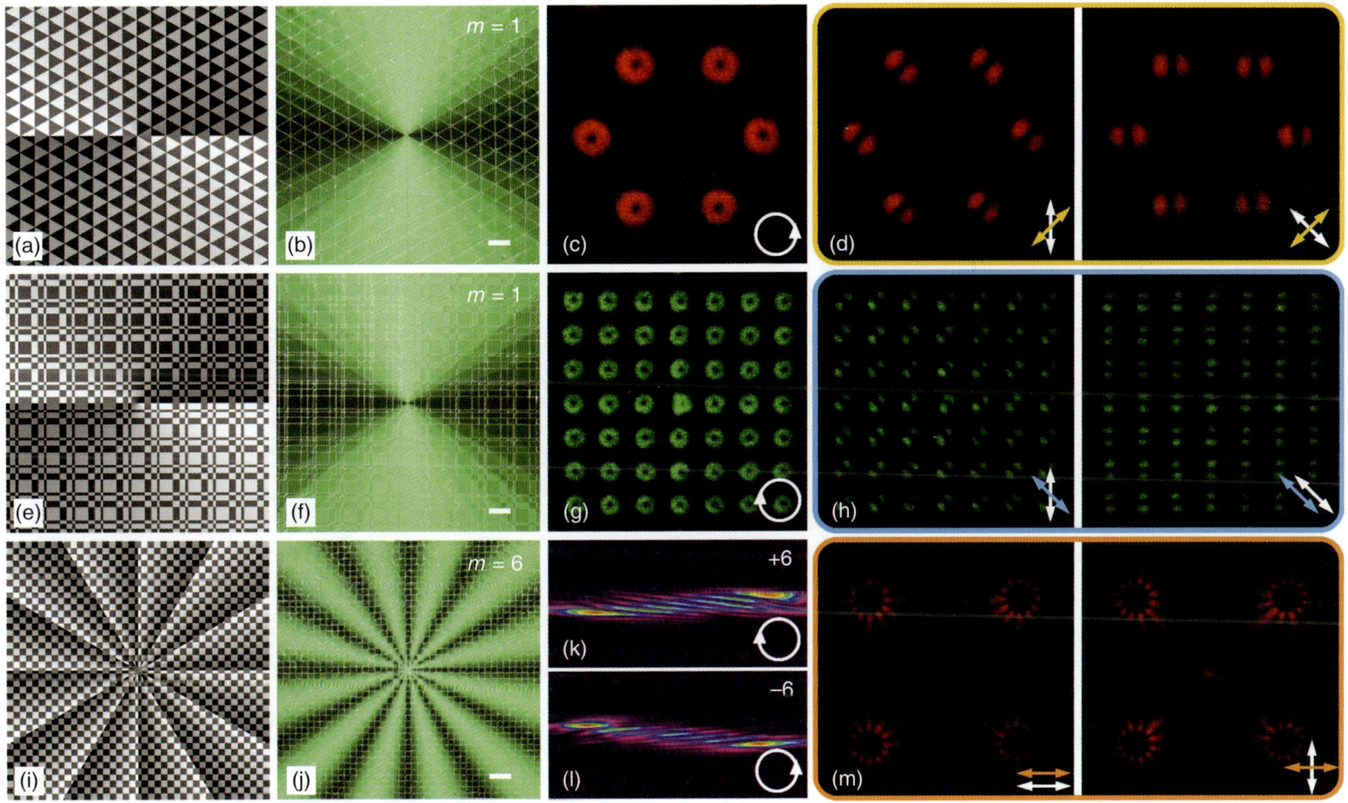

Figure 11.13 The digitalized LC domains for digitalized geometric phase modulation. Source: Chen et al. 2017 [135]. Copyright 2017. Reproduced with permission of American Chemical Society.

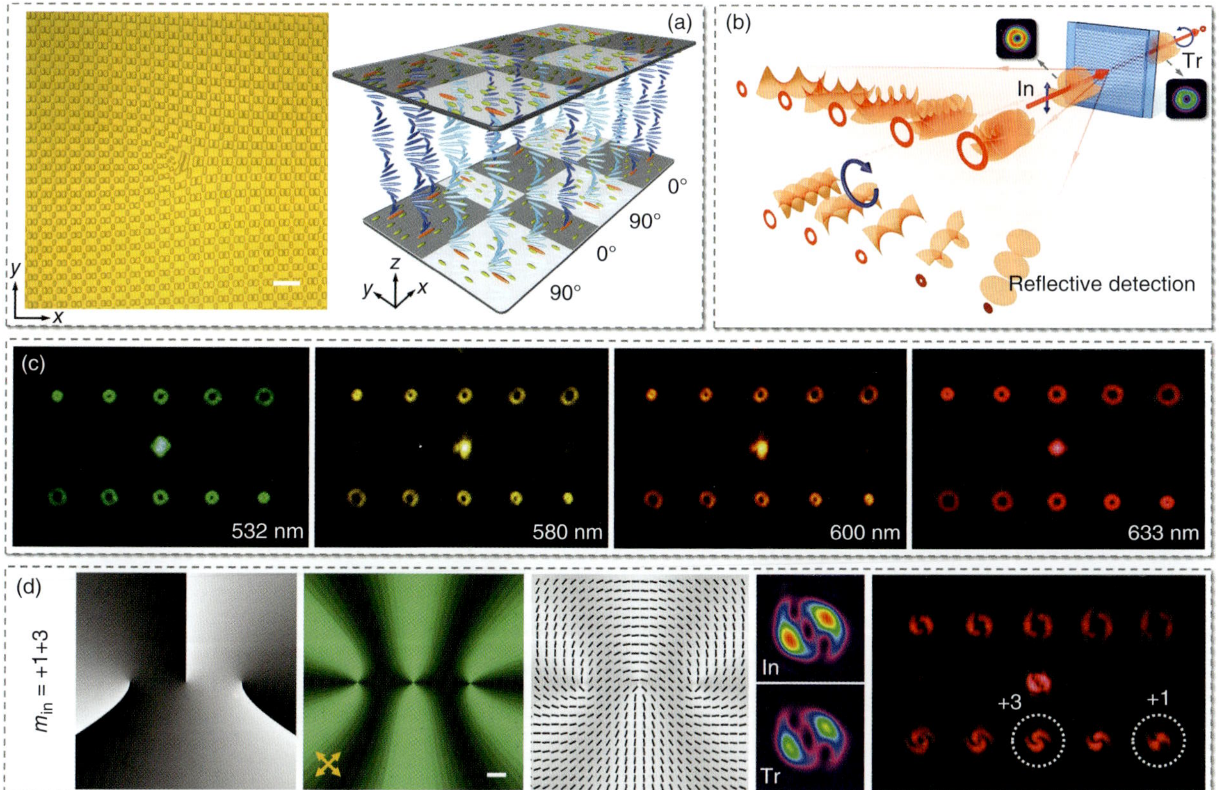

Figure 11.14 Digitalized CLC chiral superstructures for OV processing. Source: Chen et al. 2018 [136]. Copyright 2018. Reproduced with permission of John Wiley and Sons.

11.6 Discussion and Conclusion

In addition to visible range, devices with structured LC domains are demonstrated in the telecomm band [117] and THz range [139, 140] as well. Thanks to the broadband birefringence (visible, infrared to terahertz [141, 142]), further exploitation of the beam shaping extra visible range is still of great significance [143–145]. As the elements here are all induced by azimuthal angle control of LCs, the absorptive electrodes are avoidable; therefore the optical damage threshold will be drastically increased. In previous experiments, no damage was observed after 600 pulses (0.5 J cm^{-2}, 1064 nm, 10 ns, 1 Hz) received [115, 146]. The value is one order higher than the damage threshold of commercial SLM. This means the LC domain engineering based optical devices are promising for short-pulse intense-light applications, such as light bullets and industrial laser processing.

Thanks to the rewritability of the photoalignment agent, the LC orientation can be reconfigured, enabling dynamic beam shaping. Besides photoinduced LC PA domains, TN domain based LC polarization converters, which consist of one uniformly aligned substrate and a counter one with space-variant alignment, are suitable for arbitrary VB generation [22, 33, 147–149]. For most mentioned elements, traditional nematic LCs are used. In fact, the technique is compatible to various LCs, such as cholesteric, smectic [150], ferroelectric, and blue phases [151, 152]. Attempts toward new LC phases and materials is another meaningful topic.

In this chapter, recent researches about optical field control via photoinduced LC domains are briefly introduced. Some powerful photopatterning technologies are presented, which are very suitable for implementing accurate and arbitrary azimuthal orientation control of LCs. Not only binary, but also space-variant LC domains are demonstrated for the generation, manipulation, and detection of novel optical fields with high quality, high efficiency, and with unprecedented flexibility. The objective LC orientations of the discussed photoinduced LC domains can also be obtained via external-field-driven [153], self-assembled LC microdroplets or topological defects [106, 107, 154]. Compared to other techniques, the photoinduced microstructured LC devices exhibit merits of compact size, easy fabrication, and low cost and are suitable for mass production. The technique drastically enhances the capability of optical beam shaping and steering, and settles a fundamental requirement in optics and photonics. Researches on this subject may further drive the need for optical field manipulation, and bring new opportunities in the fields of optical manipulation, high-resolution imaging, OAM based informatics and quantum optics, and even some unchartered territories.

References

1 de Gennes, P.G. and Prost, J. (1993). *The Physics of Liquid Crystals*. Oxford.

2 Khoo, I.C. and Wu, S.T. (1993). *Optics and Nonlinear Optics of Liquid Crystals*. World Scientific.

3 Yang, D.K. and Wu, S.T. (2006). *Fundamentals of Liquid Crystal Devices*. Wiley.
4 Yeh, P. and Gu, C. (1999). *Optics of Liquid Crystal Displays*. Wiley.
5 Schadt, M. (2015). Nematic liquid crystals and twisted-nematic LCDs. *Liq. Cryst.* 42: 646.
6 Kim, J., Li, Y., Miskiewicz, M.N. et al. (2015). Fabrication of ideal geometric-phase holograms with arbitrary wavefronts. *Optica* 2: 958.
7 Chen, P., Lu, Y.Q., and Hu, W. (2016). Beam shaping via photopatterned liquid crystals. *Liq. Cryst.* 43: 2051.
8 Franke-Arnold, S., Allen, L., and Padgett, M. (2008). Advances in optical angular momentum. *Laser Photonics Rev.* 2: 299.
9 Yao, A.M. and Padgett, M.J. (2011). Orbital angular momentum: origins, behavior and applications. *Adv. Opt. Photonics* 3: 161.
10 Zhan, Q. (2009). Cylindrical vector beams: from mathematical concepts to applications. *Adv. Opt. Photonics* 1: 1.
11 Siviloglou, G., Broky, J., Dogariu, A., and Christodoulides, D. (2007). Observation of accelerating Airy beams. *Phys. Rev. Lett.* 99: 213901.
12 Allen, L., Beijersbergen, M.W., Spreeuw, R., and Woerdman, J. (1992). Orbital angular momentum of light and the transformation of Laguerre-Gaussian laser modes. *Phys. Rev. A* 45: 8185.
13 Yalizay, B., Soylu, B., and Akturk, S. (2010). Optical element for generation of accelerating Airy beams. *J. Opt. Soc. Am. A* 27: 2344.
14 Beijersbergen, M., Coerwinkel, R., Kristensen, M., and Woerdman, J. (1994). Helical-wavefront laser beams produced with a spiral phaseplate. *Opt. Commun.* 112: 321.
15 Machavariani, G., Lumer, Y., Moshe, I. et al. (2007). Efficient extracavity generation of radially and azimuthally polarized beams. *Opt. Lett.* 32: 1468.
16 Polynkin, P., Kolesik, M., Moloney, J.V. et al. (2009). Curved plasma channel generation using ultraintense Airy beams. *Science* 324: 229.
17 Cao, R., Yang, Y., Wang, J. et al. (2011). Microfabricated continuous cubic phase plate induced Airy beams for optical manipulation with high power efficiency. *Appl. Phys. Lett.* 99: 261106.
18 Lei, T., Zhang, M., Li, Y.R. et al. (2015). Massive individual orbital angular momentum channels for multiplexing enabled by dammann gratings. *Light Sci. Appl.* 4: e257.
19 Yu, N. and Capasso, F. (2014). Flat optics with designer metasurfaces. *Nat. Mater.* 13: 139.
20 Yulevich, I., Maguid, E., Shitrit, N. et al. (2015). Optical mode control by geometric phase in quasicrystal metasurface. *Phys. Rev. Lett.* 115: 205501.
21 Zhou, J., Liu, Y., Ke, Y. et al. (2015). Generation of Airy vortex and Airy vector beams based on the modulation of dynamic and geometric phases. *Opt. Lett.* 40: 3193.
22 Ren, H., Lin, Y.H., and Wu, S.T. (2006). Linear to axial or radial polarization conversion using a liquid crystal gel. *Appl. Phys. Lett.* 89: 051114.
23 Luo, D., Dai, H., and Sun, X. (2013). Polarization-independent electrically tunable/switchable Airy beam based on polymer-stabilized blue phase liquid crystal. *Opt. Express* 21: 31318.

24 Ge, S.J., Ji, W., Cui, G.X. et al. (2014). Fast switchable optical vortex generator based on blue phase liquid crystal fork grating. *Opt. Mater. Express* 4: 2535.

25 Lin, H.C., Chen, M.S., and Lin, Y.H. (2011). A review of electrically tunable focusing liquid crystal lenses. *Trans. Electr. Electron. Mater.* 12: 234.

26 Kitzerow, H.S. (1994). Polymer-dispersed liquid crystals from the nematic curvilinear aligned phase to ferroelectric films. *Liq. Cryst.* 16: 1.

27 Ngcobo, S., Litvin, I., Burger, L., and Forbes, A. (2013). A digital laser for on-demand laser modes. *Nat. Commun.* 4: 2289.

28 Forbes, A., Dudley, A., and McLaren, M. (2016). Creation and detection of optical modes with spatial light modulators. *Adv. Opt. Photonics* 8: 200.

29 Kim, J.H., Yoneya, M., and Yokoyama, H. (2002). Tristable nematic liquid-crystal device using micropatterned surface alignment. *Nature* 420: 159.

30 Wen, B., Petschek, R.G., and Rosenblatt, C. (2002). Nematic liquid-crystal polarization gratings by modification of surface alignment. *Appl. Opt.* 41: 1246.

31 Honma, M. and Nose, T. (2003). Polarization-independent liquid crystal grating fabricated by microrubbing process. *Jpn. J. Appl. Phys.* 42: 6992.

32 Chen, J., Bos, P.J., Vithana, H., and Johnson, D. (1995). An electro-optically controlled liquid crystal diffraction grating. *Appl. Phys. Lett.* 67: 2588.

33 Stalder, M. and Schadt, M. (1996). Linearly polarized light with axial symmetry generated by liquid-crystal polarization converters. *Opt. Lett.* 21: 1948.

34 Schadt, M., Schmitt, K., Kozinkov, V., and Chigrinov, V. (1992). Surface-induced parallel alignment of liquid crystals by linearly polymerized photopolymers. *Jpn. J. Appl. Phys.* 31: 2155.

35 Schadt, M., Seiberle, H., and Schuster, A. (1996). Optical patterning of multi-domain liquid-crystal. *Nature* 381: 212.

36 Hu, W., Srivastava, A.K., Lin, X.W. et al. (2012). Polarization independent liquid crystal gratings based on orthogonal photoalignments. *Appl. Phys. Lett.* 100: 111116.

37 Blinov, L. (1999). Liquid crystal orientation by holographic phase gratings recorded on photosensitive Langmuir-Blodgett films. *Liq. Cryst.* 26: 427.

38 Provenzano, C., Pagliusi, P., and Cipparrone, G. (2006). Highly efficient liquid crystal based diffraction grating induced by polarization holograms at the aligning surfaces. *Appl. Phys. Lett.* 89: 121105.

39 Li, Y., Kim, J., and Escuti, M.J. (2012). Orbital angular momentum generation and mode transformation with high efficiency using forked polarization gratings. *Appl. Opt.* 51: 8236.

40 Tabiryan, N.V., Nersisyan, S.R., Steeves, D.M., and Kimball, B.R. (2010). The promise of diffractive waveplates. *Opt. Photonics News* 21: 40.

41 Culbreath, C., Glazar, N., and Yokoyama, H. (2011). Note: automated maskless micro-multidomain photoalignment. *Rev. Sci. Instrum.* 82: 126107.

42 Wu, H., Hu, W., Hu, H.C. et al. (2012). Arbitrary photo-patterning in liquid crystal alignments using DMD based lithography system. *Opt. Express* 20: 16684.

43 Miskiewicz, M.N. and Escuti, M.J. (2014). Direct-writing of complex liquid crystal patterns. *Opt. Express* 22: 12691.

44 Guo, Y., Jiang, M., Peng, C. et al. (2016). High-resolution and high-throughput plasmonic photopatterning of complex molecular orientations in liquid crystals. *Adv. Mater.* 28: 2353.

45 Ichimura, K., Suzuki, Y., Seki, T. et al. (1988). Reversible change in alignment mode of nematic liquid crystals regulated photochemically by command surfaces modified with an azobenzene monolayer. *Langmuir* 4: 1214.

46 Chigrinov, V.G., Kozenkov, V.M., and Kwok, H.S. (2008). *Photoalignment of Liquid Crystalline Materials: Physics and Applications*. Wiley.

47 Pedersen, T.G., Ramanujam, P., Johansen, P.M., and Hvilsted, S. (1998). Quantum theory and experimental studies of absorption spectra and photoisomerization of azobenzene polymers. *J. Opt. Soc. Am. B* 15: 2721.

48 Pedersen, T.G. and Johansen, P.M. (1997). Mean-field theory of photoinduced molecular reorientation in azobenzene liquid crystalline side-chain polymers. *Phys. Rev. Lett.* 79: 2470.

49 Yaroshchuk, O., Kiselev, A., Zakrevskyy, Y. et al. (2003). Photoinduced three-dimensional orientational order in side chain liquid crystalline azopolymers. *Phys. Rev. E* 68: 011803.

50 Blinov, L. (1996). Photoinduced molecular reorientation in polymers, Langmuir–Blodgett films and liquid crystals. *J. Nonlinear Opt. Phys. Mater.* 5: 165.

51 Gibbons, W.M., Shannon, P.J., Sun, S.T., and Swetlin, B.J. (1991). Surface-mediated alignment of nematic liquid crystals with polarized laser light. *Nature* 351: 49.

52 Chigrinov, V., Pikin, S., Verevochnikov, A. et al. (2004). Diffusion model of photoaligning in azo-dye layers. *Phys. Rev. E* 69: 061713.

53 Hasegawa, M. (1999). Modeling of photoinduced optical anisotropy and anchoring energy of polyimide exposed to linearly polarized deep UV light. *Jpn. J. Appl. Phys.* 38: L457.

54 Nishikawa, M., Taheri, B., and West, J.L. (1998). Mechanism of unidirectional liquid-crystal alignment on polyimides with linearly polarized ultraviolet light exposure. *Appl. Phys. Lett.* 72: 2403.

55 West, J.L., Su, L., and Reznikov, Y. (2001). Photo-alignment using adsorbed dichroic molecules. *Mol. Cryst. Liq. Cryst.* 364: 199.

56 Akiyama, H., Kawara, T., Takada, H. et al. (2002). Synthesis and properties of azo dye aligning layers for liquid crystal cells. *Liq. Cryst.* 29: 1321.

57 Ji, W., Lee, C.H., Chen, P. et al. (2016). Meta-q-plate for complex beam shaping. *Sci. Rep.* 6: 25528.

58 Wei, B.Y., Hu, W., Ming, Y. et al. (2014). Generating switchable and reconfigurable optical vortices via photopatterning of liquid crystals. *Adv. Mater.* 26: 1590.

59 Zheng, Z., Song, J., Liu, Y. et al. (2008). Single-step exposure for two-dimensional electrically-tuneable diffraction grating based on polymer dispersed liquid crystal. *Liq. Cryst.* 35: 489.

60 Emoto, A., Manabe, S., Shioda, T. et al. (2009). One-and two-dimensional anisotropic diffractive gratings formed by periodic orthogonal molecular alignment in a hydrogen-bonded liquid crystalline polymer. *J. Appl. Phys.* 105: 103514.

61 Hu, W., Srivastava, A., Xu, F. et al. (2012). Liquid crystal gratings based on alternate TN and PA photoalignment. *Opt. Express* 20: 5384.
62 Li, J.N., Hu, X.K., Wei, B.Y. et al. (2014). Simulation and optimization of liquid crystal gratings with alternate twisted nematic and planar aligned regions. *Appl. Opt.* 53: E14.
63 Lin, X.W., Hu, W., Hu, X.K. et al. (2012). Fast response dual-frequency liquid crystal switch with photo-patterned alignments. *Opt. Lett.* 37: 3627.
64 Srivastava, A., Hu, W., Chigrinov, V. et al. (2012). Fast switchable grating based on orthogonal photo alignments of ferroelectric liquid crystals. *Appl. Phys. Lett.* 101: 031112.
65 Sun, J., Srivastava, A.K., Wang, L. et al. (2013). Optically tunable and rewritable diffraction grating with photoaligned liquid crystals. *Opt. Lett.* 38: 2342.
66 Ma, Y., Sun, J., Srivastava, A.K. et al. (2013). Optically rewritable ferroelectric liquid-crystal grating. *EPL-Europhys. Lett.* 102: 24005.
67 Wang, X.Q., Srivastava, A.K., Chigrinov, V.G., and Kwok, H.S. (2013). Switchable Fresnel lens based on micropatterned alignment. *Opt. Lett.* 38: 1775.
68 Wang, X., Fan, F., Du, T. et al. (2014). Liquid crystal Fresnel zone lens based on single-side-patterned photoalignment layer. *Appl. Opt.* 53: 2026.
69 Wang, X.Q., Yang, W.Q., Liu, Z. et al. (2017). Switchable Fresnel lens based on hybrid photo-aligned dual frequency nematic liquid crystal. *Opt. Mater. Express* 7: 8.
70 Srivastava, A.K., Wang, X., Gong, S. et al. (2015). Micro-patterned photo-aligned ferroelectric liquid crystal Fresnel zone lens. *Opt. Lett.* 40: 1643.
71 Torres, J.P. and Torner, L. (2011). *Twisted Photons: Applications of Light with Orbital Angular Momentum*. Wiley.
72 Ma, Y., Wei, B., Shi, L. et al. (2016). Fork gratings based on ferroelectric liquid crystals. *Opt. Express* 24: 5822.
73 Dammann, H. and Görtler, K. (1971). High-efficiency in-line multiple imaging by means of multiple phase holograms. *Opt. Commun.* 3: 312.
74 Dammann, H. and Klotz, E. (1977). Coherent optical generation and inspection of two-dimensional periodic structures. *Opt. Acta* 24: 505.
75 Zhou, C. and Liu, L. (1995). Numerical study of Dammann array illuminators. *Appl. Opt.* 34: 5961.
76 Wang, X.Q., Srivastava, A.K., Fan, F. et al. (2016). Electrically/optically tunable photo-aligned hybrid nematic liquid crystal Dammann grating. *Opt. Lett.* 41: 5668.
77 Fan, F., Yao, L., Wang, X. et al. (2017). Ferroelectric liquid crystal Dammann grating by patterned photoalignment. *Crystals* 7: 79.
78 Luo, D., Sun, X.W., Dai, H.T., and Demir, H.V. (2011). Polarization-dependent circular Dammann grating made of azo-dye-doped liquid crystals. *Appl. Opt.* 50: 2316.
79 Chen, P., Ge, S.J., Ma, L.L. et al. (2016). Generation of equal-energy orbital angular momentum beams via photopatterned liquid crystals. *Phys. Rev. Appl.* 5: 044009.
80 Wang, X., Wu, S., Yang, W. et al. (2017). Light-driven liquid crystal circular Dammann grating fabricated by a micro-patterned liquid crystal polymer phase mask. *Polymers* 9: 380.

81 Liu, J., Min, C., Lei, T. et al. (2016). Generation and detection of broadband multi-channel orbital angular momentum by micrometer-scale meta-reflectarray. *Opt. Express* 24: 212.

82 Ge, S.J., Chen, P., Ma, L.L. et al. (2016). Optical array generator based on blue phase liquid crystal Dammann grating. *Opt. Mater. Express* 6: 1087.

83 Pancharatnam, S. (1956). Generalized theory of interference and its applications. *Proc. Indian Acad. Sci. A* 44: 247.

84 Bliokh, K., Rodríguez-Fortuño, F., Nori, F., and Zayats, A.V. (2015). Spin-orbit interactions of light. *Nat. Photonics* 9: 796.

85 Yin, X., Ye, Z., Rho, J. et al. (2013). Photonic spin Hall effect at metasurfaces. *Science* 339: 1405.

86 Marrucci, L., Manzo, C., and Paparo, D. (2006). Optical spin-to-orbital angular momentum conversion in inhomogeneous anisotropic media. *Phys. Rev. Lett.* 96: 163905.

87 Mei, S., Mehmood, M.Q., Hussain, S. et al. (2016). Flat helical nanosieves. *Adv. Funct. Mater.* 26: 5255.

88 Ling, X., Zhou, X., Huang, K. et al. (2017). Recent advances in the spin Hall effect of light. *Rep. Prog. Phys.* 80: 066401.

89 Kim, J. (2011). *Liquid Crystal Geometric Phase Holograms for Efficient Beam Steering and Imaging Spectropolarimetry*. North Carolina State University.

90 Provenzano, C., Pagliusi, P., and Cipparrone, G. (2007). Electrically tunable two-dimensional liquid crystals gratings induced by polarization holography. *Opt. Express* 15: 5872.

91 Gorkhali, S.P., Cloutier, S.G., Crawford, G.P., and Pelcovits, R.A. (2006). Stable polarization gratings recorded in azo-dye-doped liquid crystals. *Appl. Phys. Lett.* 88: 251113.

92 Presnyakov, V., Asatryan, K., Galstian, T., and Chigrinov, V. (2006). Optical polarization grating induced liquid crystal micro-structure using azo-dye command layer. *Opt. Express* 14: 10558.

93 Duan, W., Chen, P., Wei, B.Y. et al. (2016). Fast-response and high-efficiency optical switch based on dual-frequency liquid crystal polarization grating. *Opt. Mater. Express* 6: 597.

94 Du, T., Fan, F., Tam, A.M.W. et al. (2015). Complex nanoscale-ordered liquid crystal polymer film for high transmittance holographic polarizer. *Adv. Mater.* 27: 7191.

95 Tabiryan, N.V., Serak, S.V., Nersisyan, S.R. et al. (2016). Broadband waveplate lenses. *Opt. Express* 24: 7091.

96 Tam, M.W.A., Fan, F., Chen, H.S. et al. (2016). Nanoscopic patterned photo-alignment for electrically switchable liquid crystal Pancharatnam–Berry phase diffractive lens. *J. Soc. Inf. Disp.* 47: 599.

97 Ke, Y., Liu, Z., Liu, Y. et al. (2016). Compact photonic spin filters. *Appl. Phys. Lett.* 109: 181104.

98 Naidoo, D., Roux, F.S., Dudley, A. et al. (2016). Controlled generation of higher-order Poincaré sphere beams from a laser. *Nat. Photonics* 10: 327.

99 Slussarenko, S., Murauski, A., Du, T. et al. (2011). Tunable liquid crystal q-plates with arbitrary topological charge. *Opt. Express* 19: 4085.

100 Nersisyan, S.R., Tabiryan, N.V., Mawet, D., and Serabyn, E. (2013). Improving vector vortex waveplates for high-contrast coronagraphy. *Opt. Express* 21: 8205.

101 Gecevičius, M., Drevinskas, R., Beresna, M., and Kazansky, P.G. (2014). Single beam optical vortex tweezers with tunable orbital angular momentum. *Appl. Phys. Lett.* 104: 231110.

102 Cheng, H., Li, P., Liu, S. et al. (2017). Vortex-controlled morphology conversion of microstructures on silicon induced by femtosecond vector vortex beams. *Appl. Phys. Lett.* 111: 141901.

103 D'Ambrosio, V., Spagnolo, N., Del Re, L. et al. (2013). Photonic polarization gears for ultra-sensitive angular measurements. *Nat. Commun.* 4: 2432.

104 Cardano, F., Massa, F., Qassim, H. et al. (2015). Quantum walks and wavepacket dynamics on a lattice with twisted photons. *Sci. Adv.* 1: e1500087.

105 Cardano, F., Karimi, E., Slussarenko, S. et al. (2012). Polarization pattern of vector vortex beams generated by q-plates with different topological charges. *Appl. Opt.* 51: C1–C6.

106 Brasselet, E., Murazawa, N., Misawa, H., and Juodkazis, S. (2009). Optical vortices from liquid crystal droplets. *Phys. Rev. Lett.* 103: 103903.

107 Loussert, C., Delabre, U., and Brasselet, E. (2013). Manipulating the orbital angular momentum of light at the micron scale with nematic disclinations in a liquid crystal film. *Phys. Rev. Lett.* 111: 037802.

108 Brasselet, E. and Loussert, C. (2011). Electrically controlled topological defects in liquid crystals as tunable spin-orbit encoders for photons. *Opt. Lett.* 36: 719.

109 Barboza, R., Bortolozzo, U., Assanto, G. et al. (2013). Harnessing optical vortex lattices in nematic liquid crystals. *Phys. Rev. Lett.* 111: 093902.

110 Tzeng, Y.Y., Ke, S.W., Ting, C.L. et al. (2008). Axially symmetric polarization converters based on photo-aligned liquid crystal films. *Opt. Express* 16: 3768.

111 Chen, P., Wei, B.Y., Ji, W. et al. (2015). Arbitrary and reconfigurable optical vortex generation: a high-efficiency technique using director-varying liquid crystal fork gratings. *Photonics Res.* 3: 133.

112 Tam, A.M., Fan, F., Du, T. et al. (2017). Bifocal optical-vortex lens with sorting of the generated nonseparable spin-orbital angular-momentum states. *Phys. Rev. Appl.* 7: 034010.

113 Duan, W., Chen, P., Ge, S.J. et al. (2017). Helicity-dependent forked vortex lens based on photo-patterned liquid crystals. *Opt. Express* 25: 14059.

114 Tang, M.J., Chen, P., Zhang, W.L. et al. (2016). Integrated and reconfigurable optical paths based on stacking optical functional films. *Opt. Express* 24: 25510.

115 Wei, B.Y., Chen, P., Hu, W. et al. (2015). Polarization-controllable Airy beams generated via a photoaligned director-variant liquid crystal mask. *Sci. Rep.* 5: 17484.

116 Larocque, H., Gagnon-Bischoff, J., Bouchard, F. et al. (2016). Arbitrary optical wavefront shaping via spin-to-orbit coupling. *J. Opt.* 18: 124002.

117 Wei, B.Y., Chen, P., Ge, S.J. et al. (2016). Liquid crystal depolarizer based on photoalignment technology. *Photonics Res.* 4: 70.

118 Wei, B.Y., Chen, P., Ge, S.J. et al. (2016). Generation of self-healing and transverse accelerating optical vortices. *Appl. Phys. Lett.* 109: 121105.

119 Rafayelyan, M., Tkachenko, G., and Brasselet, E. (2016). Reflective spin-orbit geometric phase from chiral anisotropic optical media. *Phys. Rev. Lett.* 116: 253902.

120 Barboza, R., Bortolozzo, U., Clerc, M.G., and Residori, S. (2016). Berry phase of light under Bragg reflection by chiral liquid-crystal media. *Phys. Rev. Lett.* 117: 053903.

121 Kobashi, J., Yoshida, H., and Ozaki, M. (2016). Planar optics with patterned chiral liquid crystals. *Nat. Photonics* 10: 389.

122 Kobashi, J., Yoshida, H., and Ozaki, M. (2016). Polychromatic optical vortex generation from patterned cholesteric liquid crystals. *Phys. Rev. Lett.* 116: 253903.

123 Wang, L. and Li, Q. (2016). Stimuli-directing self-organized 3D liquid-crystalline nanostructures: from materials design to photonic applications. *Adv. Funct. Mater.* 26: 10.

124 Coles, H. and Morris, S. (2010). Liquid-crystal lasers. *Nat. Photonics* 4: 676.

125 Ma, L.L., Li, S.S., Li, W.S. et al. (2015). Rationally designed dynamic superstructures enabled by photoaligning cholesteric liquid crystals. *Adv. Opt. Mater.* 3: 1691.

126 Zheng, Z.G., Li, Y., Bisoyi, H.K. et al. (2016). Three-dimensional control of the helical axis of a chiral nematic liquid crystal by light. *Nature* 531: 352.

127 Li, C.C., Chen, C.W., Yu, C.K. et al. (2017). Arbitrary beam steering enabled by photomechanically bendable cholesteric liquid crystal polymers. *Adv. Opt. Mater.* 5: 1600824.

128 Li, Y., Prince, E., Cho, S. et al. (2017). Periodic assembly of nanoparticle arrays in disclinations of cholesteric liquid crystals. *Proc. Natl. Acad. Sci. U.S.A.* 114: 2137.

129 Yoshida, H. and Kobashi, J. (2016). Flat optics with cholesteric and blue phase liquid crystals. *Liq. Cryst.* 43: 1909.

130 Rafayelyan, M. and Brasselet, E. (2016). Bragg–Berry mirrors: reflective broadband q-plates. *Opt. Lett.* 41: 3972.

131 Grier, D.G. (2003). A revolution in optical manipulation. *Nature* 424: 810.

132 Yan, L., Gregg, P., Karimi, E. et al. (2015). Q-plate enabled spectrally diverse orbital-angular-momentum conversion for stimulated emission depletion microscopy. *Optica* 2: 900.

133 Milione, G., Sztul, H., Nolan, D., and Alfano, R. (2011). Higher-order Poincaré sphere, Stokes parameters, and the angular momentum of light. *Phys. Rev. Lett.* 107: 053601.

134 Milione, G., Evans, S., Nolan, D., and Alfano, R. (2012). Higher order Pancharatnam–Berry phase and the angular momentum of light. *Phys. Rev. Lett.* 108: 190401.

135 Chen, P., Ge, S.J., Duan, W. et al. (2017). Digitalized geometric phases for parallel optical spin and orbital angular momentum encoding. *ACS Photonics* 4: 1333.

136 Chen, P., Ma, L.L., Duan, W. et al. (2018). Digitalizing self-assembled chiral superstructures for optical vortex processing. *Adv. Mater.* 30: 1705865.

137 Wang, J., Yang, J.Y., Fazal, I.M. et al. (2012). Terabit free-space data transmission employing orbital angular momentum multiplexing. *Nat. Photonics* 6: 488.

138 Willner, A.E., Huang, H., Yan, Y. et al. (2015). Optical communications using orbital angular momentum beams. *Adv. Opt. Photonics* 7: 66.

139 Wang, L., Lin, X.W., Hu, W. et al. (2015). Broadband tunable liquid crystal terahertz waveplates driven with porous graphene electrodes. *Light Sci. Appl.* 4: e253.

140 Ge, S., Liu, J., Chen, P. et al. (2015). Tunable terahertz filter based on alternative liquid crystal layers and metallic slats. *Chin. Opt. Lett.* 13: 120401.

141 Wu, S.T. (1986). Birefringence dispersions of liquid crystals. *Phys. Rev. A* 33: 1270.

142 Wang, L., Lin, X.W., Liang, X. et al. (2012). Large birefringence liquid crystal material in terahertz range. *Opt. Mater. Express* 2: 1314.

143 Ge, S., Chen, P., Shen, Z. et al. (2017). Terahertz vortex beam generator based on a photopatterned large birefringence liquid crystal. *Opt. Express* 25: 12349.

144 Wang, L., Ge, S., Hu, W. et al. (2017). Graphene-assisted high-efficiency liquid crystal tunable terahertz metamaterial absorber. *Opt. Express* 25: 23873.

145 Ge, S.J., Shen, Z.X., Chen, P. et al. (2017). Generating, separating and polarizing terahertz vortex beams via liquid crystals with gradient-rotation directors. *Crystals* 7: 314.

146 Zhu, Z.H., Chen, P., Sheng, L.W. et al. (2017). Generation of strong cylindrical vector pulses via stimulated Brillouin amplification. *Appl. Phys. Lett.* 110: 141104.

147 Ko, S.W., Ting, C.L., Fuh, A.Y.G., and Lin, T.H. (2010). Polarization converters based on axially symmetric twisted nematic liquid crystal. *Opt. Express* 18: 3601.

148 Chen, P., Ji, W., Wei, B.Y. et al. (2015). Generation of arbitrary vector beams with liquid crystal polarization converters and vector-photoaligned q-plates. *Appl. Phys. Lett.* 107: 241102.

149 Wang, X., Xu, M., and Ren, H. (2014). Polarization converting textures of nematic liquid crystal in glass cavities. *J. Appl. Phys.* 115: 023111.

150 Ma, L.L., Tang, M.J., Hu, W. et al. (2017). Smectic layer origami via preprogrammed photoalignment. *Adv. Mater.* 29: 1605671.

151 Zheng, Z.G., Yuan, C.L., Hu, W. et al. (2017). Light-patterned crystallographic direction of a self-organized 3D soft photonic crystal. *Adv. Mater.* 29: 1703165.

152 Chen, C.W., Hou, C.T., Li, C.C. et al. (2017). Large three-dimensional photonic crystals based on monocrystalline liquid crystal blue phases. *Nat. Commun.* 8: 727.

153 Loussert, C., Kushnir, K., and Brasselet, E. (2014). Q-plates micro-arrays for parallel processing of the photon orbital angular momentum. *Appl. Phys. Lett.* 105: 121108.

154 Son, B., Kim, S., Kim, Y.H. et al. (2014). Optical vortex arrays from smectic liquid crystals. *Opt. Express* 22: 4699.

12

Azobenzene Polymers as Photoactive Materials for Shape Changes of Micro/Nano-objects

Régis Barillé[1], *Ewelina Ortyl*[2], *and Sonia Zielinska*[2]

[1] *Université d'Angers/UMR, MOLTECH-Anjou, CNRS 6200, 2, Bd Lavoisier, 49045, Angers Cedex 01, France*
[2] *Wroclaw University of Technology, Faculty of Chemistry, Department of Polymer Engineering and Technology, ul. Wybrzeze Wyspianskiego 27, 50-370 Wroclaw, Poland*

12.1 Why Azobenzene-Based Photoactive Nano-objects?

One of the most important goals in nanosciences and nanotechnology is greater miniaturization of information systems. This miniaturization of components for the construction of systems and machines used in everyday life is currently carried out by a top-down approach. This approach that leads physicists to manipulate or gradually transform material into small pieces by photolithography and related techniques has been carried out so far in a remarkable way. It is becoming increasingly obvious, however, that the top-down approach faces drastic limitations for dimensions smaller than 100 nm [1]. Even if this dimension appears very small for most experimenters, it remains very large at the level of atoms and molecules. An alternative and the most promising strategy to be exploited in science and technology to build nanostructures at the nanometer scale is the use of a bottom-up approach. One of the ways to achieve the realization of the nanometric structure by the bottom-up technique is the controlled assembly or the self-assembly of nano-objects and the modification of these nano-objects. Nature allows the assembly of molecules in an elegant and efficient way by multiple noncovalent interactions. The bottom-up approach is the mainstay in the field of nanoscience and nanotechnology. This is why chemists who are able to manipulate atoms and molecules are in an ideal position for the development of nanosciences and nanotechnology. The realization of surfaces containing micro/nanoscale objects or patterns offers the possibility to develop at a large scale new modifications and new possibilities of properties developed at the nanoscale with micro/nano-objects. Toward that goal, polymer micro- or nano-objects can be used as building blocks for self- or control-assemblies that find applications in many fields of research including nano-sensors, nanomedecine, or nano-energy [2].

Photoactive Functional Soft Materials: Preparation, Properties, and Applications, First Edition.
Edited by Quan Li.

In terms of nano- or micro-objects, we refer to micro- or nano-scale objects as nanoparticles (NPs), colloidal spheres, nanorods, nanotubes, microwires, three dimensional posts, microspherical cap array, needles, polymeric strip, micro disks, etc. (Figure 12.1). The lithography technique used electrons (Ebeam), which is not suitable to transform the shape of polymer micro/nano-objects or to pattern the whole curved polymer surfaces of microspheres, especially the bottom parts of the surfaces. Further modifications of the surfaces of nanospheres are even less possible. So, conventional lithography techniques such as beam lithography methods or chemical methods, which are usually used to modify planar surfaces, are not very supportive to be used to pattern colloidal surfaces or similar size nano-objects. Polymeric materials containing azobenzene and its derivatives (azochromophores) have been actively studied and developed in recent years [3]. We show how small objects representing different shapes and dimensions (0D, 1D, 2D) can be individually modified at the nanoscale leading to surface or shape changes.

One primary strategy to achieve the functionalization of a surface is to induce a pattern or to apply "a dimension-controlled elevation" [4]. For the purpose of preparing functional polymer nano-objects in terms of modification of surfaces, the light is used as a stimulus for photoinduced surface changes. The whole shape or part shape modification of surfaces is challenging and appears as a top

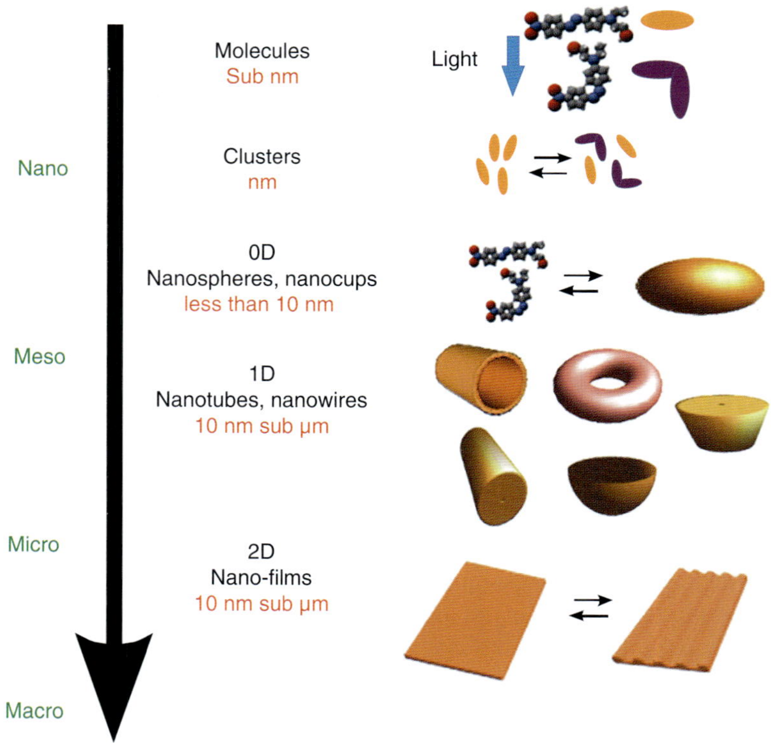

Figure 12.1 Graphical representation of the scales involved in the design of azopolymeric nano/micro-objects.

research subject due to their potential scientific and technological applications such as drug delivery, chemical and biological sensors, or mechanical actuators. The modified surface could be used as a biological sensor. The detection method is based on the modulation of the diffracted light, which is caused by the interaction between the light illuminating the biological analytes and the nanopatterned surface used as test platforms. Diffraction gratings on the surface can be prepared either by nanopatterning an azopolymer-responsive material into a periodic array or by using this photopatterned surface as a duplication for different biocompatible materials. When the grating surfaces interact with the target molecules, either absorbed onto the surface or internalized within the gratings, the geometrical parameters of the gratings and/or the refractive index contrast is changed, which results in measurable changes in the diffraction efficiency (DE). The measurements rely on the values of the ratio between the diffraction intensity in the first order with and without the analyte to be tested on the azopolymer grating. This methodology has been successfully used in the sensing of biomolecules such as human immunoglobulin-G (H-IgG) detection [5]. The quantitative detection of H-IgG was achieved by detecting the modulation of the DE when the gratings were exposed to antigen solutions. The gratings showed high sensitivity and specificity for H-IgG detection. Another potential application was the possibility to develop a new glucose-sensing strategy by using diffraction-gratings of a stimuli-responsive hydrogel bearing phenylboronic acid groups [6]. The gratings showed a fast and highly sensitive response to glucose in buffer solutions measured by the increase of the DE of the incident light. The saturated value was linearly correlated to the glucose concentration of the solutions. Authors claimed that this approach can be further explored to prepare various grating sensors for saccharide recognition and detection.

Azopolymer nano-objects contribute also to the design and development of low-cost sensors that can either be embedded or surface coated in engineering structures to continually provide quantitative information. In particular, an example of azobenzene contribution is the measurement of moisture which is an important parameter in many applications such as agriculture, food quality assessment, and materials processing [7]. In these activities it is important to be able to continually access information on the moisture content. One example is the fabrication of an FBG (Fiber Bragg Grating) sensor using a custom-synthesized azobenzene-based polymeric gel.

The information is acquired from the investigation of the strain behavior of the sensor in response to water absorption and desorption. The changes of strain and Bragg peak shift as a function of water ingresses by the gel. The application of the sensor for moisture detection in various materials could also be achieved.

Potential applications of polymeric nano-objects in the field of drug delivery have come into particular focus recently. The site-specific drug release at the pathological location is highly important for cancer chemotherapy. If triggered drug release can be achieved in response to specific *in vivo* conditions in the case of tumors, which could improve drug deliveries and decrease side effects, with local treatments the challenges can be realized. The degradation by gastrointestinal microflora of different azopolymers was studied by performing permeability experiments on isolated films and solution viscosity measurements before and after incubation [8]. The conclusion was that polyurethane was found to be degraded in a culture of intestinal flora with the azo group. To reconfirm this, drug pellets containing water-soluble drugs were undercoated with (carboxymethyl)(ethyl)-cellulose and overcoated with a segmented polyurethane containing azo aromatic groups in the main chain in order to examine the drug-releasing profiles in the culture of intestinal flora. Colon-targeted drug delivery and circumventing drug resistance are extremely important approaches for colon cancer chemotherapy. Azopolymeric micelles for colon-targeted dimethyl fumarate (DMF) delivery in colon cancer therapy has been designed based on the pharmacological properties of DMF and azo bond in olsalazine chemical structure. DMF exhibited anti-tumor activity on colon cancer cells. Azopolymeric micelles have been synthesized as a star-shape amphiphilic polymer with azo bond and the fabricated DMF-loaded.

Nano-energy is an important research field with large applications in the reversibly storing solar energy. Azobenzene-based nanostructures can be used to carry on solutions with molecular photoisomers, known as solar thermal fuels. A review is given by Zhang et al. [9]. The efficient capture, conversion, and storage of solar energy in the form of chemical bonds have potentially great interests. Recently it was shown that solar thermal fuels, with metastable forms by photoisomerization is capable of releasing close to 100% of the stored energy as the heat source under an external stimulus along with the reversion to stable forms [10]. A reversible transformation can transfer energy from and to the environment transforming the light into heat or a thermal storage of sunlight, respectively. This effect can increase the low storage capacity and short-time storage of sun energy. A nano-template with covalent attachment of new azobenzene chromophores on graphene is used as solar thermal fuels. The intermolecular hydrogen bond and proximity-induced interaction, result from high functionalization density and inter-planar bundling interaction, and lead to an improvement of both the storage capacity and lifetime. This nanoscopic template exhibits a high energy density up to 112 $\mathrm{W\,h\,kg^{-1}}$ and long-term storage with a half-life of more than one month (33 days). Novel hybrid solar thermal fuels have also been prepared through covalent linkage of carbon nanotubes (CNTs) with azobenzenes, as demonstrated both theoretically and experimentally [11].

In the field of nano-actuators for the purpose of deriving a machine or a robot in order to perform specific functions, azobenzene nano or micro-objects can be stimulated by light to provide power and information [12]. Light is used as a stimulus to remotely supply energy and transport control information to the azobenzene-based nano-actuator or robot and is suitable to mode transform from light energy to mechanical energy (Figure 12.2). It was shown that a single film of a liquid crystal (LC) network containing an azobenzene chromophore can be repeatedly and precisely be bent along any chosen direction by using linearly polarized light. This is a very smart way of controlling the direction and degree of orientations of mesophase formation using the azobenzene photoisomerization. This important photomechanical effect results from a photoselective volume contraction and is useful in the development of micro- or nano-actuators for microscale or nanoscale applications, for example, when light can propel

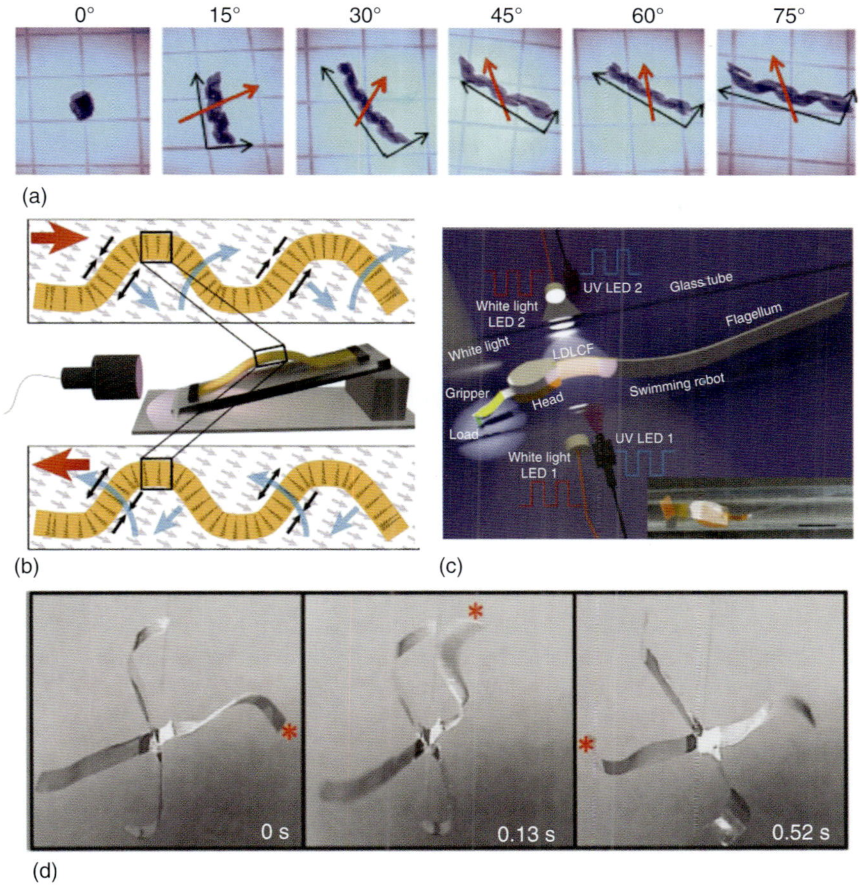

Figure 12.2 Azobenzene- based Liquid crystal micro-objects used for photo-actuation and photomotility. (a) Thin strip in the twisted nematic geometry forming a spiral ribbon. Source: Wie et al. 2016 [13]. Reprinted with permission of Springer Nature. (b) Photoactive polymer films that exhibit continuous, directional, macroscopic mechanical waves under constant light illumination. Source: Gelebart et al. 2017 [14]. Reprinted with permission of Springer Nature. (c) Micro swimming robot. Source: Huang et al. 2015 [12]. Reprinted with permission of Springer Nature. (d) Light-driven plastic mill based on hydrazone liquid crystal networks. Source: Vantomme et al. 2017 [15]. Reprinted with permission of Elsevier.

microrobots in medicine or optical catapult [16]. Most of the objects have the shape of strips and the light reversibly actuates the bending but some results have also demonstrated the possibility to induce movements with spiraling objects or rolling objects.

Recently, a micro swimming robot was designed with a head and a flexible long flagellum with the whole length of 2.6 cm moving in a liquid-filled tube with a low Reynolds number [13]. The swimming mechanism is similar to microorganism with a swing flexible oar-like flagellum, for example spermatozoons. Additionally, a gripper was added on the robot head to introduce a function to grab and release loads. Another study explored further in the design of LC polymer films incorporating azobenzene molecules for the fabrication of stimulated light materials [14]. They modified the photophysical properties by incorporating azobenzene derivatives with fast *cis-to-trans* thermal relaxation into LC networks in order to accelerate the relaxation after photoisomerization. They showed that photoactive polymer films with this new molecular design can exhibit continuous, directional, macroscopic mechanical waves under constant light illumination, with a feedback loop that is driven by self-shadowing. They provide a mechanism of wave generation to the azopolymer thin film and demonstrate the potential application of the photoactive films in light-driven locomotions and self-cleaning surfaces.

The photoinduced motion called photomotility of thin strips of a photoresponsive polymer is made with an azo-liquid crystalline network (LCN) composed of 20 wt% of an azobenzene cross-linker that transforms flat structures into spiraled ones under irradiation [17]. The material is anisotropic and the intrinsic granularity of the actuation mechanisms is done by a *trans–cis* photoisomerization at the molecular level. Owing to a twisted nematic orientation, the variation in the director profile offsets the alignment of the director to the principal axes of the strips and results in the formation of spiraled shapes. The directionality of the photomotility is programmed by the orientation of the anisotropy to the principal axes of the specimens. The motion occurs without modulating or multiplexing the actinic light source. The material does not require any other special conditions other than its own capability to be actuated by directly transducing photons into motions.

The first light-driven plastic mill was demonstrated [15]. It converts the incoming light directly into a continuous rotation. This device is composed of four blades made of azo-based LC polymer films able to bend under focused light irradiation and create a force causing the rotation of the mill. The study has shown that by accelerating the rate of the thermal Z to E back-isomerization, macroscopic deformations with fast strain rate can be obtained. The rapid transformation of the film under light is the key factor to perform the continuous rotatory motions of the mill.

Other experiments have also observed the photomotility with other type of materials as crystals. The deformation is due to a change in crystal dimensions combined with local melting and can elicit crystal creeping across a surface, such as that observed with azobenzene crystals [18]. The motion is supposed to be driven by crystallization and melting at the front and rear edges of the crystal, respectively, via photochemical conversion between the crystal and liquid phases induced by the isomerization of azobenzenes. These crystals can crawl for around 15 μm in 10 minutes when they are exposed to 365 nm light. Another experiment involves objects floating on a liquid/air interface able to create motions [19]. The experiment is similar to the well-known one with a camphor boat which is a self-propelled system in which dissolution of camphor molecules from the solid attached to the boat generates spontaneous motion of the boat on the water surface. In this experiment, a crystal of a simple azobenzene (4-methoxyazobenzene) moves on a water–air interface. The motion is triggered by irradiation with UV light. The maximum velocity of the boat was 5 cm s^{-1}. The phenomenon is explained by the photoinduced crystal–liquid phase transition and dissolution of the *cis*-isomer into water. The motion of the crystals or the boat continued even after the UV irradiation is stopped meaning that the motion is not caused by the heating effect.

Nanoparticles composed of a polymer network of azobenzene moieties with LC, capable of reproducible, photoinduced mechanical actuation was demonstrated. A 20% height contraction of nanoparticles adsorbed on a substrate was obtained [20].

The advantages of liquid crystalline actuators are their ability to act similarly in dry and wet environment as well as their reversibility of mechanical actuations. The bending direction can be adjusted in two directions. However, the design of biodegradable azobenzene liquid crystalline actuators is not easy and is limited for *in vivo* applications.

Until now, we have shown that the control of azobenzene nano- or micro-objects can be moved on a glass surface or on the surface of a liquid. However, an important situation is the possibility to control the direction of nano-objects in a 3D environment and in particular in a liquid even if this result has been partly shown with the directional control of azobenzene liquid crystal strip with a long flagellum The directionally controlled motion of nanosized polymer particles in an aqueous medium, as observed by fluorescence microscopy imaging was demonstrated [21]. These NPs are specifically designed to reduce or eliminate the Brownian motion dominating the mechanical behavior in liquid. The nanosized particles consist of azobenzene-dye-decorated polymer nanoparticles in the 16 nm-diameter range. The phenomenon in liquid with these nanoparticles is similar to the mass transport of the azopolymer when the surface is illuminated by interferences of a laser beam. The molecules move to the dark region and an accumulation of matter is observed corresponding to modulation of the photoinduced surface. The light-induced isomerization of attached molecular motors leads to a motion toward the dark regions of the modulated illumination of the sample. The photomechanical energy conversion with optical control through light-intensity gradients corresponding to the interferences illuminating the sample is transformed by the isomerizable dyes grafted on the particle surface as molecular propellers. The Brownian motion is reduced by the gradient force and can be used to explore active illuminated areas of the 3D medium efficiently, therebyleading to a directed motion. A power-tunable velocity, over tens of micrometers toward the dark regions of the optical gradient can be obtained.

Kaneko et al. have demonstrated the phototactic behavior of micrometer-sized oil droplets in a far-from-equilibrium state using photoresponsive surfactants [22]. The controlled motion of droplets was explained by heterogeneity in the interfacial tension at the droplet surface that was induced by the molecular photoactive conversion of system components with the molecular structure of the photoresponsive surfactant containing an azobenzene. The negative phototaxis was very sensitive and exhibited a directional change within 0.5 second. Such sensitivity can find applications in the spatial arrangement of micrometer-sized objects and also as transporters.

Another strategy is to create soft materials in which macroscopic motions of molecular assemblies are actuated by small amounts of molecular machines. A Japanese group has also reported reversible and spatially controlled photoinduced macroscopic motions of oleate assemblies containing smaller amounts of azobenzene with submillimeter lengths [20]. They prepare mixtures of oleic acid and oleate. The mixture exhibits a highly elastic response to mechanical stress because of the highly ordered stacking structure of oleate molecules in the assembly. They put azobenzene derivatives dissolved in oleate assemblies. Irradiation with 366 nm light induced a slight increase in the size of the vesicles. The vesicles were restored to the original size upon irradiation with 435 nm light. This reversible change in size is associated with the mechanical behavior of the azobenzene derivatives. The inner motion exhibited by the helical assemblies and the presence of an induction period strongly suggest that there are cooperatively propagating mechanisms such as domino effects occurring inside or at the surface of the supramolecular assemblies. It was assumed that the change in effective volume of azobenzene units by photoisomerization causes the propagation of expansion or contraction of inter-molecular geometry among oleate molecules.

The mechanisms associated with azobenzene can be used to probe different phenomena, in particular, with the field of plasmonics. The fabrication of hybrid core-shell nano-assemblies involving a nondoped azophotochromic core coated with a dense shell of gold nanoparticles with a diameter of 140 nm was reported for investigating the influence of localized plasmons onto the azo-core photoisomerization [23]. Such coupling causes a partial quenching of the azophotoisomerization but does not impact the thermal back relaxation.

Azobenzene is a powerful molecule able to photomechanically modify nanodevices and nano-objects. A brief summary of the main description of this molecule is then given in the following text.

12.2 Azopolymer as a Photoactive Material

Azobenzene used in the previously described experiments is an aromatic molecule where an azo linkage (—N=N—) joins two phenyl rings. The name is often referred as "azobenzenes" or simply "azo". A large class of compounds based on this molecule can be obtained by substituting the aromatic rings with various substituents, to change its geometry and electron donating/withdrawing character. The most interesting behavior common to all azo-based material is the efficient and reversible photoisomerization, which occurs upon absorption of a photon within the absorption band. Azobenzenes have two isomeric states: a thermally stable *trans* configuration, and a meta-stable *cis* form. Under irradiation, a fraction of the transazobenzenes will be converted to the *cis* form, which will thermally revert to the more stable *trans* on a timescale dictated by the chemically engineering molecule substitution pattern.

One of the most interesting properties of the materials is the photoinduced mass transport [24]. When azopolymer films are illuminated by an interference pattern of coherent laser beams in the absorption band, a surface relief grating (SRG) can be formed on the film surface at temperatures well below the glass transition temperature (T_g) of the azopolymer. The photoinduced mass transport is caused by the repeated *cis–trans* isomerization of azochromophores under polarized light irradiation, which forces the chromophore to continually change their orientations. The deformation is then caused by the variation of the orientation order and conformational change of the polymer backbones triggered by the photoisomerization of azobenzenes. The mass transport ends up with stabilization of the molecular movements in a direction perpendicular to the polarization. The photoinduced surface patterns or SRG in most of the cases are stable below the T_g and can be erased by heating samples with a temperature above the T_g or with an illumination with a circularly polarized laser beam. The SRG formation is neither a thermally driven process nor an ablation of the surface and can be attributed to different phenomena depending on the azo materials and consisting in: (i) internal pressure gradients caused by an isomerization-driven free volume expansion, (ii) a force based on the dipolar interaction of the azochromophores with the optically induced electric field gradient, (iii) a translational worm-like diffusion caused by photoisomerization of the azobenzene chromophores, and (iv) a mean-field force caused by molecule alignments.

The photoinduced orientation and related properties have been extensively investigated for various polymeric materials, such as azochromophore-doping polymers (guest–host systems), azo-functionalized liquid crystal polymers, and azo side chain to azo main chain amorphous azopolymers. Among them, photoinduced deformations observed in different forms have boosted considerable research interest because of their unique nature and potential applications where small objects can transform their properties when it is remotely modified with a light stimulus. The deformations and modifications can be induced for any kind of objects at the nanoscale.

Grebel-Koehler et al. [25] introduced azobenzene photochromes into the core of rather spherical dendrimers, resulting in (quasi) isotropic nano-objects that display significant changes in a hydrodynamic volume upon photoirradiation, with contractions as high as $\Delta Vh = 37\%$ as evaluated by a gel permeation chromatography (GPC).

A recent study has shown that a photomechanical effect occurs in thin films of azopolymers called photofluidization appearing as a new possible mechanism to explain the formation of an SRG on azopolymer films [26]. The photoinduced pattern implies that the light force could drive the azopolymers to move at least within the sub-micrometer scale. The photo-softening effect of azopolymers upon light exposure was measured by *in situ* monitoring force displacement of a tip penetration [27]. The study revealed that azopolymers behaved like elastic solids without light irradiation and became viscoelastic under illumination. This photofluidization is anisotropic and depends on the polarization of the incident light. Another study found a photo-expansion effect inducing reversible as well as irreversible changes in the mechanical properties of azopolymers. *Photo-expansion* based on repeated *cis–trans* photoisomerization generates a free volume in the surrounding polymer matrix. After multiple *trans–cis* photoisomerization, the total volume of azopolymer films is increased. The experiment was confirmed using neutron reflectometry [28]. Two competing photomechanical effects in azopolymer systems were demonstrated: one that causes expansion of the polymer matrix, and one that induces contraction of the material. In this experiment it was able to measure the expansion and contraction effects since neutron reflectometry probes the physical density rather than the optical refractive index. Later, a photoisomerization induced a softening of the polymer matrix followed by a polymer chain migration which well agrees with a study of nanoindentation with an AFM used for the acquisition of both surface time-dependent and time-independent results. Indeed, their photoinduced surface patterning provides evidence of a plastic (or viscoplastic) photosoftening far beyond what thermal effects could be responsible for. The experiments on PMMA-DR1 films [29] confirmed that the decrease of the irreversible viscosity by 1 order of magnitude is observed under illumination indicating that light strongly reinforces the viscoplastic behavior of the film. This was confirmed by indentation measurements. Additionally, the significant reduction of the composite hardness under illumination indicates a significant decrease in the hardness and/or plastic threshold (yield stress) of the material.

Different models have been tested to account for the molecular motion phenomena for amorphous polymers and liquid-crystalline side-chain polymers to explain the long-term stability of *trans–cis* isomerization. We can refer to models to study SRG formation in azopolymer films based on the interaction between molecular dipoles [30]: the semi-implicit moving-particle model [31], random-walk model [32], a stochastic model [33], and a model based on Fick's law of diffusion as a phenomenological description of the molecular motion [34]. This last model is used to explain the self-organization of photoinduced azopolymer surfaces when the light illuminates the samples. This model is based on Fick's law of diffusion that scales with light intensity. Further, light is trapped in the sample and couples into counter propagating guided modes, which interact with the azopolymer. Using this approach, it is possible to explain the self-patterning of the azopolymer film illuminated with a single beam and the surface profile and intensity evolution of azopolymer thin films during exposition to a single laser beam. The model is based on both the coupled mode theory and a phenomenological description of the molecular motion.

A unified modeling framework for the light actuation on azo-based materials was explained allowing for the understanding of surface deformation and bulk deformation of polymer films that are controlled by different laser beam polarizations: linear or circularly polarized light or a vortex shape of the laser beam [35]. It is shown that dipole forces strongly respond to polarized light in contrast to higher order quadrupole forces that are often used to describe SRG deformation through a field gradient constitutive law. Dipole forces strongly influence the evolution of *trans* and *cis* azobenzene states within monodomain and polydomain polymer networks. The coupling between the polymer network and azobenzene leads to significantly different photostrictive deformations.

12.3 Fabrication of Photoactive Nano-objects

12.3.1 Fabrication of Photoactive Nanospheres

Colloidal spheres can be prepared by methods such as controlled precipitation, emulsion polymerization, and aggregation of amphiphilic block copolymers. Amphiphilic polymers are also well known for their ability to form other types of structures through self-assembly processes. The incorporation of azopolymers into colloidal spheres, allows demonstrating that the colloidal spheres can show the photoinduced anisotropy function of the polymers and other interesting properties related with colloids.

Different techniques are used to fabricate azopolymer nanospheres. The first experimental demonstration of the photoinduced deformation of azopolymer nanospheres was done by Wang and coworkers [36]. The colloidal spheres were prepared by a method similar to those used to obtain micelles and other self-assembled aggregates of amphiphilic polymers. The method of preparation of the colloid nanospheres is based on the technique of Zhang and Eisenberg [37] and consists in first dissolving the azopolymer in a good solvent for the chosen polymer, and then adding a suitable amount of milli-q water dropwise into the solvent solution. When the water content reached a critical value, uniform colloidal spheres were obtained. The polymeric chains started to aggregate and form colloidal spheres. When the water content reached 40–50% (vol%), the polymeric chains collapsed gradually as the water content increased further. When micellization occurs, the nanospheres are dropped on a glass substrate. Density of the nanospheres on the surface can vary depending on the concentration in the solution. A scanning electron microscope (SEM) image of the nanospheres dispersed on the surface of a microscope glass slide is presented in Figure 12.3A. The external diameter measurement by a statistical analysis of the spheres can vary from nm to micron meter. With this method, different azopolymer spheres have been prepared and tested under laser illumination. An amphiphilic random copolymer functionalized with branched azo side chains has been prepared [41]. The microphase transition and structure formation are closely related with the water content in the systems. Other materials have been prepared such as four-arm-shaped amphiphilic azobenzene compound [42] epoxy-based azopolymers [43], methacrylate-based azohomopolymers [44], azopolyurethanes bearing push–pull type azochromophores [45], an amphiphilic random azocopolymer (PEAPE), and a polydispersed azohomopolymer (BP-AZ-CA), which were functionalized with azobenzene type and pseudo-stilbene-type chromophores, respectively [46], or amphiphilic azobenzene-containing random copolymer, poly(2-[4-(phenylazo)phenoxy]ethyl acrylate-*co*-acrylic acid) (PPAPE) [47]. Another study [48] shows that the electron-withdrawing groups on the azochromophores have a significant influence on the photoinduced deformation behavior of the azopolymer colloids mainly through their modulation to the absorption band position. The colloid deformation can be more efficiently induced by irradiation with the laser beams at the intermediate wavelengths between the maximum and the absorption band tails of the azochromophores. Furthermore, the electron-withdrawing groups can also affect the photoinduced deformation behavior through their influence on the hydrophilicity of the polymers. Another technique that produced azopolymeric microspheres used a glass capillary microfluidic device in which the outer phase can hydrodynamically focus on the inner phase, so as to result in droplet breakup in a dripping mode and subsequent generation of highly monodispersed oil-in-water single emulsions [49]. They used two differently sized injection orifices (20 and 100 μm in diameter) in microfluidic devices with a collection

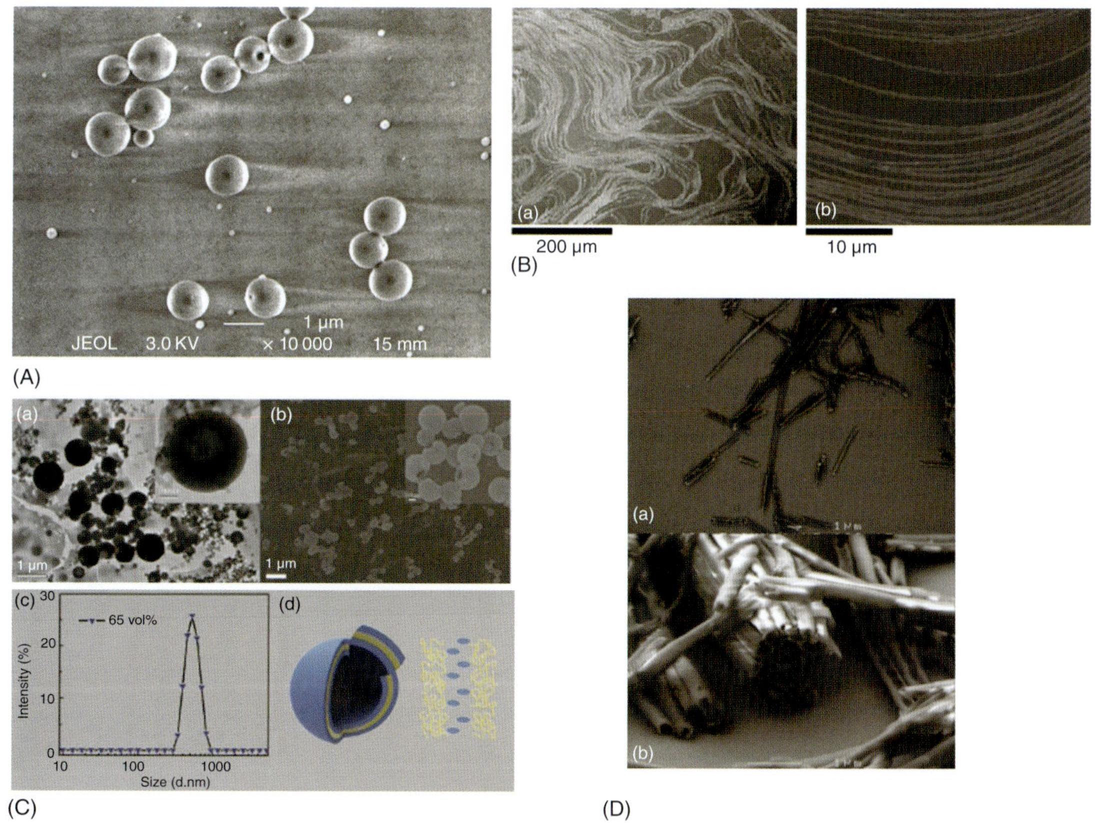

Figure 12.3 Different topologies of micro/nano-objects made with azopolymer: (A) microspheres and (B) microwires made with a soft lithography. Source: Liu et al. 2006 [38]. Reprinted with permission of American Chemical Society. (C) Hollow-core nanospheres. Source: Zhang et al. 2014 [39]. Reprinted with permission of Royal Society of Chemistry. (D) Azopolymer nanotubes. Source: Barillé et al. 2009 [40]. Reprinted with permission of AIP Publishing.

orifice 200 µm in diameter, wherein the inner flow rate (i.e. of the oil phase) was fixed at 300 µl h^{-1}, and the outer flow rate (i.e. of the water phase) was varied. A typical azopolymer poly(disperse red 1 methacrylate) (pDR1) was selected for study. Two compatible oil-phase solvents of toluene and chloroform were chosen for comparison with each other. After droplets of pDR1 emulsion were collected, the oil phase was allowed to evaporate at room temperature. They obtained highly uniform and smooth pDR1 microspheres considering the various microfluidic conditions (polymeric concentration within droplets, flow rate, and fluid).

A novel method based on noncovalent bond self-assembly technology provided hollow nanospheres with azo functional groups [49] (Figure 12.3C). Azocomplex nanospheres with a diameter of 300–1000 nm were obtained by the H-bond (noncovalently) self-assembly of poly(acrylonitrile-stat-4-vinylpyridine) (PANstat-P4VP) and azobenzene *m*-methyl red (MMR) in aqueous solvents. Polymeric hollow nanospheres with azobenzene chromospheres were formed in the THF–H_2O mixed solvent with 65 vol% H_2O in mass. Azocomplex nanospheres with a diameter of 300–1000 nm were obtained. These hollow nanospheres exhibit potential as photoinduced deformation due to the azobenzene group.

The fabrication of complex architectures of the nano- or microspheres is also interesting and in particular micro- or nanoparticles possessing noncentrosymmetric architectures like Janus particles (JPs) (Figure 12.4a). The name originates from 'Janus,' the Greek god with two faces.

The JPs consist of two distinct parts whose volume ratio is about 1 : 1, which is very close to the original feed ratio. Wang and coworkers have shown that JPs with the red part is the PAZO–ADMA phase and the transparent part is the PMMA phase [39]. Owing to high incompatibility between the two polymers, the Janus nanoparticles with symmetric compartment volumes are obtained, meaning that two volumes are self-assembled together forming two faces.

It is also necessary to reduce the interfacial tension between the dispersed droplets and dispersion medium with a suitable surfactant for forming snowman-like particles. Azobenzene-containing layer-by-layer films and microcapsules have been prepared with a range of cationic and construction of photosensitive devices. This review summarizes the syntheses of azobenze [51].

Azopolymer microspherical cap arrays have been fabricated through soft-lithographic contact printing using colloidal arrays as masters and a duplicated PDMS elastomer as stamps [52]. The size of the microspherical caps can be feasibly altered by adjusting the size of polystyrene (PS) colloids and the "ink" concentration. After irradiation of the linearly polarized laser single beam, the microspherical caps can be elongated along the polarization direction.

Most of the experiments presented until now show that the nanoparticles are well dispersed on a surface of glass substrate. However, the attractive capillary force can cause the colloidal spheres to organize into hexagonally close-packed 2D arrays or monolayers. One method is frequently used: the vertical deposition method in the goal to realize the capillary-force-driven self-assembly. In the vertical deposition method, a substrate, usually glass, is immersed vertically in

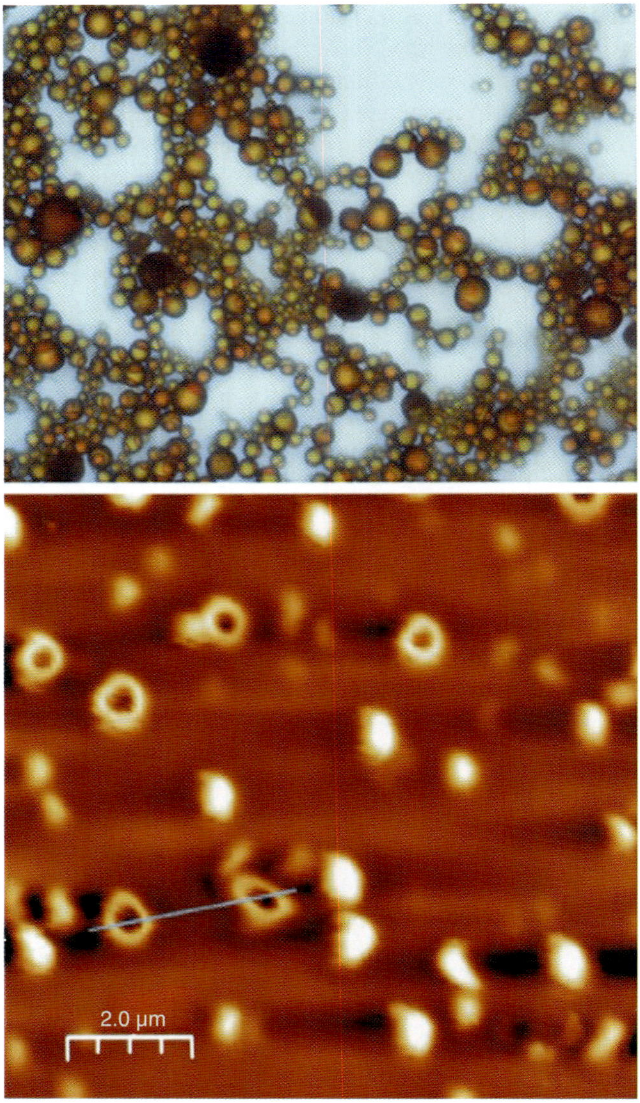

(a)

(b)

Figure 12.4 Exotic nano-object made with azopolymer (a) Janus nanoparticles made with azopolymer and PMMA and (b) doughnut-shaped nano-objects obtained with a solvent-induced dewetting technique. Source: Dubrovkin et al. 2014 [50]. Reprinted with permission of Elsevier.

a suspension containing colloidal spheres, which in our case, are azopolymer nanospheres [53]. By evaporation of the dispersion medium or lifting of the substrate out of the suspension at constant speed, the surface of the liquid moves down and the colloids are deposited onto the substrate during the decline of the surface. The average size of the spheres was estimated to be 180 nm. Dynamic light scattering (DLS) has shown monomodal distribution for the sizes of the colloidal spheres.

These assemblies of colloids on a surface can be used to design porous surfaces. The assembly of colloids can be used as a reverse surface [47]. Polymeric porous structures are very important for possible applications in selective transportation, as biosensors and as photonic opal materials. It was demonstrated that the ordered mesoporous films can be obtained by *in situ* sphere-pore inversion of 2D arrays of the azopolymer spheres. The mesoporous films were obtained by placing the 2D colloidal sphere arrays in an enclosed chamber for solvent annealing for 6–9 hours. Before solvent annealing, the colloidal surfaces are composed of hexagonally close-packed spheres. After the solvent treatment, the spheres are transformed to the mesoporous structure directly, which maintains the character of a hexagonal close-packed array. The average pore diameter is 150 nm.

12.3.2 Fabrication of Nanotubes and Nanowires

Azopolymers can also be used to adjust the shape and size of polymeric nano/microwires (Figure 12.3B). However, it is difficult to fabricate nano/microwires through a feasible approach. It was demonstrated that microwires with a uniform sub-micrometer-scale transverse dimension and macroscopic longitudinal dimension, made of an epoxy-based azopolymer can be fabricated through a soft lithographic approach [38]. The microwires can be processed by a laser beam in order to alter the transverse dimension of the microwires or inscribe regular surface relief structures on the microwires. First, elastomeric stamps made with PDMS (poly(dimethylsiloxane)) were prepared by replica molding. Then the azopolymer solution was deposited on the surface and the excessive liquid on the surfaces was removed.

Sheng et al. report the hierarchical self-assembly of a cationic gemini amphiphile azobenzene compound in a composition gradient solution generated using solvent evaporation [54]. As the gradient solution is formed, the azobenzene compound forms nanorods in the lower region of the solution. Depending on solvent composition, these nanorods can be developed into nanofibers, which can then intertwine to form double helices and other types of nanohelices in the upper region of the solution. Macroscopic wire bundles can be formed by the fusion of nanohelices. These bundles exhibit photoresponsive properties that affect the deformation.

Tubular azopolymer nanostructures with controlled diameters were prepared by wetting porous anodic alumina membrane templates consisting in an array of parallel and straight channels [40] (Figure 12.3D). The alumina membrane filters have a typical diameter of 13 mm and thickness of 60 μm. The individual cylindrical pores can be viewed as a beaker in which a piece of the desired material can be synthesized. The pores in the membrane have monodispersed diameters of 200 nm. A nanocylinder of azopolymer can be obtained from each pore. The nanotubules were prepared by filtration of a solution of the azopolymer into the pores of the template membrane. About 1 ml of the azopolymer solution was dropped into a 1 cm^2 area on the top surface of the template membrane. The membrane was left at ambient conditions until the solution was infiltrated through the membrane. It was then placed in an oven for solvent evaporation during one hour at 60 °C temperature. Nanotubules were freed by dissolving the membrane by soaking in 3 M aqueous NaOH for 10 minutes. The difficult part was to eliminate the NaOH from forming crystals on the surface. Toward achieving this, azopolymer nanotubes were extracted from the resulting mixture by three 30-second-cycles of centrifugation at 8000 rpm, followed by two more cycles in distilled water. Nanotubules were then dispersed into water by sonication. The water suspension of dispersed nanotubules was then dropped on the surface of clean microscope glass slides and dried in an oven for water evaporation (Figure 12.3D).

12.3.3 Fabrications of Other Different Nano- and Micro-objects

Rigid-rod polymers incorporating azobenzene photoswitches in a poly(para-phenylene) (PPP) backbone were fabricated [55]. The rod-like polymers have shown reversible and efficient shrinkage that have been demonstrated and quantified by static and dynamic light-scattering measurements. The deformation is accompanied by an unprecedented decrease in hydrodynamic volume. Furthermore, the design of multi-azobenzene rods with reduced solubility have led to the formation of spherical aggregates in organic solvents, which could reversibly be disrupted by alternating irradiation with UV and visible light.

Exotic nano-objects can also be designed by illumination of a pre-formed surface [50]. Doughnut-shaped nanostructures at the tailored surface of an azopolymer film have been fabricated by an incoherent unpolarized light illumination (Figure 12.4b). The method chosen is a solvent-droplet-induced dewetting of thin azopolymer films on glass substrates. The contact of a thin polymer film to a solvent droplet reduces the glass transition temperature to below the room temperature as the solvent molecules penetrate into the film matrix. The stable polymer thin film is destabilized by the introduction of polar interactions and nano-holes are created. Then an incoherent light is used for growing the nanostructures and simultaneously directing the final shape of a nanostructure by the initial seed of tiny nanoscale holes rather than for optically erasing. The sample was illuminated for 30 minutes. The volume of the doughnut has shown to grow to reach a mean value of 0.065 μm^3.

12.4 Results

SRG or photo-nanostructuration can be induced on the azopolymer surface of any nano-object as microspheres (Figure 12.5A), wires (Figure 12.5C), and thin films. The surface of the microspheres can be photopatterned with illumination by interferences of a laser beam [56] (Figure 12.5D). The observation showed a grating on the surface of the microspheres and a contraction of the pores. It was shown that azopolymer nanotubes with a diameter of 100 nm present different surface deformations as a function of the laser beam polarization [40].

After irradiation of an individual nanotube with a transverse electric field excitation, the width is observed to shrink. Using an AFM profiling, the shrinkage is estimated to be about 60% of the initial width. With a p-polarization the same

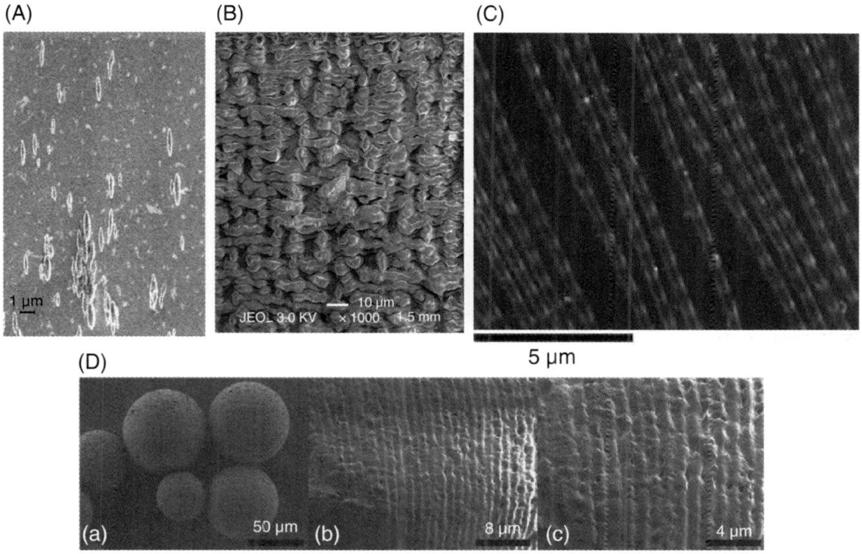

Figure 12.5 Photoinduced effect on different azopolymer objects showing the possibility to pattern and control the shape with the polarization of the laser beam. (A) photoinduced effects on nanospheres. The nanospheres are elongated in the direction of the light polarization. Source: Reprinted with permission of [57]. (B) photoinduced effects on an assembly of azopolymer colloids, and (C) photostructuration on microwires. Source: Reprinted with permission of [38]. (D) Surface relief grating observed on the surface of microspheres. Source: Reprinted with permission of [56].

experiment was repeated. At the beginning of the process, the surface profile of the measured nanotube in the longitudinal direction is flat. After 30 minutes irradiation, the nanotube presents a quasi periodically structured surface.

The force exerted on azopolymer nanospheres to induce a deformation was calculated using the Hertz theory, which considers the contact deformation of elastic spheres under normal loads in the absence of adhesion and friction. A force of 1 μN was detected. It is an order of magnitude larger than typical forces needed for the deformation of hollow polyelectrolyte microcapsules using an atomic force microscope [58]. Moreover, the stress that is exerted on a polymer during SRG formation given by Toshchevikov et al. [59] was calculated. It yields a force of $F_Z \sim 2$ μN onto the nanospheres, readings that are rather close to the estimation with the Hertz theory.

Ortyl et al. demonstrated that an assembly of colloidal azopolymer nanospheres could accelerate the process of SRG formation [57] (Figure 12.5B). One of the reasons is that the surface is already pre-formed and the process of photostructuration starts on small nano-objects. Only anisotropic photofluidization by polarized light deforms the nanospheres along the polarization direction. The phenomenon is amplified by the self-lensing effect of the azopolymer nanospheres absorbing the laser wavelength. The single beam SRG formation is accelerated by the initial natural roughness of the colloidal assembly, suppressing the long incubation time usually observed in the spontaneous grating organization. The subduction mechanism of nano-spheres pushing each other along the polarization direction enhances the grating modulation amplitude. Altogether, exceptionally large and fast SRG formation gets self-organized using a single beam irradiation in the colloidal nanosphere assembly. This method is a generic method and can be used for any material already tested during SRG formation on thin films. The preformed pattern on the surface of the azopolymer can also be used to photo-accelerate the structures of what was called nano-doughnuts

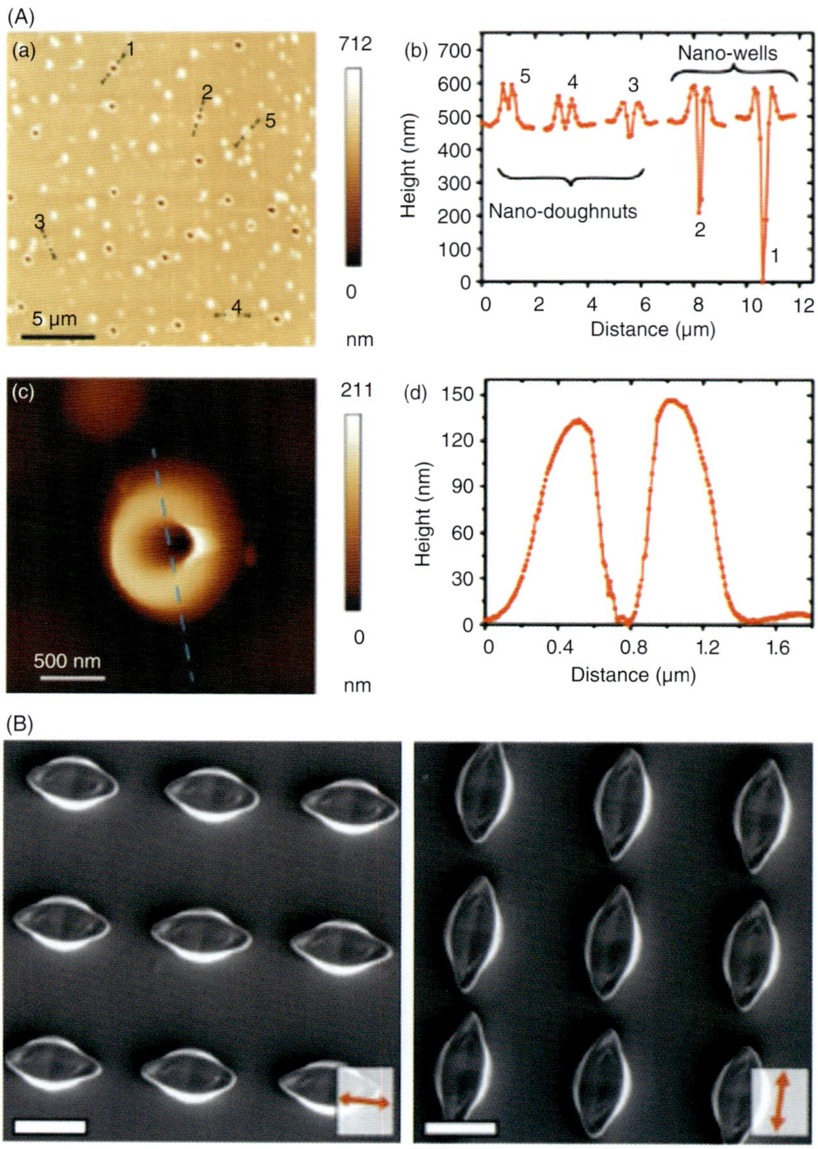

Figure 12.6 (A) Photoinduced fabrication of nano doughnut on the surface of an azopolymer Source: Dubrovkin et al. 2014 [50]. Reprinted with permission of Elsevier. (B) Light deformation of micropillar used for the control of the wettability of a surface. Source: Oscurato et al. 2017 [60]. Reprinted with permission of American Chemical Society.

(Figure 12.6A). The light amplifies the structures and can modify the geometry with a near-field set-up.

Ambrosio and coworkers have used light-driven mass migration in order to reconfigure a pre-patterned tridimensional superficial texture consisting in an array of azopolymer micropillars so as to induce a controlled directional wetting anisotropy [60] (Figure 12.6B). The wetting state of a water droplet onto the pristine cylindrical pillar array is recognized as a result of the pinning effect of the triple-phase contact line (TCL) at the sharp pillar edges. The experiment therefore shows the potential of using light parameters with the aim to design light-driven two-dimensional wetting pathways onto the surface of a single pre-patterned azopolymer pillar array. The last experiment shows that the directional anisotropic wetting is transformed into a unique direction of the liquid spreading by a superficial hydrophilization treatment of the surface.

12.5 Summary and Outlook

We have demonstrated the huge capabilities of nano-objects made with azobenzene or azopolymers. In particular, we have shown that the mechanism of the photo-orientation of azo molecules can induce mechanical actions. These mechanical actions can be further used for directional movements of micro/nanosystems or display significant changes in volume [61]. We have reviewed the different nano/micro- objects obtained up to now including azobenzene in their designs. In terms of nano- or micro-objects we have presented the fabrication and use of micro- or nanoscale objects as nanoparticles, colloidal spheres, nanoparticles, nanorods, nanotubes, nano/microwires, three dimensional posts, microspherical cap arrays, needles, polymeric strips, micro disks, etc. All these nano/micro azobenzene based objects can find applications in different media. Large possibilities of applications can be developed and can, in particular, open the way for future applications concerning the future of nanotechnology in sensors, nano-medicine or nano-robotics. We can imagine, for example the reconfiguration of subdevices with light like metamaterials to match different optical properties such as chirality or can imagine nanorods as nano reconfigurable barcodes. The inclusion of emitting compounds in azopolymer nanotubes can be tuned by opto-mechanically changing the volume of the nanotube.

References

1 Wang, L. and Li, Q. (2018). Photochromism into nanosystems: towards lighting up the future nanoworld. *Chem. Soc. Rev.* 47: 1044–1097.
2 Ionov, L. (2015). Polymeric actuators. *Langmuir* 31: 3015–3024.
3 Huck, W.T.S. (2008). Responsive polymers for nanoscale actuation. *Mater. Today* 11: 7–8.
4 Hendricks, M., Schenning, A.P.H.J., Debije, M.G., and Broer, D.J. (2017). Light-triggered formation of surface topographies in azo polymers. *Crystals* 7: 231.

5 Ye, G., Yang, C., and Wang, X. (2010). Sensing diffraction gratings of antigen-responsive hydrogel for human immunoglobulin-G detection. *Macromol. Rapid Commun.* 31: 1332–1336.
6 Egawa, Y., Miki, R., and Seki, T. (2014). Colorimetric sugar sensing using boronic acid-substituted azobenzenes. *Materials* 7: 1201–1220.
7 Moniruzzaman, M. and Rock, J. (2016). Azobenzene-based gel coated fibre Bragg grating sensor for moisture measurement. *Int. J. Polym. Sci.* 8471903.
8 Ma, Z.G., Ma, R., Xiao, X.L. et al. (2016). Azo polymeric micelles designed for colon-targeted dimethyl fumarate delivery for colon cancer therapy. *Acta Biomater.* 44: 323–331.
9 Zhang, X., Hou, L., and Samorı, P. Coupling carbon nanomaterials with photochromic molecules for the generation of optically responsive materials. *Nat. Commun.* 2015 (7): 11118.
10 Luo, W., Feng, Y., Cao, C. et al. (2015). A high energy density azobenzene/graphene hybrid: a nano-templated platform for solar thermal storage. *J. Mater. Chem. A* 3: 11787–11795.
11 Kolpack, A.M. and Grossman, J.C. (2011). Azobenzene-functionalized carbon nanotubes as high-energy density solar thermal fuels. *Nano Lett.* 11: 3156–3162.
12 Huang, C., Lv, J., Tian, X. et al. (2015). Miniaturized swimming soft robot with complex movement actuated and controlled by remote light signals. *Sci. Rep.* 5: 17414.
13 Barrett, C.J., Mamiya, J.-I., Yager, K.G., and Ikeda, T. (2007). Photo-mechanical effects in azobenzene-containing soft materials. *Soft Matter* 3: 1249–1261.
14 Wie, J.J., Shankar, M.R., and White, T.J. (2016). Photomotility of polymers. *Nat. Commun.* 7: 13260.
15 Gelebart, A.H., Mulder, D.J., Varga, M. et al. (2017). Making waves in a photoactive polymer film. *Nature* 546: 632–636.
16 Tsoi, S., Zhou, J., Spillmann, C. et al. (2013). Liquid-crystalline nano-optomechanical actuator. *Macromol. Chem. Phys.* 214: 734–741.
17 Vantomme, G., Gelebart, A.H., Broer, D.J., and Meijer, E.W. (2017). A four-blade light-driven plastic mill based on hydrazone liquidcrystal networks. *Tetrahedron* 73: 4963–4967.
18 Uchida, E., Azumi, R., and Norikane, Y. (2015). Light-induced crawling of crystals on a glass surface. *Nat. Commun.* 6: 7310.
19 Norikane, Y., Tanaka, S., and Uchida, E. (2016). Azobenzene crystals swim on water surface triggered by light. *CrystEngComm* 18: 7225–7228.
20 Kageyama, Y., Tanigake, N., Kurokome, Y. et al. (2013). Macroscopic motion of supramolecular assemblies actuated by photoisomerization of azobenzene derivatives. *Chem. Commun.* 49: 9386–9388.
21 Abid, J.-P., Frigoli, M., Pansu, R. et al. (2011). Light-driven directed motion of azobenzene-coated polymer nanoparticles in an aqueous medium. *Langmuir* 27: 7967–7971.
22 Kaneko, S., Asakura, K., and Banno, T. (2017). Phototactic behavior of self-propelled micrometer-sized oil droplets in a surfactant solution. *Chem. Commun.* 53: 2237–2240.

23 Snell, K.E., Mevellec, J.-Y., Humbert, B. et al. (2015). Photochromic organic nanoparticles as innovative platforms for plasmonic nanoassemblies. *ACS Appl. Mater. Interfaces* 7: 1932–1942.
24 Mahimwalla, Z., Yager, K.G., Mamiya, J.-I. et al. (2012). Azobenzene photomechanics: prospects and potential applications. *Polym. Bull.*
25 Grebel-Koehler, D., Liu, D.J., De Feyter, S. et al. (2003). Synthesis and photomodulation of rigid polyphenylene dendrimers with an azobenzene core. *Macromolecules* 36: 578–590.
26 Karageorgiev, P., Neher, D., Schulz, B. et al. (2005). From anisotropic photo-fluidity towards nanomanipulation in the optical near-field. *Nat. Mater.* 4: 699–703.
27 Harrison, J.M., Goldbaum, D., Corkey, T.C. et al. (2015). Nanoindentation studies to separate thermal and optical effects in photo-softening of azopolymers. *J. Mater. Chem. C* 3: 995.
28 Yager, K.G., Tanchak, O.M., Godbout, C. et al. (2006). Photomechanical effects in azo-polymers studied by neutron reflectometry. *Macromolecules* 39: 9311–9319.
29 Sorelli, L., Fabbri, F., Frech-Baronet, J. et al. (2015). A closer look at the light-induced changes in the mechanical properties of azobenzene-containing polymers by statistical nanoindentation. *J. Mater. Chem. C* 3: 11055–11065.
30 Pedersen, T.G. and Johansen, P.M. (1997). Mean-field theory of photoinduced molecular reorientation in azobenzene liquid crystalline side-chain polymers. *Phys. Rev. Lett.* 79: 2470–2473.
31 Barada, D., Itoh, M., and Yatagai, T. (2004). Numerical analysis of photoinduced surface relief grating formation by particle method. *J. Appl. Phys.* 96: 4204–4210.
32 Bellini, B., Ackermann, J., Klein, H. et al. (2006). Light-induced molecular motion of azobenzene-containing molecules: a random-walk model. *J. Phys. Condens. Matter* 18: S1817–S1835.
33 Juan, M.L., Plain, J., Bachelot, R. et al. (2008). Stochastic model for photoinduced surface relief grating formation through molecular transport in polymer films. *Appl. Phys. Lett.* 93: 153304.
34 Leblond, H., Barille, R., Ahmadi-Kandjani, S. et al. (2009). Spontaneous formation of optically induced surface relief gratings. *J. Phys. B: At. Mol. Opt. Phys.* 42: 205401.
35 Bin, J. and Oates, W.S. (2015). A unified material description for light induced deformation in azobenzene polymers. *Sci. Rep.* 5: 14654.
36 Li, Y., He, Y., Tong, X., and Wang, X. (2005). Photoinduced deformation of amphiphilic azo polymer colloidal spheres. *J. Am. Chem. Soc.* 127: 2402–2403.
37 Zhang, L. and Eisenberg, A. (1995). Multiple morphologies of "crew-cut" aggregates of polystyrene-*b*-polyacrylic acid block copolymers. *Science* 268 (5218): 1728–1731.
38 Liu, B., He, Y., and Wang, X. (2006). Fabrication of photoprocessible azo polymer microwires through a soft lithographic approach. *Langmuir* 22: 10233–10237.
39 Zhang, T., Jin, C., Wang, L., and Yin, Q. (2014). One-step synthesis of hollow polymeric nanospheres: self-assembly of amphiphilic azo polymers via hydrogen bond formation. *RSC Adv.* 4: 36882–36889.

40 Barillé, R., Tajalli, P., Zielinska, S. et al. (2009). Surface relief grating formation on nano-objects. *Appl. Phys. Lett.* 95: 053102.
41 Deng, Y., Li, Y., and Wang, X. (2006). Colloidal sphere formation, H-aggregation, and photoresponsive properties of an amphiphilic random copolymer bearing branched azo side chains. *Macromolecules* 39: 6590–6598.
42 Wang, J., Wang, S., Zhou, Y. et al. (2015). Fast photoinduced large deformation of colloidal spheres from a novel 4-arm azobenzene compound. *ACS Appl. Mater. Interfaces* 7: 16889–16895.
43 Zhou, Y., Tang, B., and Wang, X. (2015). Photoinduced deformation behavior of a series of newly synthesized epoxy-based polymers bearing pushepull azo chromophores. *Polymer* 60: 292–301.
44 Liu, B., Zhou, Y., Zhou, X., and Wang, X. (2017). Photodeformable microspheres from methacrylate-based azo homopolymers. *Macromol. Chem. Phys.* 218: 1700020.
45 Zhou, Y., Liu, B., and Wang, X. (2017). Microspheres of polyurethanes functionalized with push-pull type azo chromophores and their photoinduced deformation behavior. *Polymer* 111: 229–238.
46 Deng, Y., Li, N., He, Y., and Wang, X. (2007). Hybrid colloids composed of two amphiphilic azo polymers: fabrication, characterization, and photoresponsive properties. *Macromolecules* 40: 6669–6678.
47 Li, Y., Deng, Y., Tong, X., and Wang, X. (2006). Formation of photoresponsive uniform colloidal spheres from an amphiphilic azobenzene-containing random copolymer. *Macromolecules* 39: 1108.
48 Liu, J., He, Y., and Wang, X. (2010). Influence of chromophoric electron-withdrawing groups on photoinduced deformation of azopolymer colloids. *Polymer* 51: 2879–2886.
49 Yeo, S.J., Park, K.J., Guo, K. et al. (2016). Microfluidic generation of monodisperse and photoreconfigurable microspheres for floral iridescence-inspired structural colorization. *Adv. Mater.* 28: 5268–5275.
50 Dubrovkin, A.M., Barillé, R., Ortyl, E., and Zielinska, S. (2014). Photoinduced doughnut-shaped nanostructures. *Chem. Phys. Lett.* 599: 7–11.
51 Zhou, X., Du, Y., and Wang, X. (2016). Azo polymer Janus particles and their photoinduced, symmetry-breaking deformation. *ACS Macro Lett.* 5: 234–237.
52 Akiba, U., Minaki, D., and Anzai, J.-I. (2017). Photosensitive layer-by-layer assemblies containing azobenzene groups: synthesis and biomedical applications. *Polym.* 9: 553.
53 Liu, B., He, Y., Fan, P., and Wang, X. (2007). Azo polymer microspherical cap array: soft-lithographic fabrication and photoinduced shape deformation behavior. *Langmuir* 23: 11266–11272.
54 Sheng, Y., Chen, Q., Yao, J. et al. (2015). Hierarchical assembly of a dual-responsive macroscopic insulated molecular wire bundle in a gradient system. *Sci. Rep.* 5: 7791.
55 Bleger, D., Liebig, T., Thiermann, R. et al. (2011). Light-orchestrated macromolecular “accordions”: reversible photoinduced shrinking of rigid-rod polymers. *Angew. Chem. Int. Ed.* 50: 12559–12563.

56 Zhou, X., Du, Y., and Wang, X. (2016). Azo polymer microspheres with photomanipulated surface and topographic structure. *Macromol. Chem. Phys.* 216: 765–772.

57 Ortyl, E., Zielinska, S., Barille, R. et al. (2016). Instantaneous photoinduced patterning of an azopolymer colloidal nanosphere assembly. *Opt. Mat. Exp.* 69: 2925–2932.

58 Lulevich, V.V., Radtchenko, I.L., Sukhorukov, G.B., and Vinogradova, O.I. (2003). Deformation properties of nonadhesive polyelectrolyte microcapsules studied with the atomic force microscope. *J. Phys. Chem. B* 107: 2735.

59 Toshchevikov, V., Saphiannikova, M., and Heinrich, G. (2012). Theory of light-induced deformation of azobenzene elastomers: influence of network structure. *J. Phys. Chem. B* 113: 5032.

60 Oscurato, S.L., Borbone, F., Maddalena, P., and Ambrosio, A. (2017). Light-driven wettability tailoring of azopolymer surfaces with reconfigured three-dimensional posts. *ACS Appl. Mater. Interfaces* 9: 30133–30142.

61 Barillé, R., Tajalli, P., Kucharski, S. et al. (2010). Photoinduced deformation of azopolymer nanometric spheres. *Appl. Phys. Lett.* 96: 163104.

13

Light-Controlled Encapsulation and Release Enabled by Photoresponsive Polymer Self-Assemblies

Jesús del Barrio[1], Milagros Piñol[2], and Luis Oriol[2]

[1] *Instituto de Nanociencia de Aragón, Campus Río Ebro. Edificio I+D. C/ Mariano Esquillor, 50018 Zaragoza, Spain*
[2] *Departamento de Química Orgánica, Facultad de Ciencias, Instituto de Ciencia de Materiales de Aragón, Universidad de Zaragoza-CSIC, C/Pedro Cerbuna 12, 50009, Zaragoza, Spain*

13.1 Introduction

The process of encapsulation involves the incorporation of active ingredients into a relatively small capsule or particle. Such systems are employed for various reasons such as enhancing the stability of specific active ingredients by delaying their degradation, protecting species from substances that could prematurely react with them, improving the processability of materials by altering their physicochemical properties, or even for safe handling of toxic components [1, 2]. Encapsulation technologies are also available for the production of systems incorporating controlled-release properties. In general, controlled release refers to the presentation of compounds in response to stimuli at a desired time, and at a specific rate (delayed, sustained, or even pulsed release). A timely and targeted release improves the effectiveness of the actives, ensures optimal dosage, and broadens the application range of ingredients. In many occasions, localized release allows reduction of the total amount of active ingredient that must be delivered to achieve the desired effect. Controlled release is pursued in a number of areas including agriculture, food sciences, and pharmaceuticals.

Externally triggerable encapsulants have been engineered to address these challenges and a wide palette of inputs can be used as triggering events including changes in pH, temperature, and magnetic fields, and the presence of specific chemical species. In this regard, light is an efficient and clean stimulus. Light-controlled substance release, as compared to other mechanisms, is attractive in many ways. It is a versatile stimulus, which can act in a noninvasive fashion and provides excellent spatiotemporal control by means of modulation of properties such as wavelength, intensity, and polarization. Light also shows a high degree of orthogonality toward a great deal of chemical species and functionalities [3]. Indeed, light has gained significant interest as a trigger in drug encapsulation and delivery on account of its clinical relevance (light has a number of therapeutic uses and it is currently used in photodynamic therapy [4]).

Photoactive Functional Soft Materials: Preparation, Properties, and Applications, First Edition.
Edited by Quan Li.

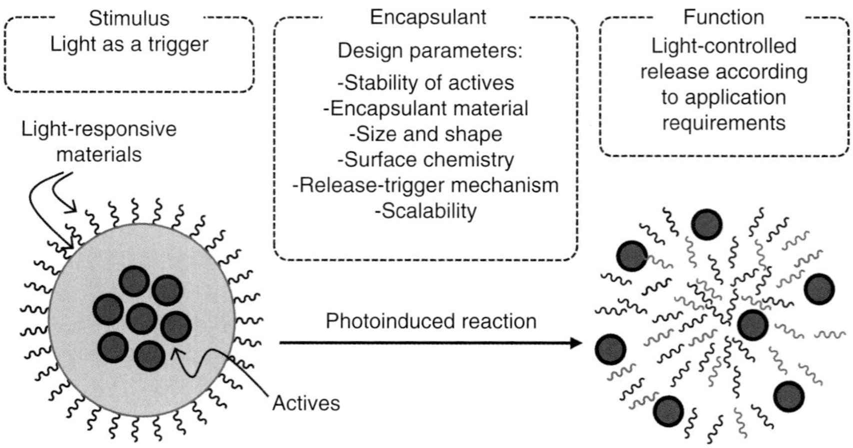

Figure 13.1 Schematic design flow diagram for a light-controlled encapsulant system.

As illustrated in Figure 13.1, several parameters need to be considered when formulating encapsulated products, which also applies to light-controlled systems. The physicochemical characteristics of the encapsulated species and its functionality in the final product, the appropriateness of the encapsulating process with respect to the stability of the actives, the physicochemical properties (size, shape, surface chemistry, etc.) and storage conditions of the final product, the release-trigger mechanism and scalability, and cost constrains are all important aspects of encapsulating systems. A further step in the design flow (Figure 13.1) involves the comparison and contrast of design elements and application specification. In such a process, a set of characteristics, like the type of encapsulant material, the concentration of the active ingredient, and the type of process to produce the formulated product, need to be selected for the preferred system. A wide range of encapsulation processes exist such as atomization, cocrystallization, emulsification (liposomes, interfacial polymerization, etc.), and inclusion complexation, and each of them has its own set of limitations and advantages and disadvantages. On the materials side, it is possible to differentiate between inorganic and organic materials including surfactants, polymers, oligomeric, and macrocyclic species. In many instances, the applicability of a specific encapsulation method or cargo-release mechanism is constrained to a very limited set of materials, which evidences the complex interdependencies between processes and materials.

There exist excellent recent reviews on photocontrolled delivery systems [5–8]. Considering the framework of this book, we highlight several polymer-based encapsulating systems exhibiting light-controlled release capabilities. In this chapter, rather than being comprehensive, we focus on those aspects and attributes of photoresponsive polymer materials that make them particularly attractive for the delivery and controlled release of active ingredients, mainly pharmacological formulations. Our goal is to provide a general sense of current research and to assist the reader in the identification of trends and challenges facing the development of practical photoresponsive polymer-based delivery systems. We begin this chapter by discussing some generalities of the design of light-responsive polymeric encapsulants (Section 13.2). Looking toward the application of light switches and photoinduced reaction in polymeric encapsulants, the families of polymer micelles and vesicles (based on amphiphilic block copolymers, BCs), capsules, microgels, nanogels, nanoparticles (NPs), and unimolecular systems will each be discussed in turn (Sections 13.3.1–13.3.6).

13.2 Photoresponsive Groups

In the framework of this chapter, the realization of light-controlled substance release requires the polymeric materials, e.g. amphiphilic BCs, which are typically capable of undergoing micellization in aqueous media, that they include at some point in their macromolecular architecture at least one photoresponsive chromophore [9–11]. Changes or reorganizations of the chromophore structure can be selectively induced with light (UV, vis, NIR), which result in temporally and spatially controlled substance delivery. This process depends, more commonly, on the light wavelength, intensity, and irradiation time. In general, three major photoinduced changes may be induced in polymeric nanocarriers. These include (i) a local polarity change (typically associated with a change of the overall hydrophobic/hydrophilic balance of the encapsulant [see Figure 13.2]), (ii) a photoinduced degradation of the polymeric structure via selective cleavage of photolabile bonds, and (iii) the photocrosslinking of specific polymeric components (see Figure 13.2). Furthermore, these photoreactions can be reversible or irreversible [12, 13].

The reversible *trans–cis–trans* photoisomerization of azobenzenes has been widely used for the preparation of photoresponsive materials due to the high synthetic versatility and stability of these compounds. As the *cis* isomer has a bent geometry and a higher polarity than the *trans* isomer, photoisomerization is accompanied by local polarity changes in the system which, without necessarily achieving a complete distortion the self-assembly, increases the level of molecular mobility and induces cargo-release. UV light is used more commonly although the photoisomerization of azobenzenes with visible light has also been described [14–16]. The photoinduced reversible ring opening/closure of spiropyrans can also alter the hydrophilic/hydrophobic balance of a given system, sometimes to a larger extent, as the open merocyanine form is zwitterionic in nature. Irreversible photoreactions such as the Wolff rearrangement of diazonaphtoquinone (DNQ) or the photocleavage of *o*-nitrobenzyl, perylenylmethyl or coumarinyl esters can also induce substantial local polarity changes. Furthermore, chromophores as DNQ or coumarins can also be efficiently stimulated via two-photon processes using NIR light (650–900 nm), which is interesting for systems of biomedical relevance, as NIR light exhibits deeper tissue penetration (up to 10 cm in some cases) than UV light (c. 1 cm).

The light-induced degradation of a polymer system typically relies on photocleavable chromophores, which can specifically serve as a bond connecting two blocks of different nature (as in BCs) or randomly distributed along a polymer backbone. *O*-nitrobenzyl derivatives are usually selected for this purpose due to

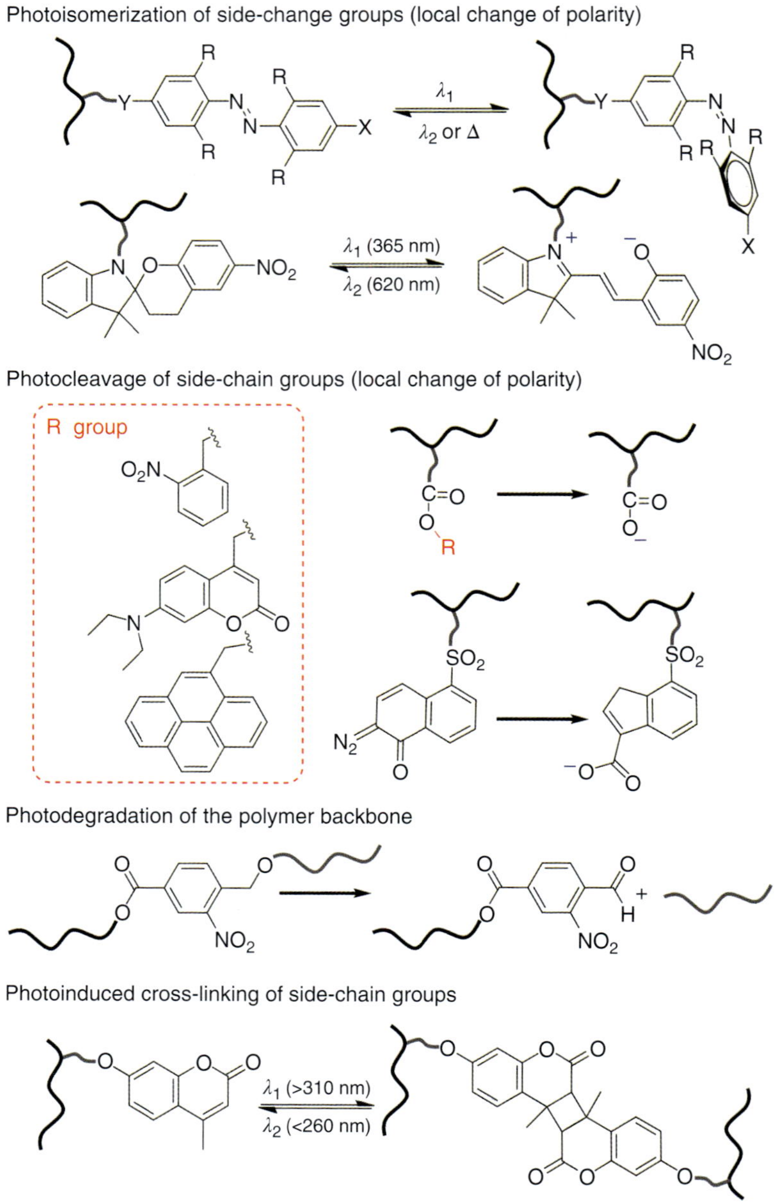

Figure 13.2 Photoinduced reactions of polymeric encapsulants.

their high synthetic versatility. Finally, some photocrosslinkable moieties as coumarins or cinnamates can be used to induce changes on the macromolecular self-assembly and stability that occur due to cross-linking-induced shrinkage. Coumarins are particularly useful for this purpose as they undergo reversible photocrosslinking using UV light. Other chromophores such as dithienylethenes have been evaluated in the search for photoactive macromolecular materials, including applications as nanocarriers [17]. When looking toward biomedical applications, other aspects must be considered in addition to the light-response of the chromophore such as the cytotoxicity of the polymer, which is particularly relevant in the case of photolabile and degradative moieties. Indeed, the generation of nitroxobenzaldehyde derivatives may induce cell death as it has been reported for *o*-nitrobenzyl-containing polymer micelles exhibiting phototriggered cytotoxicity [18].

13.3 Photoresponsive Polymer Self-Assemblies for Encapsulation and Release

13.3.1 Polymer Self-Assemblies from Linear Amphiphilic BCs

One of the most common materials for the production of polymeric encapsulants are amphiphilic BCs, which are defined as polymers consisting of at least two immiscible blocks (frequently, one hydrophobic and one hydrophilic). When a BC is dispersed in a liquid that is a "good" solvent for one of the blocks and, at the same time, a precipitant for the other, the BC associates in a reversible fashion forming self-assemblies (micelles and vesicles) of various shapes and dimensions, which to some extent are similar to those of low molecular weight surfactants. BC micelles consist of a collapsed core of the insoluble block and a corona of the soluble block. When water is used as a solvent, and depending on the hydrophobic/hydrophilic balance of the BC, spherical, ovoid, lamellar, worm-like and rod-like micelles have been found to form spontaneously. In such species, encapsulation of hydrophobic cargo at the micellar core can be readily achieved. Polymer vesicles are particularly relevant self-assemblies for encapsulation applications. They are composed of a bilayer of amphiphilic species enclosing an aqueous compartment. Hydrophilic actives can obviously be trapped inside the interior lumen, but there also exists the possibility of encapsulating hydrophobic cargo substances at the vesicle wall (a combination of both modes of encapsulation is also possible). In addition to the hydrophobic/hydrophilic ratio, there are other parameters that play a role in determining the self-assembly morphology and structure including the establishment of secondary interactions (hydrogen bonding, π–π stacking, etc.) or the crystallization of the core-forming block. Amphiphilic BCs exhibit substantially higher molecular weight as compared to lipids, thus leading to a less dynamic, more robust and mechanically tougher membrane than surfactant-forming micelles. In addition, BCs can be easily functionalized at either the polymer ends or side chains providing a useful approach to the modulation of specific properties such as membrane permeability or the incorporation of stimuli-responsive features, including photoresponsive moieties. Regarding the preparation of polymer self-assemblies, a variety of different methods have been explored. Two of these are preferentially used and are the film rehydration and the solvent switch methods. In the latter, the BC is initially dissolved in an organic solvent that is "good" solvent for both the hydrophilic and hydrophobic blocks, followed by the addition of an additional solvent (typically water), which selectively induces the collapse or nanoprecipitation of the solvophobic polymer block. Several parameters such as the type of organic solvent, solvent to water mixture ratio, and mixing rate may affect specific characteristics of the final system including the morphology and permeability of the hydrophobic domains (core of a spherical polymer micelle; and wall, in the case of a polymer vesicle). A final purification step is generally performed where the residual organic solvent is removed by dialysis or is allowed to evaporate. The use of organic solvents is undesirable for *in vivo* applications or for the encapsulation of specific bioactive compounds. The film rehydration is, to a certain extent, a method which minimizes the use of organic solvents. Here, a thin polymer film is rehydrated in aqueous media in bulk or by means of electrical methods (electroformation). Again, the method of choice will depend upon factors including size as well as the nature of the polymer and the active ingredient. Alternatively, micellization can also be achieved by less conventional methods including pH and temperature switches, and through the complexation of polyelectrolytes. The use of microfluidic, nanoimprinting and centrifugation technologies has also been described.

On amphiphilic BC self-assemblies, the response to light depends on the type of transformation induced by light and the location of the photosensitive unit into the BC. The chromophore is usually located at the blocks junction or as a pendant moiety of one of the blocks, usually the hydrophobic one, either as a repeating unit or as a side-chain. There is a wide variety amphiphilic BC designs and all combinations of hydrophilic and hydrophobic monomers are possible. The number of hydrophobic blocks studied to construct amphiphilic BCs is rather large; nevertheless, considering potential *in vivo* applications, degradability is a critical aspect [19]. Poly(ethylene glycol) (PEG) is most commonly used for the hydrophilic segment as, in addition to being biocompatible, it is resistant to immunological recognition, thereby improving circulation times. Alternative hydrophilic and responsive polymers, such as thermoresponsive poly(oligo(ethylene glycol) methyl ether methacrylate) or poly(*N*-isopropylacrylamide), may also be considered for the design of nanocarriers operating under multiple combined stimuli. Systems that are able to respond to more than one stimulus provide the opportunity to selectively regulate the release of various cargoes independently or induce a burst release by combining the effect of several of them, which greatly enhances the versatility of the systems in practical applications [20]. Indeed, temperature is a convenient stimulus that can trigger a reversible hydrophobic-to-hydrophilic transition and has been frequently combined with light stimulation.

One approach consists of the utilization of a photocleavable unit to covalently link the hydrophobic and hydrophilic segments of an amphiphilic BC (Figure 13.3). An irreversible photolysis at the linker breaks up the copolymer into their individual constituent blocks, which is accompanied by the destabilization of the self-assemblies and cargo-release. The vast majority of the reported examples have used *o*-nitrobenzyl- based linkers. Under UV or two-photon NIR absorption, the *o*-nitrobenzyl moiety cleaves at the benzylic site and disrupts the polymer micelles or vesicles. The performance of amphiphilic poly(acrylic acid)-*b*-poly(methyl caprolactone) diblock copolymer (**1** in Figure 13.3) vesicles as phototriggerable nanocarriers was tested using hydrophilic fluorescent probes and proteins [21]. Vesicles disintegrated into small micelles upon exposure to

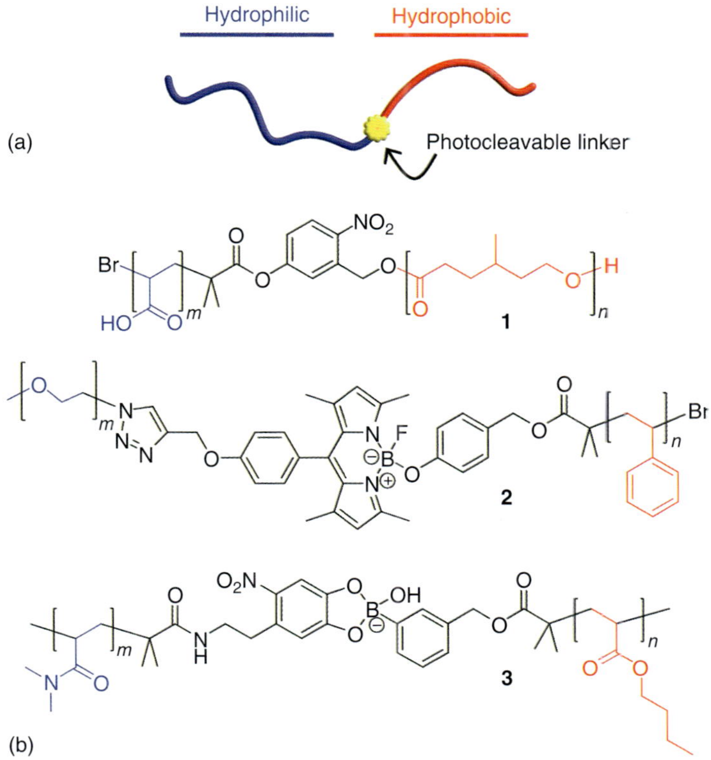

Figure 13.3 Schematic representation of an amphiphilic BC with a middle photocleavable linker (a) and some selected examples (b).

light ($\lambda = 365$ nm) concomitantly to cargo-release. The authors also demonstrated a light intensity dependent release, which enabled some level of control over the release process. In an attempt to achieve multi-stimuli-responsiveness, ABC-type triblock copolymer combining a *o*-nitrobenzyl unit with a disulfide bond, one at each block junction, were described to generate micelles in aqueous media where a combination of reducing agent and UV light induced significant disruption of the micelles [22]. With this system, the release profile of Nile Red was tuned depending on the sequences in which the stimuli were applied. An analogous approach was followed for the design of light-responsive vesicles based on a 2-nitrophenylalanine linker connecting a block of polycaprolactone with a block of PEG. After analyzing the encapsulation and release properties of the vesicles in aqueous media using biocytin, it was suggested that copolymer photodissociation induced a gradual collapse of the vesicle membrane which resulted in the precipitation of polycaprolactone aggregates [23].

Photoremoval of *o*-nitrobenzyl functionalities was used to activate self-immolative processes due to end-to-end backbone depolymerization [24, 25]. The principle was applied to ABA-type amphipilic triblock copolymers of PEG-*b*-poly(ethyl glyoxalate)-*b*-PEG with *o*-nitrobenzyl units at the junction blocks. Nile Red and doxorrubicin (DOX) were used to demonstrate the encapsulation and release properties associated with the light-triggered disintegration of the micellar self-assemblies [26, 27]. The same principle applies to the concept of self-immolative vesicles using amphiphilic diblock copolymers having a hydrophobic block whose cascade depolymerization is triggered in this case by a pyrenyl or an *o*-nitrophenyl photocleavable terminal group [28].

A few recent examples have demonstrated the utility of boron-containing photolabile linkers. Patil et al. have reported the synthesis of a PEG-*b*-polystyrene amphiphilic BC with a photolabile boron dipyrromethene linker (**2** in Figure 13.3), which exhibits a clear absorption band in the visible range. Disassembly of the micellar self-assemblies was monitored by encapsulation of Nile Red, and its triggered-release capabilities under light ($\lambda = 470$ nm) irradiation was demonstrated [29]. A boronate ester linker, from a boronic acid and nitrocatechol end-functional polymers (**3** in Figure 13.3), was used by Coumes et al. to impart multiresponsive (pH, UV light, and sensitivity toward carbohydrate molecules) properties to the BC self-assemblies. The potential of the self-assemblies as drug delivery nanocarriers was evaluated using Nile Red [30].

Insertion of photocleavable groups along the main chain of the hydrophobic block induces the breakdown of the hydrophobic backbone into smaller fragments allowing for the release of cargo, which has been proposed as a more efficient strategy to achieve a burst-release effect [31]. Han et al. have reported the synthesis and self-assembling properties of a series of ABA amphiphilic triblock copolymers, where the B block contained multiple photolabile *o*-nitrobenzyl repeating units along the polymer backbone (**4** in Figure 13.4). The triblock copolymers had the capability to undergo self-assembly in aqueous media yielding core–shell micelles, which could be degraded upon exposure to UV light. The encapsulating capabilities of the systems where initially evaluated with hydrophobic fluorescent probes [32, 33]. Han and coworkers also prepared DOX-loaded micelles that showed enhanced anticancer activity against tumor cells after UV illumination [34]. In addition, use of light was shown to be a more effective method to induce release than pH. Micelles disruption triggered by main-chain photodegradation of a polyurethane hydrophobic block has also been reported using DOX to demonstrate the encapsulation and efficient release of hydrophobic payloads [35]. Pasparakis et al. have reported the synthesis of

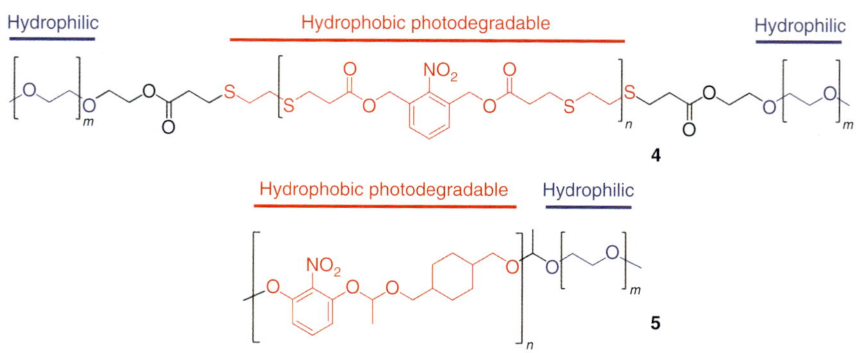

Figure 13.4 Selected examples of amphiphilic BCs with hydrophobic photocleavable repeating units.

polyacetals and polyketals, which can undergo main-chain photocleavage with very low energy inputs. These systems were tuned to make them sensitive to visible light through the derivatization with 2-nitroresorcinol moieties, which are in close proximity to the acetal moieties (**5** in Figure 13.4) [36]. BC micelles based in these polyacetals were capable of encapsulating and releasing camptothecin and hematoporphyrin via two-photon absorption photolysis (λ = 532 nm) in a process that was accelerated at low pH.

Another general approach consists of the attachment of light-sensitive chromophores to the backbone of a polymer chain or segment of a BC.

The polarity of these side-chain moieties can be switched from hydrophobic to hydrophilic by using light irradiation. Light-induced changes of the pendant chromophores can switch the polarity of the segment from hydrophobic to hydrophilic (or *vice versa*), inducing the destabilization of the self-assembly and the release of any encapsulated active. This strategy has been vastly explored because it is possible to easily combine several units at the hydrophobic block creating multiresponsive polymeric carriers where release can be regulated by applying either one or a combination of two or more stimuli. These polymers can be synthesized via direct polymerization of functional monomers or by post-polymerization functionalization of macromolecular precursors. The second approach requires a polymeric skeleton with chemical groups that are suitable for efficient functionalization with several chromophores. This offers versatility versus the polymerization of functional monomers. Several of these systems consist of a PEG-*b*-poly(meth)acrylate BC where the hydrophobic poly(meth) acrylate carries a number of light-responsive moieties (Figure 13.5). Upon exposure to light, the pendant photocleavable groups, including pyrenyl [37], *o*-nitrobenzyl [38], and coumarin [39], undergo a photoinduced reaction that leaves a highly polar carboxylic acid/carboxylate residue at the polymer chain. Thus, micellar self-assemblies with the core formed by the hydrophobic block are unable to retain self-assembly and the release of any physically entrapped

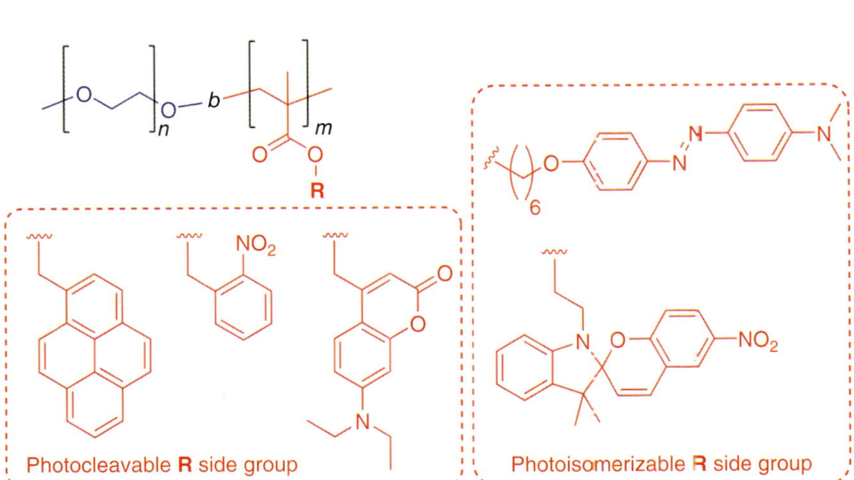

Figure 13.5 PEG-*b*-poly(meth)acrylate BCs with photoresponsive side-chain groups.

active can take place. Florescent Nile Red has been used to verify the controlled release of payloads.

Recently, the development of polymeric nanocarriers that are triggerable at the visible or the NIR wavelength range has been identified as a key step forward toward the realization of systems with relevant biomedical applications. In this sense, the photolysis of *o*-nitrobenzyl moieties is possible via one-photon UV absorption or two-photon NIR absorption; however, the rate of release is slowed down in the latter case due to inefficient two-photon absorption [15]. Micellar systems using coumarin with larger two-photon absorption cross-section can also be used as NIR light-triggered systems due to disruption of micellar self-assemblies after photolysis [16]. Additionally, coumarins can dimerize by reversible photo[2 + 2]cycloaddition activated by UV and visible light. The process affords stabilized micelles and controlled-release capabilities by photoregulated micelle core cross-linking–decross-linking [40].

Azobenzene is one of the most popular photoswitches for light-triggered applications. As the photoinduced process is reversible, micellar dissociation may be partially reverted. Additionally, azobenzenes with electron-donor or electron-acceptor groups enables the production of micelles that are sensitive to visible or NIR light [41]. Similarly to azobenzene, spiropyran can undergo reversible photoisomerization, which may lead to the disruption and regeneration of polymer micelles [42]. Encapsulation, UV light stimulated release, and visible light stimulated re-encapsulation by amphiphilic BC micelles have been demonstrated using coumarin 102.

In the pursuit of multi-stimuli responsive systems, Yang et al. have reported the synthesis of a poly(*N*-isopropylacrylamide)-*b*-poly(2-nitrobenzyl methacrylate) BCs where the hydrophilic poly(N-isopropylacrylamide) is responsive to temperature and the hydrophilic poly(2-nitrobenzyl methacrylate exhibits photolabile groups. Micelles based in this BC exhibited release capabilities, which could be tuned in a two-way manner by changing either the temperature or the irradiation conditions [43]. Light- and temperature-response groups can be both introduced in the same hydrophobic poly(meth)acrylate block by copolymerization of suitable monomers, such as *N*-isopropyl acrylamide [44, 45], or post-polymerization functionalization of a preformed thermoresponsive hydrophobic block. Indeed, Chen et al. have decorated a PEG-*b*-poly(oligo(ethylene glycol) methacrylate) with DNQ moieties that converts into 3-indenecarboxylic acid upon exposure to UV light [46]. The biocompatibility of these micelles was evaluated on HepG2 and HUVEC cell cultures and controlled-release properties were confirmed using the hydrophobic model coumarin 102. An analogous system was reported by Li et al. which uses poly(dimethylaminoethyl methacrylate) (PDMAEMA), a temperature- and pH-responsive polymer, as a hydrophilic block [47]. Multi-stimuli responsive micelles capable of encapsulation release coumarin 102 under both UV and NIR light stimulation were obtained through the partial functionalization of the PDMAEMA block with DNQ moieties. Cargo-release was enhanced under the combined stimulation of pH and light.

The vast majority of BCs micelles include one hydrophobic poly(meth)acrylate polymer chains in their structure, which is appropriate for proof-of-concept studies. However, the lack of biodegradability significantly hampers any potential *in vivo* applications. This limitation is stimulating research to develop photoresponsive BC nanocarriers based on biodegradable polymers such as polycarbonates [48–50] or polypeptides [51–54].

Unlike micelles, vesicles are capable of transporting hydrophilic and hydrophobic cargo, either packaged inside the lumen or embedded within the vesicle wall. In these systems, light-responsive hydrophobic blocks that enable the regulation of the permeability and cargo-release capabilities is an attractive strategy [55]. Wang et al. reported a couple of studies of amphiphilic diblock copolymer vesicles with a PEG shell and a spiropyran or an *o*-nitrobenzyl-functionalized polymethacrylate-based wall. Concurrent light-triggered reactions at the polymer-forming wall enabled the release of both hydrophilic and hydrophobic actives. The release capabilities of the systems were also evaluated *in vitro* and *in vivo* [56, 57].

13.3.2 Polymer Self-Assemblies from Linear-Dendritic and Branched BCs

Interest in linear-dendritic block copolymers (LDBCs) arises from the combination of the easy synthesis and the excellent processability of conventional linear polymers with the well-defined branched structure of dendrons, which is characterized by a precise and high number of peripheral ending groups [58]. Their unique macromolecular architecture and the derived hierarchical self-assembly properties are responsible for them gaining significant interest in bionanotechnology, e.g. drug or gene delivery [59–63]. Furthermore, the chemical modification of the peripheral groups allows to introduce the exact number of functional moieties, which facilitates providing functional properties to the resulting materials. Limitations in the development of these BCs arise from their challenging synthesis. However, the synthesis of functional LDBCs may be achieved via direct coupling of preformed blocks. This coupling strategy requires implementation in two complementary clickable groups, one at one end of the linear block (or both ends in dendritic-linear-dendritic BCs), and another at the focal point of the dendron [64].

Our group reported the synthesis of a series of amphiphilic LDBCs by coupling end-functionalized alkyne PEGs to 2,2-bis(hydroxymethyl)propionic acid (bis-MPA) dendrons of different generations. The dendrons were decorated with 4-cyanoazobenzene moieties at their periphery and an azide group at their focal point (see Figure 13.6). The amphiphilic LDBCs undergo self-assembly in aqueous solution and micellar structures of different morphologies were obtained. These include nanofibers, sheet-like micelles, and vesicles. It was shown that the self-assembly behavior of the BCs was dictated, in part, by the hydrophobic/hydrophilic balance (associated with the length of the PEG block and the generation number of the dendron) of the polymer. Light-induced breakup of the vesicles was only observed under somewhat high intensity light irradiation conditions [65]. The morphology and photoresponse of the self-assemblies were later theoretically studied by Lin et al. using dissipative particle dynamics [66]. The authors concluded that the photoisomerization of azobenzenes arranged in the membrane

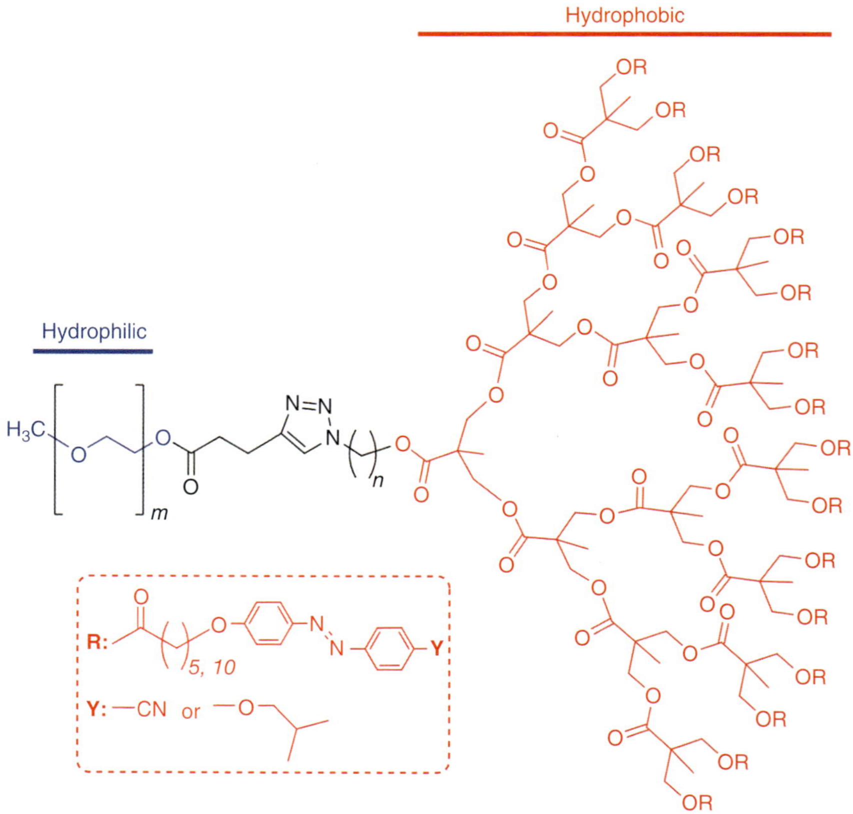

Figure 13.6 Selected example of an amphiphilic photoresponsive LDBCs.

of the LDBC vesicles should lead to distortion and eventually the rupture of the membrane consequently increasing the permeation of the nanocapsules.

The theoretical studies of Lin et al. (*vide supra*) prompted the design of a new series of LDBCs based on 4-alkoxyazobenzene peripheral moieties, which leads to a less intense packing of chromophores in the inner part of the membrane and a higher modification of polarity by photoisomerization with respect to 4-cyanoazobenzene analogues. The derived vesicles were then easily stimulated (λ = 365 nm, at 2.6 mW cm^{-2}) and their application as light-responsive nanocarriers was evaluated using both hydrophobic and hydrophilic fluorescent probes [67]. Furthermore, changes induced by light to the polymeric vesicles were modulated by tuning the functionalization of the periphery. For example, in one series of materials the decoration of the periphery of the dendron was achieved by using a combination of azobenzene moieties and linear aliphatic chains, toward tuning the membrane stability and the payload-release capabilities of the self-assemblies. In particular, vesicles with a content of around 50% of azobenzene peripheral units exhibited significant morphological changes when they were exposed to light and an almost immediate release of their cargo [68].

García-Juan et al. described the preparation of dual responsive LDBCs combining thermo- and light-response. A thermoresponsive copolymethacrylate of diethylene and oligoethyleneglycol was coupled to a bis-MPA dendron containing 50% of 4-alkoxyazobenzene moieties and aliphatic tails. The LDBC self-assembled into vesicles were able to encapsulate and release either hydrophobic or hydrophilic payloads under thermal or light stimulation [69]. Recently, Kalva and Ambade described a series of LDBCs equipped with an azobenzene moiety linking the two blocks of the BC. No light-triggered cargo-release properties were observed, as the changes before and after irradiation were too subtle to induce the disassembly of the micelle [70]. Nevertheless, the replacement of the azobenzene moiety with an *o*-nitrobenzyl linker enabled the preparation of nanocarriers that were capable of encapsulating and releasing cargo in response to pH changes [71]. Although most of the reported examples of photoresponsive LDBCs contain azobenzenes moieties, Goodwin et al. described the preparation of micellar systems based on a PEG-dendron BC (as in Figure 13.6) having peripheral DNQ moieties. The photoresponse of the DNQ moieties (*vide supra*) was associated with the release of hydrophobic payloads including fluorescent probes. Furthermore, these systems exhibited low cytotoxicity when compared with linear analogues [72, 73]. Sun et al. described a PEG-poly(amidoamine) LDBC having DNQ peripheral moieties. The authors described the cell internalization process of the LDBC micelles carrying encapsulated DOX. Triggered DOX release and cytotoxicity under NIR light irradiation using HeLa cells was demonstrated [74]. The regular branched structure of dendrons was also exploited by Sun et al. to prepare two-photon-sensitive (808 nm) and sugar-targeting nanocarriers from a poly(amidoamine) dendron having DNQ-ended linear oligo(ε-caprolactone) peripheral groups and a hydrophilic sugar moiety at the focal point. Light-triggered release of encapsulated DOX and specific binding with lectins were demonstrated [75]. They also extended this approach to the preparation of photoresponsive nanocarriers based on Janus codendrimers decorated with DNQ (hydrophobic dendron block) and lactose moieties (hydrophilic dendron block) [76]. Sugar-coated micelles prepared from these codendrimers showed binding with RCA_{120} lectin and the DOX encapsulated micelles exhibited a NIR-triggered drug release and cytotoxicity similar to free DOX. Namezi and Gillies also demonstrated the UV-triggered release of hydrophobic and hydrophilic payloads on Janus amphiphiles having a branched block with PEG linear chains combined with a photodegradable bis(MPA) dendron containing *o*-nitrobenzyl esters [77].

Polydispersed hyperbranched polymers share several of the advantageous properties of monodispersed dendrimers but their preparation is significantly less demanding, which has certainly incentivized research in this type of materials [78]. Hyperbranched polymers can function as unimolecular micelles (*vide infra*) but also via multimolecular self-assemblies as amphiphilic BCs [79, 80]. Photocontrolled response can easily be achieved through the appropriate modification of end groups. Accordingly, amphiphilic systems were prepared by grafting either DNQ or spiropyran light-sensitive motifs to the peripheral groups of a polyphosphate-based hyperbranched polymer. Encapsulation and light-triggered release were verified using hydrophobic Nile Red and coumarin 102 models [81, 82]. Alternative, hyperbranched polyglycerols with a spiropyran moiety at the focal point also form biocompatible micelles, which under UV illumination were able to release pyrene [83]. Mo et al. prepared hybrid structures consisting on hydrophilic hyperbranched poly(oligo(ethylene glycol) methyl ether acrylate) with polycaprolactone linear arms grafted at photocleavage *o*-nitrobenzyl branching points. Light-induced degradation of the polymer hybrid was accompanied by the release of Nile Red from aqueous self-assemblies [84].

Recent advances in polymer chemistry based on combining controlled polymerization methods with click chemistries have opened endless possibilities to design sophisticated polymers with precisely controlled architectures and tailored functionalities. Thus, BCs have been complemented by new amphiphilic polymer architectures, including star branched polymers. These contain three or more polymer chains emanating from a central core and their globular shape, with multiple chain ends, is attractive for drug delivery applications although applicability has been scarcely evaluated *in vivo* [85–88]. Depending on the length of the arms, star polymers may be seen as unimolecular containers or, more frequently, via self-assembly, as amphiphilic BCs with low micellar concentrations and small morphological changes. Amongst them, miktoarm polymers are asymmetric star polymers including two or more types of arms species with different chemical composition and/or molecular weights. This heteroarm segmented architecture constitutes an excellent platform to combine hydrophobic, hydrophilic and light-responsive linear blocks for the development of novel polymeric nanocarriers. Several examples of photoresponsive miktoarm azobenzene based polymers have been reported to date, and encapsulation and release of payloads from these systems have been demonstrated. Amphiphilic photoresponsive miktoarm polymers of A_3B type were described by Blasco et al. consisting of three hydrophilic PEG arms and one azobenzene polymethacrylate arm (Figure 13.7). Polymer with adjusted arm lengths formed vesicles with encapsulating and UV-triggered releasing properties for both hydrophobic and hydrophilic fluorescent probes [89]. Replacing PEG arms by temperature-sensitive poly(*N,N*-diethylacrylamide) ones, dual thermo- and photoresponsive micelles were produced where only photoisomerization of azobenzene stimulated the release of encapsulated Nile Red [90]. ABC miktoarm micelles with dual thermo- and photoresponse due to the

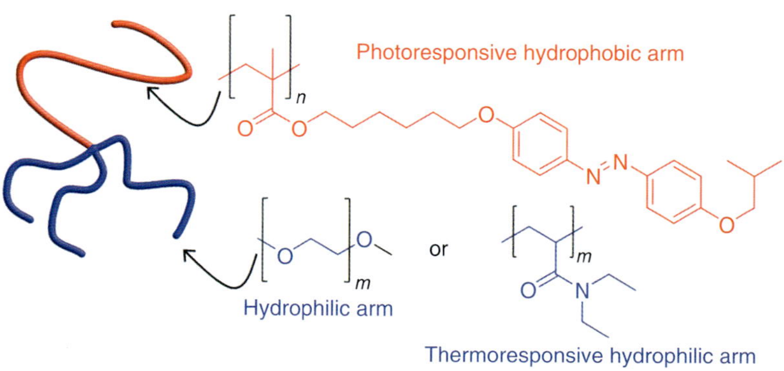

Figure 13.7 Photoresponsive amphiphilic star-polymers developed by Blasco et al. [89, 90].

combination of PEG, poly(*N*-isopropylacrylamide) and poly(*o*-nitrobenzylacrylate) where used to encapsulate Nile Red. Cargo molecules were released rapidly upon UV light irradiation [91].

13.3.3 Polymer Self-Assemblies from Supramolecular BCs

During last decades, the principles of supramolecular chemistry have been progressively combined with polymer chemistry broadening the scope for new chemically diverse polymers with controlled architectures. Following a "supramolecular synthetic approach," which in many instances exhibits key features to achieve versatility, can expedite the discovery of novel macromolecular architectures. Indeed, amphiphilic supramolecular di- or tri-block copolymers can be readily assembled by connecting individual blocks through specific noncovalent interactions (end-to-end connection, for example), combining the reversibility and tunability of supramolecular materials with the intrinsic properties of polymer chains [92]. Indeed, the incorporation of the adequate functionality either at the focal point or at some middle point of one of the individual blocks will result in systems with light responsive properties [93]. For these purposes, a wide variety of supramolecular interactions have been developed, which include host–guest interactions, hydrogen bonding, π–π stacking, and electrostatic pairing. The choice of supramolecular binding motif is conditioned by the stability of the noncovalent interaction in aqueous media, which is key requisite if applications in aqueous media are pursued [94]. In spite of this, only a limited number of studies have reported supramolecular micelles with encapsulating and light-triggered cargo-release properties.

Regarding host–guest systems, cyclodextrins (CDs) have been widely used as host molecules due to their wide availability and low toxicity. They are macrocyclic oligosaccharides with a toroidal molecular shape consisting of a hydrophilic exterior and a hydrophobic internal cavity that can form inclusion complexes with different guest molecules in aqueous media [95]. The molecular recognition between CDs and azobenzenes has been employed for the construction of photoresponsive amphiphilic BC self-assemblies. A CD-azobenzene host–guest complex can serve as a linker to connect two segments, one hydrophobic and one hydrophilic, of a BC [96]. In the vast majority of CD systems, the *trans*-azobenzene is capable of forming stable host–guest complexes, while the *cis*-azobenzene is too bulky and does not fit inside the hydrophobic cavity of CDs. Thus, the reversible coupling of individual segments of a BC can be easily regulated by alternating UV and visible light irradiation. The preparation of amphiphilic linear–linear BCs has been realized by combining azobenzene and α-CD terminated polymer chains (hydrophilic poly(acrylic acid) and hydrophobic poly(caprolactone)) as described by Yan et al. [97]. In water, this supramolecular copolymer forms tubular structures that exhibit reversible assembly properties on account of the light-controlled host-guest "bond," which connects the two blocks. The photoregulated assembly/disassembly properties imparted controlled delivery capabilities to the micellar system. Other alternative amphiphilic macromolecular species based on the host–guest recognition properties of azobenzenes and CDs include amphiphilic LDBCs [98, 99] linear-star [100], multiarm dendrimers [101], Janus hyperbranched [102], and hyperbranched–linear-hyperbranched polymers [103]. Although the self-assembling properties of these systems and their morphological transformations in response to light have been investigated in depth, clear evidence of light-controlled cargo-release properties are limited. Yao et al. have reported the preparation of a series of miktoarm AB_2 supramolecular polymers consisting of PEG and poly(*N*-isopropylacrylamide) branches; with either one or two azobenzene-CD host–guest complexes serving as linking motifs [104]. The miktoarm polymer undergoes micellization in aqueous media and the resulting micelles exhibit dual thermo- and light-responsive properties and enable the controlled delivery of DOX.

CDs have also been used to "mask" specific hydrophobic residues of polymer chains, which can alter the hydrophobic/hydrophilic balance of a system in a reversible fashion. Jin et al. have described the preparation of an azobenzene BC consisting of a PEG hydrophilic block and a hydrophobic poly(meth)acrylate carrying side-chain azobenzene moieties (**6** in Figure 13.8) [105]. Such a polymer undergoes micellization in aqueous media on account of the amphiphilic nature of the polymer, which results in the formation of stable vesicles. In the presence of CD the vesicles are disassembled as the macrocycle binds to the *trans*-azobenzene moieties rendering the BC more hydrophilic. Upon exposure to UV light, the host–guest complexes fall apart and "unmask" the relatively hydrophobic azobenzene residues of the BC (Figure 13.8). This process partially restores the hydrophobic/hydrophilic balance of the system and the vesicles regenerate. Wang et al. followed an analogous strategy by using a double hydrophobic copolymer [106]. In this case, the micelles were formed in the presence of β-CD and dissociated upon exposure to UV light. The authors demonstrated reversible self-assembly by alternating UV and visible light although no encapsulation capabilities were reported.

Supramolecular BCs based on specific hydrogen-bonding pairs linking two or more blocks have been investigated in the past by using, typically, apolar organic

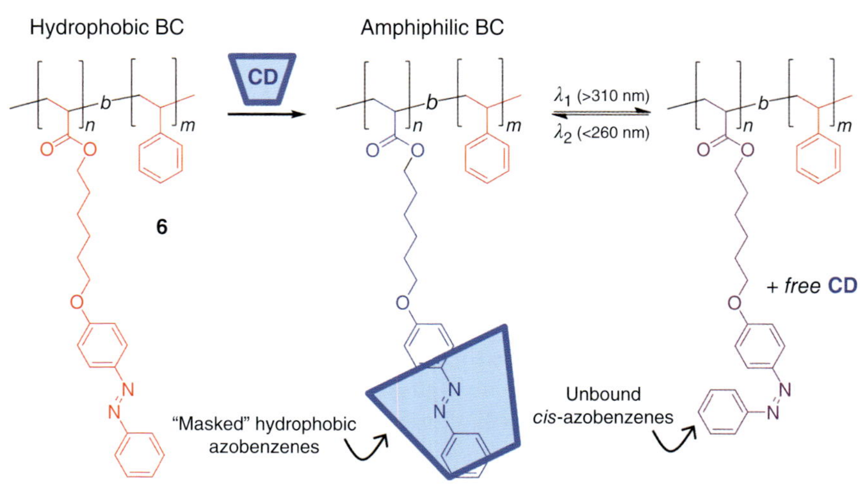

Figure 13.8 Selected example of a photoresponsive amphiphilic star-polymer.

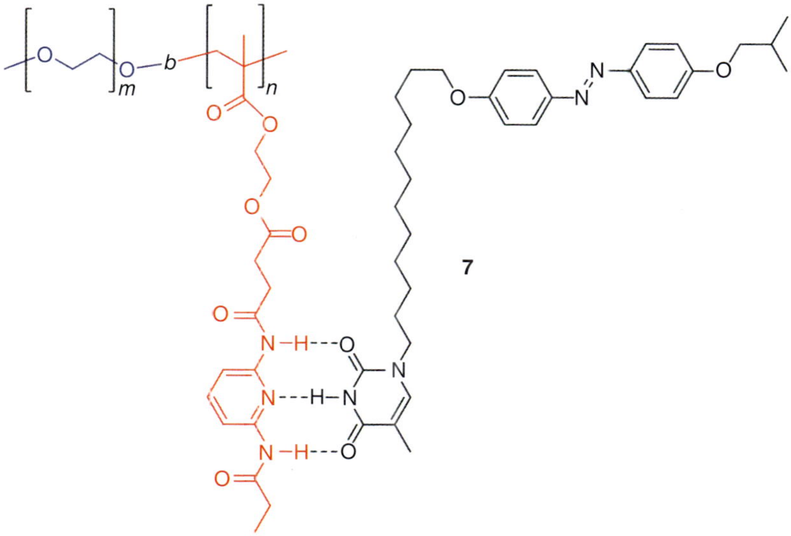

Figure 13.9 Photoresponsive supramolecular BC developed by Concellón et al.. Source: Concellón et al. 2016 [110]. Reproduced with permission of American Chemical Society.

solvents [53]. Some of these systems can also undergo aqueous self-assembly forming micelles and vesicles, which are pH-responsive; although no drug-delivery light-responsive systems have been reported to date. Light-sensitive moieties may be readily attached to a polymer chain via noncovalent side-chain functionalization reactions [107]. In one case, light-sensitive nanospheres were produced by combining a poly(4-vinylpyridine) polymer chain with a low molecular weight carboxylic acid terminated azobenzene via hydrogen-bonding interactions [108, 109]. Multiple-point hydrogen bonding motifs can establish much stronger interactions than single-point hydrogen bonding motifs such as that of pyridine-carboxylic acid, which are significantly weakened in aqueous media. In this regard, Concellón et al. used a 2,6-diacylaminopyridine, a nucleobase analogue, to impart light responsiveness to an amphiphilic BC (7 in Figure 13.9). The BC was composed of a PEG block and a polymethacrylate block with pendant 2,6-diacylaminopyridine moieties, which were functionalized with thymine-terminated azobenzene moieties, via three-point hydrogen bonding interactions. Depending on the hydrophobic/hydrophilic balance of the BC, micelles and vesicles were obtained in aqueous media. Encapsulation and light-triggered cargo-release of model compounds were also demonstrated [110].

13.3.4 Photoresponsive Polymer Capsules

In the previous sections, we have covered ways of producing polymeric encapsulants (micelles and vesicles) from amphiphilic BCs. Alternatively, non-amphiphilic structures can also be assembled into colloidal capsules (a compartmentalized pod with a polymer membrane), through appropriate manufacturing methods. In contrast to polymer vesicles (*vide supra*), which generally exhibit a bilayer membrane formed by amphiphilic macromolecular species, capsules of a much wider range of materials can be fabricated.

The production of large millimeter-size capsules is most frequently achieved through injection molding, which is a well-established low-cost technology [111]. Smaller microcapsules, with diameters ranging from tens of nanometers up to a few hundred microns, are prepared by several methods including pan spray drying, pan coating, centrifugal extrusion, emulsion-based methods, distillation–precipitation polymerization, template-based polymerization, and layer-by-layer approaches [112]. In general, each method has advantages for specific applications, which depend on a variety of characteristics of a capsule including size, shape, shell wall thickness, permeability, and composition and mechanical stability of the shell wall. A number of review articles and books have been focused on the preparation of polymeric capsules, including light-responsive systems [113–116]. Here, we present a few selected examples of photoresponsive capsules, for their preparation and mechanisms of actuation.

Light-responsive capsule formation has been achieved by the electrostatic recognition of two oppositely charged polymers at the liquid–liquid interface in an emulsified system, or through the use of layer-by-layer methodologies. These capsules typically present a thin wall and are thus rather fragile. The stability of the shell wall can be enhanced if the membrane is a covalently cross-linked network. Such covalent capsules can be produced, for example, by means of *in situ* polymerization. Here, the reactive monomers are dissolved in an aqueous phase that is mixed with an immiscible solvent and subsequently emulsified. After emulsification, the monomers are polymerized to create a covalently cross-linked shell. Cao et al. reported the synthesis of a photodegradable polymer capsule system through the one-pot free radical polymerization of *N*-isopropylacrylamide, methacrylic acid, and the photolabile 5-(methacryloyloxy)-2-nitrobenzyl methacrylate on dimethyldiethoxysilane oil-in-water emulsion templates, followed by the removal of the sacrificial core [117]. Upon exposure to UV light, the capsules degraded due to the selective photocleavage of the *o*-nitrobenzyl ester linkers in the shell wall. Photocontrolled release of coumarin 102, as a model hydrophobic cargo, was also demonstrated. Alternatively to the free radical polymerization of small monomers and cross-linkers, photoreactive groups may be directly embedded in the shell-forming material via covalent functionalization of a preformed polymer [118]. This is a relatively straightforward approach that enables the preparation of robust systems with a higher level of control over material composition and, more importantly, well-defined photoresponsive properties. Indeed, such a method has enabled the production of capsules composed of polyorganosiloxane NPs carrying nitrocinnamate residues [119]. The capsules were assembled through a water/oil/water emulsification process followed by the photocrosslinking of the nitrocinnamate groups. It was shown that the capsule content (CD, as a model guest molecule) can be released upon light irradiation. Layer-by-layer approach is another method for capsule preparation, which typically involves the use of metal oxide or polymer particles as template, and polyelectrolytes as shell-forming materials [120–123]. Park et al. prepared a series of microcapsules by coating polystyrene particles with alternating layers of negatively charged poly(sodium 4-styrenesulfonate) and positively charged poly(allylamie hydrochloride) carrying benzophenone moieties [124, 125]. After the coating process, the polymer layers are held together by electrostatic interactions and the polystyrene core particle was removed by means of selective solvent extraction, leaving behind a hollow compartmentalized capsule. With the capsules in hand, the kinetics of release of encapsulated rhodamine B were measured before and after UV light irradiation, and the photocontrol over release rates (i.e. permeability of rhodamine B through the shell) was demonstrated. It is worth noting that the photocrosslinking of benzophenone is not used as a release trigger but rather as a way to tune the rate of release. Azobenzene switches have also been exploited for light-controlled permeability purposes. Tao et al. showed that azobenzene moieties, when incorporated into the shell wall of polyelectrolyte layer-by-layer capsules, altered the permeability values of the encapsulant exposed to light [126, 127]. In a recent study, Yi and Sukhorukov developed a layer-by-layer capsule system by sequential deposition of two polyelectrolytes, i.e. negatively charged azopolymer and positively charged poly(diallyldimethylammonium chloride), on silicon dioxide microparticles [128]. Upon light exposure, the multilayer shell of these capsules underwent a structural reorganization driven by the preferential formation of azobenzene J aggregates. Such process, which is accompanied by the disruption of the capsule wall, was shown to control the release of encapsulated bovine serum albumin. Bédard et al. reported the incorporation of *meso*-tetrakis(4-sulfonatophenyl) porphyrin into the shell wall of layer-by-layer capsules [129]. The capsules were disrupted by means of laser light irradiation (λ = 532 nm, at 70 mW cm^{-2}) in the presence of an oxidizing agent, such as H_2O_2. Under these conditions, it has been suggested that the porphyrin residues have the capacity to generate reactive oxygen species (ROS) in a process that disturbs the electrostatic interactions between the polymers holding the shell together and increases the permeability of the capsule [130]. Photoreactive triphenylmethane leucohydroxide (TPML) is another photoreactive group that has been incorporated into the shell wall material of polyelectrolyte complex capsules [131]. In this case, the release rate of *p*-toluenesulfonate from layer-by-layer capsules increased under UV light irradiation, a phenomenon which was associated with the photoinduced dissociation of the TPML residues. There exist other approaches to exert controlled release including light-induced changes of ionic strength [132], and the use of specific host–guest interactions [133], diazonium compounds [134], *o*-nitrobenzyl derivatives [135], and coumarin-functionalized polymers [136].

In recent years, the area of light-regulated substance-release has witnessed several transformational advances including the use of flow focusing devices for the preparation of hierarchically structured capsules (*vide infra*) and the development of novel photocontrolled cascade chemical reactions. Gu et al. have developed a system in which a protein, caspase-3 in this case, was encapsulated in an enzymatically-degradable polymeric capsule incorporating peptidic crosslinkers [137]. The light-controlled substance-release capabilities were associated with specific residues of the cross-linkers. These residues were caged by photolabile groups, a feature that inhibited the polymer shell from degradation by the

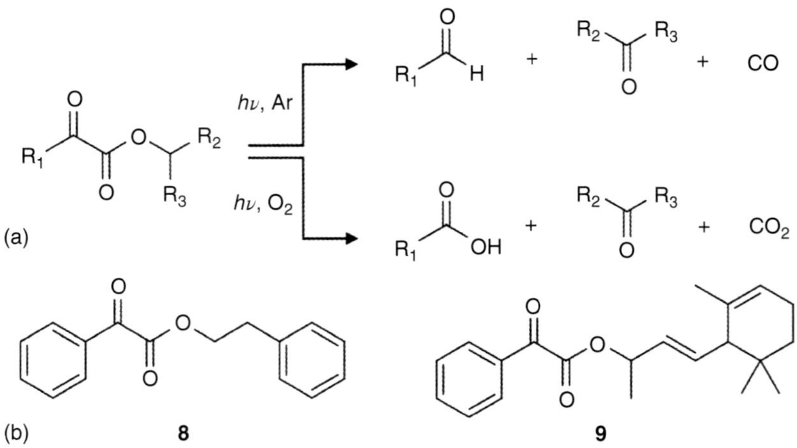

Figure 13.10 Norrish type II photofragmentation of 2-oxoacetates (a), and 2-oxoacetates derivatives (b) developed by Paret et al. Source: Paret et al. 2015 [138]. Reproduced with permission of John Wiley and Sons.

caspase-3 from within. The utility and robustness of the system was demonstrated through the UV light controlled cytoplasmic delivery of caspase-3 to HeLa cells. Another sophisticated approach has been demonstrated by the Herrmann laboratory where the encapsulation of photolabile 2-oxoacetates derivatives in core–shell capsules allowed the light-induced release of fragrances (Figure 13.10) [138]. On exposure to UV light (λ = 340–380 nm, at 3.1 mW cm^{-2}) the encapsulated 2-oxoacetates derivatives degraded to form the fragrances 2-phenylacetaldehyde and (*E*)-4-(2,6,6-trimethylcyclohex-2-en-1-yl)but-3-en-2-one; and the gases CO and CO_2 at rates that were sufficiently high to induce the breakup of the capsule.

Finally, the use of flow focusing devices for the preparation of core–shell microcapsules have allowed for a variety of liquids and actives to be encapsulated, which were incompatible with other emulsification techniques [139]. Also, capsules with multiple release triggers, including light, can be produced in a straightforward manner [140–142]. This approach was followed by the Abell and Scherman laboratories where hydrogel microcapsules were produced by combining, in the presence of the host molecule cucurbit[8]uril, a guest polymer anthracene-functionalized hydroxyethyl cellulose, and a fluorinated surfactant [143]. The electrostatic attraction between the charged polymer and the surfactant enabled the formation of a supramolecular hydrogel capsule shell in a single emulsion. Additionally, stimuli-responsive controlled molecular permeability was possible on account of the dynamic nature of the interactions holding the hydrogel shell together, which could be selectively transformed into irreversible covalent interactions (photochemically dimerized anthracene residues) by UV light irradiation. In summary, significant progress has been made toward understanding the mechanisms underlying the formation of polymer capsules, and this insight is being translated to the modulation of controlled-release, which shows great promise for the future applicability of this field.

13.3.5 Photoresponsive Microgels and Nanogels

Microgels and nanogels are swollen cross-linked polymeric particles that show a unique series of features. They exhibit a combination of the characteristics of colloidal particles (colloidal stability, high surface area, and control over particle size) with those associated with macroscopic hydrogels (three-dimensional branched configuration, structural integrity, and fluid-like transport properties) [144]. Although no clear connection between nomenclature (micro versus nano) and particle size has been established yet (the terms microgels and nanogels are often used interchangeably), particles with a size in the 20–200 nm are typically referred as nanogels. Regarding the preparation of microgels and nanogels, it is possible to distinguish between two main categories depending on whether the gel beads are produced in a single-phase or multiphase heterogeneous systems (e.g. precipitation polymerization or microemulsion). The production method of choice is dictated by the potential application of the final system, which in many occasions is related to the controlled loading and release of compounds. Light-controlled release has been proposed in the context of microgels and nanogels [145], and some representative examples are highlighted here. The photoinduced response of microgels and nanogels is predominantly associated with irreversible transformations. Only a few examples of reversible reactions have been reported to date [146, 147]. Irreversible photoinduced degradation of nanogels has been achieved by exploiting the chemistry of *o*-nitrobenzyl [148–150], and 7-amino coumarin derivatives [151]. Nevertheless, only a few selected studies have demonstrated systems with substance-release capabilities. Klinger and Landfester have reported the preparation of poly(2-hydroxyethyl methacrylate-*co*-methacrylic acid) microgels with *o*-nitrobenzyl-type photolabile linkers moieties [152]. The gel beads were successfully loaded with myoglobin, in a process assisted by the establishment of electrostatic interactions between the positively charged protein and the negatively charged polymer. In addition to the pH-controlled properties of the system (associated with the carboxylic acid residues of the polymer network), light-induced (UV light in the range of $\lambda = 315–390\,\text{nm}$, at $17–60\,\text{mW}\,\text{cm}^{-2}$) network degradation accompanied by substance-release was demonstrated by inducing the photolytic cleavage of the *o*-nitrobenzyl cross-linkers. A similar approach was followed by Azagarsamy et al. for the encapsulation of photocontrolled ($\lambda = 365\,\text{nm}$ light, at $10\,\text{mW}\,\text{cm}^{-2}$) release of alkaline phosphatase (as a model protein) from nanogels labeled with cell-penetrating peptides [153]. These studies clearly demonstrated that biocompatible nanogels are capable of encapsulating proteins, maintaining their activity and releasing proteins in response to light. The chemistry of *o*-nitrobenzyl linkages has also been combined with orthogonal triggers, and nanogels which respond to temperature, pH, and UV light have recently been developed [154].

Zan et al. have reported the introduction of NIR light dyes into polymeric nanogels, which exhibit controlled-release based on the photothermal effect [155]. In this case, the nanogels were composed of two complementary polymers: a side-chain hydrophilic random terpolymer carrying adamantine residues and a β-CD-functionalized poly(amidoamine) dendrimer. The assembly of the polymers into nanogels relied on the formation of host–guest interactions between the adamantine and the CD residues. The encapsulation of substances in these systems, which is concomitant to the nanogel formation, can happen directly in aqueous media. The simultaneous loading of indocyanine green (an NIR dye in tumor-specific imaging [156], and photothermal applications) and the chemotherapy drug DOX were the basis to demonstrate the triggered DOX-release capabilities (via photothermal effect and host–guest complex dissociation) under NIR light irradiation (λ = 805 nm, at 1.5 W cm^{-2}). Khatun et al. have developed an alternative system based also on the photothermal effect [157]. In this study, DOX-functionalized graphene was embedded in a polymeric nanogel composed of disulfide-cross-linked hyaluronic acid. It was shown that the photothermal effect associated with the graphene enhanced the release of DOX from the nanogel at pH = 5.0–7.4 upon NIR light (λ = 670 nm, at 1.0 W cm^{-2}) irradiation. Additionally, combined thermo-chemotherapy was investigated both *in vitro* and *in vivo*, and the controlled generation of heat leading to photoablation of cancer cells was demonstrated.

13.3.6 Other Miscellaneous Photoresponsive Polymeric Encapsulants

In addition to the classical families of *soft* polymer encapsulants (capsules, BC micelles, nanogels, and microgels) reviewed in Sections 13.3.1–13.3.5, there exists a large miscellaneous collection of compartmentalized polymeric species that have been explored in the context of light-controlled substance-release [13, 158, 159]. These include core-cross-linked NPs, knedel-like shell-crosslinked NPs, uncrosslinked homopolymer-based NPs, multiparticle crosslinked species and unimolecular systems, to name just a few. Given the wide variety of NP structures and compositions, rather than being exhaustive, we present some selected examples of a few recent and well-referenced studies and associate them with mechanisms of actuation.

Light-sensitive degradable polymers that can be triggered to degrade through multiple light-sensitive groups along the polymer backbone have been broadly explored in NP formulation (see Figure 13.11). Pioneering work by the Almutairi laboratory was based on self-immolative quinone-methidine system **10** (Figure 13.11) [24]. The release of Nile Red, as a model cargo, from particles based on polymer **10** was observed upon light exposure (λ = 750 nm with ~100 fs pulse widths and 80 MHz repetition rate, at c. 2.6 W). Control experiments showed that no release occurs when the particles are protected from light exposure or when the dye is encapsulated in poly(lactic-*co*-glycolic acid). Of those systems developed at the Almutairi laboratory [24, 25, 159–161], it is also worth highlighting **12** (Figure 13.11), which is a random terpolymer with sensitivity toward both UV–vis (λ = 320–480 nm, at c. 100 mW cm^{-2}) and NIR light (λ = 800 nm, at c. 82.7 W cm^{-2}), via two-photon absorption [161]. NPs of **12** encapsulating Nile Red were successfully produced by an emulsification process, and the fluorescence of the encapsulated Nile Red was monitored upon exposure to light. Fast and complete fluorescence quenching was observed when UV–vis light was employed, a phenomenon which was associated with

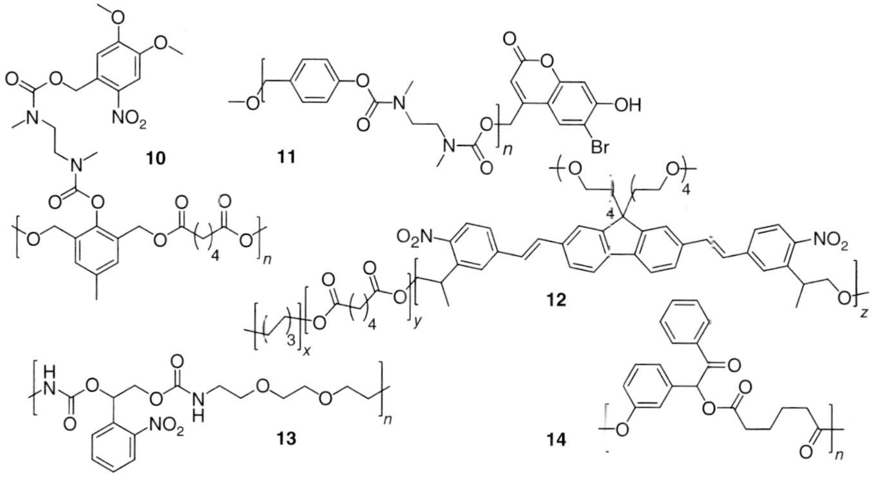

Figure 13.11 Selected photodegradable polymers for the preparation of polymeric NPs.

the efficient degradation of the particles and release of the payload. The same effect was achieved with NIR light, at the expense of prolonging the irradiation time. Lv et al. have evaluated the cargo-release capabilities of polymeric NPs based on photodegradable polyurethane **13** (Figure 13.11) [162]. Additionally, light-triggered ($\lambda = 365$ nm, at 11 mW cm^{-2}) burst release of fluorescein diacetate and Tagalsin G in cells was also demonstrated. While the majority of light-triggerable polymeric NPs systems rely on the chemistry of *o*-nitrobenzyl esters and their analogues, an alternative system based on benzoin derivatives has recently been reported by Englert et al. [163] (**14** in Figure 13.11). Traditionally used as protecting groups [164], benzoins undergo a photochemical rearrangement to yield 2-phenylbenzofurans and, in polyester **14**, serve as photolabile repeating units. Englert et al. formulated **14** as NPs and, as a proof-of-concept, the encapsulation and photoinduced release of Nile Red from the particles was demonstrated. It was shown that exposure times as short as 10 seconds were sufficient to induce significant release. An additional system has been designed at the Liu laboratory, where photoactive polymer NPs were designed as gene delivery vectors [165]. In this case, nucleic acids (and, potentially, therapeutic DNA) can interact electrostatically the photoresponsive polymer in a process that results in the formation of stable NPs in aqueous media. As the polymer carries both photosensitizer moieties, which are capable of generating ROS upon light irradiation, and ROS-sensitive linkers, the NPs degrade when exposed to light (white light, at 50 mW cm^{-2}) and release the DNA cargo. Lin et al. have recently reported the development of hypoxia-activated photoresponsive drug-release system based on a nitroimidazole-locked coumarin phototrigger [166]. In essence, the drug, etoposide, in this case, was locked by the phototrigger (coumarin), which was again locked by the hypoxia regulator (nitroimidazole residue); and where the whole system was conceived as a cleavable linker between the drug and the polymer NP. In aerobic environments the nitroimidazole relaxes the photoexcited coumarin and effectively avoids any photoinduced bond cleavage. However, under hypoxic conditions (such as those found in solid tumors), the nitroimidazole residue is reduced to aminoimidazole, which allows the bond cleavage to take place. It is worth highlighting that the drug is not only physically entrapped in the NP (as in the vast majority of previous examples), but also covalently bound to the polymer carrier and the release of the drug is achieved by selectively cleaving the linker between the active and the encapsulant via light irradiation.

Photoresponsive cinnamic acid residues have been proposed as stimuli-responsive cross-linkers to control the stability and morphology of polymeric NPs. Shi et al. reported that the photoinduced [2 + 2] cycloaddition reaction of *trans*-cinnamic acids, which is associated to reversible changes of NP size, could control cargo-release too [167]. Tan et al. have recently reported the preparation of porous crosslinked polymeric NPs through a continuous assembly of polymers strategy [168]. In this study, polymeric macromonomers carrying azobenzene, α-CD and norbornene groups were embedded into mesoporous silica templates; and subsequently crosslinked by ring-opening methathesis polymerization (irreversible bonds) and CD-azobenzene host–guest inclusion complex formation (reversible bonds). The etching of the silica templates resulted in porous polymeric NP incorporating reversible crosslinkers, which displayed light-induced burst-release capabilities. UV light (λ = 365 nm) was shown to induce burst release of encapsulated dextran, with c. 50% releases with one second of irradiation. An alternative way of inducing cargo-release from polymeric NPs is related to a specific photocontrolled switch of the hydrophobic/hydrophilic balance of NP. Cui et al. have prepared polymeric chitosan particles by mixing chitosan hydrochloride and 4-oxo-4-(pyren-4-ylmethoxy) butanoic acid, which are capable of interacting through ionic pairing [169]. The attachment of the hydrophobic pyrene residues to chitosan yields a complex with a relatively high hydrophobic/hydrophilic balance, which self-collapse in the form of NPs. Such species are sensitive to light irradiation on account of the photolabile pyrene residues. Using Nile Red as a model cargo substance, the controlled-release capabilities of the NPs were investigated by using UV (λ = 365 nm, in power range of 50–1200 mW cm^{-2}) and NIR (λ = 808 nm, at 140 mW cm^{-2}). A similar strategy has been followed for the development of swellable NP exhibiting light-triggered release capabilities [170, 171].

In the past decade, a significant focus has been placed on unimolecular micelles, which comprise a distinct class of encapsulants. Contrary to cross-linked core–shell particles, unimolecular micelles exhibit a distinct core and shell that are covalently bound together [172, 173]. This feature, in the context of encapsulation, provides some potential advantages over multimolecular and crosslinked systems. Unimolecular micelles possess a well-defined structure and are, therefore, monodisperse. Additionally, they provide many possibilities for specific functionalization to adapt to specific applications. In spite of the recent progress that has been made in the development of unimolecular micelles [174–176] light-responsive unimolecular encapsulants remain relatively unexplored. The inclusion of substances within photoresponsive single-molecule dendritic structures has been reported in the past [177, 178], but no cargo-release capabilities were reported. Allcock and coworkers described the synthesis of unimolecular multiarm star amphiphilic BCs of the type *star*-poly(methyl methacrylate)-poly(ethylene glycol)methyl ether methacrylate) [179]. These polymers were grown from a multivalent initiator with photolabile *o*-nitrobenzyl linkers, following an atom transfer radical polymerization (ATRP) procedure. The study showed that the polymers have the ability to undergo self-collapse in aqueous media producing unimolecular micelles, which subsequently self-assemble into large micellar species. The stability of the polymer micelles decreased significantly after UV light irradiation (as shown by the increased critical aggregation concentration values), a phenomenon which was associated to the dissociation of the arms of the polymers from the core. The encapsulation and photoinduced-release of pyrene from the star BC micelles were studied in order to demonstrate the potential of the system. Johnson et al. have reported the loading of unimolecular polymer brushes with drugs such as DOX and camptothecin [180, 181]. The actives were attached to the polymers via photocleavable nitrobenzyloxylcarbonyl linkers. The release of the drugs could be controlled by exposure to UV light. For instance, DOX and camptothecin-loaded polymer brushes exhibited significantly higher toxicity levels to human cancer cells after photoinduced drug release. Despite these examples, light-induced release from unimolecular polymeric carriers remains largely unexplored. We believe there are excellent opportunities in this area by combining, in a single species, macromolecular building blocks imparting both water-solubility and high-loading capacities, with light-responsive host–guest complexes [182, 183], which may provide protection to actives and triggered substance-release capabilities.

13.4 Conclusions

Light is a convenient "tool" for the isomerization of specific functional groups and the formation and cleavage of covalent bonds and supramolecular interactions. In addition, it is orthogonal to other reaction conditions and may be applied in a remote fashion, which makes it an attractive trigger for the design of photoresponsive encapsulants, including polymeric vesicles, NPs, and nanogels. The light-triggered release of actives from polymer carriers can be based, for example, on the photochemical decomposition of a capsule wall containing photolabile moieties (*o*-nitrobenzyl derivatives), or on photoisomerizable moieties (such as spiropyrans or azobenzenes). A considerable amount of work has been done in the area of light-responsive polymeric encapsulants, which has been complemented by the advent of powerful polymerization (ATRP, for example) and polymer derivatization techniques (i.e. Huisgen 1,3-dipolar cycloaddition and thiol-ene Michael addition). While the vast majority of systems are responsive to UV light, the latest developments pursue actuation at increasingly longer wavelengths. This trend is driven by the shortcomings exhibited by UV light in the context of biologically relevant applications (limited tissue penetration and potential tissue damage associated with high energy inputs). There is currently a substantial interest to develop red-shifted chromophores, which could be exploited by combining them with biocompatible polymers; and design visible and NIR-light-controlled encapsulants. Other relevant systems are based on two-photon excitation methodologies and photothermal effects. Although these techniques have not been discussed here in great detail, the increased pace of research activity is apparent and many exciting developments are expected in this area.

References

1 Sather, A.C., Lee, H.G., Colombe, J.R. et al. (2015). Dosage delivery of sensitive reagents enables glovebox-free synthesis. *Nature* 524: 208.
2 Gibbs, B.F., Kermasha, S., Alli, I., and Mulligan, C.N. (1999). Encapsulation in the food industry: a review. *Int. J. Food Sci. Nutr.* 50: 213.
3 Klán, P., Šolomek, T., Bochet, C.G. et al. (2013). Photoremovable protecting groups in chemistry and biology: reaction mechanisms and efficacy. *Chem. Rev.* 113: 119.
4 Moan, J. and Peng, Q. (2003). An outline of the hundred-year history of PDT. *Anticancer Res.* 23: 3591.
5 Xiao, P., Zhang, J., Zhao, J., and Stenzel, M.H. (2017). Light-induced release of molecules from polymers. *Prog. Polym. Sci.* 74: 1.
6 Linsley, C.S. and Wu, B.M. (2017). Recent advances in light-responsive on-demand drug-delivery systems. *Ther. Deliv.* 8: 89.
7 Bansal, A. and Zhang, Y. (2014). Photocontrolled nanoparticle delivery systems for biomedical applications. *Acc. Chem. Res.* 47: 3052.
8 Karimi, M., Zangabad, S.Z., Baghaee-Ravari, S. et al. (2017). Smart nanostructures for cargo delivery: uncaging and activating by light. *J. Am. Chem. Soc.* 139: 4584.
9 Gohy, J.-F. and Zhao, Y. (2013). Photo-responsive block copolymer micelles: design and behavior. *Chem. Soc. Rev.* 42: 7117.
10 Liu, G., Liu, W., and Dong, C.-M. (2013). UV- and NIR-responsive polymeric nanomedicines for on-demand drug delivery. *Polym. Chem.* 4: 3431.
11 Huang, Y., Dong, R., Zhu, X., and Yan, D. (2014). Photo-responsive polymeric micelles. *Soft Matter* 10: 6121.
12 Zhao, Y. (2012). Light-responsive block copolymer micelles. *Macromolecules* 45: 3647.
13 Fomina, N., Sankaranarayanan, J., and Almutairi, A. (2012). Photochemical mechanisms of light-triggered release from nanocarriers. *Adv. Drug Deliv. Rev.* 64: 1005.
14 Dong, M.X., Babalhavaeji, A., Samanta, S. et al. (2015). Red-shifting azobenzene photoswitches for *in vivo* use. *Acc. Chem. Res.* 48: 2662.
15 Dong, M., Babalhavaeji, A., Collins, C.V. et al. (2017). Near-infrared photoswitching of azobenzenes under physiological conditions. *J. Am. Chem. Soc.* 139: 13483.
16 Weis, P. and Wu, S. (2018). Light-switchable azobenzene-containing macromolecules: from UV to near infrared. *Macromol. Rapid Commun.* 39: 1700220.
17 Bertrand, O. and Gohy, J.-F. (2017). Photo-responsive polymers: synthesis and applications. *Polym. Chem.* 8: 52.

18 Anilkumar, P., Gravel, E., Theodorou, I. et al. (2014). Nanometric micelles with photo-triggered cytotoxicity. *Adv. Funct. Mater.* 24: 5246.
19 Guo, X., Wang, L., Wei, X., and Zhou, S. (2016). Polymer-based drug delivery systems for cancer treatment. *J. Polym. Sci., Part A: Polym. Chem.* 54: 3525.
20 Schattling, P., Jochum, F.D., and Theato, P. (2014). Multi-stimuli responsive polymers – the all-in-one talents. *Polym. Chem.* 5: 25.
21 Cabane, E., Malinova, V., Menon, S. et al. (2011). Photoresponsive polymersomes as smart, triggerable nanocarriers. *Soft Matter* 7: 9167.
22 Xuan, J., Han, D., Xia, H., and Zhao, Y. (2014). Dual-stimuli-responsive micelle of an ABC triblock copolymer bearing a redox-cleavable unit and a photocleavable unit at two block junctions. *Langmuir* 30: 410.
23 Katz, J.S., Zhong, S., Ricart, B.G. et al. (2010). Modular synthesis of biodegradable diblock copolymers for designing functional polymersomes. *J. Am. Chem. Soc.* 132: 3654.
24 Fomina, N., McFearin, C., Sermsakdi, M. et al. (2010). UV and near-IR triggered release from polymeric nanoparticles. *J. Am. Chem. Soc.* 132: 9540.
25 de Gracia Lux, C., McFearin, C.L., Joshi-Barr, S. et al. (2012). Single UV or near IR triggering event leads to polymer degradation into small molecules. *ACS Macro Lett.* 1: 922.
26 Fan, B., Trant, J.F., Wong, A.D., and Gillies, E.R. (2014). Polyglyoxylates: a versatile class of triggerable self-immolative polymers from readily accessible monomers. *J. Am. Chem. Soc.* 136: 10116.
27 Fan, B. and Gillies, E.R. (2017). Poly(ethyl glyoxalate)-poly(ethylene oxide) nanoparticles: stimuli-responsive drug release via end-to-end polyglyoxylate depolymerization. *Mol. Pharm.* 14: 2548.
28 Liu, G., Wang, X., Hu, J. et al. (2014). Self-immolative polymersomes for high-efficiency triggered release and programmed enzymatic reactions. *J. Am. Chem. Soc.* 136: 7492.
29 Patil, N.G., Basutkar, N.B., and Ambade, A.V. (2015). Visible light-triggered disruption of micelles of an amphiphilic block copolymer with BODIPY at the junction. *Chem. Commun.* 51: 17708.
30 Coumes, F., Woisel, P., and Fournier, D. (2016). Facile access to multistimuli-responsive self-assembled block copolymers via catechol/boronic acid ligation. *Macromolecules* 49: 8925.
31 Yan, Q., Han, D., and Zhao, Y. (2013). Main-chain photoresponsive polymers with controlled location of light-cleavable units: from synthetic strategies to structural engineering. *Polym. Chem.* 4: 5026.
32 Han, D., Tong, X., and Zhao, Y. (2011). Fast photodegradable block copolymer micelles for burst release. *Macromolecules* 44: 437.
33 Han, D., Tong, X., and Zhao, Y. (2012). Block copolymer micelles with a dual-stimuli-responsive core fast or slow degradation. *Langmuir* 28: 2327.
34 Jin, Q., Cai, T., Han, H. et al. (2014). Light and pH dual-degradable triblock micelles for controlled intracellular drug release. *Macromol. Rapid Commun.* 35: 1372.
35 Tian, M., Cheng, R., Zhang, J. et al. (2016). Amphiphilic polymer micellar disruption based on main-chain photodegradation. *Langmuir* 32: 12.

36 Pasparakis, G., Manouras, T., Vamvakaki, M., and Argitis, P. (2014). Harnessing photochemical internalization with dual degradable nanoparticles for combinatorial photo-chemotherapy. *Nat. Commun.* 5: 3623.

37 Jiang, J., Tong, X., and Zhao, Y. (2005). A new design for light-breakable polymer micelles. *J. Am. Chem. Soc.* 127: 8290.

38 Jiang, J., Tong, X., Morris, M., and Zhao, Y. (2006). Toward photocontrolled release using light-dissociable block copolymer micelles. *Macromolecules* 39: 4633.

39 Babin, J., Pelletier, M., Lepage, M. et al. (2009). A new two-photon-sensitive block copolymer nanocarrier. *Angew. Chem. Int. Ed.* 48: 3329.

40 Jiang, J., Qi, B., Lepage, M., and Zhao, Y. (2007). Polymer micelles stabilization on demand through photo-cross-linking. *Macromolecules* 40: 790.

41 Dong, R., Zhu, B., Zhou, Y. et al. (2013). Reversible photoisomerization of azobenzene containing polymeric systems driven by visible light. *Polym. Chem.* 4: 912.

42 Lee, H., Wu, W., Oh, J.K. et al. (2007). Light-induced reversible formation of polymeric micelles. *Angew. Chem. Int. Ed.* 46: 2453.

43 Yang, F., Cao, Z., and Wang, G. (2015). Micellar assembly of a photo- and temperature-responsive amphiphilic block copolymers for controlled release. *Polym. Chem.* 6: 7995.

44 Feng, Z., Lin, L., Yan, Z., and Yu, Y. (2010). Dual responsive block copolymer micelles functionalized by NIPAM and azobenzene. *Macromol. Rapid Commun.* 31: 640.

45 Jochum, F.D. and Theato, P. (2010). Thermo- and light responsive micellation of azobenzene containing block copolymers. *Chem. Commun.* 46: 6717.

46 Chen, C.J., Liu, G.-Y., Shi, Y.-T. et al. (2011). Biocompatible micelles based on comb-like PEG derivates: formation, characterization, and photo-responsiveness. *Macromol. Rapid Commun.* 32: 1077.

47 Li, Q., Cao, Z., and Wang, G. (2018). Diazonaphthoquinone-based amphiphilic polymer assemblies for NIR/UV light- and pH-responsive controlled release. *Polym. Chem.* 9: 463.

48 Shen, H., Xia, Y., Qin, Z. et al. (2017). Photoresponsive biodegradable poly(carbonate)s with pendent *o*-nitrobenzyl ester. *J. Polym. Sci., Part A: Polym. Chem.* 55: 2770.

49 Hu, D., Peng, H., Niu, Y. et al. (2015). Reversibly light-responsive biodegradable poly(carbonate) micelles constructed via CuAAC reaction. *J. Polym. Sci., Part A: Polym. Chem.* 53: 750.

50 Xia, Y., Zeng, Y., Hu, D. et al. (2017). Light and pH dual-sensitive biodegradable polymeric nanoparticles for controlled release of cargos. *J. Polym. Sci., Part A: Polym. Chem.* 55: 1773.

51 Wong, S., Shim, M.S., and Kwon, Y.J. (2014). Synthetically designed peptide-based biomaterials with stimuli-responsive and membrane-active properties for biomedical applications. *J. Mater. Chem. B* 2: 595.

52 He, C., Zhuang, X., Tang, Z. et al. (2012). Stimuli-sensitive synthetic polypeptide-based materials for drug and gene delivery. *Adv. Healthc. Mater.* 1: 48.

53 Liu, G. and Dong, C.-M. (2012). Photoresponsive poly(S-(*o*-nitrobenzyl)-L-cysteine)-*b*-PEO from a L-cysteine *N*-carboxyanhydride monomer: synthesis, self-assembly, and phototriggered drug release. *Biomacromolecules* 13: 1573.

54 Kumar, S., Allard, J.-M., Morris, D. et al. (2012). Near-infrared light sensitive polypeptide block copolymer micelles for drug delivery. *J. Mater. Chem.* 22: 7252.

55 Hu, X., Zhang, Y., Xie, Z. et al. (2017). Stimuli-responsive polymersomes for biomedical applications. *Biomacromolecules* 18: 649.

56 Wang, X., Hu, J., Liu, G. et al. (2015). Reversibly switching bilayer permeability and release modules of photochromic polymersomes stabilized by cooperative noncovalent interactions. *J. Am. Chem. Soc.* 137: 15262.

57 Wang, X., Liu, G., Hu, J. et al. (2014). Concurrent block copolymer polymersome stabilization and bilayer permeabilization by stimuli-regulated 'traceless' crosslinking. *Angew. Chem. Int. Ed.* 55: 3138.

58 Gitsov, I. (2008). Hybrid linear dendritic macromolecules: from synthesis to applications. *J. Polym. Sci., Part A: Polym. Chem.* 46: 5295.

59 Dong, C.-M. and Liu, G. (2013). Linear-dendritic biodegradable block copolymers: from synthesis to application in bionanotechnology. *Polym. Chem.* 4: 46.

60 Whitton, G. and Gillies, E.R. (2015). Functional aqueous assemblies of linear-dendron hybrids. *J. Polym. Sci., Part A: Polym. Chem.* 53: 148.

61 Mohammadifar, E., Kharat, A.N., and Adeli, M. (2015). Polyamidoamine and polyglycerol; their linear, dendritic and linear-dendritic architectures as anticancer drug delivery systems. *J. Mater. Chem. B* 3: 3896.

62 Gheybi, H. and Adeli, M. (2015). Supramolecular anticancer drug delivery systems based on linear-dendritic copolymers. *Polym. Chem.* 6: 2580.

63 Fan, X., Zhao, Y., Xu, W., and Li, L. (2016). Linear-dendritic block copolymer for drug and gene delivery. *Mater. Sci. Eng., C* 62: 943.

64 Wurm, F. and Frey, H. (2011). Linear-dendritic block copolymers: the state of the art and exciting perspectives. *Prog. Polym. Sci.* 36: 1.

65 del Barrio, J., Oriol, L., Sanchez, C. et al. (2010). Self-assembly of linear-dendritic diblock copolymers: from nanofibers to polymersomes. *J. Am. Chem. Soc.* 132: 3762.

66 Lin, Y.-L., Chang, H.-Y., Sheng, Y.-J., and Tsao, H.-K. (2012). Photoresponsive polymersomes formed by amphiphilic linear-dendritic block copolymers: generation-dependent aggregation behavior. *Macromolecules* 45: 7143.

67 Blasco, E., del Barrio, J., Sanchez-Somolinos, C. et al. (2013). Light induced molecular release from vesicles based on amphiphilic linear-dendritic block copolymers. *Polym. Chem.* 4: 2246.

68 Blasco, E., Luis Serrano, J., Pinol, M., and Oriol, L. (2013). Light responsive vesicles based on linear-dendritic block copolymers using azobenzene-aliphatic codendrons. *Macromolecules* 46: 5951.

69 Garcia-Juan, H., Nogales, A., Blasco, E. et al. (2016). Self-assembly of thermo and light responsive amphiphilic linear dendritic block copolymers. *Eur. Polym. J.* 81: 621.

70 Kalva, N. and Ambade, A.V. (2017). Synthesis and tunable thermoresponsive solution morphologies of bisMPA dendron-azobenzene-poly(*N*-isopropyl acrylamide) copolymers. *Polym. Int.* 66: 1084.
71 Kalva, N., Parekh, N., and Ambade, A.V. (2015). Controlled micellar disassembly of photo- and pH-cleavable linear-dendritic block copolymers. *Polym. Chem.* 6: 6826.
72 Goodwin, A.P., Mynar, J.L., Ma, Y. et al. (2005). Synthetic micelle sensitive to IR light via a two-photon process. *J. Am. Chem. Soc.* 127: 9952.
73 Mynar, J.L., Goodwin, A.P., Cohen, J.A. et al. (2007). Two-photon degradable supramolecular assemblies of linear-dendritic copolymers. *Chem. Commun.* 2081.
74 Sun, L., Zhu, B., Su, Y., and Don, C.M. (2014). Light-responsive linear-dendritic amphiphiles and their nanomedicines for NIR-triggered drug release. *Polym. Chem.* 5: 1605.
75 Sun, L., Yang, Y., Dong, C.-M., and Wei, Y. (2011). Two-photon-sensitive and sugar-targeted nanocarriers from degradable and dendritic amphiphiles. *Small* 7: 401.
76 Sun, L., Ma, X., Dong, C.-M. et al. (2012). NIR-responsive and lectin-binding doxorubicin-loaded nanomedicine from Janus-type dendritic PAMAM amphiphiles. *Biomacromolecules* 13: 3581.
77 Nazemi, A. and Gillies, E.R. (2014). Dendrimersomes with photodegradable membranes for triggered release of hydrophilic and hydrophobic cargo. *Chem. Commun.* 50: 11122.
78 Wang, D., Zhao, T., Zhu, X. et al. (2015). Bioapplications of hyperbranched polymers. *Chem. Soc. Rev.* 44: 4023.
79 Jiang, W., Zhou, Y., and Yan, D. (2015). Hyperbranched polymer vesicle: from self-assembly, characterization, mechanisms, and properties to applications. *Chem. Soc. Rev.* 44: 3874.
80 Wang, D., Jin, Y., Zhu, X., and Yan, D. (2017). Synthesis and applications of stimuli-responsive hyperbranched polymers. *Prog. Polym. Sci.* 64: 114.
81 Chen, C., Liu, G., Liu, X. et al. (2011). Photo-responsive, biocompatible polymeric micelles self-assembled from hyperbranched polyphosphate-based polymers. *Polym. Chem.* 2: 1389.
82 Chen, C.-J., Jin, Q., Liu, G.-Y. et al. (2012). Reversibly light-repsonsive micelles constructed via a simple modification of hyperbranched polymers with chromophores. *Polymer* 53: 3695.
83 Son, S., Shin, E., and Kim, B.-S. (2014). Light-responsive micelles of spiropyran initiated hyperbranched polyglycerol for smart drug delivery. *Biomacromolecules* 15: 628.
84 Mo, B., Lui, H., Zhou, X., and Zhao, Y. (2015). Facile synthesis of pholabile dendritic-unit-bridged hyperbranched graft copolymers for stimuli-triggered topological transition and controlled release of Nile red. *Polym. Chem.* 6: 3489.
85 Ren, J.M., McKenzie, T.G., Fu, Q. et al. (2016). Star polymers. *Chem. Rev.* 116: 6743.
86 Yang, D.-P., Oo, M.N.N.L., Deen, G.R. et al. (2017). Nano-star-shaped polymers for drug delivery applications. *Macromol. Rapid Commun.* 38: 1700410.

87 Wu, W., Wang, W., and Li, J. (2015). Star polymers: advances in biomedical applications. *Prog. Polym. Sci.* 46: 55.
88 Kuckling, D. and Wycisk, A. (2013). Stimuli responsive star polymers. *J. Polym. Sci., Part A: Polym. Chem.* 51: 2980.
89 Blasco, E., Schmidt, B.V.K.J., Barner-Kowollik, C. et al. (2014). A novel photoresponsive azobenzene-containing miktoarm star polymer: self-assembly and photoresponse properties. *Macromolecules* 47: 3693.
90 Blasco, E., Schmidt, B.V.K.J., Barner-Kowwollik, C. et al. (2013). Dual thermo- and photo-responsive micelles based on miktoarm star polymers. *Polym. Chem.* 4: 4506.
91 Huo, H., Ma, X., Dong, Y., and Qu, F. (2017). Light/temperature dual responsive ABC miktoarm star terpolymer micelles for controlled release. *Eur. Polym. J.* 87: 331.
92 Yang, S.K., Ambade, A.V., and Weck, M. (2011). Main-chain supramolecular block copolymers. *Chem. Soc. Rev.* 40: 129.
93 Yan, X., Wang, F., Zheng, B., and Huang, F. (2012). Stimuli-responsive supramolecular polymeric materials. *Chem. Soc. Rev.* 41: 6042.
94 Krieg, E., Bastings, M.M.C., Besenius, P., and Rybtchinski, B. (2016). Supramolecular polymers in aqueous media. *Chem. Rev.* 116: 2141.
95 Kang, Y., Guo, K., Li, B.J., and Zhang, S. (2014). Nanoassemblies driven by cyclodextrin-based inclusion complexation. *Chem. Commun.* 50: 11083.
96 Peng, L., Liu, S., Feng, A., and Yuan, J. (2017). Polymeric nanocarriers based on cyclodextrins for drug delivery. Host-guest interaction as stimuli responsive linker. *Mol. Pharm.* 14: 2475.
97 Yan, Q., Xin, Y., Zhou, R. et al. (2011). Light-controlled smart nanotubes based on the orthogonal assembly of two homopolymers. *Chem. Commun.* 47: 9594.
98 Dong, R., Zhou, Y., and Zhu, X. (2014). Supramolecular dendritic polymers: from synthesis to applications. *Acc. Chem. Res.* 47: 2006.
99 Zhou, J., Guan, B., Liao, X. et al. (2009). Dual reversible self-assembly of PNIPAM-based amphiphiles formed by inclusion complexation. *Macromolecules* 42: 7465.
100 Wang, J., Wang, X., Yang, F. et al. (2014). Self-assembly of a linear-star supramolecular amphiphile based on host-guest complexation. *Langmuir* 30: 13014.
101 Zou, H., Yuan, W., Lu, Y., and Wang, S. (2017). UV light-and thermo-responsive supramolecular aggregates with tunable morphologies from inclusion complexation of dendritic/linear polymers. *Chem. Commun.* 53: 2463.
102 Liu, Y., Yu, C., Jin, H. et al. (2013). A supramolecular Janus hyperbranched polymer and its photoresponsive self-assembly of vesicles with narrow size distribution. *J. Am. Chem. Soc.* 135: 4765.
103 Zhang, D., Fan, Y., Li, H. et al. (2015). A dumbbell-like supramolecular triblock copolymer and its self-assembly of light-responsive vesicles. *RSC Adv.* 5: 47762.
104 Yao, H., Qi, M., Liu, Y., and Tian, W. (2016). Host-guest binding-site-tunable self-assembly of stimuli-responsive supramolecular polymers. *Chem. Eur. J.* 22: 8508.

105 Jin, Q., Liu, G., Liu, X., and Ji, J. (2010). Photo-responsive supramolecular self-assembly and disassembly of an azobenzene-containing block copolymer. *Soft Matter* 6: 5589.
106 Wang, S., Shen, Q., Nawaz, M.H., and Zhang, W. (2013). Photocontrolled reversible supramolecular assemblies of a diblock azo-copolymer based on β-cyclodextrin-azo host–guest inclusion complexation. *Polym. Chem.* 4: 2151.
107 Romulus, J., Henssler, J.T., and Weck, M. (2014). Postpolymerization modification of block copolymers. *Macromolecules* 47: 5437.
108 Zhang, T., Jin, C., Wang, L., and Yin, Q. (2014). One-step synthesis of hollow polymeric nanospheres: self-assembly of amphiphilic azo polymers via hydrogen bond formation. *RSC Adv.* 4: 36882.
109 Jin, C., Zhang, T., Wang, L. et al. (2014). Photoinduced deformation of hollow nanospheres formed by the self-assembly of amphiphilic random copolymers and small azo molecules. *RSC Adv.* 4: 45890.
110 Concellon, A., Blasco, E., Martinez-Felipe, A. et al. (2016). Light-responsive self-assembled materials by supramolecular post-functionalization via hydrogen bonding of amphiphilic block copolymers. *Macromolecules* 49: 7825.
111 Eith, L., Stepto, R.F.T., Tomka, I., and Wittwer, F. (1986). The injection-molded capsule. *Drug Dev. Ind. Pharm.* 12: 2133.
112 Yow, H. and Routh, A. (2006). Formation of liquid core-polymer shell microcapsules. *Soft Matter* 2: 940.
113 Bédard, M.F., De Geest, B.G., Skirtach, A.G. et al. (2010). Polymeric microcapsules with light responsive properties for encapsulation and release. *Adv. Colloid Interface Sci.* 158: 2.
114 Yi, Q. and Sukhorukov, G.B. (2014). UV light stimulated encapsulation and release by polyelectrolyte microcapsules. *Adv. Colloid Interface Sci.* 207: 280.
115 Borges, J., Rodrigues, L.C., Reis, R.L., and Mano, J.F. (2014). Layer-by-layer assembly of light-responsive polymeric multilayer systems. *Adv. Funct. Mater.* 24: 5624.
116 Ochs, M., Parak, W.J., Rejman, J., and Carregal-Romero, S. (2015). Light-addressable microcapsules. In: *Layer-by-Layer Films for Biomedical Applications* (ed. C. Picart, F. Caruso and J.-C. Voegel). Weinheim: Wiley-VCH.
117 Cao, Z., Li, Q., and Wang, G. (2017). Photodegradable polymer nanocapsules fabricated from dimethyldiethoxysilane emulsion templates for controlled release. *Polym. Chem.* 8: 6817.
118 Yi, Q., Wen, D., and Sukhorukov, G.B. (2012). UV-crosslinkable multilayer microcapsules made of weak polyelectrolytes. *Langmuir* 28: 10822.
119 Yuan, X., Fischer, K., and Schärtl, W. (2005). Photocleavable microcapsules built from photoreactive nanospheres. *Langmuir* 21: 9374.
120 Borges, J. and Mano, J.M. (2014). Molecular interactions driving the layer-by-layer assembly of multilayers. *Chem. Rev.* 114: 8883.
121 Tong, W., Song, X., and Gao, C. (2012). Layer-by-layer assembly of microcapsules and their biomedical applications. *Chem. Soc. Rev.* 41: 6103.
122 Skirtach, A.G., Yashchenok, A.M., and Möhwald, H. (2011). Encapsulation, release and applications of LbL polyelectrolyte multilayer capsules. *Chem. Commun.* 47: 12736.

123 Peyratout, C.S. and Dähne, L. (2004). Tailor-made polyelectrolyte microcapsules: from multilayers to smart containers. *Angew. Chem. Int. Ed.* 43: 3762.
124 Park, M.-K., Deng, S., and Advincula, R. (2004). pH-sensitive bipolar ion-permselective ultrathin films. *J. Am. Chem. Soc.* 126: 13723.
125 Park, M.-K., Deng, S., and Advincula, R. (2005). Sustained release control via photo-cross-linking of polyelectrolyte layer-by-layer hollow capsules. *Langmuir* 21: 5272.
126 Tao, X., Li, J., and Möhwald, H. (2004). Self-assembly, optical behavior, and permeability of a novel capsule based on an azo dye and polyelectrolytes. *Chem. Eur. J.* 10: 3397.
127 Bédard, M., Skirtach, A.G., and Sukhorukov, G.B. (2007). Optically driven encapsulation using novel polymeric hollow shells containing an azobenzene polymer. *Macromol. Rapid Commun.* 28: 1517.
128 Yi, Q. and Sukhorukov, G.B. (2014). UV-induced disruption of microcapsules with azobenzene groups. *Soft Matter* 10: 1384.
129 Bédard, M.F., Sadasivan, S., Sukhorukov, G.B., and Skirtach, A. (2009). Assembling polyelectrolytes and porphyrins into hollow capsules with laser responsive oxidative properties. *J. Mater. Chem.* 19: 2226.
130 Li, C., Li, Z.Y., Zhang, J. et al. (2012). Porphyrin containing light-responsive capsules for controlled drug release. *J. Mater. Chem.* 22: 4623.
131 Kono, K., Nishihara, Y., and Takagishi, T. (1995). Photoresponsive permeability of polyelectrolyte complex capsule membrane containing triphenylmethane leucohydroxide residues. *J. Appl. Polym. Sci.* 56: 707.
132 Xu, W., Choi, I., Plamper, F.A. et al. (2013). Nondestructive light-initiated tuning of layer-by-layer microcapsule permeability. *ACS Nano* 7: 598.
133 Xiao, W., Chen, W.-H., Zhang, J. et al. (2011). Design of a photoswitchable hollow microcapsular drug delivery system by using a supramolecular drug-loading approach. *J. Phys. Chem. B* 115: 13796.
134 Yi, Q. and Sukhorukov, G.B. (2013). Externally triggered dual function of complex microcapsules. *ACS Nano* 7: 8693.
135 Li, H., Tong, W., and Gao, C. (2016). Photo-responsive polyethyleneimine microcapsules cross-linked by ortho-nitrobenzyl derivatives. *J. Colloid Interface Sci.* 463: 22.
136 Jiang, N., Cheng, Y., and Wei, J. (2017). Coumarin-modified fluorescent microcapsules and their photo-switchable release property. *Colloids Surf., A* 522: 28.
137 Gu, Z., Yan, M., Hu, B. et al. (2009). Protein nanocapsule weaved with enzymatically degradable polymeric network. *Nano Lett.* 9: 4533.
138 Paret, N., Trachsel, A., Berthier, D.L., and Herrmann, A. (2015). Controlled release of encapsulated bioactive volatiles by rupture of the capsule wall through the light-induced generation of a gas. *Angew. Chem. Int. Ed.* 54: 2275.
139 Utada, A.S., Chu, L.Y., Fernandez-Nieves, A. et al. (2007). Dripping, jetting, drops, and wetting: the magic of microfluidics. *MRS Bull.* 32: 702.
140 Amstad, E., Kim, S.H., and Weitz, D.A. (2012). Photo- and thermoresponsive polymersomes for triggered release. *Angew. Chem. Int. Ed.* 51: 12499.

141 Zhang, J., Coulston, R.J., Jones, S.T. et al. (2012). One-step fabrication of supramolecular microcapsules from microfluidic droplets. *Science* 335: 690.

142 Liu, J., Lan, Y., Yu, Z. et al. (2017). Cucurbit[*n*]uril-based microcapsules self-assembled within microfluidic droplets: a versatile approach for supramolecular architectures and materials. *Acc. Chem. Res.* 50: 208.

143 Yu, Z., Zhang, J., Coulston, R.J. et al. (2015). Supramolecular hydrogel microcapsules via cucurbit[8]uril host–guest interactions with triggered and UV-controlled molecular permeability. *Chem. Sci.* 6: 4929.

144 Nayak, S. and Lyon, L.A. (2005). Soft nanotechnology with soft nanoparticles. *Angew. Chem. Int. Ed.* 44: 7686.

145 Klinger, D. and Landfester, K. (2012). Stimuli-responsive microgels for the loading and release of functional compounds: fundamental concepts and applications. *Polymer* 53: 5209.

146 He, I., Tong, X., and Zhao, Y. (2009). Photoresponsive nanogels based on photocontrollable cross-links. *Macromolecules* 42: 4845.

147 Jin, Q., Liu, G., and Li, J. (2010). Preparation of reversibly photo-cross-linked nanogels from pH-responsive block copolymers and use as nanoreactors for the synthesis of gold nanoparticles. *Eur. Polym. J.* 46: 2120.

148 Klinger, D. and Landfester, K. (2011). Photo-sensitive PMMA microgels: light-triggered swelling and degradation. *Soft Matter* 7: 1426.

149 Klinger, D. and Landfester, K. (2012). Enzymatic and light degradable hybrid nanogels: crosslinking of polyacrylamide with acrylate functionalized dextrans containing photocleavable linkers. *J. Polym. Sci., Part A: Polym. Chem.* 50: 1062.

150 Bian, S., Zheng, J., and Yang, W. (2014). Dual stimuli-responsive microgels based on photolabile crosslinker: temperature sensitivity and light-induced degradation. *J. Polym. Sci., Part A: Polym. Chem.* 52: 1676.

151 Huang, Q., Bao, C., Wei, J. et al. (2009). Photocleavable coumarin crosslinkers based polystyrene microgels: phototriggered swelling and release. *Macromol. Biosci.* 9: 248.

152 Klinger, D. and Landfester, K. (2011). Dual stimuli-responsive poly(2-hydroxyethyl methacrylate-*co*-methacrylic acid) microgels based on photo-cleavable cross-linkers: pH-dependent swelling and light-induced degradation. *Macromolecules* 44: 9758.

153 Azagarsamy, M.A., Alge, D.L., Radhakrishnan, S.J. et al. (2012). Photocontrolled nanoparticles for on-demand release of proteins. *Biomacromolecules* 13: 2219.

154 Cao, Z., Zhou, X., and Wang, G. (2016). Selective release of hydrophobic and hydrophilic cargos from multi-stimuli-responsive nanogels. *ACS Appl. Mater. Interfaces* 8: 28888.

155 Zan, M., Li, J., Huang, M. et al. (2015). Near-infrared light-triggered drug release nanogels for combined photothermal-chemotherapy of cancer. *Biomater. Sci.* 3: 1147.

156 Chen, W.R., Adams, R.L., Heaton, S. et al. (1995). Chromophore-enhanced laser-tumor tissue photothermal interaction using an 808-nm diode laser. *Cancer Lett.* 88: 15.

157 Khatun, Z., Nurunnabi, M., Nafiujjaman, M. et al. (2015). A hyaluronic acid nanogel for photo-chemo theranostics of lung cancer with simultaneous light-responsive controlled release of doxorubicin. *Nanoscale* 7: 10680.

158 Olejniczak, J., Carling, C.J., and Almutairi, A. (2015). Photocontrolled release using one-photon absorption of visible or NIR light. *J. Controlled Release* 219: 18.

159 Klinger, D. (2013). *Light-Sensitive Polymeric Nanoparticles Based on Photo-Cleavable Chromophores*. Springer International Publishing.

160 Olejniczak, J., Huu, V.A.N., Lux, J. et al. (2015). Light-triggered chemical amplification to accelerate degradation and release from polymeric particles. *Chem. Commun.* 51: 16980.

161 Olejniczak, J., Sankaranarayanan, J., Viger, M.L., and Almutairi, A. (2013). Highest efficiency two-photon degradable copolymer for remote controlled release. *ACS Macro Lett.* 2: 683.

162 Lv, C., Wang, Z., Wang, P., and Tang, X. (2012). Photodegradable polyurethane self-assembled nanoparticles for photocontrollable release. *Langmuir* 28: 9387.

163 Englert, C., Nischang, I., Bader, C. et al. (2018). Photocontrolled release of chemicals from nano- and microparticle containers. *Angew. Chem. Int. Ed.* 57: 2479.

164 Sheehan, J.C., Wilson, R.M., and Oxford, A.W. (1971). Photolysis of methoxy-substituted benzoin esters. Photosensitive protecting group for carboxylic acids. *J. Am. Chem. Soc.* 93: 7222.

165 Yuan, Y., Zhang, C.J., and Liu, B. (2015). A photoactivatable AIE polymer for light-controlled gene delivery: concurrent endo/lysosomal escape and DNA unpacking. *Angew. Chem. Int. Ed.* 54: 11419.

166 Lin, Q.N., Bao, C.Y., Yang, Y.L. et al. (2013). Highly discriminating photorelease of anticancer drugs based on hypoxia activatable phototrigger conjugated chitosan nanoparticles. *Adv. Mater.* 25: 1981.

167 Shi, D., Matsusaki, M., and Akashi, M. (2009). Unique size-change behavior of photo-crosslinked cinnamic acid derivative nanoparticles during hydrolytic degradation. *Macromol. Biosci.* 9: 248.

168 Tan, S., Cui, J., Fu, Q. et al. (2016). Photocontrolled cargo release from dual cross-linked polymer particles. *ACS Appl. Mater. Interfaces* 8: 6219.

169 Cui, W., Lu, X., Cui, K. et al. (2011). Photosensitive nanoparticles of chitosan complex for controlled release of dye molecules. *Nanotechnology* 22: 065702.

170 Meng, L., Huang, W., Wang, D. et al. (2013). Chitosan-based nanocarriers with pH and light dual response for anticancer drug delivery. *Biomacromolecules* 14: 2601.

171 Wang, Z., Wang, P., and Tang, X. (2013). Synthesis of light-induced expandable photoresponsive polymeric nanoparticles for triggered release. *ChemPlusChem* 78: 1273.

172 Xu, J., Luo, S., Shi, W., and Liu, S. (2006). Two-stage collapse of unimolecular micelles with double thermoresponsive coronas. *Langmuir* 22: 989.

173 Zhou, Y., Jiang, K., Song, Q., and Liu, S. (2007). Thermo-induced formation of unimolecular and multimolecular micelles from novel double hydrophilic multiblock copolymers of *N,N*-dimethylacrylamide and *N*-isopropylacrylamide. *Langmuir* 23: 13076.

174 Fan, X., Li, Z., and Loh, X.J. (2016). Recent development of unimolecular micelles as functional materials and applications. *Polym. Chem.* 7: 5898.

175 Gao, H.F. (2012). Development of star polymers as unimolecular containers for nanomaterials. *Macromol. Rapid Commun.* 33: 722.

176 Satoh, T. (2009). Unimolecular micelles based on hyperbranched polycarbohydrate cores. *Soft Matter* 5: 1972.

177 Dirksen, A. and De Cola, L. (2003). Photo-induced processes in dendrimers. *C.R. Chim.* 6: 873.

178 Puntoriero, F., Bergamini, G., Ceroni, P. et al. (2008). fluorescent guest encapsulated by a photoreactive azobenzene dendrimer. *New J. Chem.* 32: 401.

179 Liu, X., Tian, Z., Chen, C., and Allcock, H.R. (2013). UV-cleavable unimolecular micelles: synthesis and characterization toward photocontrolled drug release carriers. *Polym. Chem.* 4: 1115.

180 Johnson, J.A., Lu, Y.Y., Burts, A.O. et al. (2010). Drug-loaded, bivalent-bottle-brush polymers by graft-through ROMP. *Macromolecules* 43: 10326.

181 Johnson, J.A., Lu, Y.Y., Burts, A.O. et al. (2011). Core-clickable PEG-branch-azide bivalent-bottle-brush polymers by ROMP: grafting-through and clicking-to. *J. Am. Chem. Soc.* 133: 559.

182 Qu, D.H., Wang, Q.C., Zhang, Q.W. et al. (2015). Photoresponsive host-guest functional systems. *Chem. Rev.* 115: 7543.

183 Ryan, S.T.J., del Barrio, J., Suardiaz, R. et al. (2016). A dynamic and responsive host in action: light-controlled molecular encapsulation. *Angew. Chem. Int. Ed.* 55: 16096.

14

Photoresponsive Soft Materials Based on Reversible Proton Transfer

Yi Liao and Zhuozhi Wang

Florida Institute of Technology, Department of Chemistry, 150 W University Boulevard, Melbourne, FL 32901, USA

14.1 Introduction

Proton transfer is one of the most fundamental processes in nature and numerous chemical, material, and biological processes are driven by proton transfer. For example, biological systems use proton gradient as an energy form to synthesize ATP for their biological functions; many organic reactions are catalyzed by Bronsted acids; and many materials have pH-sensitive properties. Proton transfer is also involved in different aspects of polymer science relative to preparation, property, and processing. For example, cationic polymerizations can be initiated by proton transfer from a strong acid to a monomer. It has been reported that many polymers are sensitive to the pH (proton concentration) of the environment, and their properties including mechanical, geometrical, electric, optical, and magnetic properties, change with pH due to proton transfer between the polymers and the environment. Polymers with acid-labile bonds have been studied for drug delivery; and photoinduced cationic polymerization using photoacid-generators (PAGs) has been applied to photolithography.

An important research aims at using photoinduced proton transfer as a general mechanism to develop different types of photoresponsive polymers that reversibly alter their properties or release functional molecules upon irradiation. As described above, PAGs have been utilized to initiate cationic polymerization with great spatial control [1, 2]. Upon irradiation, these PAGs generate strong acids, which induce proton transfer and the consequent events. However, although some PAGs have reversible pathways in their mechanisms, they are generally irreversible. For the development of photoresponsive polymers, reversibility of the photoinduced property change is often required. It is desired that one can not only start (initiate) a process using light, but can also stop and/or reverse the process by turning off the light or applying a light with a different wavelength. The reversibility of the photoinduced property change implies a reversible proton transfer, which requires a reversible photoacid. The well-studied excited-state photoacids such as naphthol derivatives are reversible [3–5]. However, due

Photoactive Functional Soft Materials: Preparation, Properties, and Applications, First Edition.
Edited by Quan Li.

to the very short lifetime of the acidic excited state, they cannot achieve a high concentration of proton under irradiation, although the acidity of the excited state can be very high.

Our group discovered a new type of reversible photoacid that can be activated by visible light with moderate intensity [6–8]. This type of photoacid possesses a high-acidic metastable state generated from a reversible photoreaction. Examples of the photoacids as well as their photoreactions are shown in Figure 14.1. The lifetime of the high-acidic metastable state is long enough for accumulating a large proton concentration. The photoacids revert to the low-acidic ground state and take back the protons once the irradiation is removed. The reversibility differentiates this type of photoacid from PAG, and the large photoinduced proton concentration differentiates it from the excited-state photoacid.

Most of the proton transfer processes are driven by a high proton concentration. The lifetime of the photo-generated acidic state that is needed to generate a practically useful proton concentration can be estimated by a simple model. A sample containing a photoacid is irradiated by a light beam with a photon flux of q. The irradiation area is S and the depth of the sample is L. The fraction of the photon absorbed is f and the quantum yield is ϕ. Under irradiation, the photoacid changes from its ground state G to the acidic state A. At the stationary state, the photoreaction rate is the same as the reverse reaction rate.

$$R_\mathrm{f} = R_\mathrm{r} \tag{14.1}$$

Assuming that the reverse reaction is a first order reaction with a rate constant of k, we can obtain the Eq. (14.2)

$$fq\phi/L = k[\mathrm{A}] \tag{14.2}$$

Since the half-life of the acidic state $\tau = \ln 2/k$, Eq. (14.2) can be transformed to Eq. (14.3).

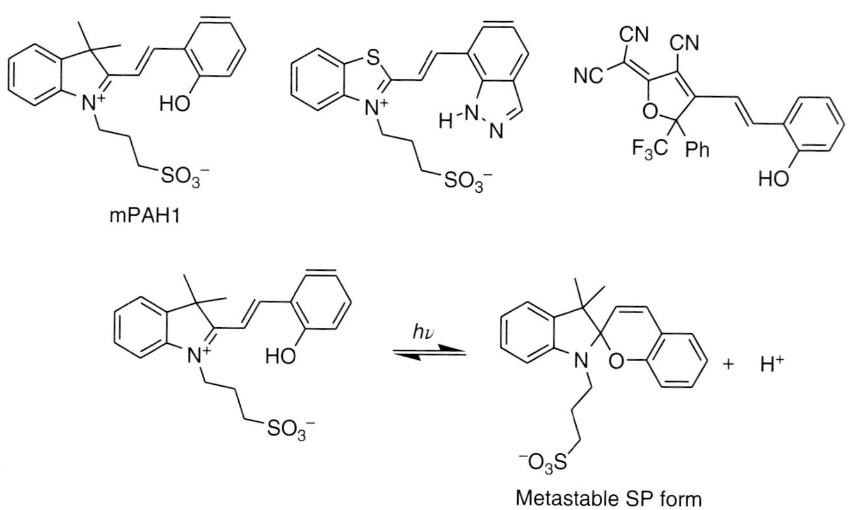

Figure 14.1 Structures and a photoreaction of some mPAHs.

$$[\mathrm{A}] = fq\phi\tau/\left(\ln 2 \cdot L\right) \tag{14.3}$$

Assuming that $q = 1\,\mathrm{mmol\,m^{-2}\,s^{-1}}$ (which is about the intensity of a strong LED or sunlight), $f = 1$, $\phi = 1$, and $L = 1\,\mu\mathrm{m}$ (which is the typical dimension for a thin film material or a cell), to obtain 1 mM of the acidic state A or higher, the half-life must be longer than 0.7 ms. In real conditions, f is often much smaller than 1. So even when a stronger light source is used and a concentration less than 1 mM is required, millisecond is still a reasonable estimate for the lower limit of the half-life. For metastable-state photoacids, the half-lives are often much longer than milliseconds. Therefore, they can generate a large proton concentration under moderate irradiation.

Since our report of the first metastable-state photoacid (mPAH) in 2011, photo-induced proton transfer using this type of photoacid has become a popular approach for the development of various photo-controlled systems. Applications in energy conversion [9, 10], sensor [11, 12], polymerization [13], micro-patterning [14, 15], nanomaterial [15–17], molecular machine [18], photochromic material [19–22], and biomedical [23, 24] areas have been reported by our group and other researchers. Regarding photoresponsive polymers, our group previously showed that a hydrogel of poly(acrylamide-*co*-acrylic acid) significantly reduced its volume under irradiation in the presence of a mPAH [6]. Protonation of the carboxylic anions by the photoacid decreased the hydrophilicity of the polymer and diminished the repulsion between carboxylic anions, which consequently lowered the water content in the hydrogel and reduced its volume. The Chumbimuni-Torres' group developed an ion-selective membrane sensor consisting of mPAHs and a calcium ionophore in a plasticized poly(vinyl chloride) [11, 12]. Upon irradiation, the protons of the photoacids in the membrane diffused into the solution. Cations in the solution penetrated into the membrane to balance the charge and thus were detected by the ionophore. The sensor could be repeatedly used due to the reversibility of the photoacids.

In this chapter, the physicochemical properties of mPAHs in polymer matrix are described first, then three photoresponsive materials recently developed by our group will be presented. The photochromic materials are based on optical property change upon protonation; the fragrant-releasing polymers described in the following section are based on acid-catalyzed reaction; and the photo-controlled nanomaterials are based on noncovalent cross-linking affected by protonation. These mechanisms together with the ion-exchange mechanism mentioned above are commonly involved in the photoresponsive materials based on mPAHs.

14.2 Photoactivity and Physicochemical Properties of Metastable-State Photoacids in Polymer Films

Photoactivity and physicochemical properties of mPAHs in solutions have been studied before [25]. The mPAH 1 in Figure 14.1 is the most well studied mPAH. It has a protonated merocyanine (MEH) structure with an absorption maximum

(λ_{max}) at 424 nm in water. Visible-light irradiation induces a *trans–cis* isomerization to form *cis*-MEH, which undergoes a nucleophilic reaction to form an acidic spiropyran (SPH) (Figure 14.1). The photoacidity is determined by the acidity of the SPH. The photoacidity of mPAH1 (pK_a under irradiation) in methanol was determined to be 3.7 using methyl orange as an indicator, and in water it was estimated to be 3.2–1.7 [20]. The dark acidity (pK_a in the dark) of mPAH1 was calculated to be 7.8 based on the pH of an aqueous solution [6]. However, the effective pK_a, which includes the effect of the equilibrium between MEH and SPH, must be measured with pH buffers. Experiment showed that the effective dark pK_a of mPAH1 is lower than 7.4 [8]. The quantum yield of the photoreaction in DMSO was measured to be as high as 0.37 [25].

The photoreaction is reversible and the metastable SP form reverts back to the MEH form after recombining with the proton it releases. The reverse reaction of mPAH1 in solutions is a second order reaction, which indicates that the rate-limiting step involves recombination of SP and proton [25]. The rate constants in water, ethanol, and DMSO are 73, 1.6, and 0.034 $M^{-1} s^{-1}$, respectively. The trend can be better explained by the hydrogen-bond donor acidity, which is related to the ability of the solvents to transfer the proton to SP. The reverse reactions of mPAH1 in organic solvent and polymers are slow. Several mPAHs with faster reverse reaction have been developed. For example, mPAH 2 (Figure 14.2) with a methoxy group on the phenolic moiety showed much faster reverse reaction rate than mPAH1 in solutions [26]. In this work, mPAH2 in polymer films was studied to understand the behaviors of metastable-state photoacid in polymer matrix [27].

Thin films of mPAH2 were prepared by spin casting solutions of mPAH2 and poly(hydroxyethyl methacrylate) (PHEMA) on glass substrates. PHEMA contains hydroxyl groups on flexible side chains, which mimic the environment of an alcohol and may assist proton transfer in the polymer. The weight ratio of the photoacid to PHEMA was 2.5%. After the solvent was completely removed by baking on a hot plate and then in vacuum, the films were very stable under ambient conditions and showed essentially the same photochemical properties after half a year. The photoreaction of mPAH2 in the polymer was monitored by UV–Vis spectroscopy. To monitor the UV–Vis absorption during the irradiation, polymer film on a glass substrate was kept at a tilted position, and the irradiation was perpendicular to the detecting beam. The light source was a 470 nm LED with a photon flux about 900 μmol m^2 s^{-1} (23 mW cm^{-2}). UV–Vis spectra were collected every six seconds during irradiation (Figure 14.2).

The MEH form of mPAH2 has an absorption maximum (λ_{max}) at 466 nm in the polymer film. As shown in Figure 14.2, upon irradiation, absorbance of MEH rapidly decreased in the first six seconds. Then the reduction of the absorbance quickly slowed down. After two minutes, the absorbance was reduced to about 66% of the original level. Although the photoreaction is formally a unimolecular reaction, the data obtained here cannot be fit to a first order kinetic equation. The rate of the reverse reaction must be comparable to that of the photoreaction. The process is further complicated by the slow proton diffusion in the polymer. Proton diffuses much slower in a polymer than in a protic solution. Upon irradiation, a local concentration of proton was built up around the photoacids. Previous work showed that the rate-limiting step of the reverse reaction is the recombina-

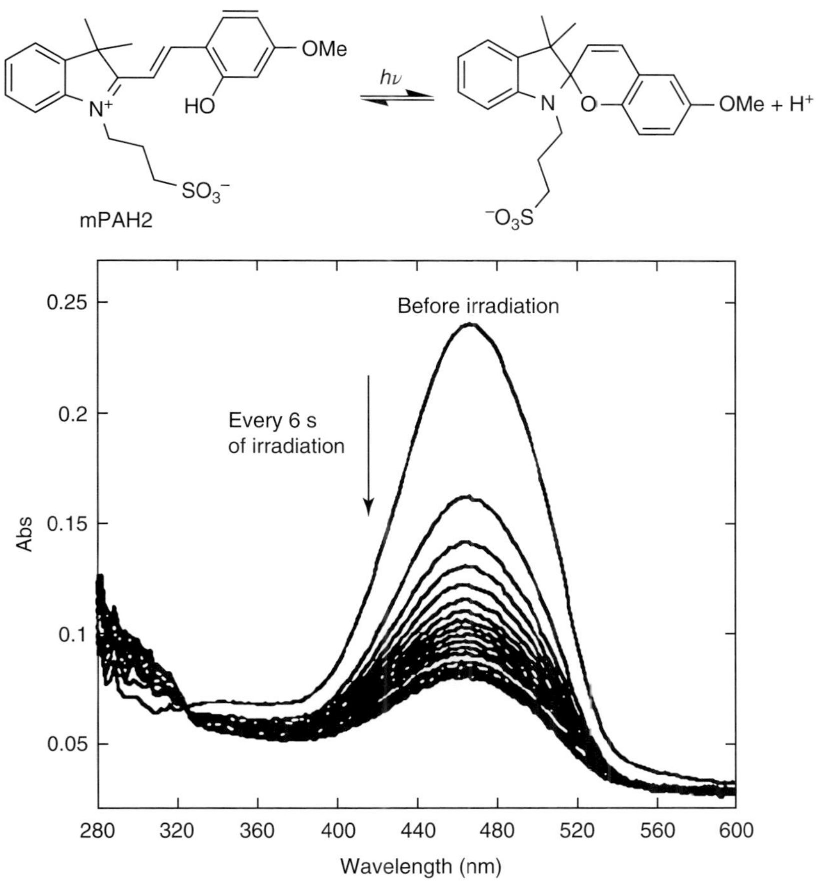

Figure 14.2 Structure of mPAH 2, UV–Vis spectra of mPAH 2-polymer film upon irradiation with 470 nm LED light source collected every six seconds.

tion of SP and H^+ [25]. In polymer, the concentration of the photoacid is high. In fact, 2.5 wt% is equal to approximately 60 mM as compared to ~0.5 mM commonly used for solution study. A high concentration of the photoacid resulted in a high concentration of SP and H^+ at the late stage of the photoreaction, which leads to a fast reverse reaction and slow overall reaction. In addition, the local concentration due to slow diffusion in polymer could be higher or lower than the average value, resulting in complicated kinetics.

The reverse reaction was studied by monitoring the recovery of the absorbance at 466 nm after irradiation. A rapid increase of the absorbance was observed within the first 20 minutes and the rate slowed down with time (Figure 14.3). Unlike the reverse reaction in solutions, the data could not be fitted into either a first order or a second order rate equation. We also tried to fit only the data in the first 20 minutes to first and second order rate equations and did not succeed. Figure 14.3 shows that, 59% of the MEH form recovered after 20 minutes, and it took 180 minutes for 90% recovery. UV irradiation (254 and 300 nm) did not increase the reverse reaction rate.

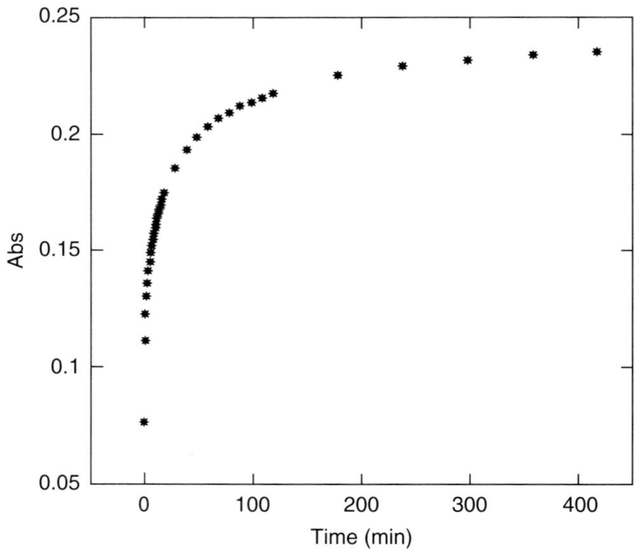

Figure 14.3 Change of absorbance at 466 nm form after irradiation.

The reversibility of the photoacid in polymer film was tested by repeated exposure to 470 nm irradiation. For each cycle, the sample was irradiated for five minutes and then kept in the dark for 10 minutes. As shown in Figure 14.4, ~80% of the photoacid reacted after each irradiation, and ~30% recovered after 10 minutes due to the slow reverse reaction. After the tenth irradiation, the sample was kept in the dark for 24 hours, 95% of MEH recovered showing good

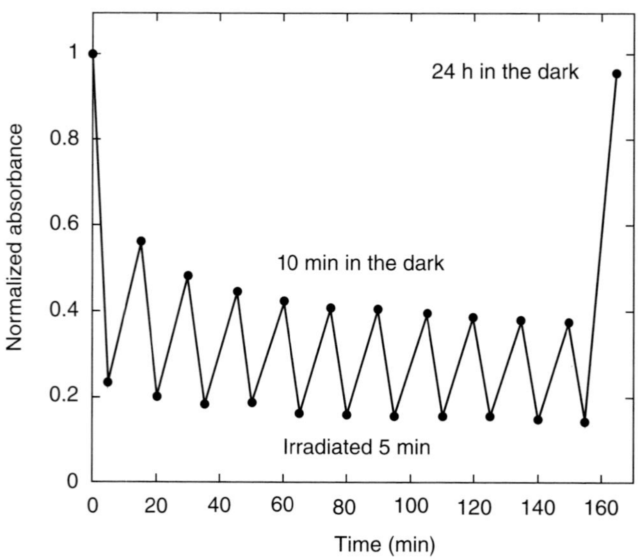

Figure 14.4 Normalized absorbance of mPAH2 in PHEMA film at 466 nm under repeated irradiating and recovery in the dark.

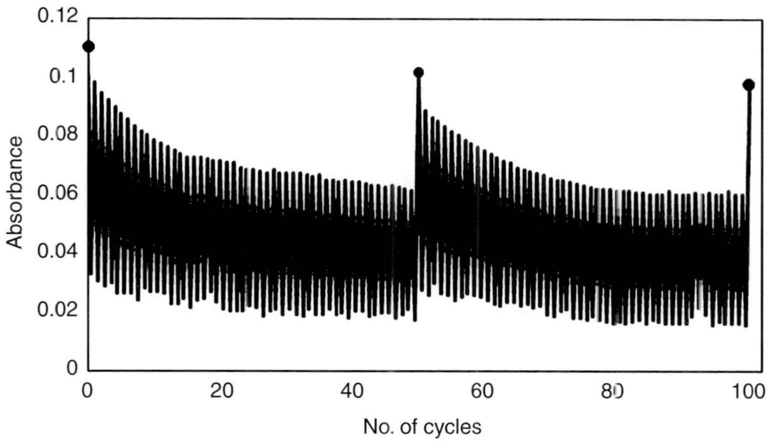

Figure 14.5 Absorbance of mPAH2 in PHEMA film with 5% of ATFMS subjecting to 100 cycles of three-minute irradiation and five minutes in dark per cycle.

photostability. Slow proton diffusion in the polymer film is one of the major reasons for the slow reverse reactions. Recombination with proton is needed for the reverse reaction. Although the initial rate for the reverse reaction could be fast due to high local proton concentration, the overall rate is slow because the protons that have diffused in the polymer cannot move back fast.

For many applications related to photoswitching, a fast reverse process is required. In order to enhance the reverse reaction rate, ammonium trifluoromethanesulfonate (ATFMS) was introduced. ATFMS has been used as a cathode additive in fuel cell systems to improve the proton conductance [28]. Therefore, 5 wt% of ATFMS was added to a polymer film and the film was tested by 100 cycles of three-minute irradiation and five minutes in the dark per cycle (Figure 14.5). About 60% recovery was observed in 5 minutes compared to ~30% recovery in 10 minutes for the films without ATFMS. After the film was tested for 50 cycles, it was kept in the dark overnight (~16 hours), at which point 92% of MEH was recovered. Then the next 50 cycles were performed. After 100 cycles, the sample was kept in the dark for 16 hours, and 90% of MEH recovered.

14.3 Photochromic Materials Based on Photoinduced Proton Transfer

Protonation of a molecule can significantly alter its electronic structure and lead to changes in electronic and optical properties. Therefore, photoresponsive electronic and optical materials can be developed using metastable-state photoacids [19–22, 29]. Photochromic materials that change their colors upon irradiation have been studied for decades. Prior photochromic materials are based on molecules that reversibly change their structures upon irradiation [30, 31]. Common photochromic molecules such as SPHs, diaryethenes, and azobenzenes have colorless or light colored forms that respond to UV light and change to colorful forms [32–36]. Some molecules show negative photochromic property. They change from colorful forms to colorless forms under visible light [37–39]. Our group has developed a new type of photochromic material based on reversible proton transfer using metastable-state photoacids.

In this work, photochromic materials were prepared by mixing mPAHs with acidochromic dyes in solutions, hydrogels, and polymers [19, 20]. Acidochromic dyes are the molecules that changes colors upon protonation. Under irradiation, the dye is protonated by the mPAH, and consequently changes its color. In the absence of the activating light, the photoacid returns to the low acidity state, takes back the proton from the dye, and the color changes back. Various color changes and the desirable stimulating wavelengths can be achieved by choosing different acidochromic dyes and photoacids.

MPAH1 was mixed with the acidochromic dye AD1, which has a pK_a of ~5 and changes from yellow to blue upon protonation (Figure 14.6). Blue and yellow are complementary colors, and have maximum color contrast. The photochromic behavior was first studied in solution. An ethanol solution of MPAH1 and AD1 quickly changed from yellow to blue upon irradiation (470 nm) in about two minutes. Before irradiation, UV–Vis spectrum showed a strong absorption at 440 nm, which is an overlap of the absorptions of both the mPAH and the dye. After irradiation, this absorption decreased and an absorption at 604 nm appeared due to the formation of the protonated dye. Once the light was removed, the blue solution changed back to yellow in about an hour. Thin films of polycaprolactone loaded with MPAH1 and AD1 were prepared by spin casting (Figure 14.7). The loading of MPAH1 and AD are both 5 wt%. In addition, 5 wt% of triethylene glycol was added to improve proton transfer in the polymer. The greenish yellow thin film changed to greenish blue under irradiation. UV–Vis spectra of the thin

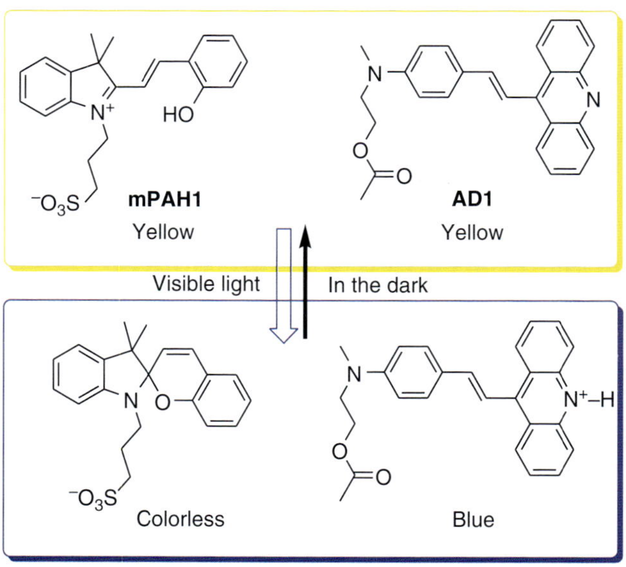

Figure 14.6 Photochromism of mPAH1 and AD1 based on photoinduced proton transfer.

Figure 14.7 Colors of a solution and a PCL film of mPAH1 and AD1 before (left) and after (right) irradiation.

film showed the decrease of absorption at 420 nm, and increase of the absorption at 590 nm. After irradiation, the UV–Vis spectrum changed back to the one before irradiation and so was the color of the film in about 10 hours. Multiple acidochromic dyes can be used together to achieve the desirable color change. For example, a hydrogel that changes from green to red upon irradiation was made by loading mPAH1, bromocresol green, and naphthyl red to a polyacrylamide hydrogel [19].

Different mPAHs can be used to tune the color, stimulating wavelength, reverse rate, etc. MPAH3 (Figure 14.8) was developed to increase the stimulating wavelength. The carbazole on the phenol side functions as a strong electron donating group. The resultant push–pull electronic configuration leads to a long-wavelength absorption with a λ_{max} at 538 nm in ethanol. A solution of mPAH3 and bromocresol green reversibly changed from brown to yellow under 525 nm yellow light. The visual effect is interesting since the color is darker in a darker condition and lighter/warmer in a lighter condition. It may be used for developing novel camouflage materials that change the darkness of their colors according to the environmental lighting condition.

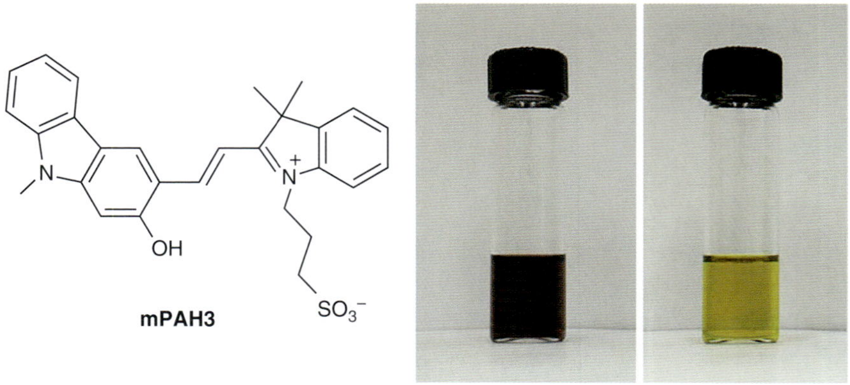

Figure 14.8 Structure of mPAH3 (left) and colors of a solution of mPAH 3 and bromocresol green before and after irradiation.

Compared to mPAH1, the TCF-type mPAH4 (Figure 14.9) possesses a faster reverse reaction rate and a lower dark acidity [20]. The former is desirable for fast recovery of the original color in the dark, and the latter allows the photoacid to pair with the acidochromic dyes with relatively high basicity. PHEMA films loaded with mPAH4 and different indicators including AD1, Congo red [pK_a (H_2O) = 4.1], Bromocreso green sodium salt [pK_a (H_2O) = 4.7], and neutral red [pK_a (H_2O) = 6.7] showed photochromic behavior under irradiation (Figure 14.9). The thin films changed color quickly under sunlight. Even indoor light can slowly change the color. The reverse reaction of mPAH4 is about 10^4 times faster than

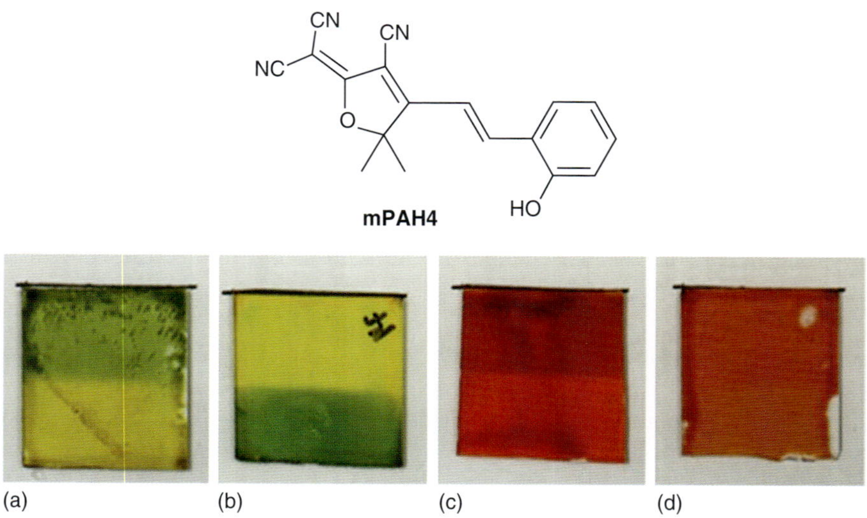

Figure 14.9 The color change of the films of mPAH4 with AD1 (a), bromocresol green sodium salt (b), Congo red (c), and neutral red (d). (The upper part of the film was irradiated and the lower part was unirradiated.)

that of mPAH1 in methanol. However, the reverse reaction is relatively slow in a polymer film. A PHEMA film containing 3 wt% of mPAH4 and 1 wt% of AD1 changed from yellow to green under 470 nm irradiation. It took about an hour for the color to change back to yellow in the dark. UV–Vis spectrum showed that ~85% of the absorbance recovered in one hour after irradiation. Heating can accelerate the color change. An irradiated film on a hot plate that was set to 80 °C changed the color back to yellow in one minute.

14.4 Photo-Controlled Fragrant-Releasing Polymer Based on Acid-Catalyzed Hydrolysis

Photo-controlled release of functional molecules has many applications in biomedical and material areas. Prior research mainly focused on grafting functional molecules via photo-labile linkage or caging functional molecules in a photosensitive polymeric or inorganic matrix. Many of these systems required UV irradiation, which could damage the materials or bio-tissues. Our group developed a new approach to release functional molecules with light, which is based on acid-catalyzed hydrolysis using mPAHs [26].

In this work, anisaldehyde, which is a commonly used fragrant molecule, is linked onto a diol-grafted poly(vinyl phenol) (P1) via an acetal group (P2) (Figure 14.10). The acetal group is thermally stable in neutral conditions, but easily hydrolyzed under mild acidic conditions. P2 and mPAH1 were dissolved in deuterated acetone and water (4 : 1). The concentration of P2 was 6.6 mg ml^{-1}, which was 8.3 mM in terms of the acetal unit. The concentration of mPAH1 was 1.0 mM. The use of deuterated solvents allowed us to study the release of anisaldehyde using NMR spectroscopy. Upon irradiation with 470 nm LED, the NMR showed the emergence of new peaks corresponding to free anisaldehyde, indicating the release of anisaldehyde from the polymer. However, when the light was turned off, the anisaldehyde peaks kept increasing, indicating that the acidity in the solution was still high enough to catalyze the hydrolysis of the acetal group. In fact, NMR also showed that the photoacid was still in the acidic state due to its slow reverse reaction.

As described in the previous sections, mPAH2 possesses a much faster reverse reaction than does mPAH1. In the 4 : 1 mixture of acetone and water, the rate constants of the second order reverse reactions of mPAH1 and mPAH2 are 3.5 and 190 M^{-1} s^{-1}, respectively. The reverse reaction of mPAH2 is 50 times faster than that of mPAH1. As described above, this type of mPAH reverses faster in water than in organic solvents. Therefore, we increased the amount of water in the solvent mixture. The reverse reaction rate of mPAH2 in a 2 : 1 mixture of acetone and water was measured to be 832 M^{-1} s^{-1}. Ideally, an aqueous solution would maximize the reverse reaction rate. However, the content of water in the solvent is limited by the solubility of the polymer.

MPAH2 and P2 in 2 : 1 deuterated acetone and water were studied by NMR. NMR spectra were taken after every five minutes of irradiation (Figure 14.11).

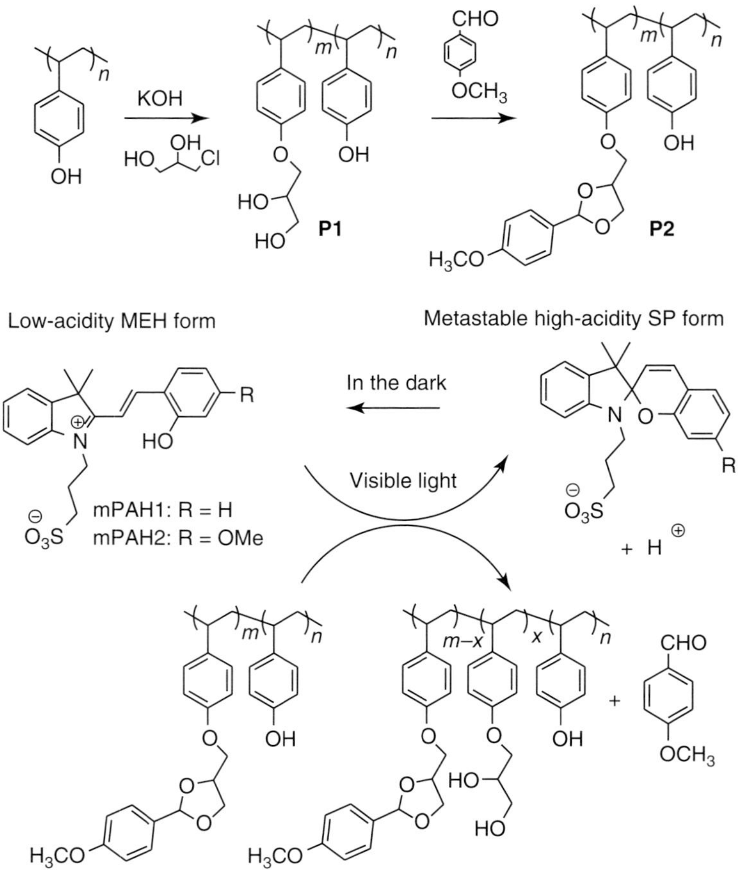

Figure 14.10 Synthesis of the fragrant carrier polymer P2 and the photoinduced release of anisaldehyde in the presence of an mPAH.

Results showed that the percentage of the released anisaldehyde increased linearly with irradiation time (Figure 14.12). After the light was turned off, the peaks of anisaldehyde did not increase with time, indicating that the release stopped immediately. No increase of the anisaldehyde peaks was observed even after the sample was kept overnight in the dark. In addition, peaks of mPAH2 was the same in all of the NMR spectra since the mPAH2 already changed back to the low acidity form before NMR was taken.

These results indicate that the release of the fragrance can be effectively controlled by turning on and off the irradiation. It is worth mentioning that the polymer carrier P1 are re-obtained after release of all the anisaldehyde, and the photoacids are reversible. They may be recycled to reduce the cost of the material. Although the method described here can only be applied to aldehydes and ketones, other types of functional molecules such as alcohols and amines may also be released by acid-catalyzed cleavage reactions using similar systems. The application may also be extended to drug delivery, etc.

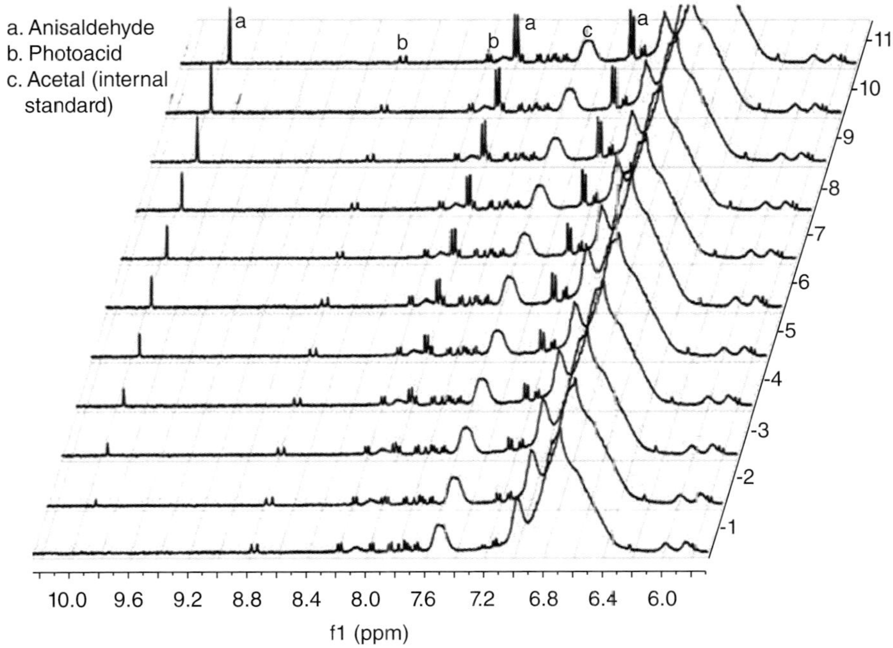

Figure 14.11 NMR spectrum of a solution of P2 and mPAH2 in 2 : 1 acetone/water before and after irradiation.

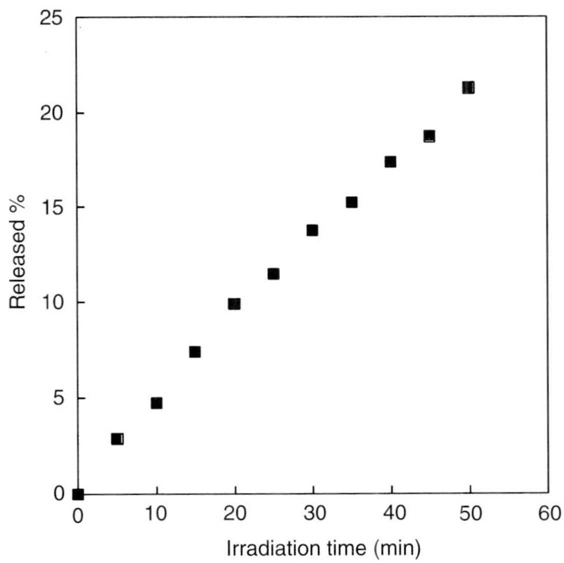

Figure 14.12 Percentage of anisaldehyde released with time of irradiation.

14.5 Photo-Controlled Reversible Dissolution/Formation of Polymer Nanoparticles

Stimulus-responsive polymer nanoparticles are polymer nanoparticles that respond to external stimuli, for example, light, pH, temperature, ionic strength, and electric field [40]. Among these stimulating methods, photo-irradiation, especially visible-light irradiation, is a convenient method to alter the properties of polymers. Polymer nanoparticles are often prepared by polymerization, precipitation, or cross-linking in a nanoconfined environment. Cross-linking of preformed polymers is commonly used to prepare stable polymer nanoparticles. Noncovalent cross-linking could be more sensitive to a stimulus than covalent bonding. In this work, we demonstrated that ionic and hydrogen bond cross-linking can be controlled by photoinduced protonation using mPAHs, which led to a polymer nanoparticle that dissolved and reformed under light and in the dark respectively [16].

The polymer nanoparticles used in this study were prepared by cross-linking poly(methylacrylate-*co*-acrylic acid) (PMA-co-PAA) with a small amount of poly(4-vinylpyridine) (PVPy) in methanol. The molar ratio between MA and AA in PMA-*co*-PAA was 4 : 1. The carboxylic group on PMA-*co*-PAA can form a hydrogen bond and an ionic bond with the pyridyl group on PVPy. Therefore, PVPy acted as a noncovalent cross-linker and only a small amount of PVPy was needed. The molar ratio of the pyridyl units on PVPy to the carboxylic units on PMA-*co*-PAA was 0.15 : 1. Addition of a higher ratio of PVPy caused quick precipitation of the cross-linked polymer. The nanoparticle suspension was stable at room temperature over days due to the negative charges of the carboxylic groups on the surface.

The noncovalent cross-linking between the two polymers can be broken by protonation of the carboxylic or pyridyl group. In either case, the two units cannot form an ionic or a hydrogen bond between one other (Figure 14.13). MPAH1 was added to the nanoparticle suspension to control the cross-linking with light. Under 470 nm LED, the cloudy suspension became transparent (Figure 14.14). The cross-linking was broken by the proton generated from mPAH1, and the two non-cross-linked polymers were both dissolved in methanol. When the light was turned off, the high-acidic state of mPAH 1 reversed back to a low-acidic state and took back the proton. The cross-linking between the two polymers formed again and so did the nanoparticle.

The size and distribution of the polymer nanoparticle were studied using a dynamic light scattering (DLS) particle size analyzer. The average size of the freshly prepared nanoparticles was 320 nm and the polydispersity index (PDI) was 0.358. After being irradiated for three minutes, the average size of the particles decreased to 20 nm and the PDI was 0.381 (Figure 14.15). The solution looked transparent. Three hours after irradiation, the polymer nanoparticles formed again and the average size was 244 nm with a PDI of 0.329. After irradiation, the nanoparticles dissolved and two polymers were evenly distributed in the solution before the nanoparticles formed again. Therefore, the reformed polymer nanoparticles became more uniform and had lower PDI.

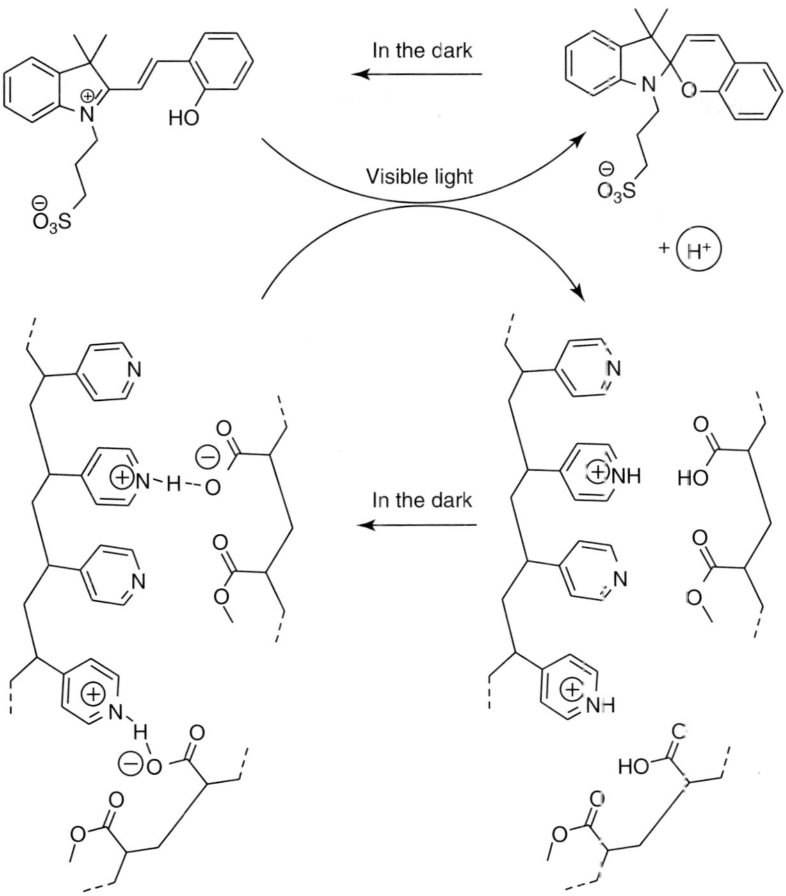

Figure 14.13 Cross-linking between PMA-*co*-PAA and PVPy, and de-cross-linking due to photoinduced protonation by mPAH1. Source: Fu et al. 2016 [13]. Reproduced with permission of Royal Society of Chemistry.

The size of the nanoparticle was further confirmed by AFM (Figure 14.14) and TEM.

The dissolution and formation cycle of the nanoparticles' suspension were studied by measuring the transmittance of the mixture using UV–Vis spectroscopy. The transmittance of a suspension prepared as described above was collected after the sample was repeatedly irradiated for three minutes and kept in the dark for two hours (Figure 14.16). Transmittance at 580 nm was monitored since mPAH1 does not absorb at this wavelength. The transmittance of the solution after irradiation was about 80%. After the irradiated sample was kept in the dark for two hours, the transmittance reduced to ~22% due to the reformation of the nanoparticles. The cycle could be repeated for many times (Figure 14.16). The material described here could be used for different applications. One

Figure 14.14 Suspension of the polymer nanoparticle before and after irradiation (left), and AFM images of the nanoparticles before irradiation (right). Source: Fu et al. 2016 [13]. Reproduced with permission of Royal Society of Chemistry.

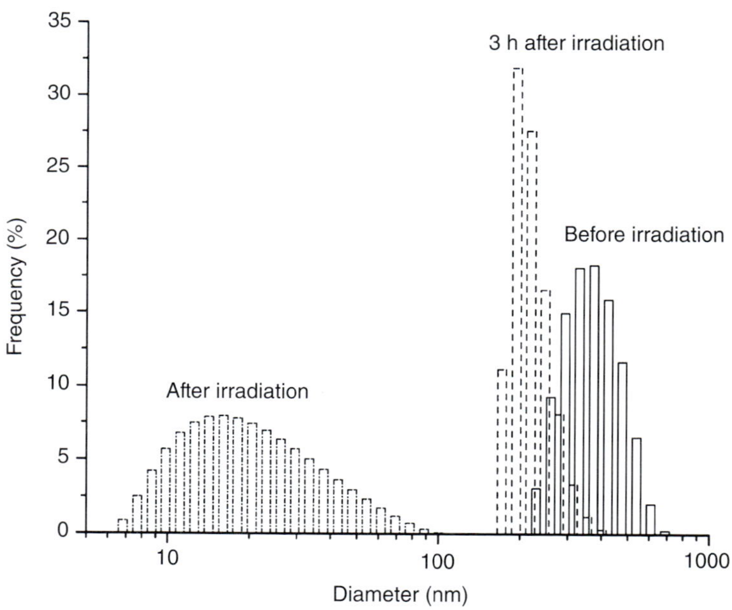

Figure 14.15 Size and distribution of the polymer nanoparticle before irradiation, immediately after irradiation, and three hours in the dark after irradiation measured by DLS. Source: Fu et al. 2016 [13]. Reproduced with permission of Royal Society of Chemistry.

potential application is as an interlayer material that is transparent during daytime and translucent during night for privacy or temperature control. In fact, we found that sunlight is strong enough to induce the photoreaction effectively. Other applications, e.g. reversible patterning and drug delivery could also be explored.

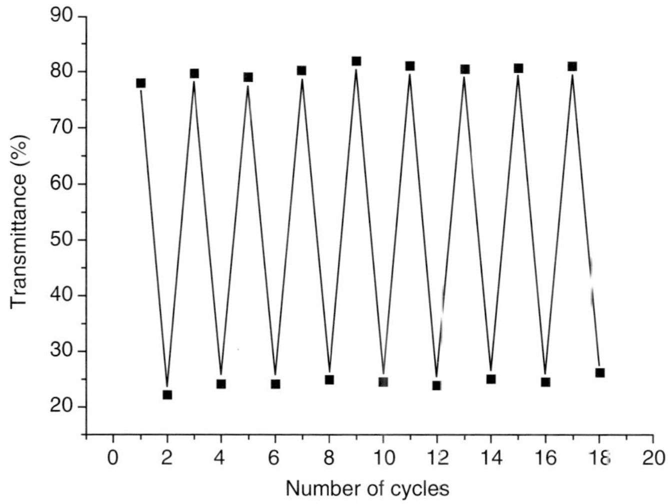

Figure 14.16 The transmittance cycles and the transmittance cycles at 530 nm measured by UV–Vis spectroscopy. Source: Fu et al. 2016 [13]. Reproduced with permission of Royal Society of Chemistry.

14.6 Conclusion

Photoinduced protonation is becoming a general approach toward different types of photoresponsive polymers. Reversible protonation of polymers can be controlled by visible light using metastable-state photoacids. This approach has shown great potential and novel photochromic materials, fragrant-releasing polymers, photoresponsive nanoparticles, volume change hydrogels, ionic sensors, etc. have been developed in the past years. Other types of photoreponsive materials may also be explored in the future. As described earlier, photoinduced protonation can affect hydrogen bonding and ionic bonding. Polymers with strong hydrogen and/or ionic bonding may show photomechanical behavior in the presence of a metastable-state photoacid. Protonation can not only affect the optical properties but also the electronic properties. Protonated polyaniline possesses high conductivity while non-protonated polyaniline is an insulator. Therefore, paring polyaniline with metastable-state photoacids could lead to photoswitchable conducting materials. However, polyaniline has very high basicity and thus a photoacid with very low dark acidity is needed. Photoacids such as mPAH1 protonate polyaniline without irradiation.

Challenges have emerged during the study of this type of photoresponsive materials in the past years. Proton transfer in polymers is often slow, which affects the photosensitivity and photoswitching rate. Development of novel neutral polymers with relatively fast proton diffusion rate is important for improving the performance of the photoreponsive materials. Covalently grafting mPAHs onto polymers could improve their compatibility with polymers and avoid leakage. More importantly, photoacid polymers could generate a localized proton concentration that is much higher than that of the environment. Given that biological systems are often in pH buffers, the capability of generating a localized pH change could have important implications for many biological applications. An effective method for synthesizing polymers covalently linked with metastable-state photoacids and demonstration of photogenerated local pH change has not been reported. Such polymeric soft materials are expected to display promising properties that would be interesting for fundamental studies and technological applications and beyond. There remains a lot to understand and more to do in this enabling and emerging frontier.

References

1 Ivan, M.G. and Scaiano, J.C. (2010). Photoimaging and lithographic processes in polymers. In: *Photochemistry and Photophysics of Polymer Materials* (ed. N.S. Allen). Hoboken: Wiley.
2 Ito, H. (2003). Chemical amplification resists: inception, implementation in device manufacture, and new developments. *J. Polym. Sci., Part A: Polym. Chem.* 41: 3863.
3 Chou, P.-T. and Solntsev, K.M. (2015). Photoinduced proton transfer in chemistry and biology. *J. Phys. Chem. B* 119: 2089.
4 Tolbert, L. and Solntsev, K.M. (2002). Excited-state proton transfer: from constrained systems to "super" photoacids to superfast proton transfer. *Acc. Chem. Res.* 35: 19.
5 Wan, P. and Shukla, D. (1993). Utility of acid-base behavior of excited states of organic molecules. *Chem. Rev.* 93: 571.
6 Shi, Z., Peng, P., Strohecker, D., and Liao, Y. (2011). Long-lived photoacid based upon a photochromic reaction. *J. Am. Chem. Soc.* 133: 14699.
7 Johns, V.K., Peng, P., DeJesus, J. et al. (2014). Visible-light-responsive reversible photoacid based on a metastable carbanion. *Chem. Eur. J.* 20: 689.
8 Abeyrathna, N. and Liao, Y. (2015). A reversible photoacid functioning in PBS buffer under visible light. *J. Am. Chem. Soc.* 137: 11282.
9 Xu, Y., Fei, J., Li, G. et al. (2017). Enhanced photophosphorylation of a chloroplast-entrapping long-lived photoacid. *Angew. Chem. Int. Ed.* 56: 12903.
10 Bao, H., Li, F., Lei, L. et al. (2014). On/off states of a microbial fuel cell controlled by an optical switching system. *RSC Adv.* 4: 27277.
11 Johns, V.K., Patel, P.K., Hassett, S. et al. (2014). Visible light activated ion sensing using a photoacid polymer for calcium detection. *Anal. Chem.* 86: 6184.
12 Patel, P.K. and Chumbimuni-Torres, K.Y. (2016). Visible light-induced ion-selective optodes based on a metastable photoacid for cation detection. *Analyst* 141: 85.
13 Fu, C., Xu, J., and Boyer, C. (2016). Photoacid-mediated ring opening polymerization driven by visible light. *Chem. Commun.* 52: 7126.
14 Maity, C., Hendriksen, W.E., van Esch, J.H., and Eelkema, R. (2015). Spatial structuring of a supramolecular hydrogel by using a visible-light triggered catalyst. *Angew. Chem. Int. Ed.* 54: 998.
15 Kundu, P.K., Samanta, D., Leizrowice, R. et al. (2015). Light-controlled self-assembly of non-photoresponsive nanoparticles. *Nat. Chem.* 7: 646.

16 Wang, Z. and Liao, Y. (2016). Reversible dissolution/formation of polymer nanoparticles controlled by visible light. *Nanoscale* 8: 14070.
17 Guo, J., Zhang, H.Y., Zhou, Y., and Liu, Y. (2017). Light-controlled reversible self-assembly of nanorod suprastructures. *Chem. Commun.* 53: 6089.
18 Yang, L.P., Jia, F., Cui, J.S. et al. (2017). Light-controlled switching of a non-photoresponsive molecular shuttle. *Org. Lett.* 19: 2945.
19 Chen, H. and Liao, Y. (2015). Photochromism based on reversible proton transfer. *J. Photochem. Photobiol., A* 300: 22.
20 Yang, C., Khalila, T., and Liao, Y. (2016). Photocontrolled proton transfer in solution and polymers using a novel photoacid with strong C–H acidity. *RSC Adv.* 6: 85420.
21 Mahvidi, S., Takeuchi, S., Kusumoto, S. et al. (2016). Gated photochromic system of diarylethene with a photon-working key. *Org. Lett.* 18: 5042.
22 Nakagawa, T. and Yokoyama, Y. (2016). All-optical fine-tuning of absorption band of diarylethene with photochromic acid-generating spiropyran. *Adv. Opt. Mater.* 4: 1350.
23 Shafaat, O.S., Winkler, J.R., Gray, H.B., and Dougherty, D.A. (2016). Photoactivation of an acid-sensitive ion channel associated with vision and pain. *ChemBioChem* 17: 1323.
24 Luo, Y., Wang, C., Peng, P. et al. (2013). Visible light mediated killing of multidrug-resistant bacteria using photoacids. *J. Mater. Chem. B* 1: 997.
25 Johns, V.K., Wang, Z., Li, X., and Liao, Y. (2013). Physicochemical study of a metastable-state photoacid. *J. Phys. Chem. A* 117: 13101.
26 Wang, Z., Johns, V.K., and Liao, Y. (2014). Controlled release of fragrant molecules with visible light. *Chem. Eur. J.* 20: 14637.
27 Abeyrathna, N. and Liao, Y. (2017). Photoactivity, reversibility, and stability of a merocyanine-type photoacid in polymer films. *J. Photochem. Photobiol., A* 332: 196.
28 Hong, S.G., Kwon, K., Lee, M.J., and Yoo, D.Y. (2009). Performance enhancement of phosphoric acid-based proton exchange membrane fuel cells by using ammonium trifluoromethanesulfonate. *Electrochem. Commun.* 11: 1124.
29 Shi, Z., Peng, P., Johns, V.K., and Liao, Y. (2012). Novel photo-switchable conducting materials based on a reversible photoacid and polyaniline. *Polym. Prepr.* 53: 125.
30 Tian, H. and Zhang, J. (eds.) (2016). *Photochromic Materials: Preparation, Properties, and Applications*. Weinheim: Wiley-VCH.
31 Wang, L. and Li, Q. (2018). Photochromism into nanosystems: towards lighting up the future nanoworld. *Chem. Soc. Rev.* 47: 1044.
32 Jin, L., Li, Y., Ma, J., and Li, Q. (2010). Synthesis of novel thermally reversible photochromic axially chiral spirooxazines. *Org. Lett.* 12: 3552.
33 Rameshbabu, K., Urbas, A., and Li, Q. (2011). Self-organized photochromic dithienycyclopentent organogels. *J. Mater. Chem.* 21: 15673.
34 Li, Y., Urbas, A., and Li, Q. (2012). Reversible light-directed red, green and blue reflections with thermal stability enabled by a self-organized helical superstructure. *J. Am. Chem. Soc.* 134: 9573.
35 Li, Y., Xue, C., Wang, M. et al. (2013). Photodynamic chiral molecular switches with thermal stability: from reflection wavelength tuning to handedness

inversion of self-organized helical superstructures. *Angew. Chem. Int. Ed.* 52: 13703.

36 Wang, Y. and Li, Q. (2012). Light-driven chiral molecular switches or motors in liquid crystals. *Adv. Mater.* 24: 1926.

37 Hatano, S., Horino, T., Tokita, A. et al. (2013). Unusual negative photochromism via a short-lived imidazolyl radical of 1,1-binaphthyl-bridged imidazole dimer. *J. Am. Chem. Soc.* 135: 3164.

38 Peng, P., Strohecker, D., and Liao, Y. (2011). Negative photochromism of a TCF chromophore. *Chem. Commun.* 47: 8575.

39 Ayub, K., Li, R., Bohne, C. et al. (2011). Calculation driven synthesis of an excellent dihydropyrene negative photochrome and its photochemical properties. *J. Am. Chem. Soc.* 133: 4040.

40 Blum, A.P., Kammeyer, J.K., Rush, A.M. et al. (2015). Stimuli-responsive nanomaterials for biomedical applications. *J. Am. Chem. Soc.* 137: 2140.

Index

Photoactive Functional Soft Materials: Preparation, Properties, and Applications, First Edition.
Edited by Quan Li.